POPULATION BIOLOGY OF TROPICAL INSECTS

POPULATION BIOLOGY OF TROPICAL INSECTS

ALLEN M. YOUNG
Milwaukee Public Museum
Milwaukee, Wisconsin

PLENUM PRESS • NEW YORK AND LONDON

Library of Congress Cataloging in Publication Data

Young, Allen M.
Population biology of tropical insects.

Bibliography: p.
Includes index.
1. Insect populations—Tropics. I. Title.
QL491.6.Y68 595.7′05248′0913 82-7562
ISBN 0-306-40843-0 AACR2

A Division of Plenum Publishing Corporation
233 Spring Street, New York, N.Y. 10013

Printed in the United States of America

"Internarse en un bosque es reconciliarse con la vida. El asombro resbala de hoja en hoja. La fecundidad telurica palpita y destella, como una libelula dorada, en los oblicuos rayos solares que se filtran por entre los densos ramajes."

....From "Bosques Y Vagabundos," by Carlos Luis Altamirano
La Nacion (seccion Ancora), San Jose, Costa Rica
November 12, 1978

PREFACE

In this book I have tried to bring together the major developments in the study of insect populations in tropical environments. In some ways, this task has been a difficult one because conceptually it is virtually impossible to limit a discussion of insect ecology to the tropics, since the same concepts, theories, and hypotheses concerning the mechanisms by which habitats support insect populations often apply both to temperate and to tropical regions. Thus one might argue effectively that a book such as Peter Price's *Insect Ecology* represents a more comprehensive treatment of insect ecology, including the tropical aspects. Yet because there has been a tremendous amount of new study on insects in the tropics in recent years, and because there has also been a strong historical interest in tropical insects, judging from early museum expeditions and medically and agriculturally oriented studies of insects in the New and Old World tropics, I believe there is a place for a book dealing almost exclusively with tropical insects. But logically so, such a book by necessity incorporates data and information from Temperate Zone studies, if for no other reason than because insights into the properties of tropical environments often emerge from comparisons of species, communities, or faunas between temperate and tropical regions. An understanding of insect populations in the tropics cannot be divorced from a consideration of Temperate Zone populations.

The book is unabashedly biased to the New World tropical studies because of my interests and more proximal awareness of research on insects being conducted in this region. But a valid attempt is made to incorporate research studies from the Old World tropics.

The basic mission of this book is to bring together various concepts and studies related to the issue of explaining the spatial and temporal patterns of insect diversity in the tropics, considering the breeding population of a species to be the major ecological unit beyond the individual within the population. The population is interacting with both biotic and physical components of the environment, within and between habitats, and regionally. Much of tropical insect population biology is therefore concerned in different ways with describing patterns of distribution of individual species in relation to resources and other environmental factors.

There is also a need for detailed descriptive data on the natural history of individual species in particular habitats in the tropics. What are some of the major attributes underlying observed patterns of distribution? Can we recognize the roles of biotic control agents in influencing the populations of insect species as well as the role of abiotic factors? We recognize that the tropical regions of the world are extremely heterogeneous in terms of climatic factors and the patterns of fluctuation of these factors. The kinds of selection pressures affecting an insect species on a mountain top in the tropics may greatly differ from those affecting the same species or a closely related species in an adjacent lowland area. Within major climatic areas, varying subclimatic conditions will be arising from localized weather patterns and effects of topography and kind of vegetation cover. Selection pressures on many insect species may be very different in a strongly seasonal region than in a relatively nonseasonal tropical region.

The fine tuning of insects and their resources, as exemplified by many plant–insect interactions in the tropics, may reflect coevolved interactions between insects and their resources, and in recent years there has been a great interest in defining the traits or properties of such associations. Emerging from such work is the concept of the individual green plant as an extremely heterogeneous resource for insects, in the sense that defense systems against such herbivores vary among different parts of the plant and as a result of highly localized environmental conditions at any one point in time and changes in the environment through ecological and evolutionary time. Herbivore pressure may be a function of the ability of insect species to adapt to the defensive chemistry of plants selected as hosts. Herbivore pressure is also a function of (1) diurnal or seasonal cycles in defensive chemistry metabolism and (2) the degree to which herbivores synchronize feeding with phases of low concentration of defensive compounds in specific plant structures targeted as food. Much of our understanding of insect–plant interactions comes from studies in the tropics.

Tropical insect studies also point to the need to consider the success of an insect species in a particular habitat or geographical region to be a function of behavioral and physiological flexibility instead of evolutionary adaptations, i.e., genetic variation in populations. Neotropical species of *Drosophila* may be capable of feeding on a variety of qualitatively different kinds of food not so much because populations of these flies are heterogeneous genetically, but because behavioral flexibility enables individual flies to effectively exploit the range of available food types or breeding sites.

Understanding the occurrence of regions in the tropics of high species richness for insect taxa is a function of the analysis of historical (biogeographical) events and ecological factors. In recent years there has been increased study of forest refugia in Central and South America in relation to periods of recent glaciation, and such studies have increased our understanding of why pockets of high species richness are surrounded by areas of relatively low richness. Ecological factors, including the occurrence of natural and human-induced changes in habitats in the tropics—the latter through agriculture, lumbering, and population expansions—also shape the distribution of insect species and the kinds of resources they exploit in a particular area. An attempt is made to examine the properties of insect populations in such man-made environments as well as the effects of natural succession on species.

Basic research studies on the mechanisms of insect–plant interactions in the tropics provide mankind with new information about the unique and regulatory properties of natural products, data that have the potential to be translated into applied use through biotechnology as related to the development of naturally based repellents, insecticides, and fungicides. Tropical floras represent a millieu of novel and diverse mechanisms of chemical defense against herbivores, and the insects exploiting such resources have developed detoxification enzyme systems and other means of countering the defenses of plants. Such interactions provide a basis for the development of new and effective methods of control for pests and pathogens associated with Temperate Zone crops and forests. The study of tropical insects and their interactions with their resources has great potential to the application of natural products to new research in agriculture, horticulture, pharmacology, etc. But the first step in reaching such applications is to determine what is out there, to examine the patterns of distributions of species or groups of species and genera, to find out what causes the observed patterns, and in the case of insect–plant interactions to elucidate the mechanisms holding such relationships together and their genetic, physiological, or behavioral basis.

I have attempted to summarize a bulk of the pertinent literature on tropical insect population biology, with an emphasis on research done within the last 50 years. This book can serve as a helpful starting point for undergraduate and first- or second-year graduate students with interests in ecology or environmental biology.

I wish to thank all of the biologists who consented to copyright permissions from their publications. Mary Joan Young offered encouragement during the preparation of the manuscript. Dr. Theresa A. Noeske provided painstaking technical assistance. Karen Heerhold did an excellent job of typing the final draft of the manuscript. The guidance and cooperation of Ellis Rosenberg and his staff at Plenum are also appreciated. Gerald R. Noonan reviewed Chapter 10. I am grateful for the encouragement and strong vote of confidence given to me by my good friend Oskar R. Zaborsky with the initial plan to write this book. My own field work experience in the tropics covers a 12-year period, with support from the Organization for Tropical Studies, Inc., the Associated Colleges of the Midwest and Lawrence University, the National Science Foundation, Friends of the Milwaukee Public Museum, and the American Cocoa Research Institute. I owe a great deal to these people and sources of funding for making my field work possible.

Allen M. Young

Milwaukee

CONTENTS

1. The Faunistic Richness of Insects in the Tropics: A Brief Overview 1

2. Individual and Population Responses to Environments **19**

2.1. Ecological Response 19
2.2. Natural Selection 22
2.3. Some Properties of Populations 22
2.4. Ecological Diversity and the Tropics 26
2.5. Some Mechanisms Underlying Tropical Insect Density . . 32
2.6. Strategies of Response to the Environment 33
2.7. Natural Selection and Properties of Populations 40
2.8. Environmental Predictability and Response Patterns 50

3. Machinery of Environmental Response Mechanisms in Insects: Key to Evolutionary and Ecological Diversification **61**

3.1. Notes on Insect Diversity 61
3.2. Comments on Insect Physiology 67
3.3. Integration of the Insect Phenotype by Hormonal Systems: A Brief Review 71
3.4. Diapause and Migration in Insect Populations: Escape in Time and Space 74
3.5. Temperature, Photoperiod, and Other Factors Related to Insect Activity: Further Comments 82
3.6. Notes on Mating and Communication in Insects 88
3.7. Rhythmicity in Tropical Insects 100
3.8. Sociality and Defense in Social Insects 104
3.9. Density-Related Behavior in Insects 111
3.10. Notes on the Physiology of Feeding in Insects 114

4. Ecological Aspects of Plant Defenses against Insects 137

4.1. Plant Defenses and Coevolution: An Overview 137
4.2. Evolution of Unpalatability and Implications for Food Chains . 148
4.3. Notes on Defoliation in the Tropics 158
4.4. Notes on Seed Predation in Tropical Habitats 160
4.5. Notes on Plant–Ant Interactions in the Tropics 164

5. Distribution Patterns of Insects in Tropical Habitats 173

5.1. The Concept of Local Adaptation 173
5.2. Examples of Habitat Differentiation in the Tropics 178
5.3. Insect Communities in Tropical Habitats 195
5.4. Plant Resources and Habitat Selection in Tropical Insects . . 201
5.5. Disturbed and Undisturbed Habitats: Some General Effects . . 207

6. Population Responses to the Environment in Tropical Insects . 217

6.1. Adaptation to Habitats 217
6.2. Demography and Fitness in Populations: Implications for Tropical Insects 219
6.3. Further Notes on Population Regulation 230
6.4. Some Examples of Census Histories in Tropical Insects . . . 239
6.5. Population Responses to Environmental Heterogeneity in Tropical Insects 245
6.6. Competition as a Selective Force in Population Responses . . 259
6.7. Spatial Environmental Heterogeneity and Demes in the Tropics 266

7. Effects of Seasonality on Insect Populations in the Tropics . . 273

7.1. Temperate Zone and Tropical Seasonality 273
7.2. Some Regional Patterns of Seasonality in the Tropics . . . 280
7.3. Some Seasonal Distributions of Tropical Insects 299
7.4. Molding of Ecological Traits in Tropical Insects by the Dry Season 322

8. Dynamics of Organization of Insect Communities in Tropical Ecosystems 335

8.1. Notes on Tropical Ecosystems 335
8.2. Patterns of Organization in Tropical Forests: Some Comments 337
8.3. Notes on the Dynamics of Tropical Insect Communities . . . 341
8.4. Some Ecological Factors Generating Patchy Distributions of Insects in Tropical Habitats 375

9. Insect Species in Agricultural Habitats in the Tropics 387

9.1. Plant–Insect Interactions and Tropical Agricultural Habitats: Some Comments 388
9.2. Insect Pollination in an Agricultural Habitat in the Tropics: The Case of Cacao 394
9.3. Grazing and Related Activities by Insects in Agricultural Habitats in the Tropics 405
9.4. Tropical Insect Pest Populations 420

10. Biogeographical and Regional Evolutionary-Ecological Effects on the Maintenance of Tropical Insect Faunas: A Brief Perspective 429

10.1. Historical Events and Ecological Processes 429
10.2. Altitudinal and Other Macrogradients in Tropical Regions . 435
10.3. Notes on Altitudinal Gradients in the Tropics 442
10.4 Biogeographical Patterns of the Origins of Tropical Insect Faunas 446
10.5. Some Mechanisms of Ecological and Evolutionary Differentiation 451
10.6. Some Examples of Differential Adaptation to Tropical Lowland and Montane Areas by Insects 455

References 461

Index 495

CHAPTER 1

THE FAUNISTIC RICHNESS OF INSECTS IN THE TROPICS
A Brief Overview

The primary purpose of this book is to discuss the various ecological factors that influence both temporal and spatial properties of insect populations in the tropics. Factors such as diet, habitat, altitude, climate, seasonality, and geography are considered to be the major determinants of both past and extant distributions of tropical insect populations. Comparisons of distribution patterns to underlying mechanisms are often made with insect populations of the temperate zones, since overall insect diversity is relatively greater in the tropics than in temperate zones. This book attempts neither to exhaustively cover all of the literature on insects in tropical environments nor to provide a complete, comprehensive unison of broad-scale historical or biogeographical explanations for insect distributions in the tropics with all explanations based upon contemporary ecological interactions. Rather, given the amount of different approaches applied to accounting for the diversity of tropical organisms in general, and insects in particular over the past three centuries, the book draws together some of the pertinent studies relating to the population structure and dynamics of individual species and how such properties of insect populations relate to the more general issue of the great diversity of insect species in tropical regions.

My approach recognizes the fusion of both conceptual and empirical studies from population genetics and population ecology, resulting in the science of population biology. As pointed out by Lewontin,[(1)] the adaptive role of genetic variation in natural populations remains largely unknown or assumed. The problem at hand is somehow to relate genotypic variation, phenotypic plasticity, and fluctuating parameters of the environment with the maintenance of enzyme polymorphisms in populations. Based primarily upon studies of certain classes of soluble enzymes there is evidence of considerable genetic variation in populations, as exemplified by the pioneering studies of Lewontin and Hubby.[(2)] The problem, therefore, is to relate natural selection to the observed highly variable loci studied in natural populations. It is also essential to determine levels of phenotypic variation and to relate these patterns to ecological factors. Enzyme polymorphisms have been conceptualized to adapt species to heterogeneous environments.[3] Different phenotypes may be better adapted to particular features of the environment in different ways, reflecting some pattern of variation in fitness among the genotypes present in the population. Genotypic differences in fitness result from phenotypic differences manifested biochemically, physiologically, or behaviorally. Such differences may be evident within the population by individual variation in life table parameters such as fecundity or developmental rate.

Population biology is viewed as one of five major component sciences of evolutionary biology, the others being biogeography, community biology, natural history, and evolutionary genetics (including analyses of developmental and physiological mechanisms underlying modes of response of organisms to the environment). I also recognize that obviously none of these components of evolutionary biology is mutually exclusive. In fact, it is somewhat difficult to suggest such a breakdown of a complex field of natural science. Yet it is my aim to draw upon primarily the major avenues of contemporary thinking in population biology to explain some mechanisms of environmental response in insect species in the tropics and to do so in such a manner that a clearer view emerges of why insect species are so diverse in the tropics. Very little of this information is my own or is new. Most of it is the result of many tropical biologists' conducting both short-term and long-term field studies, using experimental designs based upon ecological theory developed in this century.

Another major component eventually rests upon the long history of collecting and watching organisms in the tropics. The major museums around the world have extensive collections of tropical insects, a result of intensive collecting in the past. The early literature also contains some interesting natural history of specific taxa of insects in the tropics. The current emphasis in tropical biology in general is the elucidation of the temporal and spatial properties of populations and communities based upon testable hypotheses generated from contemporary ecological thinking. Yet as with most taxa of tropical organisms, there is a huge gap in our knowledge of old-fashioned natural history as well as long-term census histories of individual populations. Such data are critical in understanding the uniqueness of the tropical regions of the world and essential as a prerequisite for making sound recommendations on the conservation of tropical habitats, terrestrial and aquatic. It seems that tropical biology has passed through different stages of emphasis and approach; some sort of feedback loop may be operative. The early collecting and watching (natural history) phase gave way to detailed population studies of disease vectors in the case of insects, and today an emphasis is on experimental approaches to long-recognized issues. But as we attempt to do more experimental studies, we find the gaps in our knowledge of the natural history of the organisms being manipulated. Clearly all of these approaches are essential.

Although the science of population biology developed largely as a discipline of the Temperate Zones, the great diversity of insects and other organisms in the tropical regions of the world has been recognized for some time. A survey of Amazonian streams reveals about the same number of insect species as some Temperate Zone streams, as shown in a study by Patrick.(4) Patrick(4,5) suggested that both temperate and tropical streams contained about the same number of niches for small organisms, including insects. She explained the greater number of fish species in tropical streams as a result of a greater area being available for these larger organisms, and therefore sufficient room for a greater number of species. In the case of butterflies, the survey of Ehrlich and Raven(6) clearly indicates a far greater number of genera and species of these insects in the tropics (Table 1.1). In fact, the data also show that some families are limited solely to tropical regions. As we will examine in later chapters, the great diversity of butterflies in tropical regions is largely explained by Ehrlich and Raven and others by the expression of specialized larval host plant utilization patterns.

Another example of diversification with a single genus in the tropics is illustrated by Paulson's (7) analysis of oviposition behavior in the dragonfly genus *Micrathyria*. Six Central American species were found to have very different types of oviposition behavior in terms of how and where eggs were placed in the environment. Paulson suggests that such behaviors are related to the ecological pressure of predation on eggs and young nymphs. Emerson's classic treatises(8,9) on the zoogeography of termite genera points out that the Indomalayan, Ethiopian, and Neotropical regions support the greatest number of termite genera and that the Ethiopian region alone accounts for 47% of all genera known. The large expanse of tropical rain forest characteristic of these regions is believed to be the major factor promoting the diversity of termite genera found there. Emerson points out that tropical Africa was a major center for termite diversification but also stresses the role of ecological processes such as interspecies competition and termite dispersal by man as other factors affecting distribution. Emerson(8) suggested that introduced species of termites are usually eliminated by native species in the Belgian Congo but that the development of dwelling and agricultural habitats by man provided a new niche for such species. Peters's monograph on the taxonomic diversity of mayflies found in tropical regions(10) indicates considerable divergence within the order over large geographical areas.

Linsley(11) reports that more than 28% of the North American cerambycid beetles occur in the Neotropical region, the highest percentage for the worldwide regions surveyed, although the greatest number of species of the group occur in North America. The group exhibits considerable divergence in feeding habits, strata of forest habitat preferred, breeding conditions, disease resistance, and many other factors. There is also evidence that different species are active as adults in different seasons where wet and dry seasons occur. Watson(12) pointed out that all 30 species of Odonata known from the northwest of western Australia are of tropical affinity and consist of two groups, namely, those that breed in

TABLE 1.1
A Summary of the Worldwide Geographical Distribution of the Major Groups of Butterflies[a]

Taxon	Approximate number of Genera[b]	Approximate number of Species	Distribution
Papilionidae	24(22)	575–700	
Baroniinae	1(1)	1	Central Mexico
Parnassiinae	8(8)	45–55	Holarctic and Oriental; greatest diversity, Asia
Papilioninae	15(13)	480–640	Worldwide, mainly tropical; greatest diversity, Old World tropics
Pieridae	58(40)	950–1150	
Coliadinae	11(8)	225–250	Cosmopolitan; greatest diversity, tropics outside of Africa
Pierinae	43(29)	650–750	Cosmopolitan; greatest diversity, tropics
Dismorphiinae	3(3)	80–120	Primarily Neotropical, one small Palearctic genus
Pseudopontiinae	1(0)	1	West equatorial Africa
Nymphalidae	325–400 (ca. 202)	4,800–6,200	
Ithomiinae	30–40(10)	300–400	Neotropical; *Tellervo*, Australian
Danainae	10–12(10)	140–200	Cosmopolitan; greatest diversity, Old World tropics
Satyrinae	120–150 (ca. 70)	1,200–1,500	Cosmopolitan; greatest diversity, extratropical
Morphinae	23–26(12)	180–250	Indomalayan and Neotropical
Charaxinae	8–10(8)	300–400	Tropicopolitan, few temperate
Calinaginae	1(1)	1	Oriental
Nymphalinae	125–150 (ca. 85)	2,500–3,000	Cosmopolitan
Acraeinae	8(6)	225–275	Tropical; greatest diversity, Africa
Libytheidae	1(1)	10	Cosmopolitan
Lycaenidae	325–425 (ca. 167)	5,800–7,200	
Riodininae	75–125(17)	800–1,200	Tropical, few Nearctic and Palearctic; Metropolis, Neotropical
Styginae	1(0)	1	Peruvian Andes
Lycaeninae	250–300 (ca. 150)	5,000–6,000	Cosmopolitan; greatest diversity, Old World tropics
Total	730–930 (ca. 432)	12,000–15,000	

[a] Reproduced from Ehrlich and Raven.[6]

[b] Number in parentheses indicates number of genera for which food plant records are available.

temporary waters and those that breed in permanent bodies of water. He suggests that limits to the distributions of these species are determined largely by the specificity in larval habitats. Widespread species in his study were those occupying temporary waters, i.e., species with strong tendencies for dispersal. Species associated with permanent riverine habitats exhibit less dispersal tendency. He concludes that the distribution of some species of the region is accountable in terms of past distribution of permanent water habitats. Thus in this study the explanation of present-day distributions of dragonflies in this region takes into account relative dispersal tendencies of species utilizing two very different types of larval habitats and the interpretation of past and present changes in the distribution of these habitats. That study emphasizes the historical role of biogeographical events upon which is superimposed prevailing ecological circumstances.

Clearly, such studies indicate the role of biogeographical patterns in the general distributions of insect groups in the world. These and other studies point to the need to consider not only contemporary or ecological properties of genera or species in explaining why a particular organism is found where it is, but also the strong historical component, namely, those biogeographical events resulting in an ''early'' settling of taxa in an area and the degree of dynamic changeover at the present.

The great diversity of tropical insects makes it imperative to keep in mind both the historical and contemporary components of environmental factors accounting for the observed diversity patterns in nature. Sample sizes also confuse the picture and analyses. Apparently, the more intense the collecting or surveying of a group in the tropics, the more species or other taxa are found. Thus Park[(13)] reports a large increase in the number of pselaphid beetles from Jamaica as the direct result of increased collecting there and suggests that cases of apparently impoverished psephalid beetle faunas on other Caribbean islands may be due to poor collecting. Many other published examples of both ecological and biogeographic-type surveys account for the high generic and species abundances of insects in the tropics. But clearly the maintenance of such high diversity results from different kinds of mechanisms in different groups. If the so-called ecological niche of a species is defined as the collection of resources used by individuals in a population of a species through space and time, at least part of the explanation of why more species of insects occur in the tropics relates to divergence in the ways certain resources are exploited among closely related taxa.

Part of the justification for a book on the population biology of tropical insects comes from the well-known observation that the tropics generally contain a greater number of closely related species for many genera, and also more genera in different families of insects. Somehow, therefore, there must be relationships between the population structure and dynamics of individual species and the ability for many more species to occupy the same environment or habitat.

A great abundance of species found in a particular region can result from the following factors: (1) higher rates of speciation, (2) lower rates of extinction, and (3) the ability of tropical environments to absorb a greater number of species. The last factor implies a strong historical component in the sense that closely related species may accumulate in a particular region over a long period of time. But what is a long period of time? MacArthur[14–16] suggested that a combination of spatial differentiation of tropical forests and patchy geographical distributions of individual species promotes a greater diversity of species in the tropics, and that these effects are most pronounced in tropical environments characterized by high productivity and low seasonality. MacArthur's explanation of high species abundance in the tropics[16] indicates two major components to such diversity: a pattern of local diversity, in which some portion of the overall diversity of a group in a region is accounted for by numbers of species within and among an array of habitats, and a global pattern, in which micro- and macrobiogeographical effects add to the effects of local diversity. Thus questions of tropical diversity must be defined in terms of the area under consideration and the ability to distinguish among different kinds of habitats within an area. As will be discussed later, the issue of a relationship between the degree of environmental predictability and of ecological specialization also determines the species abundance level of a particular environment[15,16].

Historically, the lowland tropical rain forest has intrigued biologists with explanations of high species abundance as indicated in the writings of Dobzhansky[17] and Corner.[18] The observed high species abundance in such tropical environments has indeed promoted a wealth of elegant and detailed studies of tropical insects and other organisms. Thus, even though it goes beyond the scope of the tropics *per se,* Gotwald's monograph on the morphological diversification of ants,[19] with an emphasis on mouth parts and feeding, takes on particular significance since the tropics contain the greatest diversity of ants. The detailed studies on the nesting habits of *Trigona* bees by Michener[20] and Wille[21] reveal that diversification in nesting habits, rather than feeding habits, has been a major pathway of environmental modification resulting in the maintenance of high species abundance both locally and regionally. Both Fittkau[22–24] and Illies[25] have discussed the ecological diversification of aquatic biotypes in Amazonian streams, suggesting strong correlations between various biotype classes and the kinds of aquatic insects and other macroinvertebrates found in each. Their studies again provide an example of regional diversification of insects and other invertebrates in the tropics, indicating something everyone knows, namely, that geographically distinct environments sometimes have very different collections of taxonomically related organisms. Trophic relationships generally not present in Temperate Zone communities, such as the associations of certain dipterans with swarm raids of army ants studied in Central America by Rettenmeyer,[26,27] have added immensely to our understanding of just how ''ecologically saturated'' the tropical rain forest environment is in terms of unique life styles of insect species found there.

The picture that emerges is that the tropics are indeed a heterogeneous mixture of different types of environments and that within each of these environments, the many different organisms together exhibit a range of generalized and specialized life styles. I define *life style* as what the individual organism does during its lifetime, the range of such "itineraries" represented by all the individuals within a population, and the underlying reproductive capacities and breeding system in the population. A thread that weaves all of this together is how the individuals in the population allocate energy to different components of life style.

The great difference in tropical environments is exemplified by Hubbell's analysis of trees in a tropical dry forest[28] and comparisons to tropical wet forests and Temperate Zone forests. He demonstrated (see Fig. 1.1) that tree species are

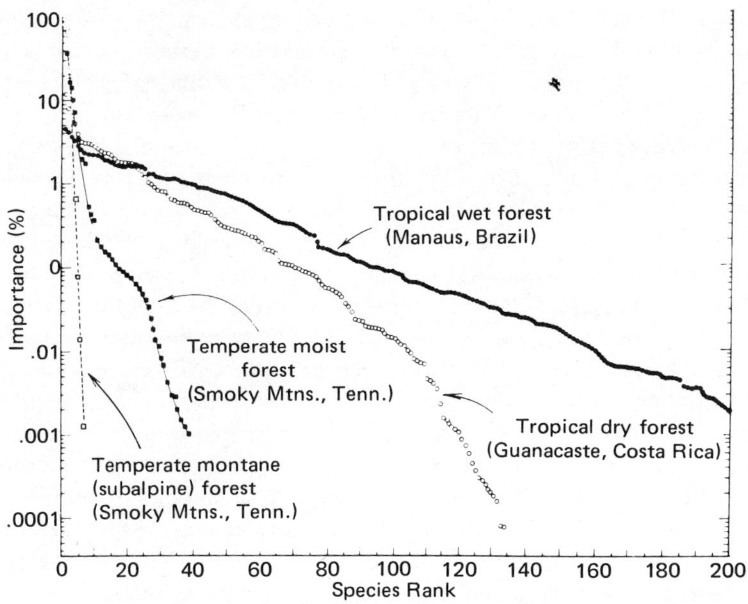

FIGURE 1.1. Ecological dominance–diversity curves for two tropical and two temperate forests. Importance values for the temperate forests are based on annual net production; importance values for Costa Rican dry forest are based on cross-sectional areas for all stems of a given species, and for the Amazonian forest from above-ground biomass. Although the comparisons are not exactly based on the same data sets, the pattern is clear: dominance of individual tree species is greater in temperate forests even though some species of tropical trees exhibit clumping. [From S. P. Hubbell, *Science* **203:** 1299–1309 (1979). Copyright 1980 by the American Association for the Advancement of Science.]

often clumped in tropical forests as they are in temperate forests but that frequent disturbances in the tropical dry forest environment create clumped distributions of many species. Comparisons of such data with tree species distribution in lowland tropical wet forests suggest that the distribution of plant resources for insects are distributed differently in different kinds of tropical forests. Such heterogeneity is expected to result in different life styles and population dynamics for insect species. As the great majority of insects are herbivores or plant predators, it is the interrelationships of plants and insects that probably account for the spatial and temporal structuring of insect populations in the tropics.

The recent studies on the coevolutionary interactions between plants and insects suggest that insects contribute in a significant way to the distribution of abundance of plant species, especially in the tropics,[6,29] but also that plants influence insect distributions as well. It has been known for some time that geographical and habitat diversification within plant taxa often result in similar shifts in distributions of pollinators such as bees. Hurd and Linsley have traced the spread of certain Cucurbitaceae and pollinating bees in the Americas.[30] The elegant studies of Gilbert[31] have also indicated that certain species of pollen-feeding *Heliconius* butterflies have acted as major selective agents in the population biology of some Cucurbitaceae. (See Chapter 3.)

The environmental heterogeneity of the tropics is also reflected in the kinds of climatic regions found there and the extent to which biologists have conducted studies in them. Thus the comparison of the recorded insect faunas of the Galapagos Islands and British Isles by Linsley and Usinger[32] indicates that a much higher number of families, genera, and species of insects are found in the British Isles. They ascribe the depauperate condition of the insect fauna of the Galapagos to accidental colonization by species from the mainland of South America. The fact that much of the Galapagos are scrub and desert may also contribute to the relatively low abundance of insect taxa found there. One might speculate that the bees studied by Linsley and Usinger were tracking the environment in the sense of changing their geographical distributions in response to the spread of both wild and cultivated cucurbits. Certain types of obligate associations between ants and plants in the tropics as studied by Janzen[33] reflect similar joint distribution patterns. Yet for other groups of insects, tropical island faunas are considerably diverse, sometimes even more so than the adjacent mainland. Paulson[34] suggests that the higher number of Odonata species in the Bahamas than in southern Florida and the Florida Keys is due to the absence of some undetermined environmental factor that typically reduces both species number and populations of these insects in the latter regions. From various studies of faunas on tropical islands, it is apparent that neither absolute island size nor area is the major determinant of faunistic size and complexity within different trophic levels. Another major component of diversity is the collection of habitats present. The interaction of regional size, habitat diversity, and climatic factors molds the communities associated with regions. In this book I hope to provide examples of the effects of climate, region size, and habitat availability on the diversity of insect faunas found in the tropics, in addition to examining the effects of these environmental factors on population ecology.

Kohn and Orians[(35)] have pointed out the need to examine the ecology of closely related species. When dealing with tropical insect faunas, it is necessary to recognize those characteristics related to taxonomic affinities and those expressed as a result of ecological conditions. Given the greater number of closely related species in the tropics, such analyses are very critical in attempting to explain both global distributions (*sensu* MacArthur[(16)]) and local distribution of species. Again, there is the need to recognize the historical component of occupying a particular environment or range of environments, and the contemporary selection pressures and nongenetic response opportunities for insects.

Because of their great diversity in the tropics, relatively short generation times, and often readily distinguishable phenotypic features, insects in general are ideal organisms for studies concerning distributions of species in relation to resources and environmental changes. The great diversification of ants in the arboreal habitats of tropical forests has provided the evolutionary opportunities for other arthropods to converge on them in terms of mimetic resemblance. An example of this phenomenon is given in the carefully documented study of ant-mimicking salticid spiders (see Reiskind[(36)]). In some groups of insects in tropical regions, there is the opportunity to study various life styles and habits of individual genera and species that indicate possible evolutionary pathways of ecological specialization and associations among closely related forms (e.g., Erwin and Erwin[(37)]). Some forms may exhibit behavior and habits intermediary in character state to other kinds of life styles. Canopy orientation, for example, has been shown to be the primary orientation mechanism underlying home range behavior in an Old World tropical ponerine ant.[(38)] Yet at least one New World tropical ponerine, *Paraponera,* is an active canopy forager unlike the Old World *Paltothyreus* studied by Holldobler, in which foraging is a horizontal or lateral activity on the forest floor. One might suggest that actual canopy foraging in a form such as *Paraponera* might be an intermediate step in more specialized forms of foraging as exemplified by *Paltothyreus*.[(38)] Tropical insect faunas overall provide the opportunity for comparative study of the behavior and ecology of closely related forms or convergent forms, because there is generally a greater diversity of insect taxa in tropical regions. Paramedical ecological studies of disease vector insects in the tropics have illustrated that local high diversity of vector dipterans such as blackflies supports a larger spectrum of disease microorganisms (e.g., Takaoka[(39)]). Sometimes the pattern of distribution of the closely related insects is defined along habitats since different forms may diverge to occupy different habitats (Figs. 1.2 and 1.3).

FIGURE 1.2. Border strip of mixed primary and advanced secondary tropical rain forest in northeastern Costa Rica. The low vegetation to the left is regenerating young secondary forest about 3 years old. The species composition and distributions of populations of individual plant species differ greatly between the two areas, providing a form of spatial environmental heterogeneity for insect communities associated with these habitats.

Perhaps an even more relevant reason for their study in the tropics is that insects are a major link in the food chains of tropical communities and ecosystems. The botanical work on the structuring and floristic composition of tropical forests cannot be divorced from the impact of insects on such systems. A bidirectional flow of information about the environment results in responses by the receiver. Thus making an attempt to account for the distribution of certain tree species in a tropical forest may not be entirely possible without examining the population biology of pollinators and seed dispersal or seed mortality agents such as certain insects. Likewise, the distribution of the insects may not be explained without an examination of the tree species acting as resources for the insects. Perhaps the greatest contribution of tropical biology in the last 20 years has been the elucidation of the interdependence between plants and animals in shaping the static and dynamic properties of tropical forests. Thus although Marston Bates studied the natural history of anopheline mosquitoes many years ago, with an emphasis on the role of these insects as disease vectors, studying these insects today may require a different approach. One might wish to examine whether the males of these mosquitoes are in fact effective pollinators of certain understory or even canopy flowers in the tropical rain forest environment.

FIGURE 1.3. Mixed primary and secondary tropical rain forest in northeastern Costa Rica. The forest patch shown here is dominated by the canopy tree *Pentaclethra macroloba* (Leguminosae) from primary forest, mixed with another dominant, *Goethalsia meiantha,* characteristic of advanced secondary forests in this region of Central America.

The role of insects as food items in tropical ecosystems cannot be disputed. Apart from the role of herbivorous and pollinating insects as major selective forces in the evolution of tropical plant assemblages,[(40–42)] many insect species are sources of nutrients for other organisms, including parasitic and predatory arthropods and many vertebrates. Just as the structuring of fruiting seasons in conjunction with the spatial distributions of individual tree species in tropical forests is clearly a major selective force in the development and maintenance of bird species diversity in tropical environments,[(43)] the availability of insects, in diverse forms, with diverse feeding and life style habits, and activity patterns compose a variegated ecological fabric exploited by other organisms. Thus it is not surprising that studies of stomach contents of crocodiles from tropical African lakes generally indicate that insects compose the bulk of food for young age classes of these animals. In such systems, the predators are active at times of the day when major foodstuffs such as belastomids and dragonflies were active or completing final ecdysis, respectively.

Hubbell[(28)] and many others have called attention to the need to study the flora of certain regions of tropical America. If we are going to study the impact of insects on plant species in the tropics, we should have reliable information on the names of the plants and their ecological distributions across regions and habitats. There is also a need for more studies of high-altitude plants and animals in the tropics. A great deal of emphasis has traditionally been on lowland systems, although the trend is changing. H. G. Baker and G. W. Frankie's analyses of flowering and pollinator phenologies in Costa Rica's Monteverde cloud forest will provide new information on the structuring of tropical vegetation from that region. A. M. Shapiro's life history and physiological studies of high-altitude pierid butterflies in the Andes will provide an explanation as to why certain groups of insects thrive there while others do not. Such studies complement nicely the earlier survey approach of detailed accounts of cloud forests such as that of Myers[(44)] in Panama.

Out of all of the above, I hope to convey the importance of various approaches to the best study of insect populations in tropical environments. First, there is clearly the need for more detailed natural-history studies, particularly on host plant relations and herbivorous predators and parasites. There is also the need to continue to develop research programs designed at examining plant–animal interrelationships and effects of climate on them, in less popular regions such as high altitudes and deserts. There is a need for more studies of aquatic ecosystems and how they are affected by habitat disturbances. And there is of course the need for the elegant conceptual and experimental studies that have developed largely through the Organization for Tropical Studies programs in the past 20 years.

A great deal of new information coming from the study of insect populations in the tropics has contributed in a significant way to our general view of the natural world and man's impact on it (Fig. 1.4). There has been a reshaping of traditional ecological concepts and a refreshing reexamination of the strategies by which insects and organisms respond to changes in the environment. A basic theme of this book, and one that does not necessarily reflect the popular viewpoint, is that we must examine what insects do in "adapting" within (I am avoiding the phrase "adapting to") the collection of intrinsic and extrinsic variables in the environment, not necessarily in terms of genetic or evolutionary response, but in terms of response plasticity at the behavioral or physiological level. Short-term responses to environmental changes might be more frequent and more the rule than evoking long-term or evolutionary responses. Another goal of this book is to emphasize the heterogeneity of the tropics. For this reason, I include a discussion of seasonality, and as with the other chapters, I draw upon specific examples to illustrate a particular phenomenon.

Tropical environments vary tremendously in terms of climatic conditions, particularly in terms of monthly distributions of precipitation and total annual precipitation. There can be considerable variation from year to year for total precipitation, and the length of a dry season may also vary considerably in

FIGURE 1.4. Clearing away of advanced secondary tropical rain forest in Costa Rica, for conversion into cattle-grazing pastures. Such activities create a dynamic condition in the wet tropics in which shifting distributions of herbivorous insect species follow major perturbations in the local or regional environment.

different years. Thus climatic heterogeneity is present within and among different tropical regions.[45] A glimpse at the precipitation regimes for Central America illustrates this point.[45] Even over smaller areas such as French Guyana, precipitation distribution varies considerably.[45] Such patterns coupled with other environmental factors impose an instability on tropical ecosystems. Further change is imposed by habitat disturbances resulting from both natural and man-made modifications in the local environment. There is also tremendous altitudinal variation with profound effects on the distribution of organisms, as discussed by Janzen.[46] Such considerations dispel the notion of a stable and constant tropics. There is a resulting uneven distribution of species over the tropics. In the next chapter I examine briefly how organisms respond to the environment and environmental change. Thus Chapter 2 is essentially an introduction to the concept of adaptation, and it forms a basis for the rest of the book.

Chapter 3 provides a brief examination of insect physiology and structure in terms of survival and reproduction within tropical environments, with an emphasis on feeding and possible diapause strategies associated with the dry season. This discussion blends into a discussion of plant–insect interactive mechanisms, including plant defenses against insects and the development of coevolved associations. As with most of the chapters, the information is topical in format, since the literature on this subject is growing rapidly and other recent publications such as Janzen[42,47] provide more in-depth and conceptual discussions. From this rather qualitative overview I examine modes of adaptation in insect populations and give examples of insect population studies from the tropics, and include a few from Temperate Zone regions that are relevant to the issues discussed.

Chapter 4 is devoted to plant–insect interactions in tropical ecosystems, with an emphasis on herbivory and pollination. There has been a flurry of research in both aspects of plant–insect interactions in the past 20 years. I also attempt, in other chapters, to examine relationships between insect diversity and habitats, and also devote one chapter to selectional and historical responses of insects to resources over habitats and regions in the tropics.

In some of these chapters, for example, Chapter 9, I incorporate some of my own research data. In this chapter, I suggest a model of cacao pollination and production, examining the question of optimal yield from the viewpoint that cacao cultivation may cause changes in pollination strategy in the wild. In Chapter 7 I include a brief discussion of different tropical regions based on climate. The book is clearly oriented toward a lot of recent work done in the New World tropics, although I attempt to incorporate studies from other regions as well. My overall theme is that complex abiotic and biotic factors mold the spatial and temporal distributions of insect species in tropical environments. We must look at mechanisms promoting abundance of individuals and high species number in both habitats and over regions. The approach is basically ecological and natural-historical *sensu stricto,* but I attempt to incorporate some aspects of evolutionary genetics whenever appropriate and comfortable for me to do so. Research on tropical insects, and their relation to other portions of the environment increases our understanding of why the tropical regions of the world contain the greater taxonomic diversity and associated patterns of ecological specialization for many groups of insects than do the other major faunistic regions of the world. Some groups of insects in the tropics may be deliberately imported to the northern latitudes as part of research on the effectiveness of introduced predators or parasites on pest species associated with food crops (e.g., Marsh[48]).

This book presents an overview of some of the major discoveries of how insect species are integrated into the complex and varied environments of tropical regions. Janzen[42] has emphasized the dynamic nature of research in tropical biology, and the future directions of it. There is a need to study in detail the ecology of individual species, especially from the viewpoint of analyzing the other organisms with which each species interacts, and the effects of seasonality and other local weather conditions on populations. Such studies are coupled with tedious natural history studies at the autecological level. A great deal of tropical biology has examined interactions of organisms over relatively short periods of time. There is a need for long-term studies on the census histories of animal and plant populations to detect cycles in abundance and other phenomena. The conservation of distinct tropical habitats (Fig. 1.5) in different tropical regions will ensure the resources essential for understanding the complex nature of food chains. At the opposite end of the spectrum, but not exclusive, is the analysis of whole ecosystems in tropical regions. It seems to me that both approaches are necessary for understanding the temporal and spatial properties of populations of component species in tropical ecosystems.

The description of population census histories, patterns of resource exploitation, and strategies of fitting into the environment for insect species in the tropics must be linked to the issue of explaining the patterns of diversity of insects within and among habitats, and over larger regions of the environment. Gradients in diversity of insects across such units of the environment are generally greater in the tropics. We expect to find habitats or regions in which the numbers of insect genera and species associated with various trophic levels in communities are considerably higher than for comparable Temperate Zone environments, and we

FIGURE 1.5. Understory of premontane tropical rain forest in Costa Rica.

also expect to find others in which such diversity is low. In the sense of Vandermeer's[49] definition of a community as a set of "operational trophic units," the collection of individuals of an insect species composing a population or subpopulation associated with a certain habitat is the fundamental component of tropical diversity patterns that warrants analysis.

The population is conceptualized as a collection of reproductively capable individuals as well as juveniles (having an age distribution) changing to varying degrees through space and time. Depending upon habitat, seasonality, trophic position, resource availability, character states, and other factors, some populations of insects might be physiologically or genetically programmed for traits favoring colonization, while others are in some sort of equilibrium with other components of the community and therefore programmed for fitting into the environment at hand. We do not expect all species within a given trophic level to be in similar states of population dynamics or degrees of integration into communities. As will be developed in this book, the degree to which insect populations are influenced by biological and physical features of environments will vary considerably among different habitats and regions in the tropics. The existence of a wide range of habitats, from both natural and man-made perturbations to local environments (Figs. 1.2–1.4), has provided the opportunity for insect species to diversify in life styles and for overall diversity of insects to increase in the tropics. Such environmental mosaics establish different sets of changing environmental conditions or challenges for insects and other organisms. We expect patterns of adaptation within populations and for whole sets of populations to vary considerably in different types of tropical habitats, and to be influenced by seasonality or a lack thereof.

We can examine some of the patterns and mechanisms of adaptation of insects to tropical environments, but we can only scratch the surface. Many species have not been studied, and the trophic organization of different communities has not been determined. We seem to have a lot more information about trophic organization at the plant–insect interface than for other levels of community organization. Extensive data on tropical litter communities are lacking, save for a few such as the interesting study in which Stanton[50] compared temperate and tropical litter communities. Stanton compared the species richness and diversity of litter mites in three habitats in Costa Rica and Wyoming and found a greater species richness and diversity of mites in forest litter at both localities. A greater diversity of mites was found in the tropical forest habitat than in all other habitats sampled, and a greater proportion of mite species in the Temperate Zone samples was classified as colonizers.[50] Such data from decomposer communities provide further support for the greater diversity and species richness for organisms in tropical forests. There is also a lack of information on the resource utilization patterns and population dynamics of parasitic wasps and flies in the tropics and on the degree to which such organisms regulate populations of herbivorous insects in tropical environments. The biogeographical studies of mammalian ectoparasites by Wenzel and Tipton[51] and others have provided some data on the broad distribution of parasitic arthropods in the New World tropics, but the omnipresent need is to examine the role of parasites in the structure and dynamics of individual communities.

Long-term studies of insects are needed to determine the degree to which insect populations are cyclic in the tropics. A plethora of short-term studies has outweighed long-term studies of 5 or more years. Data on the communities of organisms in both standing water and running water aquatic environments in the tropics are also wanting. Given the predatory role of both larval and adult Odonata and the diversity of Odonata in some tropical regions,[34] such studies might be particularly elucidating in terms of the impact of these insects on other species in terrestrial and aquatic habitats. There are many other examples, but at least one more deserves mentioning: the need to gather extensive field data on the natural history and ecology of "showy" insects from the tropics, insects that are likely candidates to end up in museum or private collections or specimen dealerships throughout the world. The riot of gaudy-colored butterflies, beetles, and other insects from the tropics has created an international market for these insects. Yet before we can establish realistic guidelines for protecting individual species of these insects, or setting aside large tracts of land for ecological preserves, we need to determine the population structure, resource requirements, and degree of "ecological flexibility" of each species in different but representative portions of its geographical range. Such a body of data is growing for some butterflies of the New World tropics but is generally lacking for butterflies in other tropical regions of the world.

This book is also not an attempt to summarize all that is known about the ecology of insects in general. One could argue that a book on tropical insects somehow excludes a great deal of knowledge on insects from other regions. I take the view that the time is right to bring together some of the interesting and informative work being done on the tropics, and a good deal of this work concerns the interactions of insects, particularly herbivores and pollinators, with other organisms. A cornerstone of this view is that the tropics represent a unique plethora of strategies concerning the ways in which organisms fit into the environment, and in some ways examining issues of population regulation, competition, effects of abiotic factors on organisms, mutualisms, and other ecological phenomena is easier than in Temperate Zone systems. This is not to say that one excludes or minimizes the importance of similar studies in the Temperate Zones. Just the opposite is true. The beauty of tropical studies is the opportunity for meaningful comparisons with Temperate Zone studies. Thus this book, drawing heavily upon some of the major directions in research on tropical insects, supplements other books concerning insect ecology, general ecology, and population biology. Realizing the magnitude of the task of bringing together the literature on tropical insect populations, it is my sincere hope that omissions of seemingly pertinent material from this book are not construed as an oversight.

CHAPTER 2

INDIVIDUAL AND POPULATION RESPONSES TO ENVIRONMENTS

2.1. ECOLOGICAL RESPONSE

The degree to which an insect or other organism becomes integrated into the environment is dependent in part upon how long the species has occupied it. The extent to which a species becomes a resident of a certain community is a function of many factors, of which the following often prevail: (1) the number and abundance of ecologically similar and/or closely related species present in the community, (2) the ability of the species to obtain sufficient resources for survival and breeding, (3) the ability of the species to withstand biotic mortality agents such as pathogenic microorganisms, predators, and parasites, and (4) the ability to cope with shifts in (1) through (3) resulting from seasonality and to cope with the abiotic changes in the environment brought into play by seasonality.

The studies of Lack[52,53] have provided ecologists with evidence that closely related species sometimes exhibit ecological isolation, achieved by geographical separation among the species, by habitat, or by feeding habits where ranges overlap. A basic assumption of such analyses is that the resources required by each species are limited but sufficient to result in competition for them. The degree to which ecological isolation plays a role in determining the geographical and habitat distribution patterns of species is also governed by the effects of seasonality on alternative food supplies or other resources in short supply. Thus although many tropical rain forest understory butterflies in Central and South America exhibit some degree of specificity for feeding on certain kinds of rotting fruits and molds, a seasonal period of high fruit abundance for particular tree species may result in a form of "ecological convergence" in which the butterfly species jointly exploit the abundant fruit supply. A similar argument can be made for some genera and species of butterflies visiting inflorescences in tropical forests (Figs. 2.1 and 2.2). The most difficult issue in terms of detecting mechanisms of ecological isolation in habitats is the degree of

FIGURE 2.1. An unidentified hesperiid butterfly visiting the inflorescences of the bush *Serjania* cf. *atrolineata* Saw. & Wright (Sapindaceae) during the dry season in the central highlands of Costa Rica. In this region and during the lengthy dry season, adult butterflies aggregate along shaded and moist strips of vegetation along streams, where they feed on nectar from flowers. Many species of plants in such habitats bloom during the dry season. Breeding tends to diminish as the dry season advances and some butterflies, and presumably other herbivorous insects associated with foliage, enter into a reproductive diapause. Under such conditions, diapause represents a physiological response to a decrease in correct oviposition substrates and larval food (i.e., young leaves). Depending upon the species of insect and foodplants, such a pattern varies considerably in seasonal tropical environments.

resolution of analysis. The literature is replete with examples of studies illustrating that one's definition of the niche of a species is a major factor in determining whether or not two or more species have niche overlap and if so to what extent.

Through detailed natural-history studies in the case of tropical insects, it will be possible to determine the temporal and spatial variations in (1) what resources are utilized, (2) what percentage of time the average individual spends on each resource, and (3) what other species use the resources and the time-frequency distribution of utilization. The issues and problems of defining the

FIGURE 2.2. The nymphalid butterfly *Metamorpha stelenes* visiting the inflorescences of the bush *Serjania* cf. *atrolineata* in the central highlands of Costa Rica during the dry season. The preponderance of blooms of this bush represents a concentrated food resource for adult butterflies at this time of the year in this region, and because breeding tends to decline as the dry season advances, adult butterfly populations become increasingly characterized by very worn individuals such as the one shown here, i.e., the age distribution of adults is skewed toward the older classes. The larval host plants of *M. stelenes,* various genera and species of herbaceous Acanthaceae, tend to cease vegetative growth and turn brown as the dry season advances, thereby dropping out of the local assemblages of potential oviposition sites and larval food sources for this species and a close ally, *Anartia fatima*.

ecological niche have been reviewed and expanded by Vandermeer.[49] How a particular insect species responds to the environment is shaped by the factors mentioned above, namely, interactions with potential or actual competitors, availability of required resources, effects of biotic and abiotic mortality factors affecting population size and structure, and the effects of spatial (within, between, habitats) and temporal (seasonal) variations in these effects of environment. The size and the age structure of the population also shapes the response of individuals to these environmental parameters.

2.2. NATURAL SELECTION

Clearly, a major approach to the issue of how organisms respond to the environment has been dominated by the theme that a great deal of such response in fact involves evolutionary change. The theory of gradual change by evolutionary processes is based largely on the premise that natural selection acts on the individual in the population. For the processes of evolutionary change, it is generally believed that the population is the most important level of biological organization. Since each population represents a gene pool with varying degrees of confinement (lack of gene flow), the evolutionary process is held to be changes in the genetic makeup of the population over time. Different populations of the same species can change in different ways through time or follow different adaptive pathways, resulting in distinct or similar adaptations or traits. Lewontin and White[(54)] demonstrated such effects in populations of the Australian grasshopper *Moraba scura*.

A major feature of populations is the distribution of individuals in space and time, described by habitat or regional surveys and census histories, respectively. The size or density of a population is determined by the interaction of birth rates, death rates, and dispersal. Changes in the environment often result in changes in these characteristics, either toward increasing or decreasing population size, if not maintaining stable conditions. Some of these effects are shown in Fig. 2.3.

Changes in the genetic makeup of a population also arise from genetic drift. Wright[(55,56)] demonstrated rigorously the effects of drift on the makeup of populations, showing that drift effects are most likely in small populations (i.e., small in size relative to other populations of that species). As each population consists of a collection of individuals or genotypes, it is the array of genotypic variation in the population that determines, under the view of evolutionary adaptation, the capacity of the species in that particular environment to cope with the environment. Such reasoning has resulted in a fusion of concepts and models from population genetics and population ecology (Table 2.1) that forms the basis of scientific study in population biology.

2.3. SOME PROPERTIES OF POPULATIONS

Two major aspects of population ecology are the census history generated by a population placed in an unlimited environment through colonization or range expansion, and the interactions of populations of two or more species existing in the same environment and at the same time. In terms of the tropics, one might expect that the observed accumulation of closely related and ecologically similar species of insects in some communities, such as indicated by Janzen[(41,47)], and the existence of a broad range of newly created secondary and

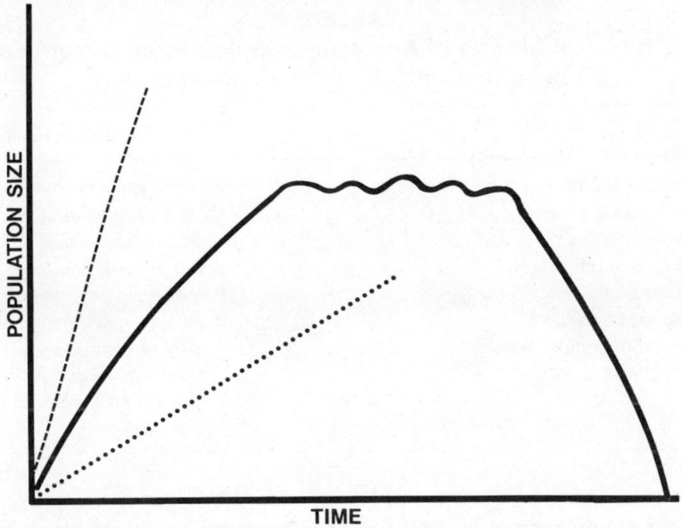

FIGURE 2.3. Relationship between time and insect population size. Depending upon the kinds of selection pressures operative on an insect population, population census histories will vary considerably. The dotted line shows a population in which increase in population size is gradual, perhaps as the result of strong selection for life history characteristics favoring a low reproductive potential (r), or perhaps as a result of intense predation of a density-related form that keeps the population from realizing the carrying capacity of the environment. The dashed line shows a population on the course for overexploiting its resources, through a high reproductive potential or as the result of relaxed biotic control mechanisms. The solid line shows the three major phases of population census history: growth, stabilization or realization of carrying capacity of the environment, and decline. If an insect population realizes the carrying capacity of the environment and produces an evolutionary adjustment, such a population may become vulnerable to extinction if selection pressures shift in such a way so as to require increased reproductive potential by the species (a higher r). All three kinds of census history are expected in the tropics, where the environment is considered heterogeneous in terms of selection pressures operative. The dashed-line ''strategy'' may be characteristic of colonizing episodes, when an insect species enters a new adaptive zone (habitat) and requires an increased rate of population growth.

agricultural habitats would indicate that interacting sets of species and colonization events are common phenomena in the tropics.[57] We return to the data on these in subsequent chapters. But returning to Fig. 2.2, it is seen that the entry of a population into a new environment with unlimited resources is characterized by four stages: (1) rapid growth of the population in the new environment, (2) a period of population stabilization, perhaps with minor fluctuations in abundance, (3) a population decline, and (4) extinction.

The vast amount of animal extinctions evidenced in the paleontological record is in itself proof of the adaptive role of extinctions, in that major changes in the environment flushes out some taxa while others are able to survive and

TABLE 2.1
The Classical Models of Adaptations in Populations, from Both Population Genetics and Population Ecology

Population genetics	Population ecology
Hardy–Weinberg equilibrium	Population parameters
Theory of inbreeding	Models of regulation
Models of selection	Density dependence
Balanced polymorphism	Density independence
Directional selection	Predator–prey systems
Stabilizing selection	Population–environment interactions
Selection in and of populations	Theory of community and regulation of diversity (species packing)
Breeding structure	Social systems

exist in altered environments. The flux of secondary growth habitats in tropical regions suggests that both local extinctions of some insect species and shifts in adaptation in others are a major component of the dynamic condition of such habitats. One might expect that changes in food availability and quantity, along with abiotic changes such as increased exposure to sunlight and precipitation, constitute some of the selective forces in newly created tropical terrestrial habitats.

The census history of a population—the record of change in size of the population through time—involves periods of population growth, stabilization, and decline. Many studies have shown that both intrinsic and extrinsic factors cause changes in the population dynamics of insect species and other organisms. It is the interaction of changes induced by the membership of the population itself and by changes from other populations or from abiotic factors that account for the observed census history. The relationship between population density and population size has been studied by many population ecologists, including Thomas Park, who has used the *Tribolium* flour beetle to study insect population dynamics (e.g., Park[58]). As population size increases, so does density and therefore intraspecific competition for a set of resources in the environment. The mortality rate in laboratory cultures of nymphs of the American cockroach, *Periplaneta americana,* increases with increasing population density, with mortality generated by periods of high population growth in which nymphs are injured and killed at ecdysis, retardation of the population growth rate due to density-related effects adversely affecting survival, and eclosion of adult roaches.[59] Dense populations are characterized by adult roaches with lower average body weight than low-density populations.[59] Pimentel[60] suggested that a genetic feedback mechanism regulates the size of populations for herbivores, predators, and parasites. Under Pimentel's model, in the plant–herbivore system, for example, a herbivorous insect species specialist on a particular plant species establishes a sphere of selection pressures influencing the defense strategy of the plant against herbivores (see Chapter 4) and through successive generations such selection pressure changes the genetic makeup of that plant population.[(60)] Such a change subsequently affects the density of the herbivore population. Pimentel envisions food chains as sets of interacting genetic feedback mechanisms.

Much of the diversity of tropical herbivorous insects is accounted for by specialist feeding behavior in which individual species tend to have restricted host plants (see Chapter 4). In one sense, the concept of coevolution between plants and animals is a restatement of the genetic feedback mechanism proposed some years ago: as herbivore density increases on a particular plant species, particularly where the herbivore in question is a specialist, and if the pressure is sustained over subsequent generations of the host plant species, selection will favor the expression of increased resistance to the herbivore in subsequent generations, and such changes will cause one of three responses in the herbivore population: (1) it will adapt to the new conditions of the host plant; (2) it will experience an ecological shift to another host plant species with a less developed defense or to qualitatively different species capable of being overcome without large losses in fitness; or (3) it will experience extinction. The evolutionary approach suggests that the outcome is largely determined by the genotypic variation in the population. Brereton[(61)] has reviewed some examples of self-regulation in animal populations, suggesting that some populations are capable of such regulation. Such self-regulation is believed to be a response to changes in the environment such as crowding and fewer resources. Deleterious crowding may induce physiological responses in individuals within a population, such as reduced average fecundity or increased tendency toward dispersal, with the result of slowing down the growth rates of populations. The entomological literature from the Temperate Zone is full of examples of the relationship between density and emigration in aphid populations associated with various plant species in seasonal regions.

The population dynamics and pattern of environmental exploitation by insects in the tropics is molded by the historical events resulting in an accumulation of ecologically similar species in a particular region, and by the more contemporary ecological factors that affect the size and distribution of populations within and among habitats. It is difficult to view both types of forces as separate entities. The centers of evolution for many groups of insects are located in the tropics (Haffer[(62)]), and superimposed upon this pattern is the existence of past forest refugia in South America during the Pleistocene in the case of the New World tropics. Such conditions have molded the general distribution of insect taxa in the tropics. Pianka[(63)] had distinguished between the historical accumulation of species in the tropics and the contemporary ecological factors that promote species abundance in the tropics. A brief discussion on the prevailing views on tropical diversity follows as a prelude to further considerations of how insects and other organisms respond to environmental conditions.

2.4. ECOLOGICAL DIVERSITY AND THE TROPICS

Pianka[63] reviews the major theories on the origin and maintenance of tropical diversity, discussing the time theory, the theory of spatial heterogeneity, the competition hypothesis, the predation hypothesis, and the climatic stability hypothesis. None of these approaches is necessarily mutually exclusive, and it is likely that the explanation of diversity must be considered for different sets of taxa at a time and that each of the above explanations may hold to some degree in each instance. The spatial heterogeneity approach professes that regions in the tropics tend to be more diverse in terms of topographic relief and number of habitats (macrospatial heterogeneity). The question of microheterogeneity is not included (within habitat diversity). MacArthur[15] has discussed the importance of habitat heterogeneity in explaining tropical diversity.

The competition hypothesis advances the notion that more species compete for limited resources in the tropics, thereby promoting ecological specialization and room for more species to fit in. The idea was put forth by Dobzhansky,[17] who suggested that competition is the most important mechanism regulating the high diversity of the tropics. The predation hypothesis, supported by data from Paine,[64] Connell,[65] and Janzen,[29] says that intense predation keeps populations of some species at certain trophic levels well below the carrying capacity of the environment, thus allowing more space for other species to fit in. Whereas the competition hypothesis requires some sort of equilibrium population sizes among potential competitors to drive selection leading to specialization, the predation hypothesis is essentially a nonequilibrium model. The climatic stability explanation advances the idea that the generally stable and benign climates of lowland tropical areas have fostered the evolution of specialized species, resulting in a greater number of species occupying various habitats. Under both the competition and time stability hypotheses, ecological specialization is thought to be the response to the environment for many species, thereby making it very difficult to distinguish between the two hypotheses by controlled experiments. The productivity hypothesis developed by Connell and Orias[66] predicts that greater production results in greater diversity, all else being equal. Presumably increased productivity provides the biomass necessary for eventual ecological diversification into many species, each exploiting a different part of the environment. A corollary of this hypothesis is that the increased temporal heterogeneity of the tropics provides different periods (seasons) to which different species in the process of evolving can adapt, thereby allowing more species to fit into a particular region. The explanation may hold in those tropical regions where there is considerable temporal heterogeneity.

This hypothesis is most interesting from the standpoint of primary production and possible ramifications for the development of insect diversity. Pianka suggests that plants use their own homeostatic mechanisms and storage capacities to increase the stability of primary production, which in turn buffers plant species against climatic variability. Under the productivity hypothesis it might be envisioned that herbivorous and pollinating insects form a major portion of the sphere of selection pressures driving the diversification of plant species in the tropics, resulting in an increased accumulation of both plant and insect species. Increases in productivity by necessity involve decreases in reproductive ability, as discussed by Cody[(67)] and others. If a greater amount of energy is allocated to vegetative structures and maintenance, less is available for the elaboration of gametes. Clutch size is a measure of reproductive ability. Examples exist of organisms of reduced clutch sizes in tropical forms. Reduced clutch sizes imply smaller populations and therefore more space for other species. Furthermore, the reduction in clutch size reduces total energy requirements and increases the chances to survive in less productive areas of the tropics (Pianka[(63)]).

MacArthur[(15)] points out that the majority of tropical environments are less seasonal and more productive, and that the regions with strong seasonality and lower productivity, such as tropical deserts and mountain tops, tend to have fewer species. Clearly such patterns reflect differences in intrinsic ability of individual genotypes or phenotypes in populations to cope with grossly different sets of environmental factors. Thus a desert or mountain top in the tropics is considered to be a marsh environment in the sense that primary production is lower due to less plant biomass present in most such regions, and that the constant or predictable severe climatic conditions establish an ecological barrier for many taxa. One might predict that the reduced faunas of such regions are in fact a select group of specialists in the broad sense, since the resource base is restricted to relatively few classes of resource types, and that such species have developed those physiological and morphological traits needed to thrive in such climatically harsh regions. MacArthur[(15)] also points out that both large regions and small areas in the tropics tend to have greater numbers of species than comparable temperate ones. He explains these patterns by suggesting that there has been a longer history of speciation in the tropics driven largely by interactions among species. The increased rate of speciation in the tropics has resulted in greater accumulation of species there; interspecies competition has precluded the development of communities characterized by one or few dominant species. Such effects are operative over small areas, resulting in great differences in faunas among different areas. MacArthur believes that marginal habitats in small areas are filled with species resulting as a spillover from the high productivity and nonseasonal conditions of the lowland tropical rain forest environment in general. MacArthur's graphical representation of the accumulation of a greater number of species in both large and small areas of the tropical environment is shown in Figs. 2.4 and 2.5. That environments with higher productivity favor the development of additional species capable of utilizing marginal resources is shown in Fig. 2.6.

Depending upon the amount of phenotypic flexibility for exploiting a range of resources by the members of an insect population, and spatial and temporal changes in the abundance of the species, island populations of insects may be more susceptible to both natural and artificially induced events that result in successful extinction of the species. Susceptibility to extinction will vary through the different stages of the taxon cycle.[68] Mainland populations of related or identical species are more likely to possess a greater tenacity for maintaining populations against environmental events promoting extinction. Presumably on the mainland a species may occupy a range of microhabitats or exploit a wider constellation of resources in some cases, although such patterns are certainly not universally true for all insects in the tropics. If the species in question has a broad geographical distribution and there are deliberate attempts to eradicate it (i.e., a

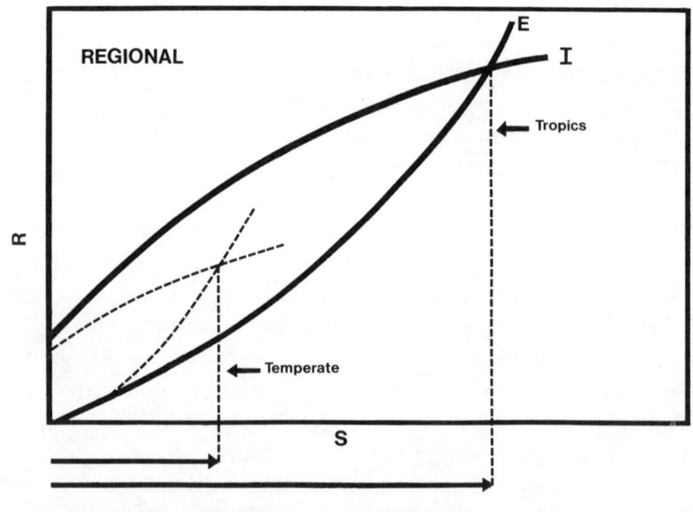

FIGURE 2.4. Relationship between area and abundance of insect species. The abundance of species in a community is the result of both regional and habitat selection pressures determining the abundance of species in an area. R is the rate of extinction plus emigration (curve E) and immigration plus speciation (curve I). Lower extinction and higher speciation rates are predicted for the tropics, resulting in a higher equilibrium number of species in an area than in a comparable Temperate Zone area (dashed line). But historical factors may prevent communities in both areas from attaining such equilibrium conditions. [From R. H. MacArthur, *Biol. J. Linn. Soc.* **1**:19–30 (1969). Reprinted with permission. Copyright by the Linnean Society of London.]

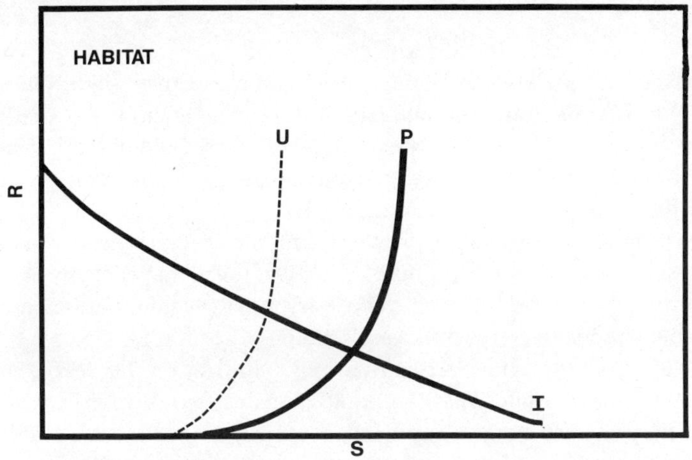

FIGURE 2.5. Relationship between productivity and insect species richness in communities: A model proposed to explain the local patterns of species accumulations in communities. In local communities, a major factor determining species richness is the productivity of the community. Recognizing unproductive (U) and productive (P) communities, the extinction rates will be different, i.e., higher in unproductive communities. Immigration (I) falls off as the number of species increases in either community as the result of competition among species. The net result is a greater equilibrium number of species in productive communities. Relatively more productive communities may be characteristic of some tropical terrestrial habitats, although a spectrum in productivity is expected for the tropics. [From R. H. MacArthur, *Biol. J. Linn. Soc.* **1**:19–30 (1969). Reprinted with permission. Copyright by the Linnean Society of London.]

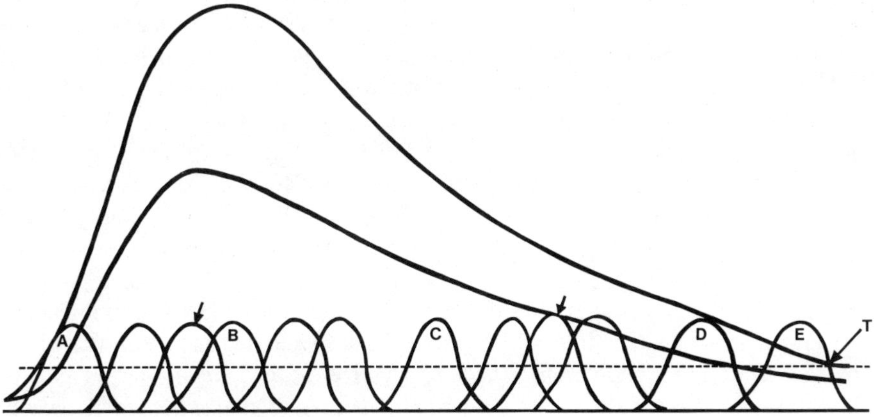

FIGURE 2.6. Utilization efficiency of various species in a community (the small bell-shaped curves) plotted against a spectrum of resources used by the species (the R axis) in a community. The large curves depict densities of resources in productive and unproductive environments; T is the threshold density needed for these species to maintain breeding populations in this community. Because the resource productivity curve is wider in a productive environment such as the tropics, more resources are available and more species can accumulate. [From R. H. MacArthur, *Biol. J. Linn. Soc.* **1**:19–30 (1969). Reprinted with permission. Copyright by the Linnean Society of London.]

pest species or disease vector) from a particular area of that range, such attempts will undoubtedly be futile, as exemplified by the attempts to eradicate the yellow-fever-carrying mosquito *Aedes aegypti* from Colombia.[69] Under such conditions, reinvasion rates into an area where an eradication program was erected will be frequent and often successful.

One of the major problems of interpreting the speculations of why a greater number of species within a particular trophic level may occupy a particular region or collection of habitats is the necessity to gather information on the extent to which the species overlap in the exploitation of resources. The considerations of MacArthur point to some overlap in resource utilization (Fig. 2.6), and classical ecological theory states that the populations of two or more species cannot coexist if they overlap completely in their utilization of the environment.

Competition for such resources will probably either cause the potentially competing species to diverge in habits or cause one species to become extinct. Obviously, partial niche overlap is expected to be a common phenomenon in natural communities, particularly in species-packed ones such as those associated with certain kinds of habitats and collections of habitats in the tropics. Thus we might predict some overlap in the guilds of host plant species utilized by different species of Hemiptera or Homoptera through an array of contiguous tropical habitats. Certainly the co-occurrence of high-density accumulations of the nymphal skins of several species of cicadas at the base of *Pentaclethra* trees in the Caribbean lowlands of Costa Rica, as reported by Young,[70,71] suggests some level of niche overlap. The overlap in the case of these cicadas might be in terms of emergence sites or feeding sites associated with the root crowns of certain tree species acting as resource patches. But what is the distribution of individuals of each of these cicada species over several different tree species in different habitats? How much overlap is there? These crucial questions form the basis of further study. If indeed certain tree species are limited resources for tropical cicadas, to what extent has competition played a role in causing the observed patterns of diversity, where diversity in this instance is defined simply as the number of cicada species and their associated nymphal skin densities associated with each tree species?

The familiar logistic equation of population growth is used as a means of conceptualizing the degree to which a species has saturated the resources of the environment. In this equation, $dN/dt = rN\ [(N-K)/K]$, dN/dt is the rate of change in population size expressed as a function of the exponential growth factor, rN, the instantaneous rate of increase (derived from life table parameters) times the existing population size, and adjusted by $(K - N)/K$, the amount of resource space available to the population now or at some future date. The asymptotic condition derived from this expression (Fig. 2.7) shows the carrying capacity, K, of the environment, the total number of individuals that the environment can support. The value for K is highly sensitive to all environmental

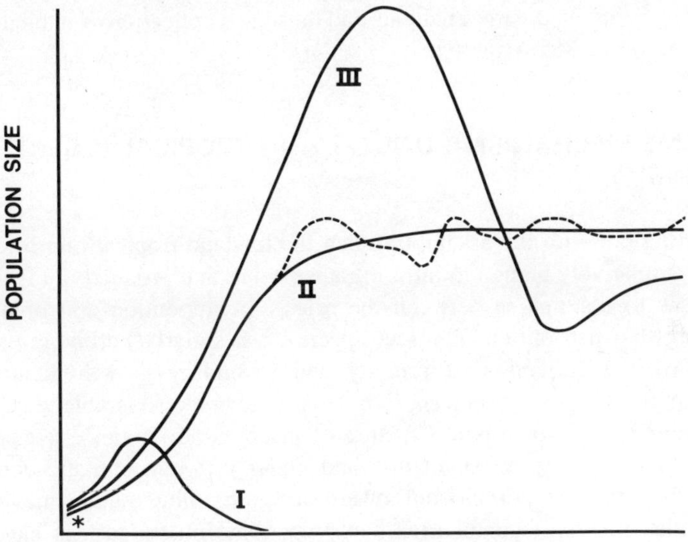

FIGURE 2.7. Alternate actions of colonization of a new habitat or adaptive zone. Many colonizations result in rapid extinction (curve I), while successful invasions may involve a period of overabundance (curve III) with eventual attainment of equilibrium following a period of oscillations. Presumably such populations are characterized by individuals programmed with very high levels of reproductive potential and where biotic control mechanisms are generally lacking. Such a species may be at an advantage if entering a community where competitive displacement can be initially overcome by having a high r, or intrinsic rate of increase. Curve II is probably least typical for such invasions, depicting a population census history pattern expected under conditions of high environmental certainty and an absence of competition from resident species. *Date of founders arriving in new environment.

factors and is expected to be in a dynamic state of flux in most habitats. Explanations of high species abundance of insects and other organisms in the tropics involving competition imply the existence of populations at or near their respective carrying capacities.

The individuals composing each population are expected to have somewhat different ways of utilizing the available resources in the environment, and this range in variation, sometimes referred to as niche breadth, is dependent largely upon the genetic and physiological/behavioral differences among the individuals, as developed below. Pianka has argued that there should be no gross differences in microspatial heterogeneity between comparable habitats in the Temperate Zones and the tropics. Yet a major issue in understanding tropical species abundance patterns is the determination of within- and between-habitat niche breadth for herbivorous insects. Such data are needed to provide information on how diversity is maintained within habitats and through a collection of habitats (i.e., a small area, *sensu* MacArthur[14]).

2.5. SOME MECHANISMS UNDERLYING TROPICAL INSECT DENSITY

Starting out with the assumption that the lowland tropical rain forest environment is relatively benign to most organisms and that productivity is high, the need arises to distinguish between the roles of competition and predation in structuring the distribution of insect species, particularly herbivores and pollinators, over a collection of primary and secondary growth habitats. The sweep-sample studies of Janzen[72] have provided considerable data on the species abundance and trophic structure of insect communities associated with different kinds of tropical vegetation and under different climatic conditions. Such studies are an important step toward designing other studies manipulating the potential effects of competition and predation. But the task at hand is immense. It would be ideal to know something about the life tables of selected herbivorous Coleoptera, Homoptera, and Hemiptera in such communities as well as the degree to which each species exploits various plant species in the habitats studied. Just as with the studies of host plant partitioning by Temperate Zone Homoptera by Denno,[73] it would also be necessary to say something about the ways each species utilized each host plant, particularly where considerable niche overlap is evident.

It would be interesting if population densities turned out to be greater within herbivorous Coleoptera for individual host plants where other species were absent. Such data would suggest competition. The ways in which a plant-feeding coleopteran views each potential host plant species might be considerably different from the ways such resources are viewed by herbivorous hemipterans and homopterans (*sensu* Janzen[72]). Thus we might expect that individuals within each taxon would respond differently to the distribution of the resources and that there might be a general response pattern for each species during a specified time period.

The following hypothetical but plausible example offers a basis for discussing how responses can be made by insects exploiting plants in tropical habitats. A light gap within lowland primary rain forest encompasses an area of about 0.25 ha and has existed for 5 years. A chrysomelid beetle species A is a specialist on one of several species of herbaceous pipers found in the light gap. The population of the beetle is generally confined to the light gap, and individual host plants are seldom defoliated, if at all. Apparently plant predation is kept in check in part by the physiological needs of the beetles and by some level of predation by ants and other organisms on the beetles. Other closely related and ecologically similar species of low understory chrysomelids occur in the surrounding forest. Suddenly the forest is cleared away, and the light gap has become a large patch of secondary growth. A second species of chrysomelid moves onto the host plant species occupied by species A. Species B beetles possess some competitive edge over species A beetles, and both exhibit a tendency to utilize the same parts of the host plant. Neither is able to handle other parts of the host plant species. Eventually the population of the host plant increases in size owing to more habitat being available, and species B beetles are able to colonize a major portion of the plants, while species A cannot. Species A beetles eventually shift to a second species of piper, and the herbivore population takes hold and grows. Eventually each beetle species enters into some sort of equilibrium with its host plant species. Ecological isolation, in the form of divergence between the two beetle species for different host plant species, has resulted presumably as a result of competitive contact between the species. Niche overlap with respect to feeding is thereby minimized as a response to an environmental perturbation. Given the dynamic state of affairs of habitat disturbance in tropical regions, such ecological character displacements may occur frequently in real systems. There is very little data on such shifts. But the documentation of such ecological shifts in the distribution of tropical herbivorous insects may provide information on response mechanisms to changing environmental conditions.

2.6. STRATEGIES OF RESPONSE TO THE ENVIRONMENT

An insect or any other organism can respond to an environmental change either genetically or nongenetically. It may ignore the perturbation and survive, or go extinct locally. The response strategies are of interest here. Natural selection is defined as the differential survival of individuals through different generations within a population of a species. Many kinds of nongenetic variation influence populations.

1. Variation in individuals through time, including age variation, resulting in different niches being utilized, as seen in the life cycles of many holometabalous insects; seasonal variations in the individual and whole generations of individuals such as different phenotypes in insects with seasons, including color forms and winged and wingless forms.
2. Social variation such as caste structure in some social insects.
3. Ecological variation, including shifts in habitat, flux of temporary habitats, some types of density-dependent variation, growth form variation, and host-induced variation.

The above sorts of variation observed in insects and other organisms need not involve changes in the genetic makeup of populations, as they are short-term adjustments to changing environmental conditions. That is not to say, however, that such responses may not eventually result in genetic changes over long periods of time. Obviously, with few exceptions such as monozygotic twins (and even here there might be somatic mutations), no two individuals in sexually reproducing species are genetically identical. Some portion of this genetic variation among individuals in a population contributes to variation in the phenotypes present in the population. Phenotype is defined as the morphological, physiological, behavioral, and developmental traits of the individual. It is the result of both genetic, somatic, and nongenetic factors influencing the structure, biochemistry, and behavior of the organism. Sometimes the variation from individual to individual within a population is rather discontinuous or abrupt with respect to the character states studied, a condition sometimes referred to as phenotypic polymorphism. Often the variation is continuous among individuals and far more difficult to detect. Discontinuous genetic variation, often a basis for observed discontinuities in phenotypes, is an intrapopulational phenomenon.

The genetic mechanisms underlying some phenotypic polymorphisms are complex, sometimes, in the case of insects controlled by large series of multiple alleles or by pseudoalleles at closely linked loci. A possible selective advantage of a large number of alleles being involved is the increased likelihood of reducing the frequency of homozygotes through successive generations. Under some environmental conditions an absence of homozygotes might be advantageous. In fact, the most frequent phenotype in a polymorphic population is by no means the result of a dominance effect of particular alleles. Genetic dominance is not equivalent to numerical dominance of a particular allele or linked alleles. In some cases the most common allele in a series of multiple alleles is a recessive, possibly maintained by a selective advantage of numerical equivalence of various phenotypes. Furthermore, what might appear "recessive" in terms of visible components of the phenotype may in fact be dominant in terms of the physiological phenotype, especially since most genes are selected for in terms of physiological effects.

Theodosius Dobzhansky and his co-workers have investigated in both field and laboratory studies the systems of chromosomal polymorphisms found in various species of *Drosophila* flies. *Drosophila* and other dipterans such as *Chironomous* and *Anopheles* possess giant salivary gland chromosomes—polytene chromosomes—which permit the direct observation of gene sequences as the result of differential staining properties of the chromosome material, leading to distinct banding patterns associated with different loci. A significant discovery from both temperate and tropical studies of natural populations of *Drosophila* species by Dobzhansky and Pavan[74] and Da Cunha and Dobzhansky[75] and others was the demonstration that many different gene arrangements can exist within a population, suggesting the maintenance of considerable genetic polymorphisms in populations. For example, the third chromosome of

Drosophila pseudoobscura has at least 16 known gene arrangements and *Drosophila simulans* more than 50. Such variation is different from variation arising from single loci, since in the former, whole gene arrangements are involved. Some of these studies revealed that central populations tend to be chromosomally polymorphic while marginal populations tend to be monomorphic. Since chromosomal inversions tend to suppress crossing-over and therefore recombination, such spatial changes in the genetic structure of these insect populations, particularly in Central and South America, were considered as highly adaptive; populations entering into or occupying marginal habitats would have the genetic resiliency (through recombination) to cope with new environmental conditions, whereas central populations existed in more or less stable conditions and locked gene arrangements in the form of chromosomal polymorphisms preserved the genetic component of the phenotypic architecture best suited for those conditions. A difficulty with such studies is the challenge to correlate specific portions of the phenotypes in such different populations with specific factors in the environments. As with loci on structurally normal chromosomes, the matter is confounded further by pleiotropic effects.

The field experiments of Lewontin and Birch[(76)] and others have shown that in both temperate and tropical environments evolutionary change does not imply the gradual perfection of a particular genotype and resulting phenotype, but rather the trends toward maintaining some level of genetic variation in the populations. Thus if some genotypes of a tropical nymphalid butterfly endow drought-resistant character states to the phenotypes, and others endow optimal traits for humid conditions and the population exists in a humid or wet forest environment, the species may have a chance to survive an unusually long dry season, particularly if the immature stages are incapable of entering into diapause. The type of environmental saturation predicted by MacArthur[(15,16)] for the formation of species to occupy marginal habitats or resources under conditions of high productivity implies a need for some sort of mechanism to allow species to shift to new resources. A popular view is that genetic variation in populations permits such events to occur. The colonization of marginal habitats is made possible by the maintenance of genetic variation or some form of phenotypic variability, perhaps nongenetic. Genetic variation arises from various sources, including the integrity of genetic factors such as alleles capable of changing in frequency and kind as the result of mutation, gene flow, and recombination, and the preservation of these sources of variation by behavioral and ecological factors as well as cytophysiological factors.

Barring inevitable consequences of small population size, maintaining some degree of phenotypic flexibility or variability within breeding populations would appear to be ecologically and biogeographically advantageous for species. It is almost trivial to say that all environments change and therefore species in them change correspondingly. Maintenance of some flexibility allows the individuals in a population to ''predict'' future environments in the sense of being preadapted in some ways to environmental perturbations. Thus we know that the colorful pierid butterfly *Perrhybris lypera* thrives as a larval specialist in the understories of some Central American rain forests. Young[77] has described some of the specialized phenotypic attributes of this species, including a highly specialized form of oviposition and utilization of host trees in the lauraceous genus *Ocotea*. Presumably such a species has experienced a long evolutionary history in this environment. If this species possesses the ability to survive some forms of environmental perturbation due to genetic characteristics, then one might predict that some level of genetic variation is maintained. That level of genetic variation might be greater than some minimal ''noise'' level that one might expect most species to possess, as the ultimate device for coping with environmental changes. Since *Perrhybris* like most other organisms orchestrates its own environment, and orchestration of the environment for a single species depends upon the activities of other species and the physical environment, it is really the individuals composing the population that possess the ability to fit into the environment wherever possible, and in such a way as to maintain a breeding population. This is another way of saying that each species, especially in those environments where species abundance is high, essentially constructs its own environment, as opposed to adapting to it *per se*.

In such cases the tools of environmental construction include a combination of genetic and nongenetic characteristics, some of which vary from phenotype to phenotype within the population. The classic models of selection summarize three basic ways that the genetic component of populations responds to environmental changes. An insect specialist can be predicted to experience strong stabilizing selection in which most individuals in the population possess the optimal phenotype (Fig. 2.8). The response of different genotypes to selection and the operation of ecological constraints on population size impose some loss of fitness to the population, but otherwise the trend is toward decreasing phenotypic variance. In terms of herbivorous tropical insects, one might anticipate stabilizing selection to be most operative in instances where a specialist insect species possesses a relatively low reproductive rate, to have good dispersal tendencies to locate patches of host plants, and to tend to utilize large patches of host plants. The latter trait lowers the likelihood of overcrowding and density-dependent selection imposing high losses of individuals as a food supply dwindles.

The case of chrysomelid beetle species A and B described earlier in this chapter provides an example of directional selection, the instance in which a species makes an ecological shift to a new resource, favoring a different phenotypic class. Environmental perturbations may result in episodes of directional selection (Fig. 2.9). There is a shift in the optimal phenotype. Thus the shift of beetle species A from its original host plant to a second species of host

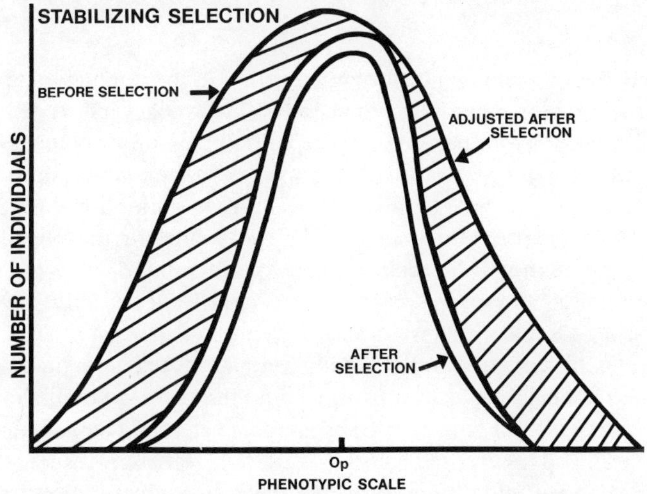

FIGURE 2.8. Relationship between stabilizing selection and phenotype variation. Stabilizing selection, operative through successive generations within a breeding population, results in a reduction of the range of phenotypic variation in the population. Some optimal phenotype, O_p, and a small range of phenotypes around it, is optimal for the environment. Stabilizing selection is expected for habitats or environments in which selection pressures have been predictable over long periods of time. Yet most populations will experience more than one general kind of selection, and the phenotype in any one generation represents a compromise of such conditions.

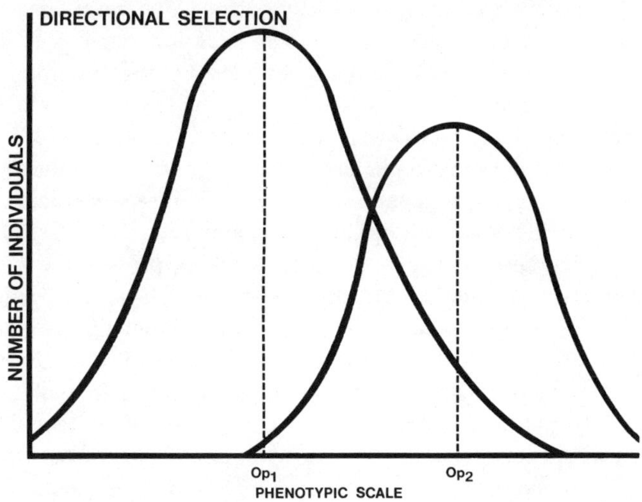

FIGURE 2.9. Directional selection. When one resource base is limited and the species can potentially utilize a second kind of resource in the same environment or habitat, assuming that the species possesses the population growth potential to exploit both kinds of resources, and the population is expanding at such a rate as to have the potential to overexploit the original resource, natural selection may result in a switch to the second resource if it is more abundant than the first and if the first resource is dwindling in supply. Thus through time, one kind of optimal phenotype (O_{p1}) is replaced by another (O_{p2}), representing an adaptation to a new resource.

plant depended upon residual phenotypic variation in the population that resulted in another phenotype becoming optimal for the new host plant species. In such cases we are talking about the sum total of those biochemical and physiological characters that allow a herbivorous insect species to feed successfully on a new resource, without appreciable losses in fitness. We examine some aspects of this topic in the next chapter. Depending upon the source of the phenotypic variability in the population, a specialist insect species may or may not be able to undergo such shifts. If *Perrhybris* is virtually totally genetically specialized to feed on *Ocotea* leaves as a larva, as a result of strong stabilizing selection, insufficient genetic variation in the population may restrict exploitation of other lauraceous and nonlauraceous host plants should the *Ocotea* population be eliminated in favor of cacao or bananas. The first response might be nongenetic rather than behavioral and physiological, but if it fails, genetic response becomes critical if there is sufficient time to adapt. If the genetic variation is not there, there would be little or no opportunity for directional selection to favor another phenotype in that habitat or region, and the species would become extinct or migrate.

An intermediate step, in the case where directional selection is possible, is the breaking down of a population structure in which one phenotype is considered optimal. This process is known as disruptive selection (Fig. 2.10). Disruptive selection can be viewed from the standpoint of plant–herbivore interactions as the initial stages of a divergence in host plant exploitation patterns, away from one host plant previously considered optimal under previous environmental conditions. If sufficient variation is present, a species might attempt to exploit several alternative resource types, a stage of disruptive selection. If suitable for maintaining a breeding population in terms of interactions with other species and the physical environment, the genetic structure of the insect population might be stabilized with a series of several optimal phenotypes, each one associated with one or more similar resource classes, together which provide the species with sufficient resources to maintain a breeding population. If not all phenotypes are optimally suitable for maintaining the population, endowing grossly different levels of survival or reproductive capacities, directional selection is expected to take over if one of the phenotypes is better than the others in the ecological shift.

Thus there might be selection for the Neotropical pierid *Dismorphia virgo* to exploit several species of *Inga* trees in a mountain forest ravine, but individuals associated with one species in particular do better than the others. Perhaps the leaves are more digestible or contain fewer defensive systems, placing less of an energetic drain on the species. The original impetus for an ecological shift might have resulted from a disappearance of the original host plant, changes in population age-structure of the original host plant, or the arrival of a new herbivorous species on the original host plant, leading to competition. All three modes of selection are operative in natural populations of insects and other organisms, and

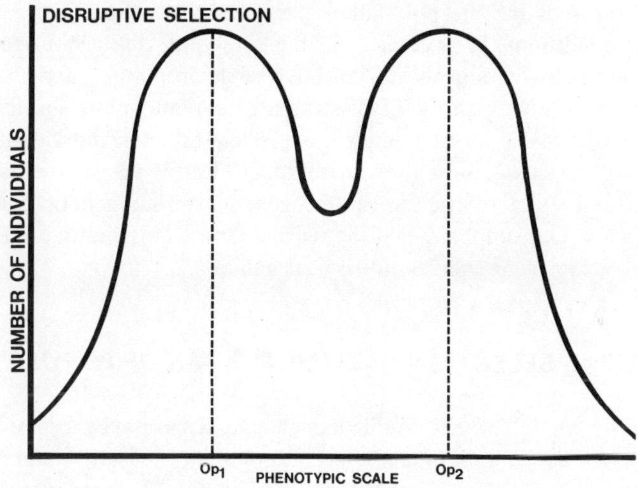

FIGURE 2.10. Disruptive selection. If a population is expanding at a rate greater than what can be thriving on a single resource base, and if there is some phenotypic variation present in the population, selection may favor the exploitation of a second kind of resource in the same environment, if it is present and accessible to the species. This disruptive selection results in a divergence into two optimal phenotypes, O_{p1} and O_{p2}, within the species, so that the population's membership exploits two different sets of resources. Disruptive selection of this sort may also result from intense competition associated with an original resource, selecting for the exploitation of a second, more cryptic resource within the same region or habitat.

because of the high species abundance of insects in many regions of the tropics, such forces undoubtedly contribute to molding the resource associations of insect species there.

But it cannot be assumed that all forms of environmental orchestration in the tropics are of this origin. A great deal of so-called adaptation is probably taking place at the level of the phenotype, without genetic change. We can say that the implied cornerstone of Darwin's theory of evolution is that some sort of variation exists within populations of species, and this spectrum of variation permits responses to environmental changes that may otherwise result in extinctions. The models of selection discussed as well as other factors such as drift affecting populations tell us that allelic frequencies, and therefore genotypic distributions, change in different generations within a population. And such changes represent significant departures from the constant proportions in allelic frequencies predicted by the familiar Hardy–Weinberg equilibrium. Where p is the frequency of one allele in a two-allele system and q is the frequency of the other, the distribution of genotypes for each generation is predicted to be

$$p^2 + 2pq + q^2 = 1$$

The model assumes infinite population size, random mating, and constant environmental conditions. It is clear from the foregoing discussions that populations and environments simply do not have such properties, and therefore the various factors affecting the genetic structure of populations discussed above provide a basis for a more complex approach to how individuals respond phenotypically to changes in the environment. Later, we will examine some effects of resource patch structure on the genetic and nongenetic response patterns of species. Our emphasis will be on the patterns of resource utilization by insects feeding on plant species in tropical habitats.

2.7. NATURAL SELECTION AND PROPERTIES OF POPULATIONS

The basic model of environmental change response by organisms is the effects of selection on individuals within a population. Such a model has these components: (1) a population consists of individuals, (2) there is genetic variation among the individuals, (3) differential fecundity and/or survivorship among the individuals biases allelic frequencies, and (4) adaptations evolve that enhance individual fitness. A model for the adaptation of whole populations as the units of selection substitutes "group" or "population" for "individual" and "community" for "population" in the above statement. Lewontin[78] has reviewed the properties of selection acting on individuals and whole populations. Others (e.g., Wade and McCauley[79]; Wilson[80]) have discussed some mechanisms on the evolution of phenotypic features at the population level and have argued that whole communities can be selected for.

Considering the above model for the response patterns of individuals within populations to environmental changes, several studies have shown that various forms of frequency-dependent selection exist in populations. Much of this work comes from studies of insect populations in general. Ayala[81,82] showed that when *Drosophila* populations become very crowded under laboratory conditions, evolutionary changes occur within the populations that allow densities to increase. In such systems the adult flies appear to have evolved an increased "tolerance" to high densities: while there are no increases in the fecundity or survival of larvae and pupae in such cultures, adults live longer and the number of flies emerging each week remains constant. Such changes in adult flies are believed to be due to a shift in gene frequencies, and therefore phenotypes (assuming that individuals within the populations are responding genetically to increased densities) as a function of population density. It is the interaction of population growth in density and natural selection that brings into play a frequency-dependent selection under these conditions. The selective value of genotypes changes with density changes in the population.

Anderson[83] has examined how the genetic structure of a population might change during the usual logistic growth pattern (Fig. 2.7) and concluded that the population during the phase of maximal growth is best described as the weighted arithmetic mean of genotype rates, and that the carrying capacity phase is the weighted harmonic mean of the individual carrying capacities of genotypes. Natural selection alters allelic frequencies so as to maximize major parameters of population growth at critical times during the census history of a population. Lewontin[84,85] has examined the effects of increasing density on viability in *Drosophila,* and such studies generally show a relationship between density and selection. Lerner[86] discusses some genetic properties that tend to mullify the effects of selection that change with changes in population density and other factors. Mechanisms such as recurring mutations and nonadditive genetic variance reduce or neutralize the effects of selection precluding the ability of the species to respond to environmental changes.

In addition to population density invoking frequency-dependent selection, the genetic structure of populations is also altered by a different kind of environmental factor, namely, interspecies competition. We envision that insect populations, at different times, and under different conditions, will be subject to the selective forces invoked by both intraspecific and interspecific competition. The degree to which these forces operate on particular populations is obviously a function of many factors, including resource availability relative to population growth parameters for a single plant-feeding species, the impact of biotic and abiotic mortality factors on the herbivore population, resource accessibility (patch structure) in time and space, phenotypic flexibility enabling the herbivore to utilize other resource types of precluding it from doing so, and the presence of competing species or potentially competing species if the environment changes.

A specialist herbivorous insect having a population structure characterized by either intense predation on immature stages or a very low reproductive rate, and exploiting an abundant food plant species having an accessible patch structure, may be occupying an essentially empty universe. Such conditions approach Anderson's[83] weighted arithmetic mean of genotype growth rates underlying overall population growth. This period of rapid population growth is analogous to the early phase of a colonizing episode.[87] If mortality is lacking and reproductive rates are high, intraspecific effects may increase, especially if dispersal ability and success likelihood are low. Assuming a sufficiently short generation time, such a condition favors selection for density-tolerant genotypes as the population approaches its carrying capacity. Thus one might argue that the ithomiid butterfly *Hypothyris euclea* in the premontane tropical wet forests of northeastern Costa Rica sometimes experiences rather intense colonizing episodes, with increased shifts in breeding populations from small, disturbed habitats in forest understories to expansive secondary habitats largely dominated in some areas by its larval host plant *Solanum rugosum*. The medium-sized tiger-striped butterfly (Fig. 2.11) places very large numbers of eggs on the host plants, and larval survival is high (Young[88,89]).

FIGURE 2.11. The Neotropical butterfly *Hypothyris* (Ithomiidae) exploiting decaying animal tissues as a food source in Central American tropical rain forests. (a) *H. euclea* feeding on a piece of trochanter muscle from a grasshopper leg and (b) feeding on the soupy tissues of a *Eurytides* pupa (Papilionidae: Papilioninae) previously gouged out by a bird.

Although there is some egg mortality, presumably from a pathogenic fungus, adult densities in secondary habitats are considerably higher than elsewhere, where smaller breeding populations occur. Within a few weeks following a concerted burst of egg-laying activity, a high percentage of the host plant individuals are severely defoliated by larvae (Figs. 2.12 and 2.13), and subsequently the butterfly population declines rapidly.[89] The system appears to be an example of high population growth ability of the herbivore, but with an absence of density-related mortality factors, eventually resulting in a density-related depletion of the resource base. No other ithomiids use this plant species as a larval host, and few other herbivores are present. The tough, sandpapery leaves may allow this plant species to drop out as a major resource for most insects, and presumably the cluster oviposition of *H. euclea* has evolved as a means of ensuring group feeding in the larvae, an activity which probably involves cooperative processing of leaf tissue before it is digested. Most other ithomiids place eggs singly on host plants and scatter them widely. Another, season-related case is that of the dry-season defoliation of the ornamental monocot *Canna edulis* (Cannaceae) by the larvae of the skipper butterfly, *Calpodes ethlius,* at Mazatlan, in Sinaloa, Mexico. Although the large final instar larvae and pupae experience severe parasitism by a chalcid wasp, earlier instars severely defoliate most patches of this ornamental during the early dry season (Fig. 2.14), and some preliminary evidence suggests a reduction in fruit set resulting from this predation on *C. edulis* (A. M. Young, unpublished data).

It was J. B. S. Haldane who suggested that when population size or density is small, intraspecific competition is greatly reduced, thus permitting the evolution of those phenotypic traits responsible for increased success in interspecific competition. The laboratory experiments of Park[58], Lerner[86], and others with *Tribolium* and other insects have pointed to the role of initial population size and genetic composition of populations as major determinants in the outcome of interspecific competition. Depending upon the amount of genetic variability present in the initial populations of each species and initial population sizes, obtaining reverse outcomes in the competition process is possible. The *Drosophila* competition studies of Ayala[90] demonstrate that the rarer species can evolve a competition advantage and reverse its density or even outcompete a formerly superior competitor. When population size or density becomes very low, perhaps as a result of intense competition with another species or from other factors, an increased competitive ability may evolve during this period when the population was on the verge of extinction.

Although the role of interspecific competition in tropical communities is not really understood, the above considerations elucidate the ways this ecological process can mold the pattern of resource utilization by closely related or ecologically similar species. Competition among species can be searched for in terms of herbivorous insects associated with plant species. As more and more tropical

FIGURE 2.12. Gregarious behavior in the Neotropical ithomiid butterfly *H. euclea*. (a) Eggs deposited in large cluster on a leaf of the larval foodplant *Solanum rugosum* (Solanaceae). (b) Third-instar larvae and gregarious defoliation of food plant leaves.

FIGURE 2.13. (a) A natural habitat of *H. euclea* in premontane tropical rain forest in northeastern Costa Rica. Low-density adult populations are associated with clearings and light gaps where scattered individuals of the food plant occur. (b) Defoliation by larval *H. euclea* typically seen in secondary growth habitats where the food plant occurs in large patches. Such defoliation is infrequently seen in forest habitats.

FIGURE 2.14. The Neotropical hesperiid butterfly *Calpodes ethlius* defoliating *Canna* lilies at Mazatlan, State of Sinaloa, Mexico during the dry season. (a) Adult, (b) dead pupa showing parasitic wasp (probably braconid) exit hole, (c) defoliated leaves, and (d) patch of larval foodplant.

FIGURE 2.14. (*Continued*)

forest habitats are altered and converted into secondary or agricultural habitats, it is possible to study the possible role, or lack thereof, of the competition process as insect species colonize resources in new habitats. In some ways this situation is analogous to the taxon cycle phenomena associated with the biotas of oceanic islands rather than habitat islands on tropical mainlands.

The taxon cycle is a succession of usually four distinct phases or stages in the change in the composition of the biota on an island and/or islands. Wilson[91] and others have developed the concept of the taxon cycle to explain the observed distributions of ants and other organisms on various archipelagos. A discussion of taxon cycles logically follows a discussion of how competition alters the genetic structure of populations, since once again we are examining the question of what forces govern the distributions of species. How insects experiencing colonizing episodes respond to new environmental conditions is a function of both the genetic makeup of the colonizing individuals and the resource heterogeneity and diversity of the native (resident) fauna of the area being colonized.

Lewontin[87] has shown that genotypes endowed with the means to promote rapid population growth make good colonizers to new environments. An array of colonizing individuals programmed for high population growth will be able to take hold of a new resource and even possibly dislodge species already using it. Thus genotypes selected for colonizing ability make initially good competitors when entering a new adaptive zone. Such genotypes are endowed with the ability to produce phenotypes capable of being generalists and taking hold of an array of resource types in the new environment. Such ecological release is characterized by population expansion into several habitats and gives the species a broad niche breadth. But such conditions are usually ephemeral, since this form of "swamping" competitive ability soon wanes as the native biota, particularly ecologically similar species on the one hand and predators and parasites on the other, respond by ecological shifts to compete with the colonizer. Thus an initially broad habitat distribution contracts, and eventually local extinctions of the colonizing species prevail. The colonizing population is conceptualized as passing through these stages of the taxon cycle: (1) expansion of the population into many habitats and perhaps over several islands, (2) differentiation of the populations in each habitat or island to adapt to local environmental conditions, (3) spotty extinctions of some populations, resulting in a fragmented distribution of the colonizing species, and (4) an expression of endemism in the surviving populations as differentiation continues and the genotypes present adapt to prevailing conditions. This final stage represents the evolution of a different kind of competitive ability, one that selects for genotypes most adaptive to the species when the population is at or near its carrying capacity, following the considerations developed by Anderson[(83)] and others such as Gadgil and Bossert.[(92)]

In terms of energy allocation, one might predict that selection favors those genotypes in the populations that enhance adaptation to the properties of a contracted niche for the species, including resource specialization and development of defensive mechanisms to discourage predation. Again, the colonizer becomes part of the environment, and the environment is no longer new to the arriving species. This is another way of saying that species themselves are the architects of their own environments. Thus fitting into a habitat does not necessarily imply that a species is adapting to the environment at hand, but rather that it responds to environmental factors in such a way as to siphon off the needed resources while avoiding mortality-generating factors that would significantly lower the size of the population to a point where extinction would occur.

It is not difficult to extend such considerations of taxon cycles to the question of herbivorous insect species in the tropics undergoing ecological shifts to either new host plant species in different habitats or to new populations of the same host plant species in new habitats. The high species density condition of some guilds of insect species associated with certain host plant species may select for regular or episodic colonization attempts when interspecific competition becomes intense. Yet from the above considerations we would expect that those genotypes present that are best adapted to competitive interactions when populations are at or near carrying capacity levels are not the genotypes best suited for colonizing episodes. Perhaps for this reason we might expect populations under such conditions to maintain some genotypes programmed for colonization, i.e., those endowing phenotypic traits such as high reproductive ability and good dispersal abilities. Such residual variation in the population, coupled with disruptive and then directional selection, could bring into play the colonizing strategy. Those genotypes that optimize the ability of individuals to overcome defensive systems of the host plant, to maximize the retrieval of nutrients from the host plant with maximal efficiency, and to have the ability to act behaviorally and physiologically in a manner that ensures survival in the face of competitive threats, represent one segment of the mixed strategy calculated to exist in most insect populations.

The other segment of the population might best be adapted to dispersal, increased energy allocation to reproduction, greater flexibility in feeding on more than one resource type, and a lack of ability to compete with ecologically similar species. When the colonizing genotypes arrive on the new resource, a taxon cycle might be set into motion, depending upon the guild of insects already associated with the resource and other environmental factors. We might also expect that the colonizing genotypes are better able to withstand new features of the physical environment.

The counteradaptation that forms the last stage of the taxon cycle, in which resident species restrict the niche breadth of the colonizer, may result in an equilibrium being established with a new species added to the community. This situation is analogous to MacArthur's idea of how new species are added to communities in the tropics. The equilibrium is established not only with other ecologically similar species but also with predators and parasites affecting population size. Sometimes, however, the counteradaptation may result in the colonizing species becoming extremely rare, in which case the selection pressures are released and the species can once again increase in abundance, and the cycle begins all over again. Such secondary expansions of species are well documented from actual archipelagos such as the West Indies, but more studies are needed to examine such cyclic phenomena in the dynamics of insect communities associated with different tropical habitats.

Janzen[(93)] has analogized the properties of colonization as suggested for islands with the issue of how insect species colonize host plants. A major aspect of the island model is the explanation of the number of closely related species found among islands in an archipelago. The discussion here has focused on the differentiation of populations either by habitat or individual host plant species within a habitat. Clearly such conditions, coupled with mechanisms of reproductive isolation, can result in speciation in the case of tropical herbivorous insects. But the discussion fares just as well in terms of such patterns representing ways that insects respond to the environment, without considering speciation. The speciation process undoubtedly involves many other factors as well, as discussed in Mayr[(94)] and elsewhere.

2.8. ENVIRONMENTAL PREDICTABILITY AND RESPONSE PATTERNS

From these considerations, it is seen that tropical terrestrial habitats, with their varying diversities of plant species and patch structure of plant species, represent complex environmental grains promoting a spectrum of adaptive pathways and response patterns in herbivorous insects, pollinating insects, and predatory and parasitic insects exploiting the former two groups. Particularly in high species density habitats such as some of those found in the lowland tropics, it is to be expected that not all individuals within a population of a particular insect species are adapting to the same features of the environment. Rather there exist mixed strategies of environmental response requiring mixed sets of behavior. Different phenotypes are interacting with the environment in different ways.

Concomitant with such diversified response patterns within a population is the development and expression of an integrated breeding system that ensures that some portion of the phenotypes reproduce each generation. Mixed behaviors may indicate a mixed pattern of adaptation to environmental factors. The tropical cicada *Zammara smaragdina* is a large, sexually dimorphic, mottled green and black insect (Fig. 2.15) with a very distinct song. Most adults emerge within primary or advanced secondary forest (more than 25 years old) habitats, and few if any emerge in more disturbed habitats (Fig. 2.16). In forest habitats males form conspicuous clusters in trees, where they chorus synchronously throughout most of the day. However, many individuals sing individually in isolated small trees in open pastures and near dwellings. The two very different singing behaviors seen in this cicada may represent a mixed response to the environment. Perhaps males in large choral groups in or near breeding habitats represent one response to the need to attract females and mate, and the disjunct and isolated distribution of others in open fields is another response to ensure that females "missed" by the first strategy are mated by the second form of singing behavior. It is noteworthy although not necessarily surprising that males emerge before females and wait for the latter to emerge. The two different forms of singing behavior may represent two different mechanisms to increase the likelihood of matings to achieve a frequency necessary for maintaining the population. The conspicuous strong sexual color dimorphism in Neotropical *Parides* butterflies (Fig. 2.17), coupled with the marked behavioral dimorphism between the sexes in which males patrol specific portions of the habitat to court females (A. M. Young, unpublished data), is another example of sex-related mixed strategies in tropical insect populations. There are other examples. In still other butterflies of the tropics, the adult sexual differences in wing coloration are only slightly different, as shown by *Morpho peleides* (Fig. 2.17), although in other species of *Morpho* such differences are more pronounced.

The sometimes tremendous size range in a collection of adult insects of one

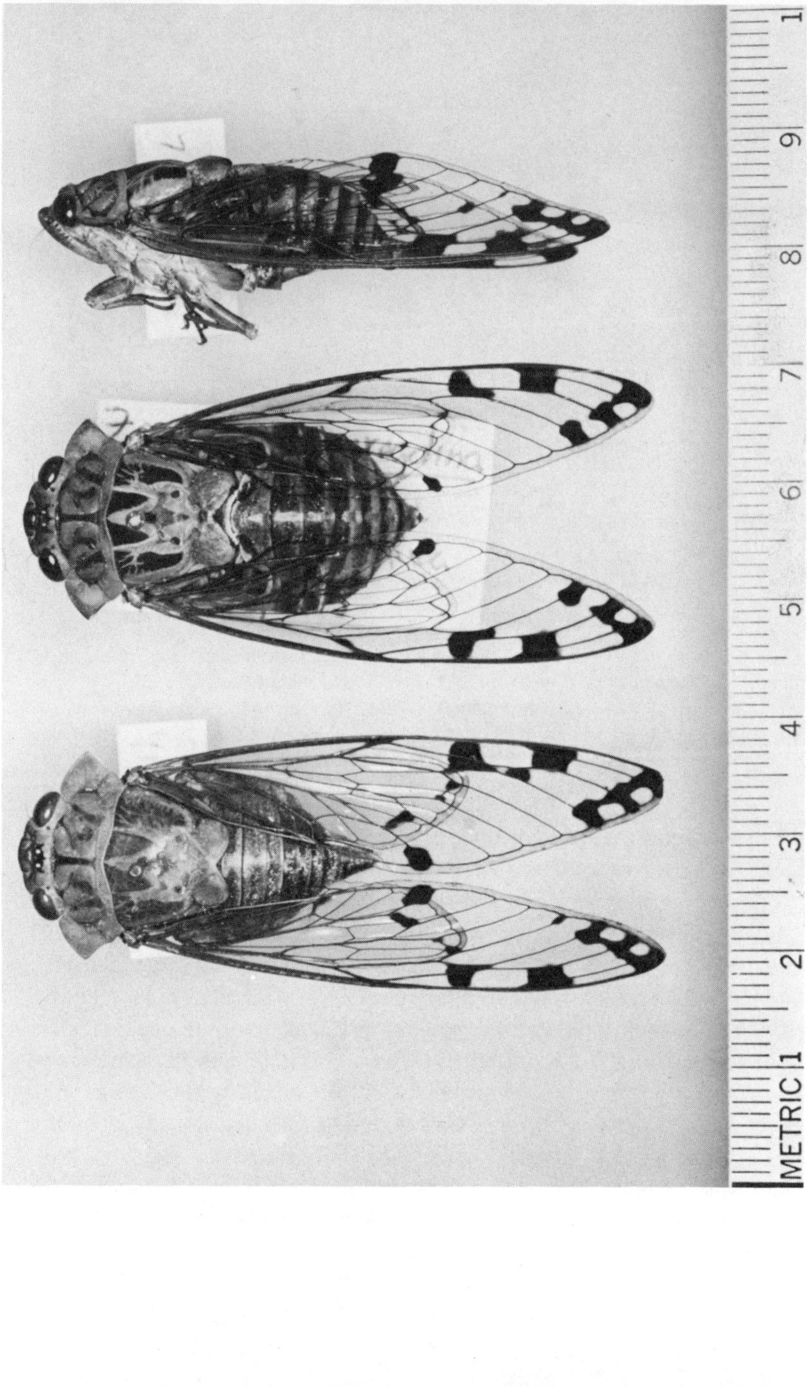

FIGURE 2.15. The Neotropical cicada *Zammara smaragdina*, from premontane tropical rain forest in northeastern Costa Rica. The species is markedly sexually dimorphic in body coloration: the male (two specimens to the right) is bright green and brown, while the female (left) is olive green and brown.

FIGURE 2.16. The Neotropical cicada *Z. smaragdina* (inset), which is associated with mixed primary and advanced secondary forest in northeastern Costa Rica as well as some areas of primary forest dominated by the canopy tree *P. macroloba* (Leguminosae). Shown here is a secondary habitat of this insect, dominated by the tree *G. meiantha* (Tiliaceae). The subterraneous burrowing nymphs of this cicada are believed to associate with the root crowns of *G. meiantha* and *P. macroloba* while secondarily exploiting intermingled root systems of other plants.

species from a habitat may reflect differences in what individuals do with the environment. In northeastern Costa Rica within a 2-week period near the end of the short dry season in one year, there was a 20% range in size differences in forewing length for male individuals of the butterfly *M. peleides* for a total of 23 individuals measured and released, and with sizes clustering into large and small individuals. One can argue that such variation reflects differences in larval nutrition or growth efficiency, but laboratory-reared *M. peleides* using one host plant species and under the same environmental conditions results in similar size variation in male butterflies. Mixed sizes may reflect mixed adaptations to mating and other activities, and such variation can be assumed to be an advantage for the species in a particular habitat.

Because insects tend to have short generation times, be small in size, and exhibit rapid increases or changes in population census histories that correlate

FIGURE 2.17. Two examples of Neotropical butterflies (Lepidoptera) exhibiting coevolved larval food plant associations in tropical forests. (a) *Morpho peleides* (female above, male below) exploits a variety of Leguminosae as larval food plants and belongs to the superspecies Achilles, which has a broad geographical range in Central and South America. (b) *Parides iphidamas,* (female above and male below) and another one to four species of the same genus, depending upon the locality, share a variety of species of *Aristolochia* (Aristolochiaceae) vines as larval food plants, although habitat differences often separate some populations in an area.

well with abiotic and biotic factors of the environment,[95–97] the observed experimental results of rapid responses to selection in laboratory studies certainly justify the assumption that insects exhibit evolutionary changes in populations in response to changing environmental conditions. Such results do not mean, however, that all observed response mechanisms in insects and other organisms represent deep-seated genetic changes in the population through time. Given the long history of migratory ability, one might expect that the changes in physiological conditions in overwintering populations of the monarch butterfly, *Danaus plexippus,* as studied by Tuskes and Brower,[98] do in fact represent stable portions of the genotype. Migratory behavior in this insect is a fairly predictable event through successive annual second-generation populations. Thus a major portion of the physiological traits accompanying overwintering behavior are believed to be directly controlled by genetic factors, resulting in a good correlation between genotype and phenotype with respect to these traits. Such reasoning may eventually prove to be false, but there is no reason not to believe that some portion of the phenotype consists of stable traits endowed by the genotype. In the case of the monarch, some additional physiological "slop" is also undoubtedly brought into play as a trigger of genetically programmed physiological changes associated with overwintering. Such additional factors may represent nongenetic responses to changing conditions in the environment that bring into play the more extensive physiological remodeling associated with overwintering.

Many other forms of environmental response by insects may operate at the phenotype level of organization rather than at the genotype level. Whole collections of species may exhibit such response systems. It is not unreasonable to suggest that short-term and rapid shifts in tropical insects from specialist to generalist feeding patterns and vice versa are commonplace and represent behavioral and physiological shifts rather than genetic changes in the population. Slobodkin and Rapoport[99] have presented the various response patterns that organisms can make following an environmental perturbation (Fig. 2.18). They distinguish between individual and population responses to the perturbation. Population responses are relatively slow responses involving the reshaping of life table parameters and genotypes in the population. Such adaptations are believed to evolve over several generations if the selection pressure arises frequently in the population and if behavioral and physiological responses do provide the quick recovery mechanisms following each perturbation. Behavioral and physiological responses enable individuals to shift habits rapidly without any appreciable loss in fitness.

Clearly there are many examples of generally species-specific behaviors in the insects, such as the one shown by the analysis of mating behavior in the dipteran genus *Drosophila* by Grossfield[100] and others. Within species, however, we might expect some degree of response plasticity, depending largely upon major shifts in types of food and other resources. It is not surprising to note that complete diet shifts in some tropical organisms from fruit to prey (Smythe[101]) represent a facultative response to shifting availabilities of fruit. Many changes in the environment represent ephemeral events rather than permanent changes, and such conditions necessitate only a short-term response by species found there. In the model developed by Slobodkin and Rapoport,[99] permanent changes in the environment may require genetic changes in the population.

Thus shifting distributions of insects associated with a certain range of

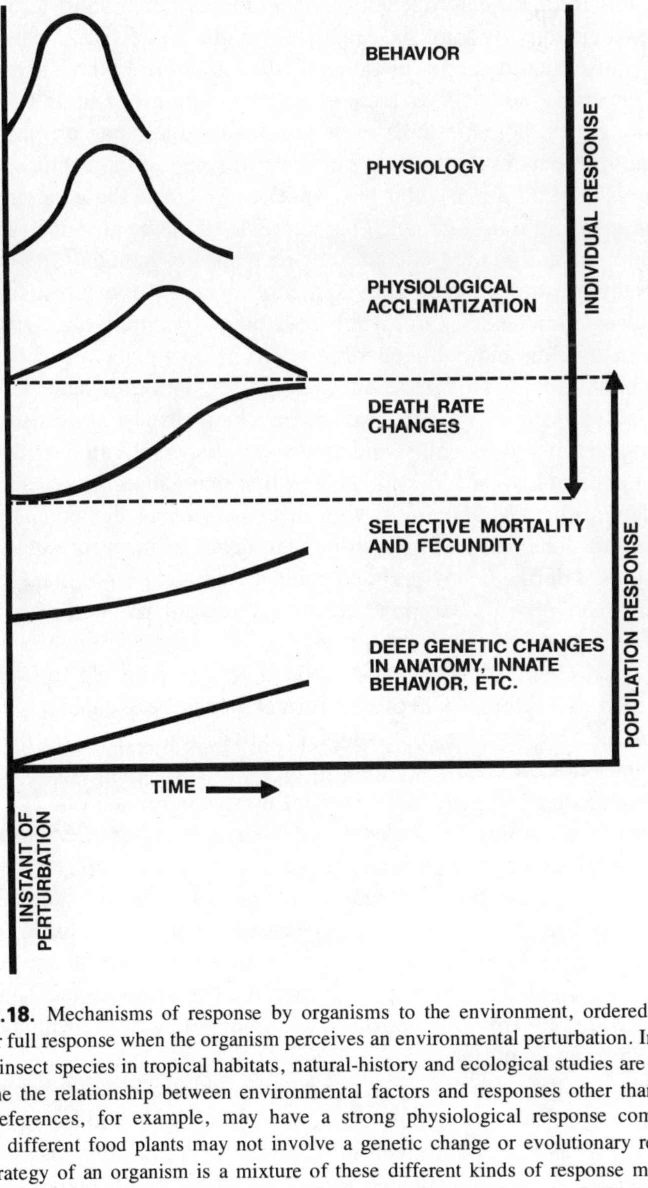

FIGURE 2.18. Mechanisms of response by organisms to the environment, ordered by the time required for full response when the organism perceives an environmental perturbation. In the study of patterns of insect species in tropical habitats, natural-history and ecological studies are often needed to determine the relationship between environmental factors and responses other than behavioral. Feeding preferences, for example, may have a strong physiological response component, and switches to different food plants may not involve a genetic change or evolutionary response. The adaptive strategy of an organism is a mixture of these different kinds of response mechanisms, a portion of which are evolutionary over a specified period of time. [From L. B. Slobodkin and A. Rapoport, *Q. Rev. Biol.* **49:**181-200 (1974). Reprinted with permission.]

solanaceous plants (genera and species) occupying a certain tropical habitat provide the environment heterogeneity that promotes rather short-term shifts in host plant associations in some insects. One might argue that a particular assemblage of chrysomelid beetles associated with solanums in the habitat consists of species generally adapted to feed on most solanums. But individual differences among each potential host plant species require some physiological or behavioral adjustments in the feeding behavior in some of the beetle species not usually associated with a particular host species. Whereas the general predilection for solanums might be a genetic characteristic of the beetle species, with the optimal phenotype within the population of each species being a "good solanum feeder" as the result of stabilizing selection, response mechanisms at the phenotype level allow species to switch host plants facultatively. Different response patterns fade in and out depending upon the host plant species attacked.

The question of physiological and genetic regulation of such feeding has been reviewed by Feeny.[102] Such response mechanisms also imply an ability to switch from generalist to specialist and vice versa. Especially in the tropics, with a plethora of distinct habitat types, it is likely that many insect species frequently encounter new resources. The ability to utilize such resources becomes a matter of possessing the behavioral and physiological means of incorporating them into niche structures. I define behavioral mechanisms with respect to plant feeding as the ability to recognize and respond positively toward particular species, and physiological mechanisms as the ability to possess the appropriate complement of digestive tract chemical agents and other factors responsible for proper processing of food. This topic is explored further in the next chapter. Geographically, one might look for cases of ecological shifts within species as the result of behavioral and physiological responses to new environmental conditions rather than as reorganizations in gene pools. Under these conditions, one can envision behavioral and physiological morphs of species rather than of genetic varieties. These considerations have overtones for studies in biogeography in which the analysis of ecological shifts within individual species or genera becomes part of arguments accounting for evolutionary divergence of groups. Not all observed divergence is evolutionary. Levins[103] has discussed the role of developmental flexibility in the manifestation of phenotypes, suggesting that such phenotypic flexibility provides a form of homeostasis within populations in response to some patterns of environmental factors.

Depending largely upon the climate and other abiotic features, some tropical regions of the world are considered to be both predictable and benign environments for organisms (Slobodkin and Sanders[104]). "Predictable" is used in this context to describe the general absence of recurring catastrophes or sudden, irregular changes in the environment. Fluctuating environments can be either predictable or unpredictable, and environments in general can be benign or harsh. Benign environments are those in which climate tends to promote biotas, whereas harsh environments preclude some biotas and promoting few. Species occupying harsh environments such as the tops of the Andes Mountains or tropical deserts can be specialists to both the abiotic and biotic features of these environments, if fluctuations are predictable through time. The rigors of harsh environments of low predictability select for more frequent extinctions of species and less speciation, so the species abundance levels in most trophic levels are relatively low. There should also be selection for dispersal or migration of species from areas of low predictability areas to those of high predictability, but not the reverse. But such invasions are probably rare or unsuccessful due to competition from resident species in predictable environments. Predictable environments tend to be saturated with species, and species abundance is high. Successful invasion of predictable environments is thought to be possible only when perturbations in the environment take place rather than as a result of ecological displacement or presence of unfilled resource patches.

Species in predictable environments tend to be specialists, while those in unpredictable environments are generalists. Generalists in such environments are believed to possess physiological and ecological buffers for the genotypes as a means of coping with a broad range of environmental conditions. Predictable environments tend to promote speciation and lower the likelihood of extinction, since most predictable environments are benign.

Such conceptualization of environments suggests that lowland nonseasonal regions of the tropics accumulate more species than seasonal regions, high-altitude habitats, and deserts. The major variable is the severity and pattern of climate, setting up selection in favor of life history traits and other phenotypic features that maximize survival in response either to physical conditions of the environment or to biotic features as in the case of species in predictable environments. Such considerations help to explain the biogeographical patterns of species in conjunction with detailed analyses of past environments and centers of origin of taxa. Successful response to harsh or unpredictable environments is more difficult physiologically than it is to benign or predictable conditions. A harsh environment might also be predictable, as we might envision habitats at the poles, but again species abundance equilibrates at lower levels than for benign and predictable environments. In an extreme case, the role of daily rising temperatures and moisture in determining the faunistic composition of biotic communities is associated with harsh environments such as that found in Antarctica and discussed by Wise and Gressitt.[105] The restricted availability of both moisture and warm temperatures lowers the biotic richness of such regions.

Insect species living in the habitats associated with high altitudes and desert regions of the tropics are believed to be generalists, since few resources are available under such conditions. There is little opportunity for speciation, and the generalist strategy, involving the expression of behavioral, physiological, and ecological buffer systems, precludes the exposure, so to speak, of genotypes to selection pressures. In one sense, unpredictable and harsh environments in the tropics relax selection pressures associated with high speciation and low extinction rates. Ecological interactions such as competition become less frequent in such environments. Being poikilotherms, insects experience such environments as being harsh in terms of temperature and moisture extremes. (This topic is explored further in Chapter 7 in a discussion of insect evolution in the tropics.) But such considerations do not reduce the importance of considering the role of short-term behavioral and physiological responses by insects in tropical habitats with high species densities. In such areas, the combined effects of adaptation at the genetic level and also phenotypic flexibility in terms of development, behavior, and physiology mold the distributions of species on resources. In harsh and unpredictable areas, there is probably less opportunity for evolutionary change to take place in taxa already established there. More data are needed on the distribution and abundance of species at high-altitude regions and deserts within the tropics.

Let us illustrate these ideas. The average carabid beetle species high up on the slopes of the Peruvian Andes might be a generalist on grass seeds, along with some level of opportunistic foraging on other resources, but the larger pool of species found in Amazonian forests exhibit somewhat species- or genus-specific feeding habits. Some species might be strictly predaceous as either ground-cover- or canopy-dwelling forms, while others forage on specific types of seeds or other plant products. The presence of benign and predictable features of the lowland nonseasonal wet forest, coupled with the existence of many habitats, both natural and manmade, promotes diversification in these beetles and other insects, so that accounting for the local diversity of the group under such conditions becomes a matter of detecting habitat distributions of species and diversification of life styles within habitats surveyed. Barring the rugged physical terrain of higher altitudes, the task of accounting for the taxonomic diversity of carabid beetles in terms of ecological conditions present would be easier since fewer taxa are present and habitat diversity is low. A complete explanation, of course, would also take into account biogeographical considerations.

The phenotype is a complex mixed system of genetic, developmental, physiological, and behavioral traits and processes that endow each individual within a population with some ability to thrive in a given environment. The environment represents a dynamic array of biotic and abiotic factors fluctuating in time and space and setting up the sphere of influences affecting phenotypes. Because the phenotype involves behavioral traits, the environment is manipulated by the individual. The energy budget and food expenditure, life table characteristics of the population, and pattern of environmental variability mold the reproductive strategy of the species in a particular habitat or collection of adjacent habitats. In most insects the combination of larval-derived and adult-derived nutrients influences reproductive effort. Thus the evolution of fecundity traits is molded by the efficiency with which both larvae and adults captivate nutrients, of which some portion will go into construction of body tissues and maintenance of body functions associated with the production of gametes. Such parental investment patterns are expected to vary between males and females.

The evolution of sexually distinct phenotypic traits imposes another issue of complexity in the manner in which individuals within a population respond to the environment. The high nitrogen requirements associated with the production of eggs in insects is a major physiological factor influencing the feeding habits of adult females. A variable environment might select for life table characteristics favoring colonization, that is, high egg production or decreased developmental time. Lewontin[87] has demonstrated that slight changes in fecundity have pronounced effects on the growth rate of populations, more so than phenotypic switches in characteristics such as developmental rate and longevity. The environment may establish selection for increased fecundity in an insect species, in response to either unpredictable conditions associated with invasions of new environments or increased predation rates within existing populations. Thus the evolution of energy budget characteristics is closely linked to the evolution of life history traits within populations. Selection is expected to be especially pronounced on females, since increases in fecundity are the most sensitive means to increase the growth rate of a population. For these reasons we also examine life history, physiological properties of insects in general, and other parameters of populations in order to understand the phenotype and range of phenotypes in a population interacting with resources for insect populations in tropical environments.

CHAPTER 3

MACHINERY OF ENVIRONMENTAL RESPONSE MECHANISMS IN INSECTS

Key to Evolutionary and Ecological Diversification

3.1. NOTES ON INSECT DIVERSITY

Any attempt to discuss the high abundance and ecological success patterns of insect species in tropical environments must, in my view, take into account two major sorts of driving forces that have contributed to the elaboration of insects as a group and to the elaboration of the multiplicity of life styles represented within the group. I define *life style* in this context as the population parameters and habitat associations of particular genera and species within families of insects. Population parameters include age-specific fecundity and mortality schedules, developmental rate, and patterns of dispersion. The habitat is that portion of the overall environment of a region that supports breeding populations of particular species. The habitat may also include a large spectrum of "transient" species that do not breed there. I assume that insect species settle into those habitats that permit them to maintain levels of population abundance that usually prevent extinctions. Much of this book is devoted to discussions of various aspects of such patterns. In the present chapter, however, we examine some of the major properties of the generalized insect phenotype that have preadapted insects in general to the patterns and structure of environments. Thus an understanding of insectan success in terms of adapting to various kinds of environments warrants an appreciation for the vast evolutionary history of the group as well as contemporaneous ecological factors affecting phenotypic diversification at generic and species levels.

Many introductory textbooks in zoology stress the tenacity and evolutionary success of the arthropods as a group, and within the group, the explosive diversification of that major terrestrial mandibulate group, the insects or hexapods. There is still considerable controversy (e.g., Schram[106]) over the probable pathways of arthropod evolution in general. The traditional view of an essentially monophyletic origin of the group is giving way to a polyphyletic view, which is growing in popularity among some invertebrate paleontologists. The new information driving such a change in view appears to be detailed analyses of embryological patterns and functional morphologies. Acknowledging the apparently obscure evolution of the arthropods, zoologists feel comfortable in stating that the insects arose from some arthropodan stock about 350 million years ago, and probably in the lower Devonian or earlier. The popular view for the origin of the insects is that they evolved from some proto-myriapod ancestor. Some evidence from morphology and embryology favors the view that the insects evolved from a myriapodlike ancestor by a form of neotony in which the rate of development of successive ontogenetic stages was slowed down, resulting in a reproductively mature adult with a larval body plan. In such a manner zoologists analogize the first instar of some Symphyla with the adult insectan body plan. A good review of these basic ideas is given by Wille[107] and in many zoology textbooks.

The paleontological record indicates that as many as 11 orders of insects, 7 of which are now extinct, were already present by early phase of the Upper Carboniferous, followed by the appearance of 19 orders in the Permian. It is noteworthy that in both periods there were some representatives of neopterous Exopterygota: insects capable of complete wing flexing and with the hemimetabolous form of postembryonic development. The two basic forms of postembryonic development in the modern-day insects, many forms of which appeared in the early Mesozoic and into the Jurassic, are (1) the more evolutionarily primitive condition in which the individual passes through three life stages—egg, larva (nymph or naiad), and adult, a pattern of ontogeny known as hemimetabolous development—and (2) holometabolous development, in which the individual passes through egg, larva, pupa, and adult stages. The addition of the pupa stage is believed to be a qualitatively new feature in insectan development, although its precise evolutionary origin is not known.[108]

In most insects with holometabolous development, there is great dissimilarity in the life styles of the larval and adult stages, a feature which permits individual species to partition resources effectively between the two stages, and which results in considerable selection in different directions within the life cycle.[109] The evolution of the neopterous Endopterygota, modern-day insect groups with holometabolous development and in which wings and other appendages develop internally rather than externally, represents the first great explosive radiation of the insects, presumably in response to selection pressures arising from the interactive effects of habitat diversification, predation, and perhaps some forms of abiotic mortality. A common point of discussion in this context is the paleontological correlation between the evolution of angiosperms and groups both of pollinating and of plant-feeding insects. Those groups of social insects such as ants and termites which usually do not enter into feeding and pollinating relationships with plants appeared relatively early in the evolutionary history of insects, about the middle of the Cretaceous, whereas other social insects such as wasps and bees, most of which depend upon nectar, pollen, and other nutrient-rich plant products, did not appear until much later, in the Oligocene. One can speculate that the high incidence of ant associations with extrafloral nectaries of many plant species in the tropics today represents a derived character state associated with ecological factors promoting further development of new or secondary feeding habits in these insects. Carpenter(110) has recently reviewed the evolutionary history of the insects.

With a modern-day group as evolutionarily and ecologically diverse as the insects, it is interesting to note that considerable diversification was present in the early history of the group, and that while whole orders have become extinct, such events were followed by an even greater diversification of orders and groups within orders. The interplay of (1) different life cycle patterns, (2) the diversification of mouthparts, digestive tract physiology, and feeding habits, (3) the evolution and elaboration of wings, and an associated musculature permitting various forms of thermoregulatory behavior, (4) defensive and courtship-related postures, and (5) sophisticated chemosensory mechanisms associated with feeding and mating and other forms of communication, superimposed upon changing arrays of habitats and climatic conditions, must have been the major features underlying the evolution of insects. To this laundry list we add the evolution of body size ranges and the ecological advantages and constraints imposed by body size and the generally poikilothermic condition of most insect species.

When examining the insects, one is tempted to reinforce the view of MacArthur(15) and others that has accounted for the high number of species in the tropics, since the high species and generic numbers of insects found in lowland nonseasonal tropical regions reflect considerable diversification in the five general types of characteristics mentioned above. The availability of many diverse habitats in the lowland tropics and the associated diversity of resources for insects in them have promoted the diversification of the group, especially at the generic and species levels. Wellington(96) has reviewed the general effects of climate on insect populations and has stressed the need to look for microclimate effects on the distribution of insects in particular regions. He suggests that while one might say a characteristic insect fauna is associated with particular regions, one must look at shifting distributions within and between regions in terms of prevailing climatic conditions, which vary considerably annually. Wellington's considerations suggest that some sort of density-independent randomness shapes

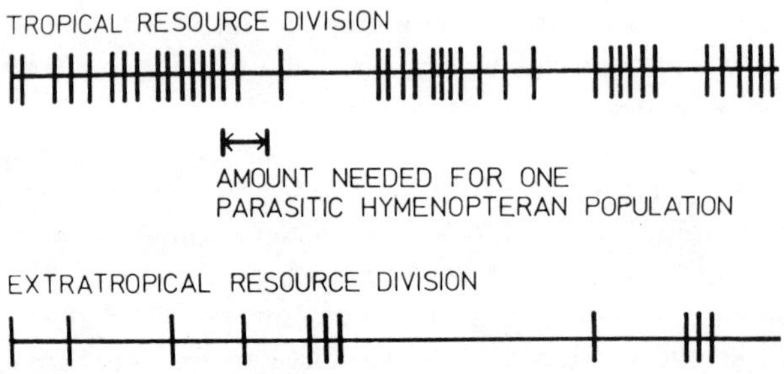

FIGURE 3.1. Patterns of resource exploitation (spaces between vertical bars) in comparable spectra of tropical and extratropical habitats. In tropical habitats there is a greater division of resources among herbivorous insect species than in comparable extratropical habitats, and the average parasitic hymenopteran requires the exploitation of more than one species of host in order to maintain its population. Since population sizes of individual herbivorous insect species are greater in extratropical habitats, fewer such species are needed to support a single species of parasitic hymenopteran. [From D. H. Janzen, Proceedings of the XV International Congress of Entomology, pp. 84–94 (1976). Reprinted with permission of the Entomological Society of America.]

the insect faunas of whole regions. The greatly shifting climatic conditions of Temperate Zone regions suggests that such effects might be less intense for tropical faunas, although little data are available to examine these effects.

When examining tropical insect faunas we must be cognizant of the prevailing climate of particular regions, especially with respect to seasonality and year-to-year trends in seasonality (see Chapter 7), and must examine the habitat diversity within a region, and the structure and diversity of vegetation within each habitat. Janzen[47] has considered the problem of explaining why there are so many species of insects, particularly in tropical regions. His suggestions

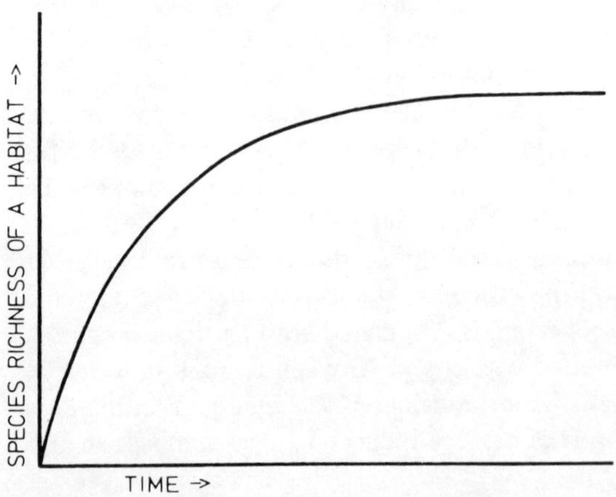

FIGURE 3.2. Correlation between time and species richness of a habitat. As resources become used the rate of increase in number of species declines, eventually leaving a community in equilibrium with available resources. [From D. H. Janzen, Proceedings of the XV International Congress of Entomology, pp. 84–94 (1976). Reprinted by permission of the Entomological Society of America.]

elucidate the roles of habitat diversity, distribution of resources, productivity, and other factors on the species richness of tropical areas.

Starting with the fact that insects are generally poikilothermic and relatively small, so that ratios of body surface area to volume in very small insects impose severe dessication problems in dry environments, physiological constraints associated with low temperatures, seasonal shifts in temperatures, and low moisture availability essentially disappear in the relatively nonseasonal lowland tropical rain forest environment. The major issues of environmental response, adaptive in both a long-term genetic sense and in the context of short-term physiological and behavioral response systems, in the lowland tropical rain forest region becomes one of adjusting primarily to biotic factors.

Janzen[47] developed a hypothesis for tropical resource division in which an absence of dominant species in tropical habitats promotes generalist feeding strategies in insects exploiting resources. Common species generally have large populations, providing a large resource base for several species in the next step of a food chain. However, because insect species in the tropics tend to have small or highly fragmented populations, it is difficult for predators or parasites on these species to be specialists. Thus parasitic Hymenoptera in the tropics are expected to be generalists (Fig. 3.1), whereas in Temperate Zone habitats the increased biomass associated with some species of plant-feeding insects in certain types of habitats provides sufficient resources for several specialist parasitic hymenopterans. But as Janzen points out, such patterns of resource utilization in the tropics will also be constrained by the degree of similarity among host species for parasite species in the habitat. The phenotypic traits of morphology, physiology, and behavior impose constraints upon predatory and parasitic members of food chains, constricting potentially broad niche breadth with respect to host species. Janzen[47] also argues that the increased productivity of some tropical habitats permits the evolution of more species of herbivorous insects and that, depending upon the patch structures of the various plant species present, herbivore populations will be disjunct to varying degrees.

In such a manner it is possible that a large number of specialist herbivore species accumulated over evolutionary time in some tropical habitats (Fig. 3.2) and that development of higher levels in food chains, such as parasitic insects, are molded largely by the structure of the herbivore community associated with the vegetation in a habitat. Add to these effects the availability of varying types of habitats in the lowland tropics in particular, resulting in a greater pool of species and genera, especially with regard to "herbivore type" species such as stem and foliage feeders, flower feeders, root feeders, and seed (fruit) feeders. Communities characterized by a high number of species will accumulate as a result of increased speciation rates promoted by a more heterogeneous environment where constraints imposed by abiotic factors are usually minimal. Janzen's models also show that environments with low primary productivity will accumulate a lower

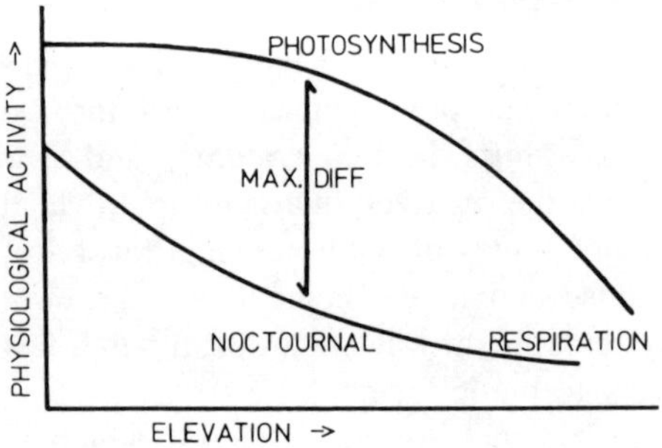

FIGURE 3.3. A hypothetical relationship between photosynthesis (upper curve) and nocturnal respiration (lower curve) as a function of increasing elevation in the tropics. The difference between the curves is proportional to the amount of harvestable productivity. The model shows that plant productivity, as a resource for insects, should be greatest at intermediate elevations in the tropics, where harvestable productivity is maximized. [From D. H. Janzen, Proceedings of the XV International Congress of Entomology, pp. 84–94 (1976). Reprinted by permission of the Entomological Society of America.]

number of insect species. Thus, higher elevations in the tropics generally have fewer insect species than lower elevations, owing to a slowing down of photosynthesis and harvestable productivity resulting from cooler days (Fig. 3.3). As harvestable productivity increases, that is, that portion of plant biomass that can be converted to nutrients by insects, the heterogeneity of resources (host plants and host plant parts) increases (Fig. 3.4) and more insect species can be supported.

Janzen[47] argues that resource heterogeneity, within and between habitats,

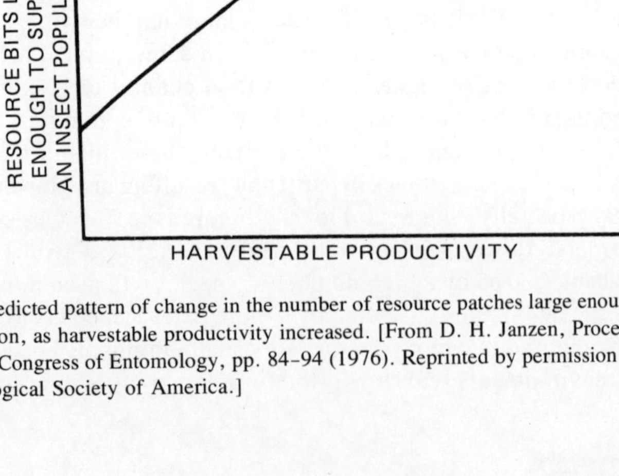

FIGURE 3.4. Predicted pattern of change in the number of resource patches large enough to support an insect population, as harvestable productivity increased. [From D. H. Janzen, Proceedings of the XV International Congress of Entomology, pp. 84–94 (1976). Reprinted by permission of the author and the Entomological Society of America.]

and the level of primary productivity are the major determinants for insect species richness. Each kind of insect is expected to view the pattern of resource heterogeneity and productivity differently, depending largely upon its lifestyle, including its energy budget and allocation of nutrients to reproductive and non-reproductive activities. Janzen argues that the small size of insects in general imposes a greater habitat and resource heterogeneity on insect species and that their ability to modify their habitats declines.

Borrowing briefly from optimization theory (e.g., Maynard Smith[(111)]), a safe assumption is that the individuals (phenotypes) within a population of an insect species will experience the prevailing environment (habitat) in such a way as to maximize their survival and reproduction (fitness). The habitat will impose a grain of resource availability and heterogeneity that will be viewed differently by different insect species. The grain of resource availability refers to the relative abundance of different types of resources available to a species, and how these resource-types are distributed temporally and spatially. That "view differential" will be largely governed by the necessity to optimize energy retrieval in a manner consistent with a host of phenotypic characters determining fitness, namely, age-specific fecundity and mortality schedules, developmental time, dispersal ability, and the amount of flexibility associated with exploitation of resources.[(112)] Thus population structure and dynamics is largely determined by productivity and distribution of different types of resources within a trophic level (resource and habitat heterogeneity). The higher number of species of insects in tropical habitats adds to these considerations the effects of competition for resources among species. Competition is expected to add, in effect, another layer of heterogeneity in the sense of promoting even greater divergence in life styles among competing species.

3.2. COMMENTS ON INSECT PHYSIOLOGY

The internal physiology of insects in general is governed primarily by environmental constraints influencing responses to temperature, moisture, food, mates, and predators and parasites. The precise problems of environmental response vary greatly within a species as a function of geographical region, habitat, prevailing local weather conditions, and resource availability and distribution. We can speak of the generalized physiology of a particular group as being very different from those phenotypic changes resulting from selection within single populations of a species. For those components of the niche of which competing species partake, we also expect selection for the development of physiological processes associated with coping with the competitors. Physiological adaptations cannot be entirely divorced from behavior, and we can view a great deal of insect behavior as resulting from physiological events or else leading to physiological changes as short-term responses to environmental perturbations (see Chapter 10).[104]

Physiological processes constitute a major set of events that buffer the genotype of an individual from the effects of the environment. The individual responds to the environment by behavioral, physiological, and genetic mechanisms. It is the joint effect of behavior and physiology that buffers the genetic system from the potential forces of selection, including outright extinction. Insects generally exhibit these classes of responses to stimuli from the environment: visual detection of resources and nonresources; sophisticated and often directional locomotor movements in response to food, enemies, shelter, changes in the physical conditions of the environment, escapes and colonizing movements to new environments; periods of rest when active responses to the environment are minimized; sophisticated chemosensory responses to odors and tastes associated with resources and nonresources; and establishment of set postures on substrates and mediated by proprioceptors.

It is beyond the scope of this book to discuss detailed aspects for all of these classes of processing of information from the environment by insects in general. It is, however, illustrative to discuss some features of some of these response classes, especially as related to how insects respond to abiotic and biotic features in habitats. Much of the discussion in some of the subsequent chapters elaborates upon the relative role of abiotic and biotic factors in molding the population structure and habitat associations of insects in the tropics.

Studies (e.g., Swihart[113]) have revealed the role of color in the visual mechanisms of some tropical butterflies and other insects. Butterfly species have been demonstrated to be visually sensitive to colors present on the wings of individuals of each species, and natural mechanisms regulating such abilities are probably rather specific for a certain set of colors. The role of ultraviolet sensitivity in the visual systems of insects has been discussed by Eisner.[114] All of these studies suggest that insects in general have evolved some rather sophisticated visual discriminatory abilities in both the visual and the ultraviolet spectra. These studies also point to the role of these mechanisms in adapting each species to the environment. The great diversity of butterfly wing color patterns in the tropics, coupled with the prevailing low population densities of many species and cooccurrence of several similar species or complexes of similar-appearing models and mimics in the same habitats, places premium selection pressure on the elaboration of highly discriminatory visual systems. Under these kinds of environmental conditions, such discrimination permits species-specific mating to take place with minimal interference among species. The joint role of visual cues in the phenotype- and species-specific odors associated with one sex or the other, functions to ensure mating frequency at some level suitable for maintaining breeding populations. Such physiological properties place constraints on the kind of spatial structure there is to the adult population of each species, compromised with constraints arising from the distribution of resources such as adult food supplies and larval host plants, in the case of many species of herbivorous holometabolic insects.

The elegant experiments of D. M. Wilson, such as Wilson and Hoy's[115] analysis of locomotory control mechanisms in the milkweed bug, *Oncopeltus*, and the beetle, *Zophobas*, have shown that the degree of relationship between visual systems and locomotor systems varies tremendously in different insects. Optomotor responses are well developed in some insects and poorly so in others. Such studies suggest that other forms of input modality exist in insects where optomotor response mechanisms are lacking. The behavioral patterns of insects involve a variety of input modality systems, including those based upon vision, odors, tastes, and tactile pressures. As will be developed later in this chapter (Sections 3.4, 3.5) and Chapter 7, the migratory ability of insects such as *Oncopeltus* depends upon the ability to discriminate changes in photoperiod. In such insects,[115] optomotor response systems are well developed and account for much of the overall behavior of the insect. Optomotor response and other forms of sensory input into the neuromuscular systems of insects constitute a significant portion of the generalized insect phenotype concerned with a multitude of short-term and long-term (migratory) responses involving walking or flying. They form the basis for the expression of many behaviors often very specialized in tropical insect populations, including mate location, territorial defense, spacing patterns associated with distribution of food and other resources, and general proneness to dispersal and movement among resource patches when patches are small and widely scattered relative to the population size of the insect species and the average-sized individual within the population.

Many studies have pointed to the effects of water stress on the ability of insects to survive under prolonged periods of dryness. Insect populations in very seasonal tropical regions are annually subjected to extended periods of water stress during the dry season. However, Janzen,[116] Frankie,[117] and Frankie *et al.*[118] have shown that the tropical dry season is a period of maximal activity for some insects, particularly those visiting and pollinating the flowers of certain tree species that flower in the dry season. Insects active in the tropical dry season tend to be large, since small insects are more vulnerable to dessication. Thus the tropical dry season acts as a sort of ecological filter in which a substantial portion of the insect (and overall arthropod) community drops out from particular habitats where the effects of dryness are paramount on plant populations and general vegetation cover. See Chapter 7 for a more detailed discussion of seasonality in the tropics.

Physiological constraints of size, cuticular structure, and ability to absorb metabolic water demonstrated by Edney[119], Fraenkel and Blewett[120], Berridge and Gupta[121], and many other workers, drive selection for mechanisms that permit species to escape in space or time from very dry conditions or permit some species to become permanent residents of perpetually dry areas such as tropical deserts and some mountain tops. The harsh nature of dry environments has been known for some time owing to laboratory studies on the effects of different humidity regimes on body weight and developmental time. Lower humidities necessitate more food to produce a certain body weight, and the developmental period is extended in the larval stage. The increased food intake at lower humidities permits a greater intake of water in response to the environmental stress imposed upon the insects. The study of Fraenkel and Blewett[120] provides a good example of such effects.

It is also evident from various studies on the fine structure of tissues associated with water uptake in insects (e.g., Berridge and Gupta[121]) that selection must have been strong in the early evolutionary history of the insects to develop efficient means of water uptake, necessitated by small body size and the invasion of various terrestrial habitats. The tropical desert today represents an example of a harsh environment in which communities are expected to have fewer insect species than comparable areas of the wet lowland tropics. Clearly a major adaptation permitting insects to colonize desert environments has been the evolution of spiracular control mechanisms that permit regulation of body water content in response to changes in humidity (Fig. 3.5). Insect species in tropical deserts are expected to be generalists toward foodstuffs that optimize water intake, metabolic or otherwise, but specialists toward being active at those times of the day or times of the year in which water stress is somewhat alleviated.

From the studies of Edney[119] and others it is evident that insects have adapted to desert environments in various ways (See also Fig. 3.5.). Desert beetles tend to have impermeable cuticles and because they tolerate high temperatures can be active on the ground surface during the day. Desert roaches, on the other hand, have thin, permeable cuticles and live beneath the sand, where they can absorb water vapor. Such traits represent ecological specialization in response to stressful abiotic features of the environment. The impact of such stress on insects in deserts promotes generalized feeding habits. The specialized adaptations, notably those that are morphological, may preclude such species from invading wetter regions of the tropics. Thus not all organisms can shift from unpredictable to predictable environments as concluded by Slobodkin and Sanders.[104] Specialized adaptations may preclude the ability of a desert-dwelling insect to respond to selection favoring ecological range expansions into less stressful wet tropical regions. Some specialization is also expected in terms of muscles and muscle physiology related to the expression of postures associated with seeking shade and rapid movements in response to predators under conditions of maximal exposure against the ground cover.

Subsequent sections of this chapter examine other physiological response

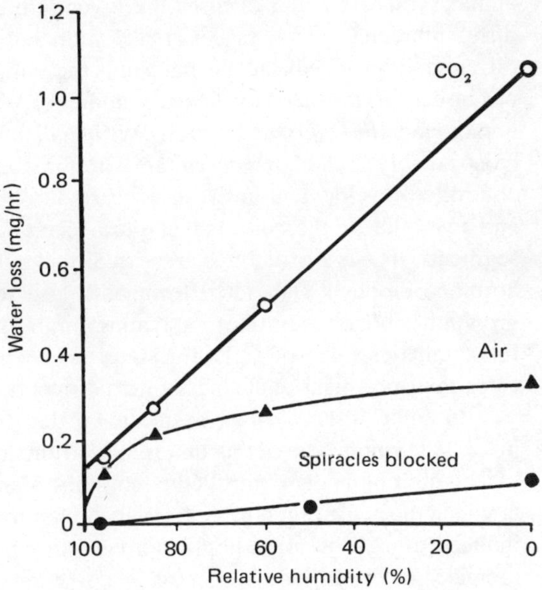

FIGURE 3.5. Relationship between water loss from tsetse flies (*Glossina morsitans*) and relative humidity when the spiracles are permanently open or blocked. The rate falls off in drier air if the spiracles are not disturbed and allowed to regulate. Insects in dry tropical environments, within certain limits, can regulate water loss by such a mechanism. [From E. B. Edney, *Science* **156:**1059–1066 (1967). Copyright 1967 by the author and the American Association for the Advancement of Science.]

systems in insects, including integration of the phenotype through hormonal control systems, feeding behavior, diapause and migration, and defense mechanisms.

3.3. INTEGRATION OF THE INSECT PHENOTYPE BY HORMONAL SYSTEMS: A BRIEF REVIEW

Many entomology and zoology textbooks discuss the processes of feedback regulation of postembryonic development in insects. A pertinent review of concepts and evidence is given by Schneiderman and Gilbert.[(122)] The early physiological studies of Wigglesworth and others have elucidated the roles of several hormones in the regulation of metamorphosis in insects. The basic model consists of the release of activating hormone from neurosecretory cells in the pars intercerebralis portion of the brain, and its subsequent storage in a gland called the corpora cardiaca in the head. The production of activating hormone and its flow to the corpora cardiaca is regulated by genetic mechanisms coupled with cues from the environment. The activating hormone undergoes timed releases from the corpora cardiaca with target effects on the prothoracic gland, an endocrine gland, which synthesizes and releases a growth hormone that induces the molt cycle at specified intervals of growth. Another endocrine gland, the corpora allata, synthesizes and releases the juvenile hormone, which is responsible for differentiation of larval tissues at each molt and the suppression of pupal or adult differentiation. The basic mechanism is the same in both hemi- and holometabolous insects, as indicated by the early studies of Wigglesworth[122a] with the tropical hemipteran *Rhodnius* and Caroll Williams's studies with saturniid moths,[122b] most notably *Hylaphora cecropia*. The three endocrine glands involved in the control of development in the insects are derived from the embryonic ectoderm, and two, the prothoracic and corpora allata, are derived from adjacent body segments. The two structures are virtually identical in cellular structure, and the hormone products elaborated from both also related chemically. The fact that growth hormone (ecdysone), activating (brain) hormone, juvenile hormone, and the synthetics farnesol and cholesterol all activate the prothoracic glands indicates their chemical relatedness in experimental studies with different insects.

In some insects it has been shown that by itself, activating hormone or juvenile hormone can act as an effective functional substitute for ecdysone and induce molting under controlled conditions. Various other studies, such as those in which the activating hormone is substituted for ecdysone in cockroach nymphs induced to molt by brain implantations following surgical removal of their porthoracic glands, provide further evidence for the common origin and similar effectiveness of the three major hormones. Simoes and Uemura[123] have reviewed the effects of these hormones on chromosomes in the dipterous insects. The giant polytene chromosomes of these insects exhibit puffing patterns associated with gene activity and protein synthesis. Puff patterns tend to be specific for different tissues and vary with different developmental stages. The experiments showing that ecdysone injections during intermolt periods change the patterns of chromosome puffing provide evidence of a feedback control mechanism on gene regulation in these insects, since puffing reflects pulses of RNA synthesis.

The studies of Krishnakumaran *et al.*,[124] have demonstrated a definite pattern of DNA synthesis in each tissue during each molt cycle of saturniid moths. Krishnakumaran *et al.* found that DNA synthesis occurred in some tissues such as Malpighian tubules and nerve tissue soon after the release of ecdysone during larval molt periods. Yet in other tissues such as imaginal wing disks and muscles, DNA synthesis during the larval period did not correlate with ecdysone secretion. Such observations point to the highly tuned system of programmed tissue differentiation during the life cycles of most insects. The factors responsible for bringing these processes into play come from the environment. Many studies have shown that the timing of molt cycles and differentiation of adults are highly synchronized with the environment, and some cue or set of cues sets up a pathway of control with ultimate effects at the molecular level. Thus the demonstrated inhibitory role of juvenile hormone or farnesoic acid derivatives that have juvenile hormonelike activity (Spielman and Skaff[125]), in the differentiation of adult insects is a function of the substances themselves but also of the integration of systems of insect development with the environment. An active area of research is the attempt to use the juvenile hormone or synthetic mimics to inhibit maturation of insect larvae as a means of biological control in natural populations.

Insect species in less seasonal and warm environments such as tropical lowlands are exposed to different seasonal-related environmental cues from species in the Temperate Zone. Studies have related the effects of change in photoperiod and temperature on postembryonic insect development. Such cues are less operative near the equator, where the length of day is fairly constant throughout the year and temperatures show little fluctuation outside of diurnal cycles. Thus one might argue that insect development in such regions is strongly genetically programmed for life cycle length and that there is little or no correlation of the timing of developmental stages with environmental changes. The sharp changes in photoperiod and temperature associated with temperate regions impose an additional and highly adaptive layer of control on insect population dynamics and natural history. The long growing season of the lowland tropical rain forest regions promotes the development of overlapping generations within each species without synchronization cues from temperature and photoperiod. In the more seasonal tropics, we might expect that the synchronizing environmental cue is precipitation or the lack of it. Thus insect communities associated with field habitats in tropical regions where the dry season is long and intense, resulting in many plant species becoming deciduous and therefore less accessible as food resources for herbivorous species, are expected to comprise species with population structures adjusted to the dry season. The tropical dry season may represent a period of reduced population growth for insect species that cannot migrate to less stressful habitats such as some riparian forests and forest habitats at higher altitudes. Such species may enter into a state of reproductive diapause as adults or as diapausing immatures.

In the lowland seasonal tropics we must look to the absence of precipitation and its secondary effects as the probable mechanisms for causing significant changes in the population structure of single species and for entire communities associated with habitats where the stressful effects of seasonality are most intense. In some ways insects experiencing such conditions are analogous to insects in the Temperate Zone, even though the environmental changes causing seasonal effects are different and the problem of surviving freezing temperatures is absent. How insect enzyme systems are affected by the tropical dry season warrants considerable study, since temperature-induced changes in enzyme activity, so characteristic of Temperate-Zone species, are absent.

Much of the successful social evolution in the Hymenoptera has been mediated by both endocrine and exocrine control systems (Wilson[(80)]). For example, Roseler *et al.*[(126)] have shown that the social hierarchy in *Polistes* paper wasps is regulated in part by differing degrees of activity of the corpora allata of females. In this system dominant females have more active corpora allata than subordinate females. Many tropical social paper wasp nests have multiple foundresses, suggesting that such a control mechanism may be particularly adaptive in wasps with multiple foundress colonies.

3.4. DIAPAUSE AND MIGRATION IN INSECT POPULATIONS: ESCAPE IN TIME AND SPACE

A cornucopia of field studies have demonstrated that insect population sizes in some tropical regions and habitats fluctuate throughout the year. Tropical regions consist in great part of very seasonal climatic conditions imposed by the interaction of prevailing winds, topography, and oceanic currents. MacArthur[(16)] summarized the major climatic regions of the Earth and the general properties of tropical regions. (For a detailed analysis of climate and weather conditions in Central and South America, consult Schwerdtfeger.[(45)]) It is clear that geographically and climatically the tropical regions show considerable variation in physical features of the environment, which in turn profoundly affect the biotas found in them. It is also clear that some tropical environments are less benign and predictable than others in terms of annual fluctuations in climate. Tropical regions thus contain quasi-temperate areas as well, which can be classified as tropical deserts, tropical mountain tops, and lowland tropical regions where for most of the five or more months of the dry season virtually no rain falls. Other tropical regions undoubtedly can be classified as harsh and unpredictable in terms of organismic response to the environment over contemporary and evolutionary time. To some degree, all three regions mentioned above provide conditions that may contain a prolonged interruption of the growing season for insect species, but such effects are most likely found in the highly seasonal lowlands.

One might expect that insects and other organisms found on tropical mountain tops and in tropical deserts experience harsh conditions throughout the year and over many successive years. In some ways, such conditions are more benign than an environment in which there is a rather gross change in the environment for about one-half of the year. Organisms on tropical mountain tops or in tropical deserts can thus evolve adaptations that permit them to maintain rather stable populations in which the average phenotype within a population is a well-buffered generalist. The challenge in the lowland seasonal environment appears to be more of an ecological challenge for species. As pointed out earlier (Section 3.4), the long interruption in the growing season in seasonal tropical lowlands is somewhat analogous to the interruption in the growing season in Temperate Zones, even though the environmental cues regulating physiological response systems are expected to be very different. To cope with tropical seasonality, insects and other organisms may develop mechanisms that allow for escape in time or space within the context of the annual alteration of seasons.

The winter season of Temperate Zone countries and the severe tropical dry season represent periods of population decline for some insect groups (Fig. 3.6). The discussion is limited in the case of the seasonal tropics to those insects that cannot physiologically remain active in lowland habitats throughout the severe

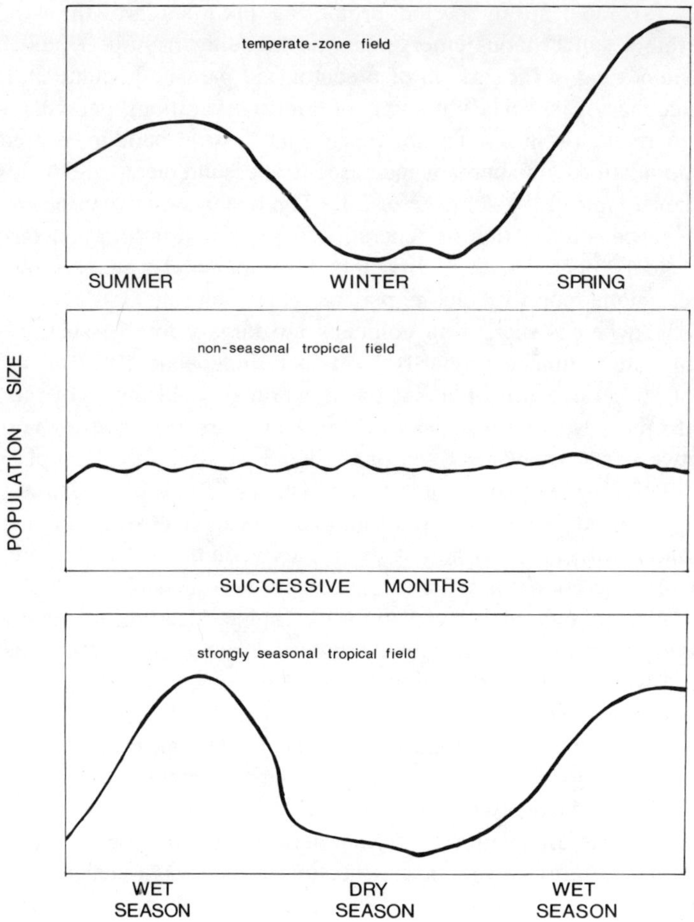

FIGURE 3.6. Predicted patterns of seasonal changes in the size of an insect population in the seasonal and nonseasonal tropics and the temperate zones. Both temperate and tropical seasonality promote rapid increases in population size at the beginning of the growing season, and enforced decline during the period of highest stress, the northern winter or tropical dry season.

dry season characteristic of tropical dry forest zones. Insect populations in the rather nonseasonal lowland wet tropics are expected to have even levels of abundance throughout the year, with minor variations in abundance superimposed by biotic mortality factors, which also tend to have constant effects throughout the year under such conditions (Fig. 3.6). Temperate Zone insect populations, especially those species that are plant feeders, may exhibit a spring pulse of moderate-to-high peak adult abundance and a second peak of moderate-to-low adult abundance later in the summer or early fall (Fig. 3.6). The interaction of a favorable spring period promoting luxuriant growth in host plant species, rather synchronous emergence and breeding in insects, and favorable weather, and a lag in the growth of predator and parasite populations, promote high abundance in species with two or more generations per year (growing season). A reduction in biomass of fresh plant tissues, buildup of predator and parasite populations, and perhaps increased dryness interact to yield a lower peak of abundance later in the summer or fall. The distance between peaks within a growing season is a function of generation time, which in turn is determined by environmental effects on fecundity and developmental time. As the summer progresses, biotic mortality factors play an increasing role in affecting herbivore populations, whereas the spring colonists are largely free of such effects and populations are regulated primarily by density-independent factors such as the pattern of successive dry or moist days, warm or cold days. In the seasonal lowland tropics, wet-season peaks of herbivores are expected to be about the same between successive wet seasons (Fig. 3.6), since the interplay of both abiotic (relatively weak) and biotic factors (strong effects of predators and parasites) are expected to be about the same each year. The size and shape of the population growth curves for insect species active in the wet season are largely a function of generation time. Perturbations in overall habitat conditions, as a result of selective clearing of vegetation, complete clearing, residual effects of insecticides, etc., are expected to disrupt such evenness in abundance in the direction toward smaller populations.

The following discussion is limited to those insect species which cannot remain active in their usual habitats during the dry season. Diapause, a shutdown in many physiological processes that promote the intake, storage, and use of nutrients for both vegetative and reproductive functions, is the mechanism that allows an insect to escape in time from a period of environmental stress so great that outright mortality of individuals occurs or fitness is reduced so much that populations experience extinctions. The environmental control of the onset and termination of diapause has been investigated extensively for insect species in Temperature Zone regions. A typical control pattern for many Temperate Zone insects is the triggering of diapause by successive days of reduced photoperiod in the fall, and a chilling below a certain critical temperature range, followed by a succession of increasingly warmer days to break the diapause. The mechanisms of photoperiod and temperature-mediated control are not the same in all insects at a given latitude. One speaks of "long-day" and "short-day" phenotypes when discussing diapause in Temperate Zone insects. Long-day conditions, such as early-to-mid-summer conditions of daily photoperiod length at the higher latitudes, usually promote rapid development to the adult stage, whereas short-day conditions, typified by late summer and fall, promote the expression of diapause in the second generation and adapt the organism for overwintering; the studies on developmental rate and timing of diapause in the Viceroy Butterfly, *Limenitis archippus,* by Clark and Platt[(127)] are illuminating. A host of secondary factors, including diet and temperature, also induce diapause in some insects.[(128)] Different strains of an insect species from different portions of its range may undergo diapause at different stages in the life cycle (e.g., Kappus and Venard[(129)]). Such studies also point to the subtle interaction of temperature and photoperiod in the induction of diapause. Wigglesworth[(130)] first demonstrated that the corpora allata were necessary for the maturation of eggs in *Rhodnius.*

Siew[(131)] showed that the adult life-span of the chrysomelid beetle, *Galeruca tanaceti,* can be divided into the physiological phases of prediapause, diapause, ovary maturation, and oviposition and that each phase is controlled by a specific level of hormone activity. At the termination of diapause, there is a marked increase in the neurosecretory system, thus promoting oocyte maturation and bringing into play the physiological and behavioral syndrome of oviposition. A fascinating study of the Argentine stem weevil, *Hyperodes bonariensis,* by Goldson and Emberson[(132)] demonstrates the selection pressures favoring the evolution of reproductive hibernatory diapause in some environments and not in others. It was shown that this insect has some populations in Argentina experiencing the diapause and others that are not. When an insect species has a wide geographical and topographical range, some populations develop diapause phenotypes when adaptive and others do not, the latter a result of mild winter conditions. Yet when the stem weevil invaded those parts of New Zealand with mild winter conditions, the diapause condition remained, presumably as a "relict" physiological trait left over from Argentinian gene pools. One might expect that such a condition would change toward a loss of this trait, depending upon the manifold effects of genes that regulate it on other aspects of the weevil's phenotype. Sims[(133)] has reported on an interesting relationship between larval host plant suitability and diapause conditions in the swallowtail butterfly, *Papilio zelicaon,* in California. The species is generally univoltine in habitats where its native umbelliferous host plant species occur, but multivoltine where it uses introduced plant species. The latter condition favors the expression of diapause in the overwintering population, and Sims demonstrated that induction of diapause is regulated by photoperiod and development, and termination by temperature. Apparently strong selection favors the exploitation of the introduced plant species (a condition of environmental change related to a biotic component of the environment and one necessitating the development of multivoltinism and associated diapause).

Field studies are needed to determine which insect species in tropical communities enter into diapause, the life stages involved, and the environmental cues triggering and breaking the diapause. Some butterfly species of the lowland Guanacaste region of Costa Rica, where the dry season lasts almost 6 months, undergo reproductive diapause as adults. Changes in moisture availability, adult nutrient availability, and larval host plant condition, which affect adult butterflies and other insects of tropical regions whose dry season is both long and severe, are worth studying. Such studies may also have implications for understanding how insects in tropical desert regions remain active under conditions of extreme and prolonged dryness. The average generation time of many insects in tropical habitats is less than 30 days, imposing a pattern of population dynamics incompatible with the longer environment oscillations resulting in seasonality. The adaptive challenge is to survive a time period in which several generations could have occurred if the dry season were absent.

In the neotropical pierid butterfly, *Eurema daira,* wet- and dry-season morphs appear to be in lowland Guanacaste in Costa Rica, as studied by Paul A. Opler (personal communication). Wet-season morphs are present in the first half of the dry season, and these individuals seek out moist forest refugia along streams and also migrate vertically into surrounding higher elevations where conditions are wetter. The dry-season morph increases in abundance in the second half of the dry season and enters a reproductive diapause, exhibiting few if any movements associated with seeking wetter conditions. Such a mixed strategy ensures against complete population decimation during the dry season, since migrants apparently return to the lowlands with the new wet season.

Short-term vertical migratory movements for many butterflies in this region of Costa Rica and in other tropical regions might be more prevalent than once believed. The energetics associated with diapause in tropical insects warrants study, since a species occupying a seasonal tropical environment must allocate energy to physiological mechanisms regulating diapause and thereby act as a sink for resources normally allocated to other aspects of the phenotype, such as reproductive physiology and defensive behavior. If overall insect species richness and trophic level complexity are less in seasonal lowland tropical habitats than in nonseasonal wet habitats,[272,137] such tradeoffs in energy allocation within the populations of individual species might be compensated, so to speak, by reduced predation and parasitism rates during periods (wet season) of optimal population growth. There is a need to determine whether season-specific changes in the xylem and phloem chemistry of plants or shifts in moisture content profiles of soils in tropical areas with distinct dry seasons provide the environmental cue for triggering of the rather synchronous emergence of some insects capable of being active during the dry season. The cicadas studied by Young[70,71,135–139] offer a system for such studies, since distinct patterns of seasonal emergence of adults are evident in many species. Similar systems might be responsible for shifts in population structure for plant-chewing insects as well.

Dingle[140–142] has shown that species of milkweed bugs, *Oncopeltus,* from nonseasonal tropical regions exhibit no diapause induced by photoperiod. The Temperate Zone migratory species, *Oncopeltus fasciatus,* undergoes photoperiodically induced diapause, as seen in some, but not all, populations. Dingle suggests that cases of photoperiodically induced diapause in some tropical species are those in which the photoperiod regime acts as a cue for other seasonal cycles, such as rainfall. Various species of tropical locusts are known or suspected to respond to small changes in photoperiod as a cue to shifting conditions of rainfall. Yet, as Dingle points out, other tropical insects enter diapause in response to cool temperatures. Periods of short photoperiods in Central America correspond to periods of reduced rainfall—dry seasons—but milkweed plants tend to be available at such localities throughout the year as a resource for *Oncopeltus*. The rugged topography of Central America and portions of northern South America create a spatial pattern of widely scattered dry areas at certain times of the year, causing some migration of the bugs and therefore a patchy distribution of the populations. Since resources are always available and the bugs are good at dispersing, they do not require a photoperiodically induced diapause to survive. Even as conditions become unfavorable in an area, there is evidence that the bugs are capable of long periods of starvation, so long as water is available. Thus one expects that the early dry season is not a period of intense migration in tropical milkweed bug populations. Such resistance gives the bugs ample time to disperse to new sites or to withstand short-term environmental perturbations such as destruction of the habitat (e.g., through slash-and-burn clearing).

Metabolically, such periods of starvation are equivalent to diapause, and the production of juvenile hormone is blocked in both. Dingle[142] suggests that the evolution of photoperiodically induced diapause in the genus, an obvious phenotypic trait adaptable to invasions of Temperate Zone environments, requires only the linking of short-day input from the environment to the neurosecretory metabolic systems. He suggests that the northern limits of the tropical species of *Oncopeltus* are set by the inability to evolve a response to photoperiodic cues. Since milkweed seeds and plants tend to be available at the end of the summer and at the time the bugs escape to more favorable latitudes, lack of food cannot be the cue. Under such conditions, an attempt to account for the additive genetic variance contributing to the age of first reproduction as a function of photoperiod would be difficult, and therefore producing diapause strains in these species from selection would be difficult. The factors contributing to this genetic variance unknown at present. Most tropical populations of milkweed bug species exhibit no photoperiodic diapause.

Migration refers to the alternate strategy of escape from unfavorable conditions of the environment. It is a gross change in those behavioral and physiological traits of the phenotype associated with the exploitation of resources in the immediate environment. By definition, migration is an escape in space, to a different habitat or geographical region where environmental conditions favor either a new burst of population growth on the part of the migrants or colonists or a period of vegetative "waiting out" and eventual return to the original environment, where new flushes of population growth ensue as environmental conditions return to a favorable state.

Johnson[(143)] maintains that migration is the most effective means of insect dispersal, the well-directed, massive movement of insects from one area to another, during which other kinds of activities cease or are held to a minimum. Migratory flight is a well-developed phenotypic trait molded to the dispersal needs of each species and is best developed in winged insects. At the age of maximum reproductive value,[(144)] young adults, especially females, are the best candidates for migrations. Johnson[(143)] refers to the "cogenesis-flight syndrome" as the temporary reduced ovarian development during the time of maximum flight activity, even though young adults will eventually be the best colonizers once they arrive in the new habitats. The interplay of individual behavior patterns and prevailing weather conditions determine the migratory routes of most insects.

Migration can be viewed as a major change in the genetic structure of a breeding population, if some portion of the population splits off and migrates to a new environment. We need to examine the correlation between genetic control of dispersal and traits that make good migrants or colonizers, such as early age of first reproduction, high fecundity, and short developmental time. Few data are available on the effect of migration on the breeding structure of insect populations and on the extent of migration of individuals within successive generations. We need to examine the factors promoting migration in different insect taxa in order to understand the biogeographic patterns of these groups. Knowing the factors that may initiate migration in a particular group may help biogeographers interpret past and present taxa distributions.

Southwood[(145)] has stressed the long-range displacements associated with migration in some tropical insects. Migration in tropical insects may be more frequent than we know and seems best documented in pest species such as some African locusts. The heterogenous topography of regions such as Central America establishes a system of climatic heterogeneity, some components of which might favor both vertical and horizontal migrations in flying insects such as butterflies, beetles, and hemipterans, those groups most associated with vegetation subject to considerable change by drought. Dingle[(142)] found that three species of tropical *Oncopeltus* exhibited far less long-duration tethered flight than the well-known Temperate Zone migrant *O. fasciatus*. He suggests that tropical habitats are less seasonally variable and that such environmental predictability should minimize the adaptive importance or tendency toward migratory flight. Strongly fluctuating habitats should result in greater frequency of migratory behavior in insects. But I believe that Dingle was referring to the lowland nonseasonal wet tropics. Lowland dry areas may select for some level of migratory ability, even if relatively short term and of short distance, in some tropical insect species. Root and Chaplin(146) found considerable short-distance movement in a mixed population of two *Oncopeltus* species in Colombia.

We need to study a few partially known cases of long-distance migration in tropical insects such as the day-flying moth *Urania fulgens*(147) and allied species. Short-distance migratory movements might be prevalent among insects occupying habitats in the lowland seasonal tropics where a dry season occurs for 5 or more months each year. Few data exist on the movement patterns of insect species in such regions. A good general working hypothesis for such studies would seem to be the degree of environmental unpredictability incurred by the dry season as a driving force in the phenotypic response of movement to new, less stressful habitats. To what extent do such movements occur in these tropical communities? Are the majority of species involved, herbivorous insects? What is the impact of such movements, if they exist, on the population structure of species left behind and upon the community structure as a whole? These seem to be among the great unanswered questions about the impact of tropical seasonality on the expression of temporal and spatial escape mechanisms in insect species.

To what extent do genotype and phenotype reorganize when insect species develop migratory movement patterns? What are the population consequences? Dingle(142) has elegantly demonstrated the interplay of density, photoperiod, and temperature regimes on the life history and population growth of *Oncopeltus*. He showed that high temperature, long photoperiod, and low population density result in a lowered age of first reproduction, an age-specific fecundity schedule highly skewed toward young age classes of adult bugs, and therefore a higher rate of population increase. Photoperiod and the effects of population density influence population growth only on the adult bugs; temperature affects the rate of development. Dingle concludes that such effects result in a phenotype highly responsive to the need for migratory movements and argues that such characteristics make this insect an ideal colonizer of milkweed patches during spring in northern latitudes. Migratory movements are induced by short days (fall), cool temperatures, and high densities, all of which delay reproduction and thereby provide the time the species needs to migrate to more suitable regions as winter approaches. Such conditions in *Oncopeltus* during the fall generations represent incipient diapause. Such data point to rather short-term changes in the genetic structure of milkweed bug populations at northern latitudes, in which spring and fall adult populations exhibit very different response patterns to the environment. Similar shifts in populations may occur in tropical milkweed bug populations where selection has favored the evolution of a dry-season form highly adaptive to short-distance migration to wetter sites. Such morphs, however, may not possess great changes in the phenotype in terms of those parameters underlying population growth, although some response is predicted if migratory ability, even in short-distance movement patterns, requires a shift in energy allocation. Alternatively, it might well be that the response mechanism resides at the phenotypic level and represents no significant shift in genetic structure of the population over a few successive generations. We need to dissect the response patterns of *Oncopeltus,* in both temperate and tropical populations, into those components resulting from gene frequency shifts within two or more generations and those arising from physiological-stage changes. The planned studies of Dingle and others should provide the necessary genetic analysis of milkweed bug populations to elucidate these effects.

The best-documented examples of migration in tropical insects come from the classic studies of African locust populations. In these cases, large shifts in the phenotypes present in migrating or nonmigrating populations clearly involve the interplay of shifting availability of food and increasing population densities, representing intraspecific competition within populations. Migration is induced by these changes in food supply and density and synchronized with seasons favorable for the growth of host plants elsewhere near the equator. Short-distance migratory movements in tropical insects might prove to be more frequent than previously documented, notably for those regions where lengthy dry seasons promote escape from prevailing environmental conditions.

3.5. TEMPERATURE, PHOTOPERIOD, AND OTHER FACTORS RELATED TO INSECT ACTIVITY: FURTHER COMMENTS

The effects of rather small changes in daily photoperiod on tropical organisms have not been well investigated. Comparisons of the effects of photoperiod at different latitudes are available. Stubblebine *et al.*[(148)] found that the tropical leguminous tree *Hymenaea courbaril* exhibits more growth in the seedling stage when grown under long-day or tropical regimes than under short-day regimes, and found good correlation between latitude of seed origin and vegetative growth. Incidence of diapause induced by photoperiod was found to vary in different geographical strains of the moth *Pectinophora gossypiella* for material from populations ranging from El Paso, Texas, to Venezuela, with the frequency of diapause being greatest in the El Paso poulation.[(149)] Changes in diet, such as lipid content, and in temperature did not alter the response to diapause or lack thereof.

Temperature is another niche component that influences the distribution of insect species in different environments. Tropical regions exhibit considerable diurnal fluctuation in ambient temperatures promoting displacement in activity periods for different groups of insects in tropical habitats. Some insects are active in the day, others at night. The development and expression of physiological mechanisms that allow activity at cooler temperatures in the tropics permit the co-occurrence of a great number of taxa in the same habitat by partitioning the environment temporally into daytime and nighttime components. Very high temperatures in the tropics such as those associated with the midday period promote lethargy and general inactivity in many insects and other organisms. The observed peak activity periods at dawn and dusk in some insects, such as certain species of cicadas and butterflies, suggest a synchronization of activity with times of least stress from excessively cool or high temperatures. Young[(150)] suggested that some tropical cicadas may have optimal chorusing at such times, borrowing from a model of optimal sound transmission developed for vertebrates.

The degree to which insects are subjected to various forms of environmental stress will shape the diurnal patterns of activity. Whitford *et al.*[(151)] found greatest foraging at the soil surface in the morning and evening in the Chihuahuan desert ants *Novomessor cockerelli* and *Novomessor albisetosus* when air temperatures ranged from 20 to 40° C. The limiting factor was demonstrated to be water, since the foraging period was extended when ants were given supplements of seeds. Such conditions of physical stress from the habitat may affect the activity patterns of various insect species thriving in high-altitude regions of the tropics as well as lowland desert areas within the tropics. Populations are expected to be more physically controlled (*sensu* Slobodkin and Sanders[(104)]) than populations in wetter regions. Lowland hot and wet habitats are expected to be the most benign in terms of physical stress conditions, even if a dry season is present.

Except when small insects are dispersed passively at high altitudes, where temperatures are cooler and the ability to be active is probably reduced,[(152)] flight initiation, speed, and direction are generally influenced by temperature and wind. Snow[(153)] demonstrated that *Anopheles* and *Culex* mosquitoes in Africa lose the ability of directed flight in winds with speeds exceeding flight speeds.

Environments which place less thermal stress on insects are usually those where the greatest diversification of species occur. The lack of such stress, and other forms of abiotic stress, promotes speciation in response to biotic features of the environment. Maximum allocation of resources goes to specialized adaptations that allow each species to utilize smaller portions of the habitat and that usually involve a reduction in population parameters promoting the growth of large populations. Downes[(154)] has argued that the habit structure and resource availability in Arctic environments make possible a greater diversity of insects than actually found there.

But the cold cannot be tolerated by the physiological mechanisms of most insects, even those species derived from Temperate Zone faunas. Downes suggests that the diversity of insects will never increase appreciably in the Arctic due to speciation in response to thermal stress. The greatest diversity of insects occurs in the tropics, where species are still well within their physiological tolerance of temperature. The well-documented, highly synchronized emergences and swarming of Arctic chironomids (see Oliver[155]) suggests another barrier to the accumulation of more species in the Arctic: the compression of activity into relatively short seasons precludes the development of environmental partitioning by different tolerable seasons.

Temperate Zone environments also require the evolution of low-temperature tolerance mechanisms to adapt to chilly fall and spring conditions. Young[156] demonstrated the rather unpredictable nature of the first killing frost date in Appleton, Wisconsin, which has a range of as much as 30 days, and the pattern of its occurrence clearly shows the instability of temperature associated with northern latitudes. Insects must be able to thrive under progressively cooler conditions as the first killing frost approaches. Such adaptations provide time for insects to prepare for the winter. The analogous issue for tropical insects would be the evolution of low-temperature tolerance at high altitudes, such as in the Andes. Mutchmor[157] has shown that insect species capable of low-temperature tolerance exhibit greater ATPase activity. Thus cold-tolerant species can maintain some level of muscle activity at temperatures that normally would cause inactivity in less tolerant species. Cold tolerance is manifested in different life stages.[158] Mutchmor and Richards[159] have shown that several kinds of insects exhibit cold tolerance by maintaining certain levels of activity of muscle apyrase but that others are incapable of such physiological regulation.

The initiation of flight in insects, analyzed by Nachtigall and Wilson[160] for dipterans, involves the stimulation of a succession of different muscles at the appropriate millisecond and the subsequent integration of the whole system. In some insects with highly waterproofed cuticle, the heat generated from activated flight muscles is not dissipated into the environment and such insects appear to have no special mechanisms for protection from overheating.[161,162] The coats of fine hairs of bees, hawkmoths, and other large insects provide effective insulation against convective heat loss during flight,[161] and one might expect such large-bodied insects in the lowland tropical forests to be active during cool periods of the diurnal cycle, since heat generated from flight can be retained. Larger insects should also be found on tropical mountain tops, where it is cool. The patterns of wing muscle activity during the characteristic warmup periods of many Lepidoptera are generally the same patterns of muscle activity exhibited during flight, although in some the burst lengths of muscle firing are shorter during the wingbeat period.[163] Such an adaptation permits a greater generation of body heat immediately prior to flight and would be advantageous in cool environments. Some butterflies exhibit regulation of the thoracic temperatures by behaviorally changing abdomen and wing positions, orientation to the sun, and perching.[164] Quiet basking in sunny spots raises body temperature in *Papilio polyxenes* far more readily than shivering.[164] Higher temperatures promote greater circulatory exchange between thorax and abdomen, dissipating heat from the thorax. Heinrich[165] has demonstrated the large amount of heat production associated with preflight warmups in the sphinx moth, *Manduca sexta* (Fig. 3.7), resulting from rapid wing vibrations.

Whereas *P. polyxenes* is active at sunny times of the day and can therefore capitalize by behavioral shifts to thermoregulate in sun patches (and cool off away from them), most sphinx moths are active in uniformly cool conditions of dusk and must rely on wing vibration movements for preflight warmups. Such behavior is a requirement for successful flight in large insects. Heinrich points out that such thermoregulation is a key factor in shaping the energetics of foraging in many flower-visiting insects. The great differences in temperature between day and night in the lowland tropics require considerable preflight warmup for larger insects that forage early in the morning or at dusk, while smaller insects are limited to intermediate periods but away from the hottest period.

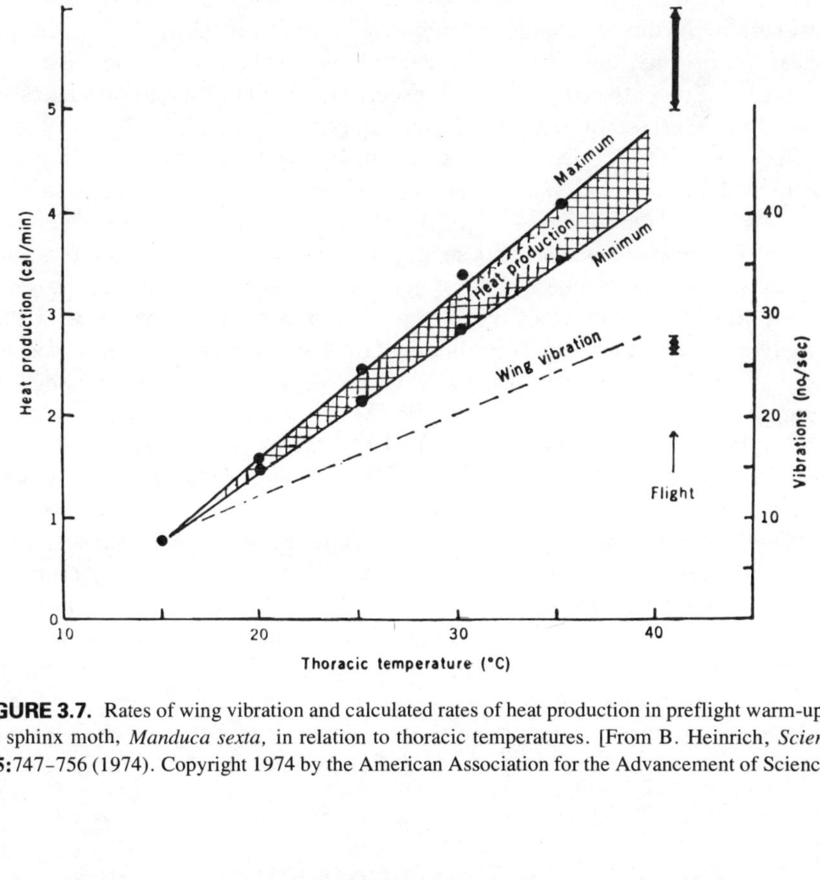

FIGURE 3.7. Rates of wing vibration and calculated rates of heat production in preflight warm-up in the sphinx moth, *Manduca sexta*, in relation to thoracic temperatures. [From B. Heinrich, *Science* **185**:747–756 (1974). Copyright 1974 by the American Association for the Advancement of Science.]

Such differences have profound effects on the flowering times of many tropical plants.[166] The flowers of some species must be synchronized to open at those times of the day when pollinating insects can fly with minimal physiological constraints. Thus hawkmoth-pollinated species have flowers which open at dusk or late afternoon, and some species pollinated by the large euglossine bees open before dawn, when the bees are foraging. The higher the muscle temperature, the greater the number of flowers that can be visited. Under conditions of large food supplies, bees may use large amounts of energy for thermoregulation. Heat produced while a bumblebee is on flowers, under conditions of moderate ambient temperatures, helps to maintain a thoracic temperature near the minimum required for flight between flowers. Such regulation breaks down at high ambient temperatures as thoracic temperature rises.[165] During the period of visiting flowers, the bee must obtain a certain number of calories per minute to remain flying. Heinrich[165] has reviewed the energetics of foraging in bumblebees. At cool temperatures, the energy investment required for foraging is usually greater, since a certain body temperature must be maintained (Fig. 3.8). The abdominal temperature, only a few degrees higher than the air temperature, is maintained by passive heat flow only. Until the air temperature reaches 25°C, however, maintaining thoracic temperature at an average of 25°C requires a greater energy investment than for intermittent flight from flower to flower. The coevolution between pollinating insects and plants in the tropics takes into account the energy requirements of pollinators.

The decreased population densities of many plant species in the lowland wet tropics is due to competition, predation, and other factors (see Chapter 8), and usually a succession of plant species is flowering throughout the year at a given habitat. The lowland wet tropics support fewer bee species than other regions such as arid environments, and most bee species found here are generalists for visiting many different kinds of flowers. *Trigona* bees, for example, are extremely opportunistic foragers on many plant species. The apparent absence of oligolectic behavior in lowland wet forest bees results from their being long-lived and multivoltine, making them switch flower species throughout the year. Speciation in bees[167] is presumably blocked in such regions by these effects, since specialized behaviors and resource utilization do not occur. Janzen[40] has discussed two different patterns of flowering in tropical plants that promote outcrossing even when population densities are low. Some tropical plants exhibit a large amount of synchronous blooming at a specified time of year, and such compressed flowering is often synchronized with the dry season. Such flowering creates a signal for many insect visitors, and some of these have high energy requirements. Janzen shows that such mass flowering tends to show up in forest canopy and open areas. In the second pattern of reproduction in tropical trees, a plant species produces a few large flowers continuously from an inflorescence that may last several months. Large-bodied pollinators, including euglossine

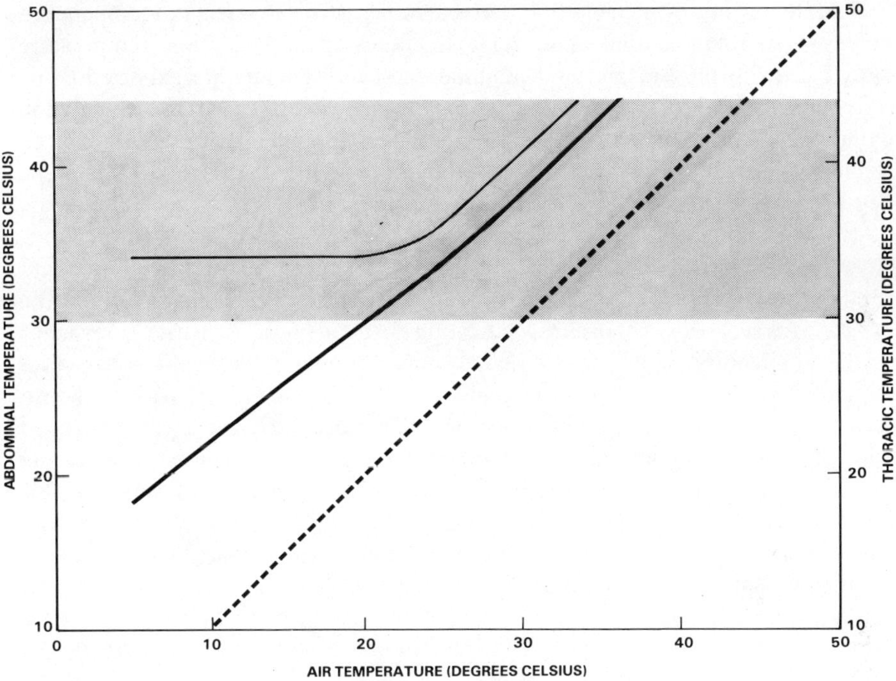

FIGURE 3.8. Patterns of energy investment necessary for a bumblebee to maintain a level of body temperature required for flight. Thick solid line: abdominal temperature, averaged in both sunlight and shade. Dashed line: Average thorax temperature, (the bee's flight muscles are in the thorax). Thin solid line: limits of thorax temperature. [From B. Heinrich, *Flight Energetics of the Bumblebee* (1973). Copyright 1973 by Scientific American, Inc. All rights reserved.]

bees, other large bees, sphinx moths, and other animals, exploit this system.[166] Each flower provides a large energy reward, which in turn provides sufficient energy for visitors to fly among them, even though great distances might be involved. The flowers are usually morphologically specialized to exclude visitors other than their pollinators, and their pollinators tend to be generalists and forage at flowers not a part of such a system. With increasing altitude in the tropics, the number of pollinating insects declines as a result of unfavorable ambient temperatures adversely affecting small insects and insects in general. Some tropical mountain tops tend to have bumblebees as pollinators.[166] Large pollinators require more energy, thereby necessitating that nectar be accessible for outcrossing to occur in a particular plant species on a tropical mountain top. Outcrossing is often impossible and apparently unnecessary, since a large percentage of the montane flora may be autogamous, a situation analogous to Temperate Zone plant species that bloom in early spring.

Many studies (e.g., Weir[168]) have demonstrated the effects of temperature on insect population dynamics. Wide fluctuations in the normal temperature daily cycles in the lowland and mountain tropical habitats are expected to be infrequent and therefore probably have no major effect on the population dynamics of most tropical insects. Owing to an absence of unpredictable wide fluctuations in ambient temperatures throughout much of the tropics, shifts in the degree of stability between predator–prey and host–parasite interactions are largely a function of changes in communities resulting from selection pressures created by habitat modification and succession. Many studies [e.g., Miller[169,170] (dragonflies); Lamb[171,172] (aphids); Wall[173] (cockroaches)] have analyzed the effects of water stress on insects, indicating that insects are sensitive to changes in the availability of water and that some forms of species shifts in tropical communities may in fact be the response to such a stress. The physiology literature is replete with examples of water stress in arthropods. The great changes in species richness and composition of insect communities associated with pasture vegetation in a very seasonal region of Costa Rica (Janzen and Schoener[134]) have shown that small insects drop out of the community in the dry season. Seasonal lowland tropical habitats are expected to impose considerable environmental stress from lack of water, even though temperature regimes may be the same between seasons.

3.6. NOTES ON MATING AND COMMUNICATION IN INSECTS

In those insect populations where individuals belonging to a particular species have a rather patchy spatial distribution owing to patchy distributions of preferred host plants or other kinds of resources, specific sets of behavioral patterns that structure courtship, copulation, and overall conspecific communication are expected to be integrated into phenotypes. Such adaptations are also expected to be present in species with clumped population distributions, and depending upon the degree to which such traits are genetically determined, such behaviors might be specific for a species as a whole rather than varying considerably among different populations of that species. Given the high species richness of insects in the lowland wet tropics, one might expect that specific forms of behavior are different in each species, particularly when examining closely related species or genera. Specialized behaviors associated with mating and other reproductive activities under general conditions of low population densities are expected for habitats where species richness is high. The mating strategy of the aposematically colored metallic green and orange (the band is orange) coreid bug, *Paryphes blandus,* entails low densities of adult bugs breeding on the small patches of *Anguria warscewiczii* (Curcubitaceae), the host plant used by de-

FIGURE 3.9. Copulating pair of the Neotropical hemipteran *Paryphes blandus* (Creidae) on the food plant of the nymphs, *Anguria warscewiczii* (Curcubitaceae).

veloping nymphs (Fig. 3.9). The host plant is patchy spatially, and breeding must be a function of bugs' successfully locating widely dispersed patches. There will be strong stabilizing selection for such traits if the species population is well integrated into the community. If the species population is not well integrated, one might predict that there will be a mixed strategy, in which there exists a set of behavioral patterns that assist integration into the existing community, and another set that is most adaptive toward shifts in habitats associated with colonizations. Strong intraspecific competition, associated with inability to disperse into other habitats, would also favor the expression of mixed behaviors as an adaptation for different segments of the population to utilize different portions of the environment.

Both interspecific and intraspecific competitive pressures arising from limited aerial space, ground space, and space for general spacing patterns are expected to mold the behavioral patterns of insects, especially in tropical habitats where species densities are high. The vertical structuring of butterflies in tropical forests (Papageorgis[(174)]) and the general observation that some species of

FIGURE 3.10. The Neotropical butterfly *Hymenitis andromica lyra* (Ithomiidae) visiting the flowers of *Eupatorium* (Compositae) at the border of a rain forest in the mountains of Costa Rica.

Morpho butterflies fly near the ground and others in the canopy,[175,176] may reflect differentiation among closely related species to utilize different portions of the environment for courtship, oviposition, and other activities.

Many species of ithomiid butterflies that dwell in the tropical wet forest understory tend to have rather transparent wings, as do some of the satyrid butterflies in the same habitats. The studies of Pliske[177–179] have revealed the role of specific chemical substances in shaping courtship-related communication between males and females in ithomiid species. The males of various ithomiid butterflies routinely visit dried or fresh flowers from which they apparently obtain precursor substances for pheromones associated with courtship (Figs. 3.10 and 3.11).

When a cluster of species associated with a habitat exhibit similar morphologies, reproductive isolation and mate recognition are determined largely by olfactory rather than visual signals between sexes. Ithomiid and satyrid butterfly species in deep forest understories might have converged in cryptic coloration to escape predators that hunt visually, and the evolution of chemical communication systems in such groups compensates for a loss of species specificity that might otherwise be associated with color and morphology under other conditions. In other insects visual cues may operate at a distance to bring potential mates into close juxtaposition, but actual courtship is then mediated primarily by olfactory and tactile cues at close range. Many naturalists know that the males of several species of *Morpho* butterflies will be attracted to the same decoy, suggesting that cues other than vision result in the species recognition process. Patchy distributions of nonfloral food sources of *Morpho* butterflies, such as fermenting sap flow wounds in trees in tropical forests, function to bring males and females together. Although conditions are generally subdued on the forest floor, where these butterflies feed in aggregate, sudden flashes of wings may function to attract others (and perhaps mates) to these feeding sites (Fig. 3.12). The cryptic undersides of the wings render the feeding and virtually motionless butterflies very inconspicuous against the forest floor. These butterflies, and several other allied groups, exploit rotting fruits and soupy fungal growths on sap flows on or near the forest floor, establishing selection pressure for cryptic coloration and behavior while feeding. High percentages of feeding morphos reveal symmetrical fragments of missing wings, suggesting attempted attacks by small insectivorous vertebrates such as lizards, even while the butterflies feed. The role of olfactory and tactile cues in butterfly species that partake of mimicry complexes is also clear.

Many investigations, such as those of Brower and Jones,[(180)] have described the olfactory secretions of male danaid butterflies. In this system a courtship-inducing substance is produced in glands of the abdominal hair pencils and deposited, through a specific behavioral sequence, into wing pockets. From there the secretion disseminates into the habitat to attract females. Vane-Wright[(181)] has discussed the phylogenetic importance of wing scent pockets and other structures in the evolution of the satyrid and morphid butterflies. Aldrich *et al.*[(182)] and Aldrich and Blum[(183)] have studied the production of courtship scents from abdominal glands in male coreid bugs. They suggest that such adaptations facilitate the colonization of new habitats if the male bugs precede females and then attract them into the new zone.

In other systems, such as the mating swarms of male lovebugs (Diptera:

FIGURE 3.11. The ithomiid butterfly *Mechanitis isthmia* visiting an unidentified bloom at the border of a coffee plantation in the central highlands of Costa Rica. Because its larvae feed on various species of Solanaceae, the conspicuous orange, yellow, and black tiger-stripe coloration of the wings is believed to be aposematic. This ithomiid is a familiar species of secondary habitats throughout much of southern Central America and also South America.

Bibionidae) studied by Thornhill[184] in Mexico, swarming by males establishes a syndrome of competition for females, which are attracted into such swarms. The swarms are established before females emerge, and the swarms are stratified vertically in terms of male lovebug sizes, with larger individuals near the bottom. Larger males near the bottom of swarms have greater access to emerging females and copulate more frequently. The conspicuous swarming behavior of many tropical dipterans, most notably of the families Culicidae, Cecidomyiidae, and Ceratopogonidae, in which swarms are dense and often columnar, warrants further study considering the findings of Thornhill. Such swarms may be highly structured in terms of the mating strategy. Yet relatively few data exist on the mating strategies of nonsocial insects, with the obvious exception of many studies of Odonata. In many species, males tend to aggregate near emergence sites or critical resources to attract females, which generally have less dense populations. It is difficult to generalize across the board for all insects, but data from butterflies, hemipterans, flies, cicadas, and others in the tropics suggest mating behaviors in which female individuals exhibit cryptic behavior and in which males exhibit single or mixed strategies of attracting females for mating.

Such patterns of adaptation are likely to be present in some groups outside of the tropics, but these might be due to the tropical origin of the groups or specialized habitat conditions favoring male competition for females.[185] In studying the distribution of males and females of tropical forest butterfly species that feed on rotten fruit on the forest floor, Young[175,186–189] and Young and Thomason[190] found that males predominate at such baits, for all species studied. Furthermore, females tend to show up at times different from periods of peak male abundance. Mating does not occur at bait sites. Females, particularly mated ones, presumably have different daily activity patterns from males, probably associated with oviposition and avoidance of predators such as birds. The observed high densities of males of some butterflies, such as *M. peleides,*[175,190] at fruit baits suggest that males aggregate in the habitat to attract females, even though this behavior might not occur at baits; males are prone to aggregate. The tropical rain forest contains a wealth of rotting fruits at different times of the year, a situation which undoubtedly promotes swarming behavior in many flies and other insects attracted to the fruit for food and oviposition. The presence of vultures in the tropics probably reduces the likelihood of courtship systems evolving in many insects that could feed on carrion.[191]

Swarming in the social insects has been reviewed by Wilson[80] and others. Naumann[192] has examined the role of trail odor marks in determining the swarming direction in *Polybia* wasps in Ecuador. *Polybia* nests are frequently raided by *Eciton* army ants, as are nests of *Polistes,*[193] resulting in the formation of "absconding swarms" to differentiate from those swarms associated with the founding of new colonies. (See Wilson[80] and West-Eberhard[194] for detailed discussions of the evolution of social behavior in insects and the analysis of

FIGURE 3.12. Cryptic feeding behavior in *Morpho* butterflies. (a) A group of several morphos feeding at a small sap wound in the exposed root of a canopy-size legume tree near Puntarenas, Puntarenas Province, Costa Rica. (b) A disturbed butterfly that has opened its wings, revealing a flash of iridescent blue in a patch of sunlight.

FIGURE 3.12. (*Continued*)

chemical communication in social insects.) In general, the high species richness of social wasps and ants in the tropics reflects considerable phylogenetic and ecological diversification in behavioral mechanisms regulating a variety of colony integrative activities such as alarm responses, recruitment, swarming, and reproduction. Such features of colonies are molded largely by prevailing ecological conditions superimposed on the historical development of character states within taxa of social insects. For example, Crewe and Blum[195] have discussed the phylogenetic significance of alarm pheromones in the Attine ants. They discovered a "common chemical heritage" based on 3-ketones and 3-alcohols for the genera studied, and these properties were found to agree with the accepted phylogeny based on other characters. Other studies (e.g., Blum *et al.*[196]) have analyzed the roles of certain chemicals in activities such as trail laying and robbing in social insects.

The behavioral repertoires of social insects must be interpreted in terms of both historical and contemporary ecological factors affecting species and genera. The great diversification of foraging behavior exhibited by ants, especially in the tropics (Carroll and Janzen[197]) reflects considerable specialization in terms of food gathering. Eisner[114] and others have stressed the convergent occurrence of similar chemical substances in different insects and other arthropods (largely the result of the substances being the most effective in carrying out a particular function, and being able to be metabolically produced directly or indirectly by the internal physiology of the organisms in question.) Thus citral is used by stingless bees in tropical forests to mark trails and by robber bees in tropical forests to disorient their victims.[196]

It is impossible to review here all of the interesting features of insects in tropical habitats in terms of behaviors associated with mating and communication. To understand the diversification of such behavior one must examine assemblages of closely related taxa of insects in a particular habitat or range of habitats and relate the observed behaviors to differences in how individuals in populations of each species utilize available resources in space and time. The types of mating and courtship behaviors that evolve will be a function of the past evolutionary history of the group and the present-day selection pressures shaping resource utilization, degree of intraspecific and interspecific competition, predation, energy budget, and population structure as related to seasonality or lack thereof.

A recent emphasis on sexual selection mechanisms in populations[185] has emphasized the importance of intermale competition for females in insects and the role of such mechanisms in the evolution of mixed mating strategies within single-species populations. We need to reexamine and describe the mating strategies exhibited by different groups of tropical insects to determine the importance of intermale fighting or competition as a driving force in providing organisms with "better fits" into their environments. In the context of this book it is important to realize that such data are ordinarily lacking in tropical insects and some kinds of tropical habitats are the places where such specialized forms of mating behavior are most likely to be found, particularly if such behaviors are related to the high densities of species usually associated with such habitats.

Outside of the social insects, butterflies, flies, and cicadas figure strongly as suitable insects through which to examine some of these ideas.

It is also necessary to consider the role of selection in determining the internal reproductive physiology of insects as a major component of reproductive behavior in general. The integration of the generalized insect phenotype in terms of the fine balance of hormonal processes regulating embryonic development[(198)] is at the very foundation of reproductive physiology in insects. Nutrition, light, and mating figure prominently as regulative factors among the wide variety of internal and external environmental factors that determine egg maturation in insects (Engelmann[(199)]). Reisen and Aslamkhan[(200)] have found that the gonotrophic cycles of some tropical mosquitoes are regulated by blood meals, whether or not females have been inseminated.

The considerations of Cody[(67)] stress the high premium placed by selection on the evolution of clutch sizes, notably in tropical organisms. If there is a greater amount of biotic regulation of insect populations in the tropics, particularly for lowland wet environments, selection will favor those clutch sizes within single populations of each species that permit these populations to remain in some fashion of equilibrium with the populations of other organisms composing the niche. Cody argues that smaller clutch sizes evolve in the tropics to ensure that populations stay at their carrying capacities. Unstable conditions tend to select for larger clutch sizes associated with changes in other population growth parameters that promote a higher rate of population growth. Yet a growing body of evidence suggests that the very instability of some tropical regions promotes a greater diversity and therefore should select for population growth parameters favoring colonization and invasions to new habitats.

The great range of habitats in the lowland wet tropics, including some that are relatively new and unstable, provides sufficient habitat heterogeneity to promote colonizing episodes by insect species, particularly and initially at the herbivore and decomposer levels of community organization; this means the elaboration of phenotypes programmed for increased clutch sizes or at least a mixed strategy favoring phenotypes of low and high clutch sizes. There appears to be too much environmental heterogeneity in the lowland tropics to suggest a uniform hypothesis of reduced clutch sizes. Simpson[(201)] has discussed the role of historical regional events in shaping areas of high diversity in South America, suggesting that large-scale environmental heterogeneity accounts for increased diversity.

A. H. Ehrlich and P. R. Ehrlich[(202)] suggest that multiple matings are probably frequent in most butterflies, including tropical species, but that multiple spermatophores are used for food rather than for mating. The existence of multiple matings in these insects seems to be indirect evidence for the existence of mating systems in which males compete for females. In some tropical butterflies, mated females exhibit specific displays designed to discourage multiple matings.

High predation rates from birds and other vertebrates on tropical butterflies are debatable. Much of the theory of mimicry and development of wing coloration in the Lepidoptera is based upon the relative roles of such color patterns and associated flight behaviors in mating and predator avoidance or evasiveness. Studies by Brown, and his associates[203–205] and others with heliconiine and ithomiid butterflies with tiger-stripe wing color patterns have shown that birds actively eat butterflies, as has the study by Brower *et al.*[206] of overwintering monarchs in Mexico. *Morpho* and other butterflies feeding on rotten fruits in tropical forests tend to exhibit symmetrical wing damage patterns at the posterior edges of the hind wings[175,188,189,207] (Fig. 3.13). Young[188] has interpreted such patterns to imply predatory attacks primarily on butterflies that are resting or feeding rather than flying. The evolution of false eyespot patterns in these fruit-sucking butterflies as an adaptation for increased crypsis for late-afternoon feeding has been discussed by Young.[188,189] Allied species active higher up in the forest tend to have reduced or no eyespots on the ventral sides of the wings, suggesting a lack of selection favoring these markings away from the forest floor.

Mating in some butterflies involves males emerging before females within a cohort and waiting for females to emerge. This behavior is well known for some species of *Heliconius* butterflies (e.g., Alexander[208]). Young[77] suggested that aggregates of freshly eclosed males of the highly sexually dimorphic tropical pierid *P. lypera* wait together in the forest understory near the emergence sites of the females and then compete for mates. This species exhibits cluster oviposition and highly gregarious larvae, conditions promoting aggregated pupation and synchronous adult emergences. Males of *Morpho amathonte* exhibit extremely predictable flight routes through forest clearings on a daily basis, perhaps as an adaptation for attracting females waiting in vegetation along such routes (A. M. Young, unpublished data). Such a system may disperse males so that inter-male competition for the same females is lowered. In the above example with *Perrhybris,* unmated males appear to exhibit a form of lek behavior by aggregating to attract females. The behavior of pupal mating is well known for some species of the Neotropical butterfly genus *Heliconius* (Fig. 3.14). Males will be attracted to the yet uneclosed pupae of females, presumably as a result of an odor produced by the latter, and mating takes place shortly after eclosion and expansion of the wings. Such behavior may reduce the amount of time required for mating in the wild and thereby increase the time available for egg laying. The selection pressure driving such behavior would be predation on the adult butterflies and perhaps predation on the immature stages as well. The adaptive significance of such behavior, coupled with the simultaneous development of the ovaries, is that females can begin to oviposit soon after eclosion. Ringo[209] discussed the lek behavior of males in some species of Hawaiian *Drosophila*. In *Drosophila grimshawi,* male displays are density dependent on frequency. As population density increases, so does the frequency of male displays. The presence of females increases the frequency of communal displays and aggression in males.

FIGURE 3.13. Eyespot markings on the undersides of the wings of *Morpho* butterflies (a) compose a form of crypsis when these insects are feeding in shaded areas of the forest floor. When light rays hit the wings (b) the vertebrate-like eyespots are enhanced and may function otherwise to discourage attack by small insectivorous vertebrates. Attacks from the rear of a feeding butterfly may be largely unsuccessful, leaving symmetrical gaps in the wings.

FIGURE 3.14. Pupal mating in the Neotropical butterfly *Heliconius sara*. (Photographs from Finca La Selva, Puerto Viejo, Heredia Province, Costa Rica, courtesy of Robert Borth.)

3.7. RHYTHMICITY IN TROPICAL INSECTS

The large change in temperature associated with day–night cycles in the lowland tropics provides a background of abiotic effects governing the expression of clear-cut diurnal rhythms of activity in many insects. Often different but closely related species with similar ecological habits exhibit different temporal activity rhythms. Insects of Temperate-Zone areas also exhibit different diurnal patterns of activity, but in the tropics, where insect diversity is great, such behavior is probably very adaptive in the sense of reducing competition among ecologically similar forms that exploit similar or identical resources. The nocturnal habits of some Temperate-Zone insects, such as many Orthoptera, may form a relict trait that was previously adaptive in tropical environments. Basically four major groups of insects, classified according to activity patterns, are seen superimposed upon the diurnal cycle in the tropics: (1) insects active in the day, (2) insects active at dawn and predawn, (3) nocturnal insects, and (4) insects primarily active at dusk. Not all of the classes are mutually exclusive, since some groups of insects, such as cicadas, can include species with individuals active at dusk as well as during the day. Young[(186,189)] found that fruit-sucking tropical butterflies in forest habitats exhibit clear daily patterns of feeding. An example is *M. peleides,* a familiar, widespread species belonging to the superspecies *achilles* of South America. Males turn up in greatest numbers between noon and 2:00 P.M. on sunny days to feed at deliberately placed baits of rotting bananas on the forest floor. Females are less predictable, turning up, always in very small numbers, at all hours of the day. On some days a burst of feeding activity in both sexes occurs near dusk. *Caligo* butterflies, on the other hand, appear in peak numbers at the same baits early in the morning and steadily decline throughout the morning.

Although Hermann[(210)] has shown that workers of the giant tropical ant *Paraponera clavata* are nocturnal foragers in Ecuador, Hermann,[(210)] Young,[(211)] and Young and Hermann[(212)] discovered considerable diurnal foraging in cloudy or overcast weather conditions in northeastern Costa Rica. In Young's studies, there was usually a prominent burst of nest exits in late afternoon and through the dusk period. Young attributed the observed patterns of foraging rhythmicity to prevailing weather conditions. There is some evidence, however, that this species and other ants possess endogenous activity patterns.[(213)] Tropical ants in general might be genetically programmed for specific endogenous activity periods, but these are modified by the influence of prevailing weather conditions experienced by nest members each day. Many tropical ants exhibit definite diurnal cycles of foraging activity, while some species are "generalist" foragers, being active through most of the day and night.

The highly synchronized nocturnal chorusing patterns of many Orthoptera is well known. In tropical forests, the greater vertical and horizontal structural complexity of vegetation in forest habitats provides a pattern of environmental heterogeneity for diversification of habits among the Orthoptera. Diversification in feeding habitats and habits is accompanied by species-specific songs and chorusing patterns. The evolutionary significance of singing in the Orthoptera has been reviewed by Walker.[(214)] There are more genera and species in most families of Orthoptera in lowland tropical wet forests, including groups such as mantids, phasmids, and roaches where sound production is less developed than in other lineages. Convergent or phylogenetically generated nocturnal activity in the Orthoptera is made possible in part by the great diversification in singing habits, coupled with ecological differences associated with feeding. There is a greater incidence of seed-eating orthopterans in the tropics. Many tettigonids are opportunistic predators on insects at lights in the tropics.

The "sundown cicada," *Fidicina mannifera* (Fig. 3.15), is a large, robust insect that exhibits short bursts of singing at dusk and occasionally at dawn. In the lowlands of northeastern Costa Rica, the males begin to sing at about 5:50 each evening and sing to about 6:05 or thereabouts. On the Pacific side of the country, the cicada sometimes sings intermittently throughout the day, as seen for one population in lowland Puntarenas Province, and possibly further north in Guanacaste. These differences in the pattern of daily singing between populations on either side of the central Cordillera suggest that abiotic factors such as daily regimes of temperature change are responsible for the timing of chorusing. A host of other cicada species, such as *Fidicina sericans, Fidicina pronoe,* and *Quesada gigas* (Fig. 3.16), also exhibit bursts of dusk singing on the eastern side of the Cordillera, but these species also chorus during the day. The males of these

FIGURE 3.15. The Neotropical cicada *Fidicina mannifera*. Lowland tropical dry forest populations in Costa Rica are characterized by smaller average adult body size and lighter markings on the body than adults from lowland tropical rain forest.

FIGURE 3.16. Some characteristic cicadas (Homoptera: Cicadidae) of lowland tropical rain forest in Costa Rica. From left to right: *Quesada gigas*, *Zammara smaragdina* (male and female, respectively), *Fidicina pronoe*, *Fidicina sericans*, and *Fidicina amoena*.

and other species position themselves in aggregates in trees[215] perhaps in such a manner so as to optimize sound transmission to attract mates. Young[150] has hypothesized that dawn or dusk aggregated chorusing in tropical cicadas represents a convergent adaptation of different genera and species to chorus at times when sound transmission is the highest. Selection for optimal mate attraction in a multispecies system has resulted in convergent chorusing, and participant species have distinct songs. Costa Rican lowland and premontane tropical rain forests support a rich diversity of genera and species of cicadas, and different species in the same area often exhibit differences in peak annual emergence period and habitat association. When several species occur in an area, there is considerable range in population densities as well.

The sweep sample studies of Janzen[72] have shown that whole groups of insects in different trophic levels of communities replace one another between day and night in various tropical habitats. Such data emphasize the role of diurnal cycles in the structure and composition of communities of insects associated with vegetation. Detailed reviews have shown the role of abiotic factors in shaping the diel rhythms of activities in many insects. The interesting consideration for the tropics is the possible selection pressure created by limited resources as a force in determining differences in peak activity periods of ecologically similar species.

Another major set of factors regulating activity rhythms in tropical insects is the response of plant species to diurnal cycles. A clear case is that of synchronized flowering in many tropical plants. The recent studies of H. G. Baker and I. Baker[216,217] and Baker *et al.*[218] have revealed diel rhythms of nectar production in several tropical plant taxa. Pollinating insects and other animals seeking nutrients from nectar will have foraging periods synchronized with peak nectar production.[218] As discussed in Section 3.5, such insects must have the physiological means to be active at these times of the day in order to obtain sufficient nutrients. In short, the very sharp temperature gradient over a 24-hr period in tropical lowlands provides a cue for those physiological mechanisms responsible for synchronized flowering in many species. These conditions in turn provide selection shaping the internal physiology and foraging behavior of pollinating insects. The existence of different flowering strategies over 24-hr periods in many tropical plants provides a source of environmental heterogeneity promoting diversification in peak foraging periods of flower-visiting and pollinating insects.

3.8. SOCIALITY AND DEFENSE IN SOCIAL INSECTS

Since it is not possible or appropriate to discuss here the biological features of social insects in depth, the reader is referred to Wilson,[80] who provides considerable information on the definition of true social behavior in insects and discusses the evolution of sociality.

Because the social insects, including both the Hymenoptera and Isoptera, reach their greatest taxonomic diversity in the tropics, some brief comments are made here concerning generalized features of these insects, possible mechanisms of diversification, and expression of defense and other forms of behavior generated at the level of the colony. In subsequent chapters we return several times to examples of the impact of social insects on tropical communities.

The tropics provide a greater structural and chemical diversity for highly adaptable social insects such as ant species to respond to in the context of ecological specialization. One need only think about the great diversity of nest site patterns exhibited in ant species in the tropics. Many ant species are specialists at arboreal nesting, making nests in hollow twigs, epiphytes, and other structures off the forest floor. Brown and Wilson,[219] for example, have traced the evolutionary changes in the foraging behavior of the Dacetine ants, suggesting shifts in both foraging habitats and degree of generalized versus specialized predatory behavior. The dacetines exhibit well-delineated patterns of morphological–ethological coadaptive systems associated with foraging in different kinds of habitats.

The general hypothesis is that the great opportunities for evolutionary diversification at subfamilial, generic, and specific levels of taxonomic organization have been concomitant with the need for niche specialization in terms of nest sites and to some extent feeding habits. Perhaps more than most other groups of social Hymenoptera, ants have exhibited the most morphological and behavioral diversification in response to diverse ecological resources. Diverse feeding habits, some generalized and others specialized, are seen in the ant groups of the tropics: scavengers, predators, exudate gatherers, nectar gatherers, pollen feeders, and parasites. Termites, another large group of social insects, apparently have less diversification, being limited to functioning as decomposers of dead wood and other cellulose detritus in tropical forests. The tremendous ecological diversification in hymenopteran social insects was initiated with a huge adaptive radiation more than 50 million years ago, endowing the groups with these three features: systems of castes and functional roles within the colony that change with adult age; elaborate systems of chemical communication that include signals of alarm, recruitment, and recognition of nest mates and non-nest mates; and elaboration of nest structure and nest positioning in the environment as means of regulating temperature and humidity inside nests.

Wilson[80] discusses the various stages in the evolution of eusocial behavior in the social insects, as well as promising directions in research that bring together the analysis of social systems with population dynamics and evolutionary genetics. According to Wilson,[80] three major traits are characteristic of eusocial insects: individuals of the same species cooperating in caring for young; reproductive division of labor, with sterile or semisterile individuals working on behalf of the female reproductive caste (queen); an overlap of at least two generations in life stages capable of contributing to colony labor, so that offspring assist parents for some portion of their life-span. The greatest diversity of eusocial insects occurs in the tropics, along with many examples of presocial behavior and colony organization in some insects. The study of tropical ants and social wasps provides the opportunity to determine the relationship between modifications in a basic design of eusociality adapted to particular ecological factors, including next habits, food resources, predation and other forms of biotic mortality on whole colonies, and influences of abiotic conditions.

It is noteworthy to appreciate the extreme ecological diversification of social insects in the tropics. Wilson[(80)] points out that the center for speciation in the vespine paper wasps is tropical Asia, where as many as 50 species occur. From here, the group has penetrated the Temperate Zones, although fewer species are present. The long growing season of the tropics has promoted the development of multiple-foundress systems in some social paper wasps, a uniquely tropical phenomenon which allows the production of large workers castes to counteract the effects of biotic mortality on workers within a given nest. Such systems are found in Neotropical species of *Polistes*. In the Vespinae, another major group, the queen is much larger in size and is the sole egg producer in the nest. In this group there is considerable sophistication in the structure of nests, with carton nests being located in the ground, on branches, or in tree cavities. Tropical wasps are basically predatory but also exhibit opportunistic foraging for nectar, extraflora exudates, rotting fruit, and other materials rich in nutrients such as carbohydrates.

Wilson[(80)] has traced the hypothetical evolutionary steps in the development of eusociality in wasps, accounting for the various forms of presocial behavior exhibited by families such as the Sphecidae and Pompilidae. The steps involve a progressively increasing care for young, food provisioning, and elaboration of nest structure to house overlapping generations of young and adults. Wilson[(80)] emphasizes that studies of wasps have yielded much of our understanding of the nutritional control of castes; trophallaxis, the process by which adults exchange food with larvae; and the mechanisms by which queens regulate the activities of the colony. The social wasps of the tropics are probably significant predators on a variety of soft-bodied insects.

Young[(220)] and Young and Moffett[(221,222)] found that vespids take heavy tolls of young lepidopterous larvae in a study of the butterfly *Mechanitis isthmia* (Fig. 3.17), and there is no reason to believe that other species are avoided as prey. Few studies have been made, however, on the impact of predation by tropical social wasps on insect populations. Given the usually patchy distribution of nests, such effects, if significant, are also predicted to be patchy, a function of nest site relative to optimal prey and other resources and to colony size. Although some preliminary evidence suggests that under some conditions *Polistes* populations can be regulated by predation on larvae and pupae by army ants,[(193)]

FIGURE 3.17. A polybiine wasp taking (in its mandibles) a second-instar caterpillar of the Neotropical ithomiid butterfly *M. isthmia* off its food plant, *Solanum lancefolium,* in northeastern Costa Rica. Note the other larvae present. Despite repeated attacks by the wasps, the larvae were unable to defend themselves, and although *Ectatomma* ants roam over the leaves and larvae, they do not protect them from such predation. But because the ants collect drops of fluid exuded from the freshly damaged leaf where the larvae are feeding, the interaction could be the initial stage of a coevolved mutualism involving the larvae and ants in response to intense predation by these wasps.

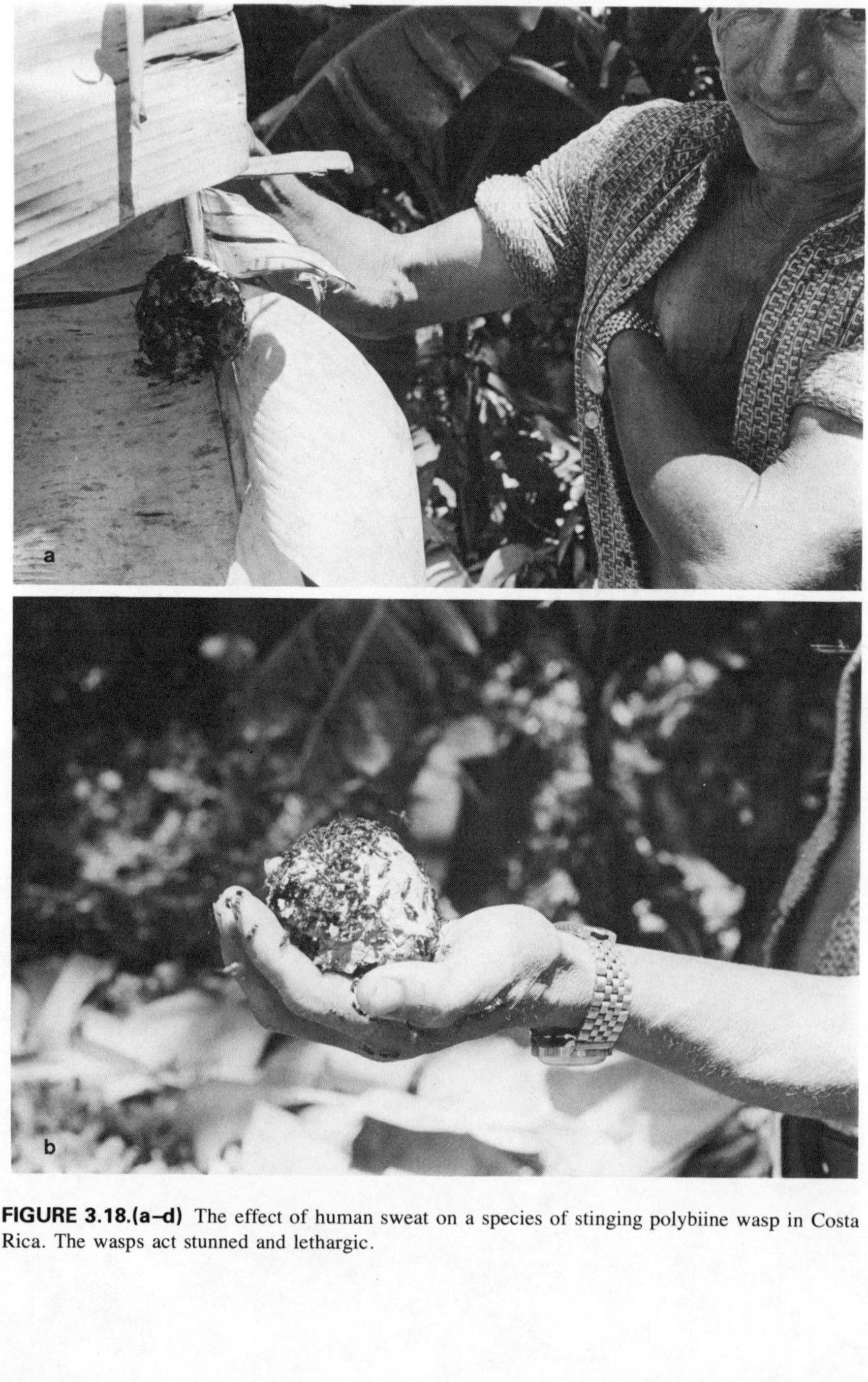

FIGURE 3.18.(a–d) The effect of human sweat on a species of stinging polybiine wasp in Costa Rica. The wasps act stunned and lethargic.

FIGURE 3.18. (a–d) *(Continued)*

experimental proof is lacking. Army ants also destroy broods of *Polybia* and other vespid wasps in the tropics.[192] Whether nesting sites for certain kinds of social insects are limiting resources in the tropics has not been studied. Given the high density of some taxa such as ant species per unit area of habitat and the patterns of ecological specialization in some of these, nesting sites may in fact be limiting resources. However, the study by Matthews[223,224] on the presocial wasp *Microstigmus cornes* in Costa Rican tropical rain forest suggests a rather high density of nesting sites per unit area of forest understory.

For most ant species, the grain of the environment with regard to nest sites is probably far more complex than for wasps in the same habitat. This is an intuitive impression based on the observation that ant species densities are greater than densities of wasp species within and between habitats in a region. But such reasoning is something like comparing bananas with coconuts until more data are available on resource specificities in various taxa. Preliminary data indicate that the giant ant *Paraponera clavata,* a ponerine, prefers to nest beneath the root systems of only certain tree species.[212] One might expect considerable nest site limitation in highly specialized nesting taxa such as ant species associated with hollw twigs, and certain epiphytes.

The studies of Rockwood[225,226] suggest that leaf-cutter ant colonies are limited by the availability of preferred tree species whose leaves are used as substrates for fungal gardens harvested for nutrients within the nests. The apparently generalized foraging habits of army ants,[26,27] in which these ants are predatory on a variety of arthropods, probably resulted from the nomadic condition of the colonies, in which there is no permanent nest sites. A nomadic lifestyle allows army ants to become supergeneralists on arthropods in general and to retrieve energy in small packet from large amounts of habitat. Once again, however, very little information exists on the impact of army ants as predators in tropical communities and whether their movement patterns impose a sort of regulation on social wasp species that tend to nest repeatedly in the same habitats over a run of years or longer.

Schmidt[227] has reviewed the great diversity of ant venoms found in the group and relates such biochemical diversity to the tremendous ecological success of ants in general. Ants produce some of the most toxic of all animal venoms known,[227] and the evolution of the sting apparatus in the Hymenoptera[228] has been a key factor in allowing such large concentrations of insects to escape from vertebrate predators. Schmidt[227] also suggests that the venoms of other hymenopterous groups such as bees and wasps might be very different from those of ants overall. Young[229] reported that human sweat appears to calm workers of a *Polybia* wasp species to the point that occupied and thriving nests can be crushed in a hand coated with human sweat (Fig. 3.18).

We need to relate the level of toxicity of hymenopteran venoms to both phylogenetic and contemporary ecological factors such as predation rates on nests by vertebrates and invertebrates along with other forms of colony defense. Even though stingless bees do not sting, they can effectively discourage attack by vertebrates through behavioral responses.[230] In stingless bees, a dominant group of social insects in the American tropics and a group of opportunistic pollinators of many plant species and nectar robbers of others, nests are hidden in tree cavities and the next entrance is a resinous tube sealed over each night to prevent other organisms from entering. When a nest is attacked, the bees will swarm and bite the attacker, and tangle up hair and feathers (A. Wille, personal communication). Wille and Michener[230] provide a review on the evolution of stingless bees and their nesting habits.

3.9. DENSITY-RELATED BEHAVIOR IN INSECTS

The density of an insect population influences the environment in several ways: (1) increasing densities coupled with a lack of dispersal that result in an overcrowding of the population's membership for the resources the species needs for survival and reproduction, sometimes resulting in population decline and extinction; (2) selecting for phenotypes suitable for dispersal to new habitats in response to increasing densities on limited resource supplies; (3) creating genetic and nongenetic changes in developmental time or other parameters affecting the growth rate of the population; (4) alternating existing spacing patterns of the population's membership. (5) changing various forms of behavior collective or singly, including mating systems and defense; (6) increasing densities that may intensify competition with other ecologically similar species in the same habitat.

The laboratory studies of Thomas Park and his students have demonstrated the deleterious effects of increasing population densities on survival and reproduction,[231] and studies (e.g., Dawson[232]; Ayala *et al.*[233]; Lewontin[84,85]) have shown the operation of density-dependent selection altering the genetic structure of insect populations. Allee[234] stressed the effects of population density on developmental time, survival, and other properties of animal species in general. Many ecology textbooks discuss the effects of density-dependent and density-independent factors on insect populations. Studies (e.g., Wharton *et al.*[59]) with cockroaches have demonstrated that increases in the population density of larval stages can result in adult cohorts of reduced mean weight of adults and that males are more sensitive to the deleterious effects of high densities than females. Low densities promote faster development and higher survival to the adult stage (Fig. 3.19).

Other studies indicate that in some insects, aggregative behavior, in which several or many individuals of a species stay together, is adaptive for feeding and defense against predators. Fraser-Rowell[235] demonstrated the role of visual stimuli arising from individual color patterns in promoting the formation of

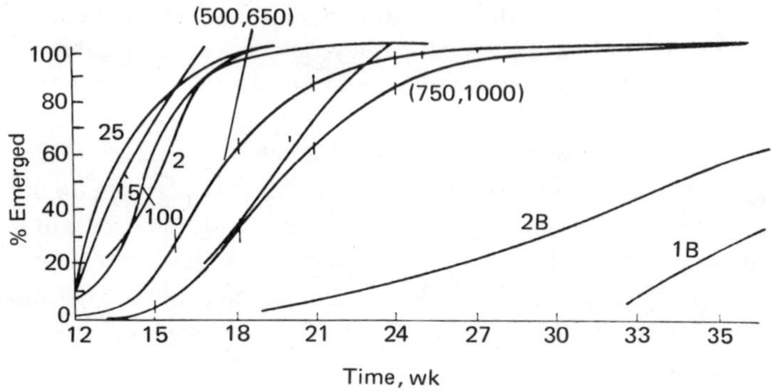

FIGURE 3.19. Egg-to-adult development curves for different population densities of nymphs grown in the same size jars (except for curves 1B and 2B, which were grown in smaller containers) expressed as percentages of all surviving insects, both nymphs and adults. The vertical lines crossing the curves show limits of the averages for the group of (500, 650) and (750, 1000). [From D. R. A. Wharton, J. E. Lola, and M. L. Wharton, *J. Insect Physiol.* **13:**699–715. Reprinted with permission. Copyright 1967, Pergamon Press.]

aggregations in the tropical grasshopper *Phymateus purpurascens* in Uganda. Aggregative behavior is most evident in the brightly colored larval stages of this species and is believed to reinforce noxious properties to predators. The larvae rest and move as groups. Myers and Smith[(236)] described the synchronous head-flicking behavior of tent caterpillars in aggregates as a means of defense against parasites. Here, the stimulus is the flight sound of a fly parasite that lays its eggs on the caterpillars. Otte and Joern,[(237)] also working with grasshopper populations, developed a model which predicts that the number of females successfully mated within a population by the most successful males (defined here as those having optimal territories) will initially increase as density increases, and that as density increases further territorial defense becomes too costly energetically and aggressive behavior then decreases. At low densities there are many more suitable territories (suitable bushes) than male grasshoppers, but at high densities territorial defense of suitable bushes increases and many males are excluded from entering the bushes and mating by the most aggressive (successful) males. Thus fewer males mate with more females as density increases, but after a point the defense of suitable bushes becomes a major energy drain and the system breaks down.

Because of the tremendous variation in the plant species composition and structural organization of tropical habitats within a particular region, most notably in the tropical lowlands, there exists the potential for various strategies of spacing patterns in insect species inhabiting different trophic levels. Two extremes within such a spectrum of habitat types might be one in which many taxonomically unrelated plants are present and in which each taxon occurs in small clumps (patches) or single individuals with considerable distances between patches. In the second case, few plant taxa are present and the abundances of each are high. The first habitat therefore is more heterogeneous spatially than the second one. An example of the first type might be a strip of advanced secondary forest bordering lowland primary wet forest, and of the second, a strip of very young secondary growth along a road in the same area. If herbivorous insect species in both habitats are assumed to be specialists, selection might have favored the evolution of population growth characteristics favoring the maintenance of population densities at or near carrying capacities for each species. If such species are regulated by predators and parasites, the populations might have high growth rate characteristics counteracted by such density-dependent mortality, and if such mortality is low or absent, the energy drain, so to speak, has been the tradeoff between parameters such as high clutch size and effective defense systems against predators and parasites. Under these conditions, and because there are more insect taxa in the first habitat within the plant-feeding community, densities of individuals on each host plant are probably low to moderate but seldom reaching levels for intraspecific competition to be invoked. The insect predator and parasite populations under these conditions are expected to be composed primarily of generalists capable of exploiting more than one kind of host, since the density and distribution of each host species make each resource rare.

The litter beneath the vegetation in such a habitat is expected to be heterogeneous physically and chemically, and it may support a greater diversity of decomposer and scavenger insects and other arthropods, including ants. This assumes that certain distinctive taste and nutritional features of each plant species are retained in some form during at least initial stages of the decay cycle. For the first habitat, therefore, little or no density dependence is expected to be within the population of specialist herbivore species but perhaps some in predator and parasite populations, since these species (tachinids, various hymenopterans) are expected to be generalists and therefore overlap in resources.

A different picture is expected for the second habitat. The population of each herbivorous insect species is expected to be large since resource patches are large, and within such a habitat predators and parasites might in effect become specialists on these herbivore populations, which can absorb such mortality from having population parameters promoting rapid population growth associated with colonizing new resources. Litter communities are expected to be less diverse in this habitat unless multiple species can exploit the same resources, since patches are large. Keep in mind that these examples are not necessarily the case for all situations in the tropics. It is dangerous to oversimplify complex systems or assemblages of species and other taxa. Habitats with many plant species may very well support mixtures of specialist and generalist herbivorous insect species, and most litter organisms may be generalists to a large degree. In general, we will have to examine the population dynamics and general ecology of individual insect species to determine what sorts of environmental conditions promote density dependence in the tropics.

In habitats where a species is a resource specialist (low niche breadth) and population growth is high, density-dependent effects will be operative, perhaps promoting frequency-dependent selection resulting in geneotypes best adapted for colonizing new habitats. Young[71] has discussed the possible interplay of forest habitat destruction, host tree specificity, and population density in promoting some cicada species to invade secondary habitats in the tropics. Data from sweep-sample studies in the tropics[72] suggest some overlap in insect faunas in different adjacent habitats, and the types of selection pressures affecting each species might be very different between habitats (see Chapter 8). Following similar lines as my discussion of the clutch size hypothesis, we can expect considerable variation in the population structure and dynamics of insect species within a trophic level, implying the differential role of density dependence, interspecific competition, and degree of resource specialization as the major factors affecting populations, particularly in the nonseasonal lowland regions. Further aspects of population dynamics and other ecological properties of insect species are discussed in Chapters 5 and 6.

3.10 NOTES ON THE PHYSIOLOGY OF FEEDING IN INSECTS

What follows is a brief introduction to the complex matter of insect feeding, including food procurement, ingestion, and digestion. My purpose is to highlight some aspects of feeding in insects that relate to the kinds of generalized and specialized interrelationships found in the tropics, and to lay a foundation for discussions of such interrelationships in subsequent chapters. Much of the emphasis in the present discussion will be on plant–insect interactions instead of on feeding by insect predators and parasites on other insects because it is becoming increasingly evident that much of the structure of tropical terrestrial communities and whole ecosystems is determined largely by the effects of insects as herbivores and pollinators on plant species and the counterselection of plant species as resources for insects, and because the bulk of descriptive and experimental data on the ecology of insects in the tropics concerns the plant–insect ecological interface. This is not to say there are no other relevant aspects to tropical insect biology. We have large gaps in our knowledge about the community biology of predators and parasites, not to mention litter insects and arthropods in general. And much more autecological or natural historical work is needed on herbivorous insects and their host plants. Excellent general discussions of feeding in the insects are found in Brues[238] and Dethier.[239–241]

Using the "feeding categories" of Brues,[238] the tropics support a greater diversity of plant-feeding, predatory, parasitic, and scavenging insects than other regions of the world. Tropical environments present complex mixtures of food shapes, colors, odors, and nutrient contents for organisms, and given the relatively small body size of most insects, the grain of food heterogeneity generated by other organisms in the tropics and by the insects themselves is great. Charles Darwin emphasized the role of food in the survival of animal species in general. The implication is, of course, that the behavioral and physiological processes that shape food procurement and processing form a major, complex sphere of selection pressures influencing both the evolutionary history of the insects and the strategies of ecological diversification within genera and species. Within each level of trophic organization, insects exhibit considerable preferences for certain kinds of food, based on both nutritive and nonnutritive components of food. Selection is expected to favor the evolution of a feeding behavior within populations of a species that maximize the retrieval of energy necessary for maximal survival and reproduction. The evolution of sophisticated visual, tactile, and olfactory sensory organs, mouthparts, digestive tract chemistry, and neuromuscular mechanisms associated with food handling and capture have shaped the evolution of feeding behavior in the insects.

The long evolutionary history of insects in the tropics has provided the opportunity for considerable ecological diversification as part of the differentiation of whole families, subfamilies, genera, and species. And although insects feed on a wide variety of animal, plant, and dead organic matter, Fraenkel[242,243] was one of the first to demonstrate that most insects have similar nutritional requirements. Most insects require sterols, B vitamins, carbohydrates, and water to varying degrees.

Food procurement involves being able to recognize and ingest the correct food items in the habitat. Insects ingest food by chewing it; fluid feeding, in which enzymes may be injected into the food to assist ingestion; bloodsucking, in which anticoagulants are injected into the blood; and prey immobilization by the injection of venom such as through a sting. A few insects, such as some ants, grow their food, examples being the fungus gardens of tropical leaf-cutting ants and the "honeydew" secretions of certain Homoptera farmed by some species of ants. In the social insects there is feeding of one individual by another in the form of a worker caste caring for the young, and such behavior is a major component of such social systems.

Insects that feed on plants are phytophagous, and if many different plant taxa are consumed, the species is polyphagous. If food consumption is limited to plants in one family or closely related families, the insect is monophagous. The term *herbivore* is used to connotate the trophic role of the insect in a community. Phytophagous feeding involves the use of visual, olfactory, and tactile sensory mechanisms, coupled with gustation. The physiology and structure of chemoreceptors and eyes in food recognition in insects has been studied extensively. Of interest is the food recognition of biting or chewing insects. When a grasshopper lands on a plant, chemotactile stimulation results in biting, a nonspecific response. Sustained feeding is due to the presence of certain attractant substances, some of which, such as sugars and amino acids, may be nutritive. Suitable gustatory stimuli are necessary for continued feeding. Yamamoto and Fraenkel[(244)] discuss the roles of olfactory and gustatory stimuli in the oviposition and feeding of the tobacco hornworm on solanaceous plants, and Dethier[(240,241)] gives a general review of the role of secondary substances in plants on the feeding specificity of insects. Dethier stresses, among other things, the specificity of certain chemoreceptor sites and secondary substances interacting with them as the mechanism for host plant selection in phytophagous insects. Since secondary plant substances apparently have no direct metabolic function in plants, they are believed to have evolved to protect plants from ingestion by herbivores (see the recent review of Dethier[(241)]). Dethier[(241)] has shown by behavioral and electrophysiological comparisons that in general the gustatory receptors of herbivorous insects possess a much wider receptor sensitivity than the gustatory receptors of nonherbivorous insects. The data suggest that herbivorous insects have evolved receptors sensitive to a wide range of secondary compounds rather than specific deterrent receptors. Dethier reasons that such generalist receptors provided the central nervous system of these insects with a large pool of information that formed the basis for the evolution within the CNS of specific deterrent and acceptance behaviors. It follows that the great secondary compound diversity present in the lowland tropical floras of the world provide the response systems for herbivorous insects to act upon, resulting in ecological diversification in the form of host plant specificities (see Chapter 4).

Visual, olfactory, and tactile stimuli form a major aspect of the evolution of animal tissue feeding and plant feeding habits in insects. Most insects use some combination of these types of cues from the environment in the food recognition process. Studies (e.g., Downes[(245)]) have demonstrated that different sexes may utilize different kinds of food. Thus female ceratopogonid midges in some genera and species feed on pollen grains, while males feed on nectar and other substances rich in carbohydrates. Pollen feeding in insects appears to be related to females' requiring amino acids from extrinsic sources for the production of eggs, and some species of *Heliconius* butterflies use pollen as a source of egg protein too.[(31)] I suspect there are many undescribed examples of similar behavior in tropical insects.

The paucity of female *Morpho* butterflies and the abundance of males at baits of rotting fruit (a concentrated source of sugars)[(175)] and the presence of about equal numbers of male and female butterflies on molds (A. M. Young, unpublished data) suggests that females may obtain different nutrients than males. Phytophagous insects appear to have feeding specificity greater than predatory and parasitic insects. Such an effect is likely the result of prey discrimination based largely upon size and not so much on taste, since most soft-bodied prey should taste about the same (save for noxious substances that could be detected by biting or sucking predatory insects).

Schoonhoven and Dethier[(246)] have reviewed the physiology of host plant discrimination in the Lepidoptera and have stressed the role of olfactory information received by the antennae and maxillae in determining discriminatory levels.[(247)] Other forms of specificity in the insects are determined by other environmental factors. Jaisson[(248)] discusses the role of early experience in the determination of nest site preferences in ants, which can be entrained by a plant stimulus to prefer a particular plant. Heinrich[(249)] and others have shown how leaf shape and size affect the feeding behavior of caterpillars, and such studies stress the role of geometrical and mechanical features of plants in determining how tissues are consumed.

Janzen[(72)] has suggested that the mechanisms of feeding specificities in plant-sucking tropical insects might be considerably different from those regulating the gustatory preferences of chewing or biting herbivorous insects. Host plant specificity may not be as well developed in plant-sucking insects as in many chewing insects, because xylem and phloem fluids may contain fewer secondary substances. Host plant selection by these forms may be mediated primarily through the use of physical and mechanical properties of plant tissues rather than by chemical defense mechanisms. Miles[(250)] has found that some plant-sucking Hemiptera secrete a salivary phenolase that may induce the production of a chemical defense mechanism in plants based upon the oxidation of phenols to quinones. Sweet[(251)] argues that phytophagy was the original feeding habit of the Hemiptera and that selection has favored the expression of feeding on nitrogen-rich plant tissues (buds, seeds, meristems). Schaefer and O'Shea[(252)] reported that three unrelated tribes of coreine hemipterns feed on Leguminosae and exploit those parts of the plants with high nutrient content, and the trait is considered a symplesiomorphic character. (A *symplesiomorphic character* refers to the presence of plesiomorphic character states in different species. A *plesiomorphic character* is the character or character state from which transformation begins in a monophyletic group.) The precise mechanisms responsible for host plant specificity in both the Hemiptera and Homoptera remain obscure. Young[(71,138,139)] has suggested that the nymphs of many tropical cicadas have preferred primary host trees in the sense of preferring to feed on the root crowns of these trees, but not necessarily to the exclusion of root systems of other plants (Fig. 3.20). Adult tropical cicadas feed on a variety of tree species (Fig. 3.21) in a habitat (A. M. Young, unpublished data; Young[(215)]), but ovipositing females of some species prefer to place eggs near particular tree species.[(70)]

Webb[(253)] found that spring populations of the South African psyllid *Acizzia russellae* were correlated in size with nitrogen concentrations in the leaves of a host plant, *Acacia karroo*. Webb also found that Homoptera that feed on the

FIGURE 3.20. Nymphal skins of the cicada *Diceroprocta* sp. on a *Tamarindus indica* tree along the northwest coast of Costa Rica and northern Central America as well. Peak emergence occurs annually between dry and wet seasons in these strongly seasonal regions of the tropics.

FIGURE 3.21. The cicada *F. pronoe* (above) feeding from a branch of the secondary habitat shrub *Veronia patens* (Verbenaceae), and *Q. gigas* (below) feeding from a trunk of the same tree species, in northeastern Costa Rica. Both cicadas are part of a complex of species widespread from Mexico through Central America and into South America.

cellular contents of plant tissues do not increase greatly in abundance on cut plants, probably because leaf nitrogen levels are no higher than in uncut plants or stump sprouts. But phloem-feeding species seem to respond to the increased flow rate of nutrients in growing tissues of cut plants, promoting increased population growth. Attempts to make statements about the host specificity of plant-sucking insects such as aphids and other Homoptera must thus take into account the physiological condition of the host plant with respect to feeding habit (phloem feeding, cell feeding). Specificities may change in response to conditions of the host plant and the population growth requirements of the insect species involved. Levins and MacArthur[254] believe that a major factor determining the choice of a host plant by an insect is the effect of such a switch or diet breadth on fitness within the population. The evolution of monophagy or polyphagy is determined by the balance between the cost of diet breadth expansion and the cost of searching for new host plant types.

The sequestering of ingested secondary substances from plants by insects refers to the ability to use these substances for defense. Because most insects are small, abundant, and full of nutrients such as proteins and lipids, they are a major food source for many animals. Thus selection within insect species may often favor the evolution of effective predator-deterring defense mechanisms,[114] one of which is the ability to be unpalatable. One avenue of unpalatability is the sequestering of plant secondary substances by some insects. In the tropics the diversity of secondary plant substances is expected to be very high given the high diversity of plants (although the relationship is not one to one), and predation rates on insects by other animals can also be high. Thus there might be considerable selection for sequestering mechanisms in tropical insects. Alkaloids are often sequestered, as shown by experiments with danaid butterfly larvae, which store high concentrations of cardiac glycosides from their asclepiadaceous host plants.[255] The stored substances, which are bitter, are passed on to the adult butterfly, which through aposematic coloration reinforces the unpaltable properties in the predator species with relatively few encounters. Whether or not flying butterflies are frequently attacked is still somewhat controversial, with evidence for both points of view.[189,204] Studies (e.g., Pliske[177–179]) have shown that some secondary substances sequestered by butterfly larvae are eventually incorporated into male pheromones. Thus the pyrrolizidine alkaloids of some plants are converted to dihydropyrrolizidine derivatives in male pheromones of danaids.[114]

The inability to sequester toxic secondary substances from host plants results in the evolution of cryptic behavior to escape predators, or mimicry to resemble noxious models. In the tropics (notably the lowland wet regions), where many groups of phytophagous insects comprise many closely related genera and species, the pattern of evolutionary acquisition of the ability to sequester may exist. For heliconiines (Benson[255]), two different genera may feed on the same larval host plant, but only the advanced genus will have the ability to sequester the secondary substances and become unpalatable. Sequestering appears to be a derived phenotypic trait superimposed upon the detoxification mechanisms of phytophagous insects. The majority of ithomiid butterflies tend to possess transparent to translucent wings with reduced markings (Fig. 3.22), and such genera and species may be palatable even though they feed on the larvae of Solanaceae, a tropical plant group well known for high alkaloid contents. These groups may lack the ability to sequester the alkaloids of these plants, even though many of the same plant species support other aposematically colored insects. A few ithomiids such as *Mechanitis* and *Hypothyris* possess aposematic coloration, perhaps reflecting unpalatability characteristics derived during the larval stage. These genera tend to feed on solanaceous species found in open secondary habitats, whereas the majority of other ithomiids feed primarily in forest understories and utilize different host plant species. In this instance the evolution of sequestering might have been induced by the transition of somc ithomiids to secondary habitats. The bright coloration of the wings may be more adaptive for

FIGURE 3.22. The ithomiid butterfly *Hypoleria cassotis* in the shaded primary forest understory in northeastern Costa Rica. Species such as this one within the Ithomiidae are typical for the shaded understories of Central and South American rain forests, where adult population densities are generally low (save for contractions of populations into moist pockets during dry seasons in highly seasonal areas) and larval food plants (various papery-leaved genera and species of Solanaceae and Apocyanaceae) are dispersed over large areas, especially within light gaps.

courtship in the more exposed habitats in addition to reinforcing unpalatable taste profiles in these insects.

From the standpoint of insects, plant populations are food resources possessing varying degrees of accessibility of nutrients. Although, as pointed out in the foregoing discussion, individuals of different plant species may be similar in terms of the quality and quantity of nutrients available for animals, including insects, accessibility of nutrients is determined largely by the lines of defense exhibited by each plant species at a particular place and point in time. Phytophagous insects are not necessarily seeking out the most nutritive plants to feed on, but rather those with defense systems that can be most efficiently overcome by the prevailing genotypic and phenotypic traits of the insect. The vast botanical literature is filled with examples of insects attacking all stages in the life cycle of plants and all structures of adult plants and seedlings. As a result of natural selection, unstable equilibrium is expected between the plant's continued abilities to defend itself against predators and the predators' abilities to overcome the defense mechanisms of host plants without appreciable losses in fitness. In artificial systems such as cultivars and associated crop plantings, the equilibrium breaks down and usually results in increased predation on the cultivar, because selective breeding often removes the defense mechanisms of the targeted plant species while enhancing other traits associated with human consumption. Weed species lie at the opposite end of the spectrum from cultivars, either possessing effective defense mechanisms or saturating the environment with such a large harvestable biomass that phytophagous insects cannot attack in a density-dependent fashion. Both strategies result in the successful occupancy of many habitats.

Levin[256] lists the following defense mechanisms in angiosperms: (1) the texture and composition of stem and leaf surfaces, (2) thorns and resin ducts, (3) an absence of nutrients, (4) a presence of hormonelike substances that alter the development of insects, and (5) the accumulation of secondary substances in plant parts subjected to attack. The evolution of secondary substances, as discussed earlier, is maintained to be a major line of defense by plants against phytophagous insects. Secondary substances are diverse, being alkaloids, terpenes, phenolics, and mustard oil glycosides, and are steroidal and cyanogenic.[256] The alkaloids enjoy a wide distribution among many taxa of angiosperms in particular. Secondary compounds can be sequestered in cells or cellular organelles and released only when the cell is broken, or they may be stored and secreted by specialized epidermal glands, sometimes located on trichomes.[256] In terms of general modes of action, terpenes, alkaloids, resins, or phenolics tend to be glandular exudates and can act as contact poisons or volatile inhibitors. The object of the game is to discourage insects from eating the plant tissues.

Levin[256] points out that the production and storage of secondary substances is often an energy drain on plants, not to mention the multitude of other phenotypic traits that are associated with release of such substances. The proportion of total resources allocated to the production and maintenance of effective chemical defense systems in plants is expected to vary with many environmental factors, including season, age of plants, relative allocations to vegetative and reproductive growth phases, and the degree of attack by insects and the types of insect feeding habits involved.

Several of the papers in Rosenthal and Janzen[257] stress that the distribution of secondary compounds or toxins in a plant will vary for different parts of the plant and with both spatial and temporal properties of the surrounding environment. Finely tuned timing of the production and release of such substances is also expected in the individual plant, a condition which may allow a herbivorous insect to bypass the toxins by feeding on that part of the plant before the substance is produced. Because of these complexities, and the fact that a plant may contain a wide spectrum of distinctive secondary compounds interacting to varying degrees with different species of insects feeding on the plant, it is extremely challenging for the biologist to confirm the precise relationship between an insect and a host plant species.

The deleterious effects of plant secondary substances on insects have been elucidated by experiments (e.g., Feeny[102,258]; Feeny *et al.*[259]; Erickson and Feeny[260]; Rehr *et al.*[261–263]). High concentrations of such substances may retard development even though rates of feeding may not be reduced.[260] Even though there are many kinds of secondary substances, most of them share the trait of being effective in deterring attack by insects. The recent tome of Rosenthal and Janzen[257] attests to current research interests regarding secondary plant substances and their effects on herbivorous animals. The field is still embryonic, with much to be done. Especially with respect to the tropics, one interesting avenue of research is understanding the long-term relationships between temporal and spatial changes in the secondary substance profiles of plant species in relation to changes in the insect herbivore guild associated with the plant species, and changes in the population structure of member species in the guild. How such interactions change as a function of habitat in the tropics is also of interest, given the high densities of plant species in some tropical habitats.

Levin[256] discusses the general means of detoxification and sequestering of secondary substances by herbivores. Detoxification refers to the ability of an insect or other organism to reduce or remove the toxic features of the substance, usually by oxidations, reductions, hydrolyses, and conjugations, depending upon the substance involved. The chemical interactions occur in the digestive tract, and the detoxified substances are usually voided in the feces following the retrieval of nutrients and water. Microsomal mixed-function oxidases appear to be the major class of detoxification enzymes in insects, and there are considerable differences in the level of microsomal oxidase activity among different

FIGURE 3.23. The life stages (a–d) of the Neotropical ithomiid butterfly *M. isthmia,* a species associated with secondary habitats in Central and South America. The larval food plants are tough-leaved, and spinose species of *Solanum* found in open habitats, including pastures and roadsides.

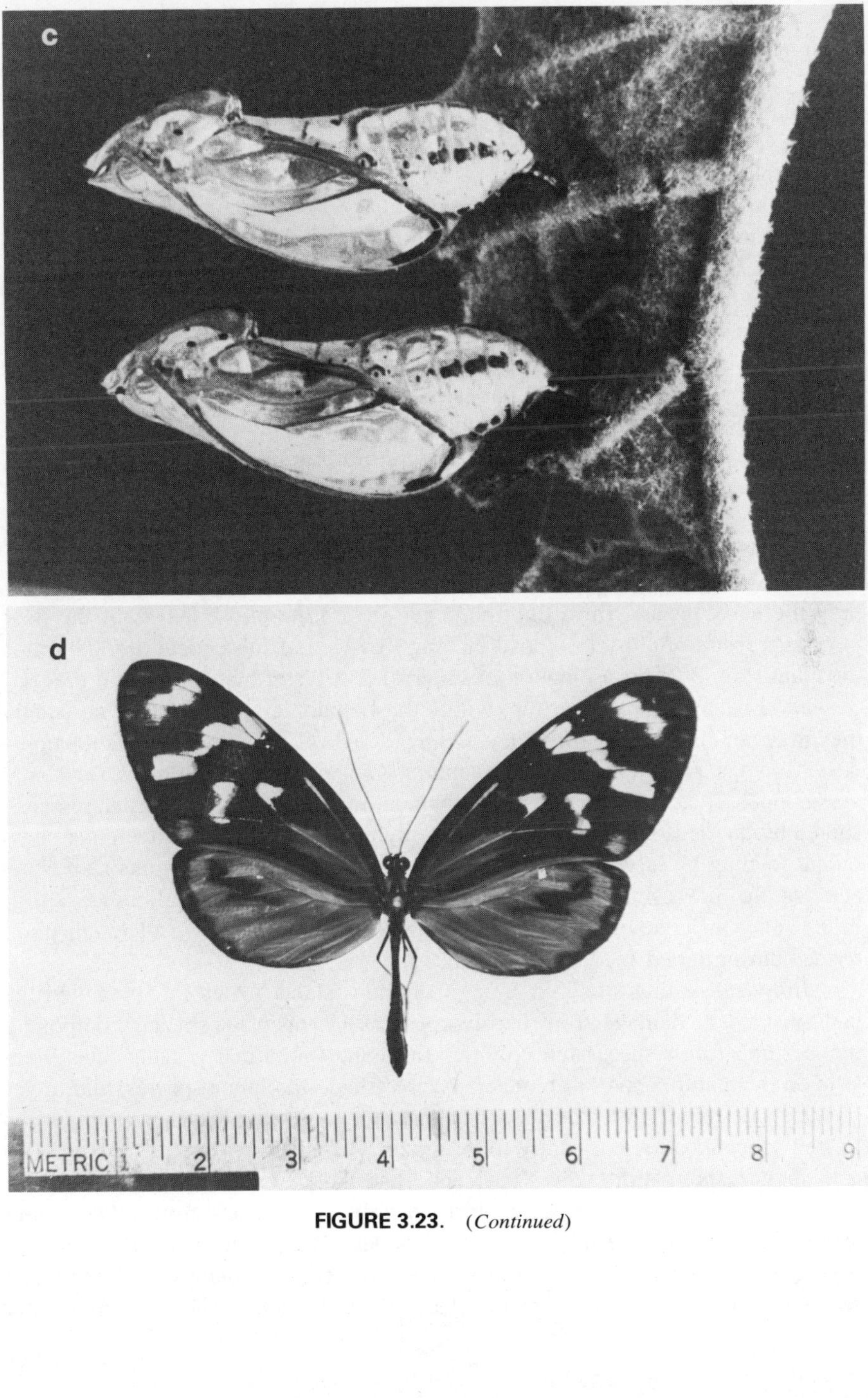

FIGURE 3.23. (*Continued*)

species of insects attacking the same plant species in a particular habitat, and differences within the same insect species in different habitats or utilizing different plants as food. Natural selection is expected to enhance the efficiency of detoxication in insect species, since the process requires high amounts of energy.[256] As expected, microsomal oxidase activity is higher in polyphagous insects than in monophagous insects.[264]

Stamp[265] discussed the existence of cluster oviposition in butterflies, and A. M. Young (in manuscript) has extended her ideas to the role of communal feeding in Lepidoptera larvae as an adaptation to host plant species presenting formidable physical barriers to successful feeding by the individual larva. Based on natural history studies of tropical butterflies, Young suggests that cluster oviposition occurs primarily when a present-day host plant, or host plant in the evolutionary past of the insect, has necessitated communal or group feeding to ingest the food. Ghent[266] studies such behavior in sawflies. Young[88] found communal feeding by the larvae of the egg-clustering ithomiid *H. euclea,* which uses a solanaceous host plant with thick, tough leaves. Young[220] and Young and Moffett[221,222] reported communal feeding in the larvae of the egg-clustering ithomiid *M. isthmia* (Fig. 3.23) and suggest that bites by multiple first-instar larvae are necessary to reach the soft tissue layer of the leaf. The larvae, particularly the early instars, form tight aggregations on the pilose leaves of the host plant as a result of cluster oviposition (Fig. 3.24), and subsequent defoliation of the plant (Fig. 3.25) is a function of group size. Cluster oviposition and gregarious larval behavior are rather rare within the Lepidoptera. Where they do occur, they may be related to a variety of ecological factors, including host plant patchiness, average patch size of food plants within an area, preferences for shade versus open areas, and degree of mechanical toughness of plant structures consumed by larvae (for example, when it is high), may select for cooperative communal feeding by larvae. Cluster oviposition and larval aggregations also characterize the life cycles of Neotropical *Battus* butterflies (Papilionidae) (Fig. 3.26). Such behavior may facilitate the mechanical breakdown of tough plant tissues during larval feeding.

In an interesting study on Barro Colorado Island, Coley[267] examined the patterns of leaf damage from herbivorous insects on plant species composing pioneer and forest succession communities. She found that grazing rates were high on the mature leaves of pioneer species (shade-intolerant species) and lower on the shade-tolerant species of forests (Table 3.1). Young leaves in both groups of tree species had significantly more grazing damage than mature leaves. The data suggest that mature leaves in general and those of climax forest habitats in the tropics have better defenses against attack by herbivorous insects than young leaves. The majority of tropical butterflies deposit eggs on or near young leaf tissue, and young larvae consume young leaf tissue. In many cases first- and second-instar cannot feed on mature leaves. These patterns are not universally

FIGURE 3.24. *M. isthmia* ovipositing on its larval foodplant (a), and gregarious second-instar larvae (b).

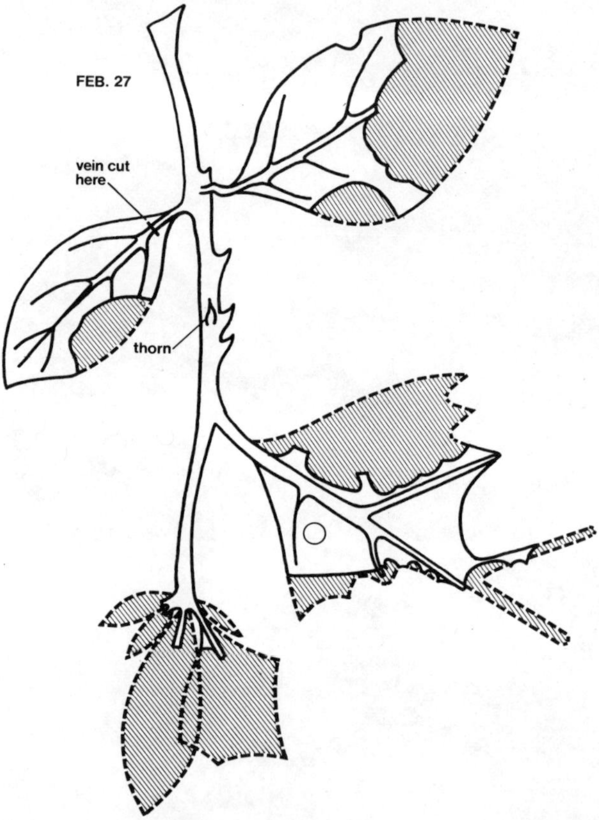

FIGURE 3.25. An assay of leaf damage to *Solanum hispidum* by a group of 20 larvae of *M. isthmia* within a 10-day period.

true, since some species (*M. peleides,* for example) differentially place eggs on mature leaves and the larvae feed successfully there. But the interactions of herbivorous insects with foliage of different ages and defensive properties may be more complex than they seem. While theory predicts that young leaves of trees should be preferred by herbivores, the Temperate Zone study of Raupp and Denno[268] on the leaf preferences of adult and larval stages of the imported willow leaf beetle, *Plagiodera versicoolora,* indicates that preference for young leaves occurs only in the adult stage. Survivorship of eggs and feeding success of young larvae were found to be greater on mature leaves. Because adults of this beetle and many other herbivorous insects prefer young leaves, survival of young larvae and eggs on young leaves in the wild is very low, owing to damage effects. In this system selection would appear to favor oviposition on mature

FIGURE 3.26. Egg cluster of the Neotropical swallowtail butterfly *Battus belus* on the underside of a mature leaf of the vine *Melothria guadelupensis* (Curcubitaceae) (a), and first-instar larvae aggregate (b). Although this butterfly is an *Aristolochia* feeder in the larval stage (Aristolochiaceae), egg clusters are sometimes placed on *Melothria* leaves entwined with the correct food plant.

TABLE 3.1
The Grazing Rates of Herbivorous Insects Associated with a Variety of Plant Species in Tropical Rain Forest on Barro Colorado Island, Panama[a]

Species	Mature leaves	Young leaves
Persistent		
Faramea occidentalis	0.002	0.069
Virola sebifera	0.002	0.108
Prioria capaifera	0.002	0.014
Swartzia simplex	0.003	2.504
Simarouba amara	0.003	0.026
Trichilia cipo	0.003	0.522
Poulsenia armata	0.004	0.027
Desmopsis panamensis	0.004	0.783
Tachigalia versicolor	0.005	0.775
Tetragastris panamensis	0.005	1.477
Protium tenuifolium	0.008	0.928
Hirtella triandra	0.042	0.120
Quararibea asterolepis	0.114	0.316
Alseis blackiana	0.136	0.096
Zanthoxylum panamense	0.136	0.701
Cupania sylvatica	0.311	1.151
Mean	0.041	0.631
Standard error	0.024	0.151
Number of species	16	16
Number of plants	120	115
Number of leaves	452	509
Pioneer		
Didymoponax morototoni	0.007	0.001
Croton bilbergianus	0.025	0.310
Zanthoxylum belizense	0.081	0.624
Spoldias radlkoferi	0.186	1.492
Miconia argentea	0.189	0.509
Ochroma pyramidale	0.191	0.178
Aichornea costaricensis	0.210	0.818
Luehea seemannii	0.456	1.108
Cecropia insignis	0.797	0.111
Trema micrantha	1.071	0.053
Cecropia obtusifolia	2.267	1.299
Mean	0.466	0.663
Standard error	0.162	0.161
Number of species	11	11
Number of plants	102	105
Number of leaves	386	398

[a]Data from Coley.[(267)]

leaves, since the loss in fitness due to laying eggs on leaves of reduced nutrient quality (mature leaves) is less than the loss from the mortality of early stages on leaves of high nutrient quality (young leaves). The data suggest the importance of examining the whole assemblages of herbivorous insects associated with both young and old leaves of host plants, and the interactions among the species. Such effects may even be more important to examine in tropical herbivorous insect assemblages associated with particular tree species, since species richness may be greater in many cases than in comparable Temperate Zone communities.

The ability for an insect to feed successfully on a mature or young leaf depends upon its ability to overcome the defensive systems of the host plant. Feeny *et al.*(259) have demonstrated that some polyphagous species of insects have a higher activity of detoxification enzymes in their digestive tracts to handle the chemical defenses from diverse plant groups. Krieger *et al.*(264) found that the mixed-function oxidases (detoxification enzymes) of the polyphagous larval armyworm can be induced by a variety of secondary plant substances. Carroll and Hoffman(269) reported that experimentally damaged leaves of *Cucurbita moschata* exhibit marked increases in certain substances at the damage site within a short period and that these substances stimulate feeding in a chrysomelid and inhibit it in a coccinelid. But the coccinelid under natural conditions cuts a trench around the damaged areas, preventing the flow of these defensive substances to the area, and it can therefore feed. Young(270) reported that later instars of the tropical butterfly *Aeria eurimedea* cuts the midrib of leaves of the apocyanaceous host plant before feeding on the distal portion of the leaf. Such behavior presumably blocks the flow of secondary substances, known to be very toxic in the Apocyanaceae,(271) to the feeding site.

Herbivorous insects associated with different host plants known to have toxic secondary substances are unable to feed successfully on one another's plant species, but a generalist species can to some degree. Gilbert and Ehrlich(272) discuss the evolutionary affinity of the Ithomiidae and Danaidae, and suggest that the exploitation of closely related plant families by the two groups is further evidence of their relatedness. Some primitive genera of ithomiids, primarily a solanaceous feeding group, use the Apocyanaceae as larval host plants, while danaids use the Asclepiadaceae. Because of the close evolutionary relationship between the Apocyanaceae and Asclepiadaceae, Young(273) has demonstrated in a preliminary way partial exchangeability in host plants between the two groups of butterflies. In the analysis of feeding preferences in herbivorous insects, it is necessary to determine whether instances of polyphagous feeding or switches to new host plants result from genetic changes in populations or from physiological induction of already present adaptations associated with gustation and digestion. Given the high overall diversity of insects and plants in the tropics, many cases where shifts in host plants by insects, in response to changes in habitat or competition pressures or other ecological factors, may not be evolutionary at all. Many such shifts, depending upon the phenotypic flexibility of the insect species, may be behavioral shifts in which new resources are encountered and found to be acceptable. Yet good evidence exists for the genetic regulation of feeding preferences in Temperate Zone Lepidoptera.[(274)] Young[(275)] has discussed the possible role of genetic and physiological processes in the evolution of polyphagous and monophagous feeding behavior in tropical butterflies.

The nutritional requirements of insects vary considerably according to the kind of insect and its habits. Insects require environmental sources of at least 10 amino acids for tissue and enzyme synthesis, and carbohydrates as an immediate source of energy. Various vitamins, inorganic salts, and small quantities of fats are also essential for normal development and reproduction. Under the working hypothesis that many tropical dicotyledonous plants have evolved effective physical and chemical defense systems against herbivores, it is clear that such defenses insulate or protect the nutritional components of the plants from being mobilized by herbivores. The concept of coevolution between plants and herbivorous insects is examined in the next chapter. The greater elaboration of plant taxa in the tropics has entailed the diversification of rich arrays of chemical defense systems, and these processes have been a source of both evolutionary and short-term behavioral and physiological challenge to insects. The development of host-location mechanisms in parasitic and predatory insects also depends in part upon the elaboration of plant odors associated with defense against herbivorous insects. Rhoades[(276)] has reviewed the evolution of chemical defenses against herbivores by plants. Toxic secondary plant substances play a major role in determining the feeding specificities of insects.[(277)] Tropical lowland habitats must present a very heterogeneous odor and tactile field for herbivorous insects. The complexity of such environments is appreciated by a consideration of oviposition behavior in addition to feeding.

Many butterfly species exhibit precise oviposition for placing eggs carefully on the correct host plant for the larvae, although[(278–280)] errors are made in which eggs are placed on incorrect plants or plants toxic to larvae. Sometimes insect eggs are placed on host plants that are suboptimal for normal development, and fitness is reduced (Fig. 3.26). Using McClure's study[(281)] of introduced scale insects in Connecticut and Coley's study[(267)] of herbivore grazing effects on tropical foliage, we can argue that selection should favor an oviposition behavior that places eggs on the correct parts of host plants as well. The studies of Gilbert,[(31)] Benson,[(282)] and Benson *et al.*[(283)] show the great degree of oviposition-site selectivity in *Heliconius* butterflies in Central and South America. In the interaction of these butterflies with their passifloraceous host plants, the oviposition behavior is defined by availability of meristems, plant size, presence of other heliconiine species, and predation by ants. The studies of Young and Muyshondt in Central America have shown that oviposition behavior can be very specialized in most groups of butterflies.[(77,221,222,270,284–288,289–319)]

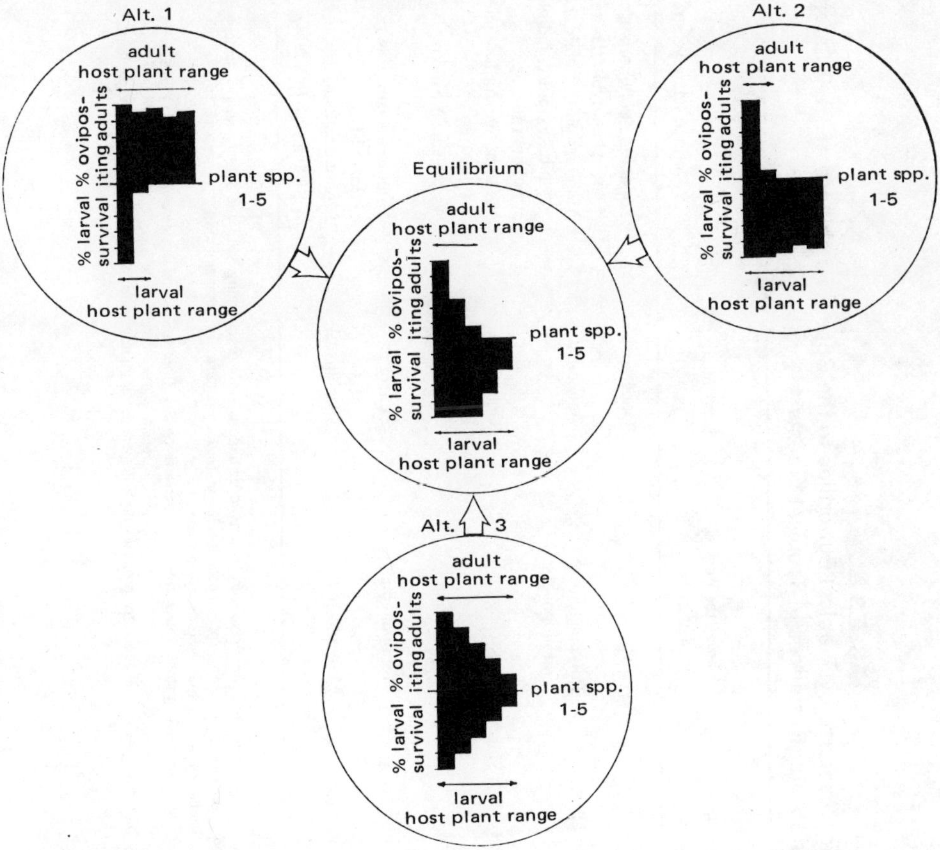

FIGURE 3.27. Hypothetical model for the coevolution of alternative (alt.) adult and larval food plant ranges for a butterfly population that has successfully colonized an island containing five equally abundant larval food plant species. [From C. Wiklund, *Oecologia* **18:**185–197 (1975). Copyright 1975 by Springer-Verlag.]

Although the larval host plants of ovipositing female butterflies are known to determine the host plants the butterfly oviposits on for some species, Wiklund(320,321) found no such correlations in tests with *Papilio machaon* using four alternative host plant species. Such data suggest a lack of dependence between larval host plant preference and that of ovipositing adults.

Whether similar preference patterns exist in tropical butterflies remains to be studied. When large numbers of closely related genera or species of butterflies occur in a tropical habitat, as is often the case, intense selection might favor specialization at both the larval and adult preference levels as a means of lower-

TABLE 3.2

Some Predicted Strategies of Adaptive Response by Tropical Butterflies to the Spatial Patchiness of Their Larval and Adult Resources in Tropical Forests[a]

Habitat	Spatial patchiness of host plants	Relative patch size	Larval feeding pattern	Adult feeding pattern	Optimal adaptive strategy
Forest canopy	Very high	Small	Polyphagous	Generalist	Population polymorphism; low phenotypic flexibility
Forest understory[b]	High	Small	Polyphagous to monophagous	Generalist	Monomorphism; high phenotypic flexibility; individual polymorphic
Forest understory	Low (palmaceous)	Large	Monophagous	Generalist	Monomorphism; low phenotypic flexibility
Forest light gap[c]	High	Small	Polyphagous	Generalist	As per forest canopy
Forest light gap	Low	Large	Monophagous	Generalist	Monomorphism; low phenotypic flexibility
Secondary[d] succession	Low	Large	Monophagous	Specialist	Monomorphism; low to high phenotypic flexibility

[a] From Young.[275]

[b] Two cases are recognized here. Woody understory trees and woody vines generally have very patchy distributions, thus converging with the forest canopy strategy for most butterflies using dicot host plants; some understories are highly palmaceous, selecting for a specialist larval feeding pattern. In both situations, as with the forest canopy, adult butterflies of the tropical forest tend to be generalized feeders of nectar, pollen, fungus, and decaying fruits.

[c] The forest light gap represents a highly transitional habitat between the highly patchy forest flora and the less patchy flora of secondary succession. Two cases are recognized here, although most light gaps very likely are mixtures of both.

[d] It is believed that reproductively active adults of forest canopy butterfly species are generally prone to high dispersal, facilitating outcrossing and reducing inbreeding. Inbreeding may be more frequent in butterfly species of secondary succession.

ing competition. Resource limitation in the tropics and high cooccurrence of many ecologically similar species in an area or habitat may select for highly coadaptive gene complexes and physiological mechanisms that influence both adult oviposition preferences and larval host plant preferences in the same direction in individual species.

Based on his studies of the butterfly *P. machaon,* Wiklund[320,321] suggests that host plant range of adults within a species and of larvae in the same population are controlled by independent gene complexes. Wiklund[321] has developed a model for the coevolution between adult and larval host plant ranges in which it is assumed that larval mortality resulting from the inability to accept a host plant as food is the only selective pressure, adult and larval host plant ranges are determined by separate gene complexes, and the host plant family is large and contains many biochemically similar species. Given three very different starting populations as alternatives (Fig. 3.27), Wiklund predicts that an equilibrium eventually will be established between adult and larval preferences in another population. Host plant preference in butterflies and other insects is determined by (1) the phenologies of both butterflies and host plants, (2) oviposition preferences of adults, (3) host plant preferences of the larvae, and (4) the combination of (1) and (3). The question of how adult oviposition preferences can be linked to, rather than be independent of, larval host plant preferences has been discussed by Young [275] for tropical butterflies. Young suggests that the evolution of polyphagous or monophagous feeding in tropical butterflies is a function of the patch structure of tropical vegetation in addition to the genetic or physiological ability to detoxify distantly related plant groups. Young believes that polyphagous tropical butterflies possess genetic systems of mixed morphs and that the population is polymorphic as a whole. Monophagous butterflies are believed to have populations of genetically similar individuals and with a high amount of genic variation within individuals. Monophagy can also result from highly monomorphic populations possessing various forms of phenotypic flexibility from behavioral, physiological, or developmental switch mechanisms. Different strategies are suggested for different kinds of tropical habitats (Table 3.2). Monophagy is predicted to occur when a host plant family is locally very abundant, i.e., has many large patches in an area. Oviposition "errors" in some Neotropical butterflies may be related to a general colonization strategy[322] or hiding eggs from egg predators.[323,324]

CHAPTER 4

ECOLOGICAL ASPECTS OF PLANT DEFENSES AGAINST INSECTS

In this chapter a review of the roles of herbivorous insects as selection pressure in the evolution of morphological and chemical systems of defense against herbivores in plants is given, with an emphasis on examples from the tropics. Additional aspects of this topic are touched upon in later chapters as well. The survey and analysis is by no means exhaustive, but rather is designed to illustrate some of the interesting defense strategies of plants against herbivorous insects and the implications for such interactions on food chains in the tropics.

4.1. PLANT DEFENSES AND COEVOLUTION: AN OVERVIEW

In Temperate Zone habitats there are many well-documented cases of severe defoliation by insects in certain tree species. Typically an infestation is initiated in the spring and steadily increases throughout the summer, and the insect species involved in such decimations of foliage are uni- and multivoltine. The gypsy moth, tent caterpillar, and spruce budworm are three herbivorous insects involved in severe outbreaks on forest and fruit trees in North America. These species tend to exploit a wide range of tree species and do not appear to be specialists. In tropical forests such outbreaks are less frequent although not entirely absent. The young and mature leaves of a particular plant species may exhibit damage from several different herbivorous insects, but individual plants are seldom stripped of their foliage.

The defoliation of a plant may involve a loss of leaves, meristem, buds, flowers, fruits, and seeds. As with other organisms the individual plant is viewed as allocating energy to vegetative and reproductive functions. Vegetative functions include those biochemical and physiological processes that promote the growth and maintenance of the individual and also the development of morphological and chemical defenses against herbivores. The simultaneous or sequential allocation of energy to reproduction promotes the differentiation of inflorescences and flowers, then seeds and fruits. The individual plant is viewed as having some sort of dynamic equilibrium with the internal and external environment, so that fitness is maximized as much as possible.

An array of phenotypic types in the population promotes different types of responses to the selection pressures of the environment. Other forms of selection pressure that also affect the modes of adaptation in plants include soil texture and nutritive properties, intra- and interspecific competition, position in the habitat relative to exposure to the sun and shade tolerance, availability of seed or fruit dispersal agents, seed and fruit dispersal methods, availability of pollinators and system of pollination, and seasonality. Other, often subtle features of the environment, such as the influence of microorganisms, are also operative, and the above list is by no means complete. The high diversity of tree species in lowland tropical wet forests (see, for example, Pires *et al.*[(325)]; Connell[(65)]; Janzen[(29)]) suggests that each plant species in such a habitat is highly specialized in some or all of the above adaptations. In the nonseasonal lowland tropics, the interplay of herbivore pressure associated with foliage, fruits, and seeds, or competition on the one hand balanced by adequate pollination and fruit or seed dispersal on the other accounts to a large extent for the observed lack of species dominance in most habitats. The role of herbivores in structuring the organization of tropical forests is reviewed in Chapter 8, when the impact of insect populations on tropical plants is approached from the view of how insects promote diversity in the tropics; other aspects of this topic, including pollination systems and fruit and seed dispersal systems, are also discussed.

Out of the many elegant studies conducted in the tropics emerges the following delineation of the modes of interactions between insects and plants:

1. Coevolutionary associations between plants and herbivores, exemplified by butterfly–host plant interactions.
2. Mutualism between ants and plants.
3. Seed and seedling predation by insects.
4. Interactions between extrafloral nectaries and ants.

The means by which a plant species adapts to selection pressure arising from herbivorous insects consists of four alternative strategies: emigration to new habitats where the selection is, or is very likely to be, less intense; elaborate specialized morphological traits such as hairs or spines, designed and strategically placed to discourage feeding and oviposition; elaborate secondary plant substances that block feeding; and mutualism with an insect partner (ant) that helps to defend against herbivores. It should be immediately apparent that these categories are not necessarily mutually exclusive. But what particular pathway of defense is expressed in a function of genotypic and phenotypic flexibility coupled with preadaptations in a population, types and intensity of herbivore pressure, and potential availability of mutualist ant species in the habitat? Since not all parts of a particular plant are equally vulnerable to attack by herbivorous insects, it stands to reason that the distribution of defense mechanisms within the plant need not be, and should not be, uniform. Rhoades[(276)] suggests that the allocation of secondary substance to a particular tissue or structure is proportional to the value of that structure to the plant. Yet one can clearly predict that other forms of physiological constraints within the phenotype of a plant will alter such effects, and generalizing about such effects and defensive compounds across the board is difficult.

On the other hand, in encountering a potential host plant species with effective defense systems the herbivorous insect species can do the following: individuals within the insect population can adapt biochemically or physiologically in terms of digestion and assimilation of detoxified secondary substances; individuals can "learn" new feeding habits to overcome the blocking of favorable gustatory cues from the plant, and perhaps feed on other parts of the plant where cues are more inducive to sustained feeding; and a portion or all of the population can invade a new habitat where a more accessible host plant population is found, thereby circumventing entirely the defensive system of the original host plant species. The development of the detoxification enzymes necessary for handling secondary plant substances, through evolutionary modifications in metabolic pathways, is a requirement for some insects to overcome the chemical defenses of a particular host plant. Such selection must ultimately be coupled with the peripheral and centrally controlled sensory physiology of the insect so that each insect species recognizes the proper host plant to feed on under most conditions. Thus behavioral diversity, at the level of high specificities in recognizing host plants, is coupled with the diversity of detoxification systems, and together they account for the expression of considerable feeding diversity, especially in tropical insects.

In the tropics, where physical conditions of the environment are generally stable, a greater diversity of coevolved relationships is expected between plants and insects and other animals. The higher specialization in feeding habits means that many insects are specialist feeders on plants, one factor contributing to the absence of outbreaks and severe defoliation. But such specialization within an insect species may entail the allocation of considerable energy to the metabolic pathways responsible for the morphological or chemical defenses of the host plant, removing resources that could potentially be used to increase fecundity or some other highly selection-sensitive portion of the phenotype concerned with increasing the rate of population growth. The tradeoff between increasing population growth and the acquisition of mechanisms for overcoming plant defenses therefore precludes outbreak conditions in most tropical insects associated with species-rich and stable habitats.

Another energy sink operative is that endowed by the patch structure of the vegetation: resources must be allocated to effective interpatch dispersal in a specialist species. It is therefore the species composition of the vegetation, coupled with the patch structure of the vegetation and the great diversity of morphological and chemical systems of defense by plants, together with the fact that the majority of insects are plant feeders, that accounts for the high diversity of insects in the tropics, notably the nonseasonal lowland tropics. Each plant species can be viewed as being in some sort of dynamic equilibrium with the biotic agents affecting it, including herbivores and pollinators. If such an equilibrium is shifted drastically by some environmental perturbation and short-term, rapid responses (of the sort discussed by Slobodkin and Sanders[104]) are ineffective, considerable fitness in that generation may be lost. Rockwood[326] created such an environmental pertubation on trees in lowland Guanacaste Province, Costa Rica, where he deliberately hand-defoliated trees to simulate intense herbivore activity. He demonstrated a significant loss of fruit and decrease in seed crop size following the defoliation. He concluded that heavy defoliation in wild trees virtually eliminates seed production for the year it takes place. Various studies have shown that growth and reproduction in plants greatly depend upon leaf surface area since most of the photosynthesis takes place here. Breedlove and Ehrlich[327] found a drastic reduction in seed set in lupine populations as a result of severe predation by the larvae of a lycaenid butterfly. Heavy defoliation reduces fitness by postponing the age of first reproduction and lowering the number of seeds and fruits produced. Varying levels of defoliation of individuals of a given plant species in a habitat are likely to be regular occurrences, and during periods of intense defoliation a plant may respond by subsequently reducing energy allocations to reproductive structures and maximizing the replacement of those vegetative structures involved in photosynthesis, a physiological shift that minimizes the depletion of food reserves to nonvegetative functions.[328]

Whether an insect species is a specialist or generalist feeder is a function of the phenotypic ability to handle more than one kind of plant species. I equate generalist with the polyphagous habit and specialist with the monophagous habit. Rosenthal and Janzen[257] provide a collection of papers concerning the many-faceted aspects of the role of secondary plant substances in the defenses of plants against herbivores, and the reader is urged to consult that volume for an introduction to much of the current thinking about this topic. One might expect that a great deal of host plant selection takes place at the level of how insects perceive and process secondary plant substances. Other substances, such as tissue water, may also determine feeding preferences. It is suggested, therefore, that water content, physical properties such as toughness and nutritive and nonnutritive substances in plants act together to determine the attractiveness of the plant to particular insects as a food source. It is also largely a matter of the cost to the insect to exploit a particular plant species in terms of possible losses in fitness in other niche components, and the availability of alternative food sources in the habitat or nearby habitat. Owing to this form of cost–benefit relationship, plant species evolve to maximize the energy investment in secondary substances with defensive roles but do so only until any gain in adapting to herbivores is balanced by the cost in fitness in doing so. But cost cannot be viewed solely in terms of current tradeoff, because the eventual penetrance of the ecological barrier created by defenses of the potential host plant[6] may endow a much greater fitness to the insect population. Thus in addition to protecting itself against many herbivorous insect species, the plant allows itself to be penetrated by one or few such species, and these insect species in turn develop protective devices against their own predators through becoming unpalatable. Unpalatability, in this context, is the result of the insect developing the appropriate mechanisms to incorporate the secondary plant substances, in some form, into its own body tissues. Thus the coevolution between plants and insects involves not only the processes by which insects become feeders on certain plants but also the development of unpalatability in insects.

Coevolution between plants and insects depends upon the processes of coadaptation between insect and plant. The concept can be applied at the familial, subfamilial, generic, and specific levels of taxonomic organization. Benson *et al.*,[283] for example, discuss both the overall and very fine features of the coevolved associations between heliconian butterflies and their larval host plants, the Passifloraceae. They emphasize the joint role of evolutionary history and contemporary ecological factors in molding the manifold specialized relationships of the butterflies to their host plants and the geographical diversification in these associations. They found a basic pattern of host plant exploitation in which restricted primitive taxa of heliconians oviposit on the mature leaves of primitive passion vines of large biomass and the more evolved species of *Heliconius* using leaves and meristems of the *Granadilla* and *Distephana* lineages of passion vines. Other interesting patterns of plant usage and local adaptations of both butterflies and plants were found. The work of Benson *et al.*[283] provides an excellent example of how the general coevolved association between heliconians and passion vines has become diversified geographically and locally, and also shows some patterns accountable from the evolutionary history of the groups as well.

Other studies of butterfly–host plant associations in the tropics (e.g., Gilbert and Smiley;[329] Rathcke and Poole[330]) suggest that the evolution of physical defenses such as trichomes follow the evolution of chemical defenses in host plants. The studies of Rathcke and Poole[330] and Young and Moffett[221,222] on the feeding behavior of the larvae of *M. isthmia* suggest that the larvae of the tropical butterfly can exhibit behavior patterns associated with avoiding or circumventing the thick layer of trichome hairs covering the leaves of the host plant. Such studies and stress the dynamic nature of the coevolutionary process between insects and plants. The relationships are not static but continually changing in response to selection pressures promoting specialized adaptations in each organism to counteract the other. In some groups of tropical butterflies there is considerable ecological specialization to feed on certain plants. A. M. Young (unpublished data) found that several genera and species of Ithomiidae in a forest habitat exhibited considerable diversification in larval host plants, with little overlap among species. Young[301] also found that the little ithomiid *Oleria zelica pagasa* utilized a vine growth form of solanaceous host plant, one not used by other ithomiids in the same habitat.

Another interesting example comes from the association of some species of *Morpho* butterflies with Leguminosae as larval host plants in tropical wet forests. I have been studying the natural history of two very similar species, *M. peleides* and *Morpho granadensis* at one site in northeastern Costa Rica. In earlier studies,[311,316,331] *peleides* was found to accept a broad range of leguminous trees and vines as larval host plants (Fig.4.1), and eggs and larvae of this butterfly have been found on these plants in the wild (Fig. 4.2). Several genera and species of papilionaceous and nonpapilionaceous legumes are involved.[316] In northeastern Costa Rica both butterfly species are sometimes found together in the same habitat. But *granadensis* is never as abundant in the adult stage as *peleides* where the two co-occur, and *granadensis* seems very restricted in terms of habitats while *peleides* occurs in several kinds of habitats. Philip DeVries has also gathered data on the geographical distribution of *granadensis* in Costa Rica, and his results tentatively show that this species is also restricted geographically, much more so than *peleides* (summarized in Young[331a]). Reciprocal feeding tests with the larvae of both species indicate that *granadensis* utilizes only one host plant, *Machaerium seemanii,* of *peleides,* and larvae do not accept others such as *Mucuna urens*. While preliminary, the data suggest that the coadapted relationship between these morphos and legumes is not the same in the two species, perhaps a result of contemporary ecological factors, including habitat type and breadth of habitat used by each species. The role of habitat diversity in insect diversity in the tropics is examined further in Chapters 5, 7, 8, and 9.

The evolution of chemical defense systems in tropical plants is also expected to vary with respect to other forms of environmental factors beyond those shaped by herbivores. Janzen[332] predicted, for example, that regions in the tropics with nutrient-poor soils, such as white sand soils, should have plant species with greater concentrations of secondary substances very effective against herbivores. This prediction comes from the observation that plant assemblages on such soils are vulnerable to extinction since replacement from severe herbivore pressures would be very difficult. McKey *et al.*[333] found that the mature leaves of common tree species in the tropical rain forest on white sand soils in Cameroon contained about twice the concentration of phenolic compounds found in similar forest vegetation on lateritic soils in Uganda, confirming Janzen's hypothesis. The implications for herbivorous insects in the tropics is clear: the evolutionary challenge for insects to exploit plant species growing in

FIGURE 4.1. Fourth-instar larva of *Morpho polyphemus* (Morphidae) from El Salvador (a), and third-instar of *M. peleides* from Costa Rica (b).

FIGURE 4.2. Seedling of *Machaerium seemanii* (Leguminosae), a larval food plant of *M. peleides* in the central highlands of Costa Rica (a) and a fifth-instar caterpillar of this species in typical resting position on the stem of the food plant (b).

nutrient-poor soils should be greater than for the exploitation of similar plant species in relatively greater nutrient rich regions of the tropics.

Overall species richness of insects as herbivores on those plant parts with high concentrations of secondary substances in communities on nutrient-poor soils should therefore be lower than in comparable communities on nutrient-rich soils. The average plant species in the nutrient-poor community cannot afford the cost of withstanding severe herbivory, the lack of required nutrients for rapid growth make recovery impossible. The metabolic cost for elaborating effective chemical defense systems is overshadowed, in this case, by the potential threat of localized extinctions resulting from severe defoliation or other forms of herbivory. Tropical plant communities occupying nutrient-rich soils should be buffered against such effects by the ability to recover with the mobilization of soil nutrients. This is not to say that plant species in such communities would not have antiherbivore defense compounds. It does say that the concentrations of such eompounds might be less than for similar species in communities on nutrient-poor soils. Thus soil quality, in terms of nutrient profile, is a fundamental geographical factor in shaping not only the composition of vegetation of a region but also the phenotypic commitment to chemical defenses within populations of each species, and the composition and species richness of the herbivorous insects found on these plants.

From studies in the tropics, it is now clear that the coevolutionary process between plants and animals, especially insects, goes beyond the interactions of herbivores with their host plants. Frankie(117) has reported on the strategies of annual flowering and pollinator activity in relation to how tropical communities are organized in time. H. G. Baker and I. Baker(216,217) stress the possible role of nectar constituents in promoting specific patterns of coevolution between plants and pollinators. Van der Pijl(334) has outlined the general diversity of pollination systems in the tropics. Coupled with the data of Frankie(117) and studies of nectar by H. G. Baker and I. Baker,(216,217) the evidence indicates that the great diversity of pollinating insect species in the tropics is promoted by the coevolution of plant species with their pollinators in specific habitats. High diversity of pollinators is also promoted by allochronic flowering patterns amoung plant species in the same community or habitat, often regulated by, or timed with, seasonal cycles of rainfall(335) (see also Chapter 7).

The energetics for maintenance of optimal foraging behavior associated with thermal and water-stress conditions (reviewed by Heinrich(165) for bees) represents the combined effects of selection within plant populations for flowering periods and nectar nutrient profiles and selection within pollinating insect populations in terms of physiological constraints of temperature and water stress, flight muscle physiology (see Chapter 3), and resource needs for maintenance and reproduction. See Chapter 8 for discussion of pollination in tropical communities.

The degree to which an abundant plant species becomes exploited by a particular herbivorous insect form depends, in some cases, on the degree to which the plant population in question occupies a heterogeneous environment and to which the insect species is capable of maintaining fitness on different kinds of environmental patches occupied by the plant species. In the tropics, particularly lowland and montane humid or wet areas, considerable topographic heterogeneity may create different microenvironmental factors influencing both plant and herbivore distributions, and the two may not always overlap. If the plant species is capable of occupying patches where the herbivore does poorly, this may function as an effective escape mechanism for the plant in space, if the average size of the antiherbivore patches and the frequency of such patches are great enough so that the plant species maintains a breeding population capable of withstanding fluctuations in the environment perhaps totally unrelated to the plant–herbivore interaction. Individual groundnut trees growing on sandy or gravelly soils in some parts of Nigeria experience severe defoliation by the larvae of the cossid moth, *Azygophleps albovittata,* while individuals on patches of lateritic soils escape such predation.[(336)] The physiological explanation of why groundnut trees on lateritic soils are less attractive to the larvae of this moth than trees with similar leaf crowns on sandy or gravelly soils is not known. The spatial patchiness of soil types keeps populations of the herbivore relatively isolated, reducing the likelihood for an outbreak of the species as a pest of groundnuts. A dry-season diapause in this herbivore also contributes to the prevention of outbreak conditions in all patches.[(336)] In the tropics, high percentages of butterfly taxa in local communities do not feed on flower by-products such as nectar and pollen. Pliske[(177,178)] has shown that various male ithomiine butterflies are attracted to plants containing pyrrolizidine alkaloids by the release of certain highly volatile esterifying acids released from the rotting plant tissues where the insects feed.

Using the chemical defense system of the Cruciferae as a model, Feeny[(102)] has stressed that herbaceous plants exposed to herbivores with only low frequency should evolve chemical defenses in small amounts. Feeny suggests that herbaceous plants in general should have relatively low concentrations of toxic secondary substances, whereas the mature foliage of trees, shrubs, and grasses have well-developed chemical defenses against herbivorous insects. In the tropics, one might therefore expect that the species richness of forest herbivorous insects is much lower than that of adjacent secondary habitats dominated by herbaceous species. The sweep samples of Janzen[(72)] from tropical lowlands suggest a greater number of both chewing and sucking herbivorous insect species in secondary growth, and the pattern is not correlated with a greater absolute diversity of plant species there. The sweep samples compared include those from forest understories, and little data are available for insects in the canopy. The forest environment may contain a number of specialized herbivorous insect species, since Feeny[(102,258)] and others predict that trees should have greater concentrations of toxic secondary substances in mature leaves, and Coley's study[267] in Panama showed that grazing rates were lower for mature leaves of forest species. Thus the overall species pool of herbivorous insects is lower in forests, but those present probably include a good many specialist species.

In contrast, secondary habitats in the tropics may select for "patch specialists" and generalist species of herbivorous insects using the following reasoning. Patch sizes of plant species may be larger in secondary habitats, perhaps even to the extent that the population of a particular herbivorous insect species is confined largely to one or few patches of a host plant. This insect is a "patch specialist." On the other hand, the presence of a pool of a lower number of plant species than what is found in an adjacent forest, but also a pool with species having less effective toxic secondary substances, promotes the development of generalist feeding habits in some insects. Many Homoptera may fit this category. Thus secondary habitats in the tropics may support mixtures of specialist and generalist insect species as a function of larger patch sizes per host plant species, generally fewer host plant species, and less toxic secondary substances found in the average plant species of secondary habitats.

Not all kinds of secondary habitats are expected to be similar in these properties. The presence in some lowland tropical forests of dominant tree species, such as *P. macroloba* in northeastern Costa Rica and elsewhere in southern Central America, provides an increased uniform biomass providing the evolutionary opportunity for some specialist insects to exhibit convergence for feeding on that abundant tree species. In this hypothetical situation, different insects are specialists on different tree species in the forest environment. But if one tree species through evolutionary time becomes a dominant, perhaps as a result of high seed and seedling survivorship coupled with colonization of light gaps and marginal areas,[337] and if the foliage or phloem fluids of this tree contain very low concentrations of secondary substances, cost–benefit considerations may promote insects to invade the tree as a new resource that is readily available in large patches and free of a large energy drain associated with detoxifying toxic secondary substances. The striking occurrence of several genera and species of cicadas on *Pentaclethra* trees in Costa Rica might be an example of such convergence, although other explanations also exist. With disjunct patch structures of plant species in the tropics, the specialist or generalist feeding strategy in herbivorous insects must be a function of both the physiological capabilities to feed on certain species and the patch structure of each species.

If the tree species of tropical forests represent very different sets of toxic defenses to herbivorous insects, that is, each species has high apparency of its defense substances, the specialist feeding behavior will evolve in the insect. Low apparency among several plant species in a habitat would promote generalism in insects. The generalist strategy is therefore most likely to turn up in insects in secondary habitats in the tropics if in fact most secondary growth plant forms have low apparency of toxic defenses. The patch specialist insect might in fact be a generalist physiologically that has happened upon an attractive resource with sufficient biomass to allow the majority of the population to exploit it as a food source. Further studies on the levels of toxic secondary compounds in plant species in different habitats in the tropics, as well as detailed natural history studies of feeding preferences in tropical insects, are needed to examine some of these ideas.

4.2. EVOLUTION OF UNPALATABILITY AND IMPLICATIONS FOR FOOD CHAINS

The foregoing discussions have emphasized the general role of the versatility of response mechanisms of insects to the environment, with the focus being how one explains the great diversity of insects in the tropics. I use the term *diversity* here not in the sense of weighting species numbers by abundances of individuals in populations, but simply to refer to the great number of genera and species of insects per unit area of tropical habitat. Reference to the high diversity of insects in the tropics usually means to regions and habitats where insect diversity is greatest, namely, the nonseasonal to slightly seasonal lowland tropical rain forests, and to the pattern of environmental heterogeneity imposed by the existence of different terrestrial habitats.

I have emphasized in a general way that the long evolutionary history of insects as a group and the increased stability of the physical environment in tropical regions have together promoted insect diversity. Of course, the important roles of biogeographic events such as expansion and contraction of lowland and mountain forest refugia in South America must also be taken into account in explaining present-day distributional patterns of tribes, genera, and species. The studies of Gilbert[(338)] and Benson[(282,339)] stress the interplay of evolutionary history of a group of tropical butterflies and contemporary ecological factors promoting diversity within the family. In this system, it is the geographical and local changes in the environment that are continually restructuring the coevolved associations between the adult butterflies and their food sources, and between the larvae and their host plants. Simberloff[(340)] points out that there are relatively few examples of the rates of local population extinctions for any group of organisms, including insects. It is necessary to document such extinctions or lack thereof, through detailed field studies, as a means of testing the generality of the equilibrium island biogeography model of MacArthur and Wilson.[(341)]

In both a broad sense and at the level of specific interactions between insects and plants, the evolution of plant communities has promoted the diversification of insects. The following pattern of events is suggested.

1. Evolution of flowering plants, followed by regional differentiation and speciation in relation to the physical environment, including land area size, soil quality, topography, and prevailing regimes of rainfall and temperature.
2. Overlapping with (1) of the evolution of insects, with an emphasis on diversification of qualitatively distinct feeding habits resulting in herbivory, pollination, etc.
3. Large-scale diversification of tropical plant communities, including the establishment of different ecosystems at different altitudes, soil types, and climatic regimes.
4. Development of high speciation rates in tropical floras in response to year-round favorable conditions of climate, competition for space, and continual natural occurrence of habitat differentiation resulting from floods, mud slides, volcanism, tree falls in forests, and early systems of agriculture.
5. Induction of high speciation rates in tropical insect faunas in response to extant floras to reduce competition.
6. Induction of selection pressures for speciation in plants through herbivory and pollination. Elaboration of specific plant parts and metabolic pathways resulting from such forms of environmental pressures. High specialization in the tropics for seed and seedling predators, herbivores of leaves and meristems, seed and fruit dispersal agents, and pollinators.

Events (1) through (6) establish equilibria in the tropical biotas in terms of species number and types of interrelationships, but superimposed upon such patterns in evolutionary time are large-scale biogeographical events such as Pleistocene glaciation that create contractions and expansions of biotas (see also Chapter 10). The result is the need to interpret present-day distributions of species and other taxa of insects in the tropics not only in terms of contemporary and past ecological factors promoting diversity, but also in terms of biogeographical events causing regional changes in biotas.

Like plants, insects do not adapt to the environment in terms of food recognition and food processing as related to fitness. Other pathways of selection pressure include that caused by the trophic position of most insects as food for other organisms. Visually hunting insectivorous vertebrates and many invertebrates create intense selection pressure in the tropics, especially where speciation in these groups has also been high. In insects, therefore, the expression of phenotypic traits that promote passive escape from predators or make insects less attractive as food for predators is linked in some ways to the interactions of insects with their host plants. Part of the great ecological diversification of insects in tropical habitats is related to the insects' ability to survive being eaten.

Yet we have no well-documented examples from the tropics of insect populations going extinct as the result of intense predation by vertebrates.

The classic examples of aposematic coloration in many tropical insects, notably butterflies, grasshoppers, beetles, and hemipterans, represent the outward manifestation of unpalatable properties to predators. Such properties are derived (e.g., Brower *et al.*[342]; Rothschild[343]; Rothschild and Kellett[344]; Rothschild and Reichstein[345]; Rothschild *et al.*[346]) from the incorporation of toxic secondary substances from plants into the bodies of insects. The ability to incorporate such substances and turn them into defenses against predators is considered to be a highly evolved characteristic.[339] Thus there are many examples from tropical biotas where several insect species may feed on a plant species with toxic secondary. Thus it is not always the case that an insect species feeding on a toxic plant will be unpalatable and possess aposematic coloration. When a toxic plant has several insect species feeding on it, individuals belonging to one or more of these species may be cryptically-colored, suggesting an absence of unpalatability. Such an insect does not have the ability to incorporate the toxic substances from the plant into its own body tissues. The ability to improve survival based upon the evolution of warning coloration depends on being able to sequester the toxic compound and then assimilate it in some effective form into the body tissues. Brown and Neto[205] have reported that some birds in Brazil successfully feed on supposedly unpalatable butterflies, so the system is not foolproof. Pough and Brower[347] discuss a form of functional palatability in which males of the butterfly *Ascia monusta* are attacked more frequently than females as a result of behavioral differences such as more frequent thermoregulation in males, even though both sexes are unpalatable. Well-studied systems include the sequestering of cardiac glycosides from Asclepiadaceae and Apocyanaceae, two tropical groups, by several different taxa of insects.

The studies of Brower and Moffitt[348] and Brower *et al.*[349–352] suggest that considerable geographic variation exists in both the concentrations of these toxic secondary substances in host plants and the degree to which insects become unpalatable as a result of feeding on them. Brower *et al.*[351] studied the cardenolide profiles and palatability in a population of the butterfly *Danaus chrysippus* in West Africa and found considerable variation in the concentration of this substance in a sample of butterflies from one area. They related these differences to the spectrum of host plant variation and differential abilities of different butterflies to sequester the substance efficiently in the larval stage. Largely as the result of the studies of Brower and Moffitt,[348] it is evident that butterflies and other herbivorous insects whose immature stages or adults feed on plant groups known to contain toxic secondary substances and sequester these substances are not always unpalatable to potential predators, because individual species of host plants may vary considerably in the content of such substances, within local populations and geographically. The studies of Duffey and Scudder[353] and Vaughn[354] demonstrate the role of cardiac glycosides in making selected Hemiptera and Coleoptera unpalatable. Duffey and Scudder[353] suggest that the occurrence of several unrelated insects with the same palatability properties, small shape and size, and similar patterns of warning coloration provides a basis for the co-occurrence of these insects throughout much of North America and the Caribbean. In an important set of experiments, Vaughn[354] demonstrated that the amount of sequestering of cardiac glycoside from milkweed seeds by the milkweed bug, *O. fasciatus,* is inversely proportional to the concentration of the secondary substance in the seeds (Fig. 4.3).

Some lowland and mid-elevation tropical rain forests support species- and

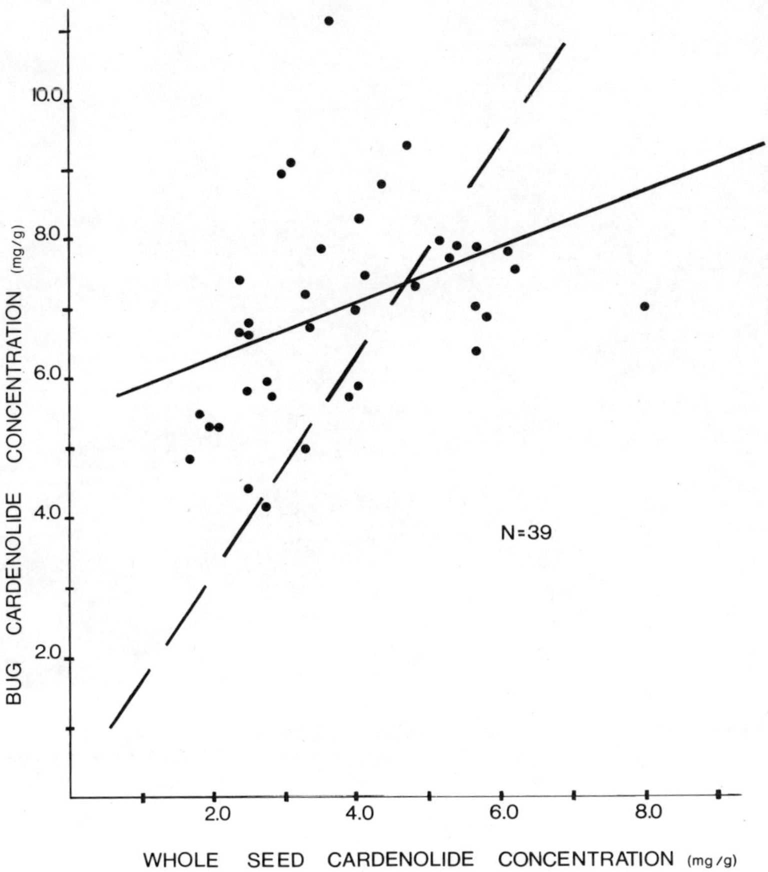

FIGURE 4.3. Patterns of increasing concentrations of cardenolides in *Oncopeltus* milkweed bugs as a function of increasing concentrations of these toxic secondary substances in whole milkweed seeds. [From F. A. Vaughan, *J. Chem. Ecol.* **5**:89–100 (1979).]

FIGURE 4.4. Young nymphs of the coreid bug, *P. blandus,* aggregated on the underside of a leaf of the food plant *A. warscewiczii* (a); the nymphs are bright orange with black bulbous antennae and legs. Older nymphs (b) are metallic green and orange. The nymphs aggregate on the food plant throughout the juvenile period and disperse as adults.

genera-rich assemblages of butterflies associated with a particular microenvironment within the forest, such as the ground cover and low understory vegetation.[186,189] In such well shaded environments, scent organs rather than strictly

FIGURE 4.4. (*Continued*)

visual mechanisms may play a major role in species-specific courtship patterns, although no inverse relation necessarily exists between color patterns and effectiveness of scent organs in courtship for tropical butterflies.[181] The solution to understanding the distribution of certain kinds of scent structures in butterflies is the examination of both phylogenetic and ecological properties of the species involved.[181]

In some Neotropical Hemiptera, the adults are migratory, large, and not aposematically colored, but their nymphal stages are sedentary, gregarious, and warningly colored. Aldrich and Blum[183] demonstrated the defensive role of aggregative behavior in the nymphs of the coreid *Thasus acutangulus* in lowland Guanacaste Province, Costa Rica, during the dry season. The nymphs form conspicuous clumps on the host tree and, when disturbed, simultaneously spray anal fluids and produce a defensive secretion covering the abdominal terga. The aposematic display is released by mechanical disturbance, and aggregation is induced by a pheromone. The behavior presumably is adaptive: the large bugs are active in the dry season and therefore under conditions of reduced foliage. Without such a defense, the bugs would be sitting targets for birds, lizards, and some amphibians. The brightly colored nymphs of the hemipteran *Paryphes blandus* form conspicuous, tight congregations on the leaves of *Anguria* (Cucurbitaceae) (Fig. 4.4), the adult bugs breed there (Fig. 3.9), and presumably these insects are unpalatable due to uptake of toxic substances from the host plant.[355]

In the tropics one encounters many examples of aggregative behavior in brightly colored insects, but further studies are needed to relate such conditions to host plant chemistry, *de novo* synthesis of defensive compounds by insects, and kinds of predators involved. There is also a need to analyze the spectrum of individual variation for such aggregating insects in terms of the spectrum of host plant species utilized. If, for example, the basis for unpalatability is derived from host plants, and if the insect is polyphagous, how is unpalatability conferred if the toxic substances are different among the host plant species in the habitat? Is it a system in which some morphs within the insect population are physiologically adapted to sequester the compounds from one or a few of the host plants? Or can any individual in the population sequester the compounds from several different host plants? These two alternative patterns of interaction imply different genetic and physiological traits. The population may consist of distinct genetic morphs each adapted to process some subset of the total food spectrum, or the population is essentially monomorphic with each individual possessing enough phenotypic flexibility (in the form of mixed-function oxidases or a similar system) to sequester the compounds from all host plants. The evolutionary implications of unpalatability in tropical insects also relate to population growth features. Unpalatable species may have, on the average, reduced population growth capacities resulting from energy being allocated to those physiological mechanisms endowing unpalatability. Yet if predation rates are high on immature stages such as eggs or larvae, new adaptations may arise to offset the loss in fitness from reduced fecundity or some other characteristic related to population growth. For example, Gilbert[31] has found pollen feeding to be a specialized behavior of some species of supposedly unpalatable *Heliconius* butterflies (Fig. 4.5), and he

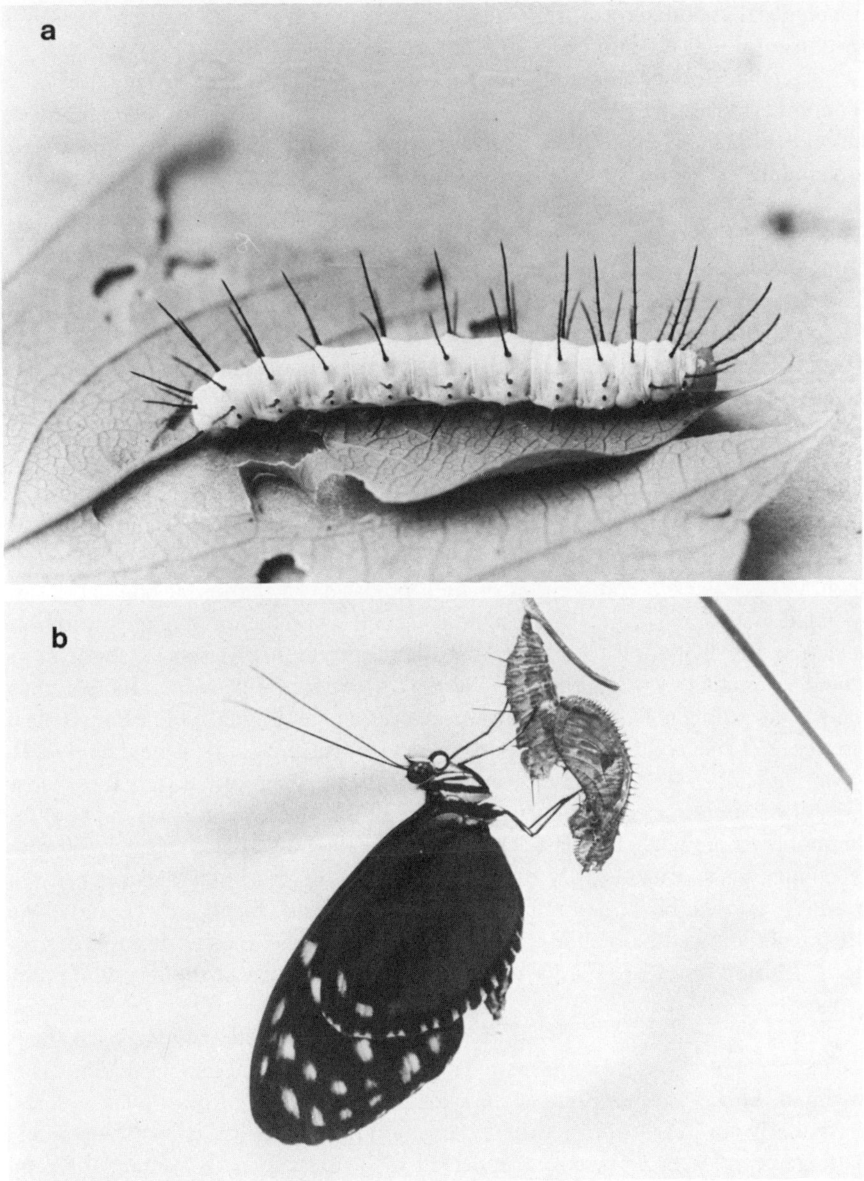

FIGURE 4.5. (a) Fifth- (final) instar caterpillar of *Heliconius hecale,* a specialized herbivore of some species of *Passiflora* (Passifloraceae) in Neotropical rain forests, and (b) the adult butterfly, clinging to pupal shell.

concludes that pollen feeding enables the butterflies to increase egg production in response to high predation rates on eggs by ants.

Detailed life cycle studies of tropical butterflies (e.g., Muyshondt[284–287]; Young[77,221–222,270,289–315]; Young and Muyshondt[316–319]) that include records of larval host plants and adult food resources provide the information necessary for developing conceptual models of how specific groups of insects, such as the major taxa of butterflies, are distributed locally (habitat wise) and geographically in the tropics. Such information also provides insights into why some tropical butterflies adapt well to man-made environments such as gardens.[356]

The double-edged feature of the concept of coevolution between plants and insects is that, on the one hand, the expression of toxic secondary substances in plants for defense against herbivores may screen out many potential feeders on that plant species, while on the other, other species may "penetrate" the ecological barrier through evolutionary time, overcome the defense system, and feed successfully on that plant. The ecological advantage of feeding on a plant with effective toxic substances for antiherbivore defense is that those species that possess the ability to detoxify or sequester the toxic compounds enter into an essentially empty universe in terms of food and absence of competitors. Intense ecological factors such as interspecific competition or high predation in the insect population associated with an "original" host plant species population probably promote selection for switching to a new host plant species in the same habitat or in adjacent habitats. The ability to make the switch would be a function of many factors, including (1) taxonomic distance between the original and new host plant species, (2) the ability of the insect to respond in the short term (behavioral, physiological) to feed on the new host plant, (3) the long-term adaptive response capabilities in terms of the genetic structure of the insect population, and (4) the amount of ecological stress associated with the new host plant species, in terms of competing species already established on it, and the kinds and intensity of mortality factors, biotic and microenvironmental abiotic factors, that could lower the population well below the carrying capacity and so low as to promote extinctions. Ehrlich and Raven[6] discuss some of these aspects of the coevolutionary process.

In plant populations, considerable variation is predicted among individuals for the concentrations of toxic secondary substances present in a particular tissue or organ. Such variation is expected since the populations of most plant species, particularly in the tropics, where selection pressures from herbivores and pathogenic microorganisms are expected to be intense, will contain different genotypes and phenotypes with somewhat different character states of various traits. Janzen[40] has stressed the adaptive role of maintaining outcrossing in plant species in lowland tropical forests, to the degree that certain plant species have evolved temporal patterns of flowering that promote long-distance pollinator movement patterns to ensure genetic exchange. Superimposed, so to speak, upon the residual variation among individuals in plant populations in the tropics is the pattern of temporal changes in concentrations of toxic secondary supstances found in individuals. Feeny,[258] working with oaks in the Northern Temperate Zone, assessed seasonal changes in the concentration of tannins in leaves and the impact of these changes on herbivorous insects. One type of pattern is the difference in concentrations of defensive compounds in young versus older tissues and structures. Shifts in concentration and distribution of defensive compounds may be matched with seasonal patterns of vegetative growth and reproduction in tropical plants. Some Neotropical vertebrate omnivores exhibit considerable shifts in diet as a function of size, sex, and age,[357] with such shifts related to changing physiological needs and availability of food sources. In tropical herbivorous insects, similar shifts may occur in terms of progressive reduction in quality of a food type during the growing season. Some Neotropical Orthoptera and carabid beetles are opportunistic predators on other insects and plant materials, respectively. The diet of a herbivorous insect must be viewed in terms of temporal changes in quality of suitable food sources and quantitative changes in food abundances. Many tropical butterflies preferentially oviposit on young leaves and meristem of host plants, and according to general theory about plant–herbivore coevolution, such tissues may have lower concentrations of toxic substances, making it easier for young larvae to assimilate nutrients efficiently. Young,[292] for example, described the preference of *Dismorphia* butterflies for placing eggs on the young leaves of *Inga*. Such studies lead us to believe that herbivorous insects exploiting the young tissues of host plants to the exclusion of older tissues will be subject to considerable resource limitation, which is a function of the vegetative growth pattern of the host plant species. The effects are expected to be pronounced for specialist herbivores.

Frankie *et al.*[118] have described some aspects of the seasonal or temporal vegetative growth patterns of trees in both seasonal and nonseasonal lowland tropical forests. An important implication for coevolutionary theory from this study is the great variation among tree species in these patterns, which must impose considerable variation in the population cycles of insects associated with these trees. Taking this one step further, those groups of butterflies and moths, and flies and beetles which make up pollinator communities in tropical forests will also be regulated in the larval stages by the availability of suitable host plant tissues, if the species are specialists on certain tree species and if they are also specialists on young tissues. The impact on plant species that require these insects for effective pollination is that limitation in the abundance of pollinators through time will be a function, in part, of patterns of resource limitation on larval food sources, which are usually different taxa of plants from those being pollinated by the adult stages of these insects. Such considerations point to the immense complexity of interactions in the tropical forest habitats, especially from the viewpoint of different selection pressures influencing adults and immature stages in insects.

4.3. NOTES ON DEFOLIATION IN THE TROPICS

In the preceding discussions it was generally concluded that the incidence of severe defoliation in tropical plants ought to be considerably less than for plant species in comparable Temperate-Zone habitats, primarily because of the inferred role of biotic mortality factors and deleterious effects of competition in keeping populations of herbivorous insects in the tropics well below the carrying capacity. Yet we must recognize that not all species of herbivorous insects in the tropics will be regulated by predators and parasites and those that are will not be regulated to the same degree. Because the essentially nonseasonal lowland wet forest tends to have the greatest numbers of insect species and plant species, incidence of defoliation in natural forest habitats may be low in these regions. As new insect species evolve in these regions, there might be a short period initially when populations are not exploited by predators and parasites. But because predators and parasites tend to be generalists, these organisms will rapidly accumulate on the new species and the species will come into equilibrium with the rest of the communities. If the new species is a good competitor and has a low population growth rate as a result of an evolutionary tradeoff with energy allocated to defensive strategies, it may remain "cryptic" as a new resource for predators and parasites for some time. But if the species consists of one or more populations with individuals programmed for rapid population growth, resulting in a swamping of a host plant species, the new species becomes a signal for predators and parasites and the response to the new resource may be rapid.

Another factor in such considerations is the degree to which different species of predators and parasites that are potentially capable of exploiting the new resource are limited enough in sufficient resources to make such a facultative switch. Then again, such "decision making" may not be an issue if the exploitation of the new host is basically a matter of simple encounter and easy access to it. The reestablishment of an equilibrium between the new herbivore species and its host plant mediated by regulation from predators and parasites reduces the likelihood of outbreak conditions and subsequent defoliation. If the new species enters into a competitive adaptive zone with other species, the tradeoff in energy allocation associated with niche restructuring to lower competition will also keep the population below outbreak levels.

Other factors may influence the degree to which predators and parasites can regulate herbivore populations. Campbell and Duffey[(358)] have discovered that tomatine, an alkaloid toxin in tomato plants, is passed on from the body of the herbivore, a pest of tomatoes, to the parasitic wasp, and the substance is toxic to the wasps. Such a mechanism lowers the ability of the wasp to regulate the herbivore population. The wasp is an endoparasite of the herbivore and therefore ingests the tissues laden with tomatine. Thus the effectiveness of parasitic wasps and flies in regulating herbivore populations will be a function of the amount of toxicity of the secondary substances accumulating in the body tissues of a potential host species and the ability of the parasite to buffer the toxic effects of the secondary substance. Given the diversity of plant species and secondary substances in tropical floras, such selection might be intense in the tropics and constitute yet another level of coevolutionary associations, namely, between parasites and predators of herbivores, and the defensive compounds of the host plants, for those herbivores that sequester these substances in their diets.

Although defoliation seems to be an ecological phenomenon of Temperate Zone communities, I believe many examples of it will be discovered in high-altitude communities in the tropics, lowland forest light gaps expanding into secondary growth habitats, a variety of disturbed habitats including agricultural systems and horticultural habitats, and very seasonal habitats. Kenyi[(360)] described the defoliation of *Capparis tomentosa* by the larvae of the butterfly *Belenois aurota* in Uganda and found it to be a seasonal event each year with larvae selectively browsing young leaves. The observed defoliation is thought to reduce the host plant population to the extent that other plant species in the same park environment increase in abundance. Defoliation of a host plant species will also depend upon the amount of defensive chemistry variation among individuals of the host plant population and the ability of the insect population to handle that spectrum of variation in host plant defenses. The extent of herbivore attack will also depend upon the spatial arrangement of individuals in the plant population (e.g., Vandermeer[(359)]). The isolated individuals may have a significantly lower probability of being attacked than an individual in a large clump. Yet such patterns will vary considerably with the kind of herbivore, its dispersal ability, and variation in attractiveness of host plant individuals.

If temporal patterns of growth and overall diversity of plant species are low in high-altitude regions in the tropics such as parts of *altiplano* in the Peruvian Andes, there might be periods of defoliation corresponding to those periods in which vegetative growth of host plants is high and herbivore or omnivore populations are somewhat synchronized to the flushes of these resources. Physically controlled environments are believed to have a greater incidence of insect outbreaks as a function of synchronized seasonal emergence patterns and reproductive cycles during the limited growing seasons of such environments. It does not necessarily follow that highly seasonal lowland tropical areas will also contain communities characterized by seasonal outbreaks of herbivorous insects and defoliation. Lowland seasonal environments represent old ecosystems with considerable diversity of organisms and sufficient evolutionary time to permit species in different trophic levels to adjust population structures to seasonal fluctuations in resources. As with nonseasonal lowland environments in the tropics, such interactions are finely tuned so that herbivores, predators, and parasites follow in close synchrony the seasonal changes in the environment. The major departure from all of this reasoning is, however, the consideration of agricultural systems in the tropics and their proneness to insect outbreaks and defoliation. See the discussion of this important topic in Chapter 9.

Ferns make up a major group of tropical plants that are not intensely exploited by insects as food, and the reasons are not obvious.[(361)] The evidence suggests that toxic defensive substances in ferns cause low predation rates by herbivorous insects, but interestingly, about 75% of all polyphagous insects feeding on ferns are also associated with woody angiosperms,[(361)] a condition suggesting that under-utilization of ferns by insects is largely the result of other suitable food supplies being readily available in the same habitats. The explanation, therefore, may be an ecological one, related to the local distribution and abundance of angiosperm host plants versus ferns as hosts.

4.4. NOTES ON SEED PREDATION IN TROPICAL HABITATS

The studies of Janzen[(41,362)] demonstrate the role of host-specific seed and seedling predation in determining the species composition of tropical forests. Working primarily with various species of Neotropical leguminous vines, Janzen[(363,364)] and Rehr *et al.*[(261,263)] have shown that the large seeds of these plants contain high concentrations of toxic secondary substances that make the seeds immune to attack by certain seed-eating insects. For example, the seeds of the common vine *Mucuna* contain high concentrations of free L-dopa, which when isolated and combined with a synthetic diet leads to substantial mortality and deformities in the southern armyworm larvae.[(261)] Rehr *et al.*[(261)] surveyed a variety of seeds from Central American legume species and conducted similar feeding tests demonstrating that these seeds contain high concentrations of "insecticidal" amino acids that repel or kill armyworm larvae and presumably have similar effects on many seed-eating insects in tropical forests. The various insect species that attack such seeds in the wild may be under limiting conditions of other food sources, and considerable selection might favor the expression of new coevolved relationships between these insects and leguminous seeds.

The studies of Janzen[(365)] have shown that nonleguminous trees in tropical forest have similar coevolved systems. From his study it is clear that seeds of tropical trees may contain toxic secondary substances and other kinds of toxic substances that through evolutionary time have filtered out a potentially large assemblage of seed-eating insects, at the same time resulting in a few species' overcoming the defenses and successfully attacking the seeds in annual seed crops. Janzen[29] has shown that seeds are rapidly discovered on the ground beneath the parent tree or vine and that discovery rate and intensity of attack on seeds drop off considerably as one moves away from the tree. The system is most evident for tree individuals within forest communities, but sometimes the entire trees or seed crop of vines in secondary succession habitats escapes from predators, presumably because the crops escape detection by insects. Mammals such as squirrels and monkeys are the primary dispersal agents of seeds, carrying off pods and then dropping some seeds away from parent trees or vines. The interaction of host-specific seed predators, annual patterns of seed crop production, seed characteristics such as size, shape, and toxicity,[362] and mammalian dispersal agents largely determine the distribution of adult and juveniles of tree species in tropical forests. Other processes, such as differential rates of seed survival and germination on different soil patches within a forest, inhibitory effects of other plant species on the germination or seedling growth process, and other microen vironmental factors also help determine the spatial distribution of plant species in tropical forests. Janzen[29] also discusses the impact of predation on seedlings in causing virtual loss of seed crops beneath parent trees and reviews the theory and evidence of the ecological processes shaping plant distributions in tropical forests. The model developed by Connell[65] and Janzen[29] predicts that each tree species in a tropical forest will have its own set of host-specific herbivores attacking seeds and seedlings. Different kinds of insects are involved in attacking seeds and vegetative structures. Herbivorous insects such as some Lepidoptera may be associated with the canopy of such trees, but larvae will also attack seedlings growing beneath the tree. The canopy crown is envisioned to act as a signal to attract herbivores while filtering out many others through defensive compounds present. Cicadas may constitute another group of insects attacking forest trees in the tropics and using the canopy crown to search for the appropriate root crown necessary for nymphal development.

A dominant tree species of lowland tropical forest in the eastern watershed of southern Central America is the legume *P. macroloba,* whose population structure and dynamics has been studied by Hartshorn.[337] Unlike most trees of these forests, this tree species is abundant per unit area of habitat, and adults and seedlings are often found tegether. At least three genera and six species of cicadas emerge in large numbers beneath individual adult trees of this species year after year, and the same trees are full of cicada nymphal skins each year.[70,71] Even though the forest understory is full of emergence sites, the nymphs appear to emerge directly beneath the crown shadow of *Pentaclethra,* suggesting a possible feeding association of nymphs with the root crowns of the tree. But given the abundance of intermixing root systems from other tree species, the hypothesis is that *Pentaclethra* adults, and their cohorts of seedlings beneath them, act as signals for mated female cicadas to oviposit within the patch. Young[70,71] reports frequent oviposition by *Z. smaragdina* in *Penta-*

clethra seedlings beneath adult trees. *Pentaclethra* xylem or phloem fluids may have relatively fewer toxic substances than other tree species in the same habitat and therefore represent an accessible food source for cicada nymphs on its root crowns. Thus there might be strong selection for cicada species to converge and exploit the root crowns of this tree species, and the preference is mediated through the oviposition behavior. White and Strehl[366] discuss the effects of low-nutrient plant fluids on the evolution of developmental time in cicadas and other insects, and because of such an effect, one might predict that cicadas prefer host plant species where nutrient assimilation is most efficient, thus speeding up developmental time, even if only in small amounts. *Pentaclethra* trees, because of a hypothesized lack of toxic defenses in internal fluids, may be an optimal resource for cicadas in tropical forests. Because patch size is relatively large, considering the combined root crowns of adults and juveniles, not to mention the biomass available for oviposition sites, the environment can support populations of several genera and species.

This is not to imply that Neotropical cicadas in geographical zones where such a tree species occurs will feed only on this tree species. The opposite is probably true, namely, that cicada nymphs will attempt to feed on most roots encountered within a certain stratum of soil. What is being suggested, however, is that while *Pentaclethra* may have a primary host root crown. Other tree species are exploited in direct proportion to the frequency of encounter by burrowing nymphs. Disease and predation associated with egg masses and young nymphs undoubtedly keep cicada populations below the carrying capacity of the habitat. Adult cicadas are also attacked frequently by birds such as oropendulas, which successfully locate singing males.

Janzen[29] discusses the high survivorship of *Pentaclethra* seeds and seedlings beneath parent trees based upon the immunity of seeds to attack by seed-eating insects. It should be pointed out that the patterns discussed above for cicadas in tropical forests can also be explained by other hypotheses. For example, an alternative explanation for the apparent convergent association of cicadas with *Pentaclethra* is that this tree species is the most abundant resource type in the forest habitat and therefore can be exploited by several species without competition. This hypothesis implies nothing about the toxic compound chemistry of internal fluids being ineffective relative to other tree species. The combination of large patch size and relative ease of exploitation together may account for associations of cicadas with certain tree species in tropical forests.

A similar pattern seen in advanced secondary growth forests in the same region is the association of some cicadas such as *Z. smaragdina* and *F. sericans* with large patches of *Goethalsia meiantha* (Tiliaceae).[71,215] But one must also examine the evolutionary history of cicadas and broad patterns of host tree exploitation to account for present-day distributions and associations. Young[137] reported that the majority of species in Costa Rica are associated today with leguminous tree species in general and that legumes might have constituted the original host plant group during the early adaptive radiation of cicadas in the tropics. Thus, an overall preadaptation to legumes in general, coupled with contemporary ecological factors, accounts for the associations of cicadas with particular habitats in the tropics and for association of nymphs with the root crowns of certain tree species in these environments. Ecological shifts to disturbed habitats may occur in cicada species best adapted to marginal habitats and may involve shifts to nymphal feeding on different plant families, especially for those plant species with large patch sizes and low concentrations of defensive compounds in xylem or phloem fluids. Some recent Temperate Zone studies with plant hoppers show that individuals of a single species preferentially select individual host plants or parts of host plants containing the highest concentrations of amino acids in phloem or xylem sap.[367] The feeding on plants containing high concentrations of nitrogen-containing nutrients increases the fitness of some sap-feeding Homoptera.[368,369] Such studies reveal that sap-feeding insects in the Temperate Zone exploit host plants in many different families, although any one species may exhibit a coevolved interaction with a single host plant species. Since the tropics contain a greater array of Leguminosae in a variety of terrestrial habitats than comparable areas of the Temperate Zone, a greater evolutionary opportunity exists for some sap-feeding forms such as cicadas to coevolve with these plants, presumably as the result of the high availability of these plants in the tropics and the presumed higher concentrations of nitrogen-containing nutrients in the xylem fluids of roots in these plants. Outside of the Leguminosae, one might expect to find strong selection for cicadas to exploit plant species with mycorrhizal associations that provide additional nutrients to the root crowns of these plants; a large portion of plant species in tropical forests may have such associations.[370]

Studies by Sgarbieri[371] have demonstrated the considerable variation in protein content of legume seeds from many plant species in Brazil, variation both among and within species. Similar patterns of nutrients and nonnutritive constituents may also occur in phloem fluids of tropical legumes, providing considerable heterogeneity in resource quality for cicadas and other insects. Thus preadaptation to legume feeding in cicadas in tropical forests is followed in evolutionary time by the acquisition of abilities to feed successfully on legume species with different nutritive and nonnutritive profiles. Other phenotypic characteristics, such as the attractiveness of bark and pith to ovipositing females, will also influence host selection in cicadas. Our overall knowledge of the natural history of tropical cicadas is fragmentary at best, but the interactions of these organisms with tree species in tropical forests may shed some light on the original plant species composition of tropical forests.

Various examples of the chemical and functional diversity of antiherbivore defenses of seeds in tropical floras are reviewed by Janzen.[41,42] A noteworthy point about these studies is that fewer data are available for seed defenses in mountain floras in the tropics. It is difficult to generalize about the antiherbivore defenses of plants and their role in shaping the species richness of tropical habitats, but an even greater problem is predicting changes in such interactions in mountain forests. Whether or not the large seeds of legume vines such as *M. urens* are attacked by species of bruchid weevils along similar lines as studied in lowland habitats remains to be investigated in high-elevation regions. The selective forces affecting a vine population high up on the fertile slopes of a volcano in southern Central America might be very different from the lowlands in the same region. For example, a preliminary examination of seed mortality and vine recruitment levels in a forest patch containing several adult *M. urens* vines at about 1600 m elevation indicates little evidence of attack by seed-eating insects or high predation on "milk stage" seeds by squirrels. Rehr *et al.*[261] predict that *Mucuna* seeds in general should be free from insect attack. Data from a highland population of *M. urens* seem to confirm this. But the high level of predation by squirrels in this population results in low recruitment of new vines within this habitat. Presumably the squirrel population is confined to the small patch of forest, and foraging for milk stage seeds is unusually high. Such foraging represents a facultative response by squirrels to restricted habitat and easy access to seed crops maturing on several vines.

4.5. NOTES ON PLANT–ANT INTERACTIONS IN THE TROPICS

An earlier discussion in Chapter 1, pointed out the great taxonomic and ecological diversity of ants in the tropics. The evolution of eusociality, tremendous diversification in nesting habits, and diversification in feeding habits figure as the prime factors accounting for the great ecological success of ants in the tropics. From the poles to the equator, the number of subfamilies, genera, and species of ants increases gently. Lowland tropical rain forests tend to have the most ant taxa within the tropics. The pattern of environmental heterogeneity both horizontally and vertically in lowland tropical rain forests has provided ants with considerable opportunities for ecological diversification. I define ecological diversification in ants as the specialization within different taxa for particular nesting sites in different microhabitats within the forest environment and the associated specialization in feeding habits in many of these taxa. (See Wilson[80] for a complete review of ecological diversification of ants.) While many tropical ants are opportunistic scavengers and predators, others combine such behaviors with specialized foraging activities, while still others exhibit completely specialized foraging habits. From an initially terrestrial or cryptozoic existence, ants have been able to exploit different foods and nesting sites arboreally and otherwise in tropical forests in response to the great diversity of available resources. Extensive surveys of tropical insects will probably leave undescribed many taxa of forest canopy ants and of ants exhibiting behaviors adapted to canopy existence.

Some forms of ecological specialization in tropical ants result from the structural and organismic complexities of lowland tropical forests. The long growing season, generally favorable climate, high diversification of plants and various animals, promote diversification in ants, and one can envision such a process laying the fundamental pattern of speciation in tropical ant faunas. Different patterns of ant speciation are associated with tropical islands.[91] But just as taxon cycles in some ant faunas of tropical islands have molded the geographical distribution of ant species in archipelagos, so could phenomena similar to the stages of a taxon cycle govern the diversification of ant faunas on tropical mainlands. The invasion of ant species into secondary successional stages and disturbed habitats may follow such patterns.

Another form of ecological specialization in tropical ant faunas is characterized by the entering of particular taxa into obligate mutualism with certain taxa of plants. Such associations do not form a major portion of tropical ant faunas, but another form in ant–plant interaction, namely, the feeding by ants on exudates of extrafloral nectaries of plants, constitutes a fairly common form of opportunistic feeding exhibited in many different ants. In the mutualism between plants and ants, both partners exhibit specialized physiological, behavioral, and morphological adaptations or phenotypic features designed to promote the interaction. Perhaps the best known examples come from the classic studies of Janzen[33] on the obligate mutualistic actions between certain species of *Pseudomyrmex* ants and *Acacia* shrubs in Central America. Although the association of these ants with acacias was known for some time, the experiments of Janzen[33] confirmed and proved the nature of the mutualism. Janzen, for example, was able to demonstrate that young swollen-thorn acacias deliberately deprived of their *Pseudomyrmex* founding colonies exhibited a reduced fitness compared with ant-occupied plants in the same habitats. Janzen also demonstrated the detailed adaptive traits in both ants and plants for the mutualism. Janzen[372] describes the geographical distribution of swollen-thorn acacia species in Central America, illustrating that the group is mainly adapted to agricultural habitats such as pastures. The major traits of the mutualism are summarized in Table 4.1. Keep in mind that these phenotypic characters have developed in response to selection pressures favoring the evolution of mutualism between *Pseudomyrmex* ants and acacias. Presumably a pre-swollen-thorn acacia population experienced considerable herbivore damage to vegetative structures and the mutualism evolved as a means of reducing predation on the plants. Janzen shows that the ants, a complex of several species in Central America, protect the vegetative parts of acacias by removing herbivorous insects. The ants possess a powerful sting and are mobilized rapidly on the leaves and stems of the plant in response to mechanical disturbance. The acacia provides

TABLE 4.1
The Major Phenotypic Traits of Ant Acacias[a]

General features of acacias of importance to the interaction	Specialized features of swollen-thorn acacias (coevolved traits)
Woody shrub or tree life form	Woody but with very high growth rate
Reproduce from suckers	Rapid and year-round sucker production
Moderate seedling and sucker mortality	Very high unoccupied seedling and sucker mortality
Plants of dry areas	Plants of moister areas
Ecologically widely distributed	Ecologically very widely distributed
Leaves shed during dry season	Year-round leaf production[b]
Shade intolerant, sometimes covered by vines	Shade intolerant and free of vines
Stipules often persistent	Stipules longer persistent, woody with soft pith
Bitter-tasting foliage	Bland-tasting foliage
Each species with a group of relatively host-specific phytophagous insects, able to feed in the presence of the physical and chemical properties of the acacia	Each species with a few host-specific phytophagous insects, able to feed in the presence of the ants
Foliar nectaries	Very enlarged foliar nectaries[b]
Compound unmodified leaves	Leaflets with tips modified into Beltian bodies[b]
Flowers insect pollinated, outcrossing	Flowers insect pollinated, outcrossing
Seeds dispersed by water, gravity, and rodents	Seeds dispersed by birds
Lengthy seed maturation period	Lengthy seed maturation period
Not dependent upon another species for survival	Dependent upon another species for survival

[a] From Janzen.(33)
[b] Essential to the interaction.

domicile in the form of hollow swollen thorns, and food in the form of nutrient-rich secretory structures (Beltian bodies) on the leaves. Ants actively remove vegetation surrounding the young acacia, increasing its likelihood of surviving in the canopy of secondary growth vegetation. Acacias deprived of ant colonies experience intense attack by herbivorous insects and may die. The swollen stipular spines and foliar nectaries of ant acacias are adaptations to acacia ants.

Rehr *et al.*(263) report on the chemical defense in non-ant acacias in which leaves of these plants were found to be cyanogenic and thereby toxic to herbivorous insects. Leaves of ant acacias are not cyanogenic. An absence of evolutionary contact between *Pseudomyrmex* and some acacias, or the lack of phenotypic flexibility to develop a mutualism with ants if such contact is made, promotes selection for the expression of toxic compounds as defense against herbivores. Janzen(373) also demonstrated that occupancy of *Barteria* (Passifloraceae) shrubs by *Pachysima* ants in Nigerian rain forest enhances the vegetative growth of the plants. The ants reduce herbivore damage and attack plants adjacent to their host shrubs and clean *Barteria* of various debris. A similar interaction has been studied in Central America and the Caribbean islands.[374,375] Risch *et al.*[376] describe the mutualistic association of *Pheidole* ants with three species of *Piper* (Piperaceae) in lowland tropical rain forest in Costa Rica. The ants occupy hollowed-out stems and petiolar cavities of the plants and feed on lipid-rich secretions from secretory bodies in the petiolar cavities. In addition to removing nearby vines, the ants deposit debris in the hollowed stems which is then absorbed by the plant as an additional source of nutrients.

Bentley[377] found that individual flowering plants of *Bixa orellana* in Costa Rica with higher frequencies of visits by ants to extrafloral nectaries have higher levels of mature fruit production than other plants with reduced visits by ants. She concludes that the presence of many ants on a plant reduces the activities of herbivorous insects and thereby increases fitness by increasing yields of mature fruit. The daily pattern of nectar production in extrafloral nectaries suggests a peak in late afternoon; ants also are active at this time (Fig. 4.6). The interaction is so well tuned that peak nectar production in extrafloral nectaries occurs at the time mature buds are produced (Fig. 4.7). Studies (e.g., Keeler[378]; Keeler and Kaul[379]; Koptur[380]) indicate that extrafloral nectaries produce nutrients for ants and are strategically placed on plants to maximize protection against herbivorous insects. The elaboration of extrafloral nectaries in many plant species is a form of facultative mutualism in which the plant's phenotype includes ant-attracting substances to increase the general presence of one or more ant species on leaves and

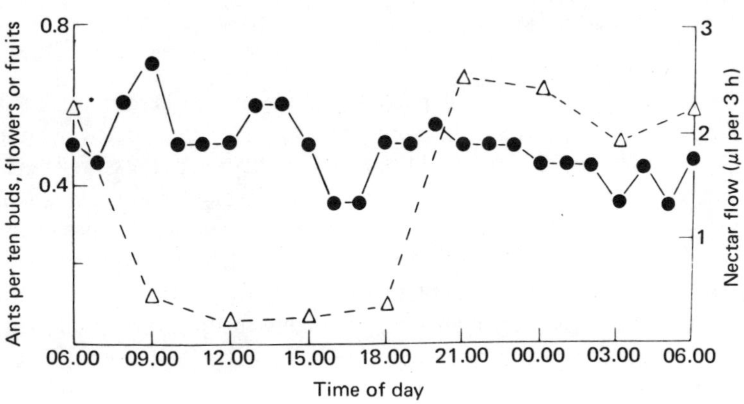

FIGURE 4.6. Ant activity (●), as the mean number of ants per 10 buds, flowers, or fruits over successive hours, and hourly nectar production (△) from extrafloral nectaries of the Neotropical tree *Bixa orellana*. Nectar production or flow was measured as the mean volume of nectar secreted in 3 hr by 10 inflorescences over three 24-hr periods. [From B. L. Bentley, *J. Ecol.* **65**:27–38 (1977). Copyright 1977 by the British Ecological Society.]

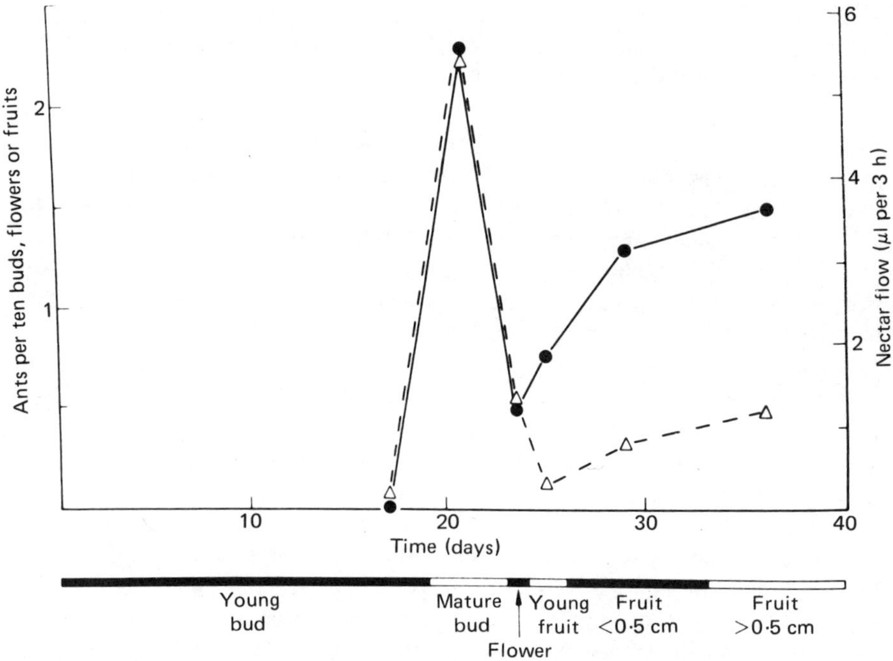

FIGURE 4.7. Ant activity (●) and nectar flow (△) for the six developmental stages of buds, flowers, and fruits of the Neotropical tree *B. orellana*. Ant activity is the mean number of ants per 10 buds, flowers, or fruits averaged over 24 hr, based on hourly observations, and nectar flow is the mean volume of nectar secreted in 3 hr by 10 buds, flowers, or fruits. [From B. L. Bentley, *J. Ecol.* **65:**27–38 (1977). Copyright 1977 by the British Ecological Society.]

stems, which in turn makes the plant less attractive to herbivores. Although extrafloral nectaries are found in plant groups outside the tropics,[379] they are probably more abundant in tropical floras, where herbivore pressure is great and where there is a large pool of ant species that could enter into such associations opportunistically. There might be a correlation between the ability of a plant to produce toxic secondary compounds as effective antiherbivore deterrents and the degree of phenotypic commitment to other defense traits such as extrafloral nectaries and their associated phsyiology: plants incapable of producing such secondary substances are more likely to develop extrafloral nectaries to attract ants, and within this group a small percentage may evolve into obligate mutualisms. But facultative mutualisms between plants and ants are expected to be more common.

Foraging responses in some tropical ants to the exudates of extrafloral nectar ies is most likely the result of encounter frequency with such resources for most cases of facultative associations. Carroll and Janzen[197] discuss shifts in the foraging behavior of ants in response to differential needs of colonies in terms of breeding. Opportunistic foraging coupled with periodic phases of different nutritive requirements within nests determine the exploitation of extrafloral nectaries by each ant species. The exudates of extrafloral nectaries may contain substances such as lipids and carbohydrates, not found in sufficient concentrations in other food resources such as prey or debris animal tissues. Intake of lipids, for instance, might be important for proper nourishment of larvae in nests, although there are no data on such requirements. The observed patterns of extrafloral nectary exudate production being timed with the reproductive cycles of bearer plants suggests that these plant species are vulnerable to intense herbivory that could lower fitness in terms of seed crop production. The highly timed production of exudates within a population of a particular plant species may be so narrow that it becomes an unpredictable resource for foraging ants. Thus examples of facultative mutualism are expected to consist of ant species with flexible foraging habits. In the Neotropics, ant groups such as *Ectatomma* typify such opportunistic forms of behavior. The investment return for a plant species to allocate resources to elaboration of extrafloral nectaries and associated exudates will depend in part upon the synchrony of production within the population: the greater the synchrony, the greater the reinforcement for worker ants and the greater the protective role of the ants. Strong stabilizing selection should be influencing those traits governing the development and timing of exudate elaboration among different individuals within the population. Presumably ants at extrafloral nectaries reduce attacks by herbivores. The adaptive role of these structures and their secretions is seen in some Passifloraceae that have evolved extrafloral nectaries that mimic the eggs of heliconiine butterflies. Females in search of oviposition sites will avoid placing eggs on meristems with egg mimics[338] (Fig. 4.8), reducing attacks of these vines by heliconiines.

The presence of ants with or without extrafloral nectaries does not necessarily mean that herbivores will be absent from the plant. Counterselection or short-term behavioral responses induced by ants is predicted in these insects. The extremely specialized egg placement behavior in some tropical butterflies, in which eggs are placed on tips of tendrils and other structures where ants cannot reach them, may be responses to ant predation. Other factors, including the proximity of ant colonies and foraging habits of ant species in a habitat, will also contribute to the probability of ants being present on a plant with extrafloral nectaries. The evolutionary commitment on the part of plant species to evolve extrafloral nectaries will be a function of (1) the genetic variation in the population for those phsyiological and morphological control mechanisms that would be involved in the development of these structures and (2) the long-term intensity of predation by herbivores and the reduction in fitness resulting from such damage.

Because selection pressures arising from herbivory are expected to vary

FIGURE 4.8. Some examples of extrafloral nectaries in the Passifloraceae of Neotropical forests. (a) The cuplike structure at the base of the folded leaflet is an extrafloral nectary. (b) The light round, dotlike areas on the leaf of *Passiflora* aff. *capsularis* are extrafloral nectaries.

considerably within and among plant species in the tropics, the production of toxic secondary compounds in plants may be facultative responses during periods of intense predation. Thus once a tree or herb is attacked, the physiological machinery of the plant produces more of the substance to increase effectiveness as a feeding deterrent. Such facultative chemical defense systems, as hypothesized here, can work in concert with other lines of defense of plant species, such as leaf toughness, water content, and phenological patterns of leaf production. The possession of facultative response mechanism of this sort, along with other more stable phenotypic characteristics, may be less of an energy drain on the plant species in the sense that the machinery is geared up only as needed. Such a system might be very adaptive for plant species in seasonal regions of the tropics where facultatively controlled defenses are brought into play during the periods of peak herbivore activity, such as the first half of the rainy season.

CHAPTER 5

DISTRIBUTION PATTERNS OF INSECTS IN TROPICAL HABITATS

As discussed in Chapters 1 and 2 various authors (e.g., MacArthur[(15,16)]) have considered the joint effects of within- and between-habitat diversity being the major determinants of observed patterns of high species richness in the tropics. This approach implies that much of the adaptation within species occurs as a result of evolutionary responses to patterns of local environmental heterogeneity, and such patterns of resource distributions and biotic interactions among organisms are very diverse in some regions of the tropics. Such an approach complements considerations of broad-ranging adaptive responses such as observed patterns of geographical variation in phenotypic characters and gene frequencies within a species. It is well known that studies of geographic variation within single species and for closely related species resulted in the formulation of the concept of speciation by allopatry or geographical differentiation. But in this chapter, we are concerned with adaptation in local environments rather than over a broad geographical range.

5.1. THE CONCEPT OF LOCAL ADAPTATION

One can speak of the adaptive responses of individuals within a single population of a species as adaptation to the local environment. Many studies have demonstrated that such local patterns of adaptation within populations may differ among different populations of a single species and that major determinants of local adaptation include (1) spectrum of potential resources available for a species, (2) biotic interactions with other organisms, and (3) local effects of the physical environment. These three sets of factors impose patterns of environmental heterogeneity upon organisms, and adaptation is often a function of the distribution of these ecological factors in time and space. Such distributions are often patchy, and the collection of patches associated with a particular ecological factor is said to have a grain pattern, where grain refers to the arrangement of the patches as experienced by individuals of the species.[3,381] Furthermore, the abundance of patches and the relative fitness of individuals in each patch determine the manner in which a species exploits resources in a particular habitat. Either within a habitat or between two or more habitats in a region, the distribution of resources for a species may be coarse grained, resulting in an individual remaining in a single patch for all of its prereproductive and reproductive life. The result is that different genotypes will adapt to different patches, and considerable microspatial variation in the patterns of adaptation can result. Such microspatial variation is expected to be most evident if there is reduced gene flow among patches. This type of response mechanism to environmental heterogeneity results in differentiation within a species at the local habitat level; as a means of evolutionary change in a species, it is therefore a very different mechanism from geographical changes.

Heterogeneous environments establish a mosaic of different selection pressures that in some cases will favor the maintenance of genetic variation in populations occupying such environments. In such situations, disruptive selection maintains the polymorphism, with different genetic morphs adapted to different niches or portions of the environmental mosaic. Habitat selection promotes the maintenance of alleles advantageous to the species in different patches of the environment, and in some cases deleterious alleles can also be maintained in a population by this mechanism.[382] The maintenance of such an allele represents an evolutionary compromise when limited resource patch types are present, and the population requires the exploitation of all patch types to maintain a population size sufficient to prevent local extinction.

In species responding to available resources in a fine-grained fashion, that is, in which resource patches and different habitats may be indistinguishable in terms of differential fitnesses of genotypes, most or all individuals in the population will utilize the same resources and therefore be generalists rather than a mixture of specialists as developed above for coarse-grained environments. Genetic differentiation among populations or subpopulations, in fine-grained environments, will be nonexistent. Such considerations apply to the analysis of niche differentiation, or lack thereof, within a single species and in a particular habitat, or along a series of habitats in the same region. Considering the tropics, we would expect the great species richness of plant species within some lowland habitats in particular to provide the potential for a great deal of niche differentiation within a herbivorous insect species, assuming for the moment an absence of intense interspecific competition and the availability of patch sizes sufficiently large to support a breeding population of that species. The same reasoning can be extended to summed floristic diversity of several adjacent tropical habitats.

Several studies with *Drosophila* have demonstrated that insect species exhibit considerable microspatial preferences in resource utilization,[383,384] and revealing along these lines are studies on the specialized oviposition habits of these flies.[385] Habitat selection is a major behavioral mechanism generating within-habitat spatial variation in the distribution of *Drosophila* species on the island of Hawaii.[385] The analyses of thermal preferences in Puerto Rican *Drosophila* by Levins[386] illustrate considerable diversification among species for different microhabitats. Other studies (e.g., Powell[387]; McDonald and Ayala[388]) indicate that microenvironmental heterogeneity influences the genetic structure of *Drosophila* populations, so that observed phenotypic differences correspond to genotypic differences within populations.

There is now considerable evidence that genetic variation in populations is promoted by the complexity of the environment. Within biological limits, one might say that as the environment becomes increasingly complex, genetic variation in the population of a given species should also be high so that the species can thrive in the environment. I use the term *thrive* to mean the capacity to maintain a breeding population at or near the carrying capacity or to counteract severe biotic or abiotic mortality so that the population can continue to increase even though well below the carrying capacity. In complex environments, it is believed that one kind of allele for a given locus is favored in one kind of microhabitat, another allele in a different microhabitat, and so on. The utilization of many different microhabitats within a particular habitat is maintained by natural selection, since such a pattern of resource exploitation allows the species population to lower the possible detrimental effects of both intra- and interspecific competition.

Field studies with *Drosophila* have demonstrated, in some cases, a positive relationship between levels of genetic variation and ecological complexity of the environment. DaCunha and Dobzhansky[75] demonstrated, for example, a clear relation between the incidences of chromosomal polymorphisms and floral diversity in *D. willistoni*. Taylor and Powell[389] showed that morphs of *D. persimilis* collected over a region of field and forest habitats in the Northern Temperate Zones retain habitat associations when marked, released, and subsequently recaptured. But other studies of different species do not show a clear relationship between the kinds of chromosomal polymorphisms present in a population and habitat affinities.[390] On the basis of various kinds of electrophoresis, there have been attempts to determine if different enzyme-genotypes within a population are associated with different microhabitats, the basis of selection being suitability of microhabitat occupancy at the biochemical level.

Given such approaches, the question of habitat selection distills down to the determination of whether there is a pattern of genetic variation in a population and the degree to which different genotypes exhibit preferences for certain microhabitats present in the habitat. The population is seen as consisting of several subsets of different genotypes, each exhibiting a preference for different microhabitats, with breeding largely confined to those genotypes associated with a particular microhabitat. Thus many subdemes may exist, and the result is a preservation, along with the effects of selection, of the microhabitat affinities of each genotype.

In the tropics, where there is often a great deal of host plant specificity in many phytophagous insects, selection might result in different genotypes exploiting different host plant species in the same habitat, or over several habitats if the entire breeding population is distributed over more than one habitat, and breeding occurs among individuals in different habitats. The selection for ecological divergence in such insect populations can arise from interspecific competition for the same plant species as resources or during periods following colonizing episodes in which population density might be very high, resulting in considerable intraspecific competition. Both sorts of situations can create selection favoring microhabitat divergence where each microhabitat type is a different host plant species.

The studies of Bush[391] on Temperate Zone, fruit-inhabiting flies in Wisconsin indicate that such divergence may lead to partial reproductive isolation and the beginnings of speciation. It is tempting to speculate that the high diversity of insects in the tropics is the result of considerable microhabitat divergence on resources with high harvestable productivity, initially giving rise to different genotype sets associated with certain resources, then leading to speciation. Such a system would fit the lowland tropics well, where resource diversity for insects is high. Thus one can envision habitat selection for a given species of insects in a lowland tropical field as being the association of the breeding population with that habitat, and also on a fine scale, in that microhabitat selection within the habitat promotes ecological divergence, i.e., different insect species using different plant species. Depending upon the genetic systems associated with feeding behavior and physiology, and the amount of genetic variation in the population for alleles involved in the regulation of feeding, polyphagy or highly flexible monophagy can result.[275] Which type of expanded feeding evolves, assuming the initial condition to be traditional monophagy, would also be a function of the grain of the environment in terms of plant species diversity and the patch structure of the plant species found in that habitat? Given the great taxomonic diversity of lowland tropical floras, high harvestable productivity, and great structural diversity of plant species in the tropics, the response surface available to insects is expected to be more heterogeneous and textured in the tropics than for most comparable habitats in Temperate Zone regions.

Habitat selection can be viewed as the pattern of niche differentiation exhibited by individuals of a particular species and in which the resulting spatial isolation works in concert with geographical isolation in causing the local differentiation of populations. It is the mechanism by which the individuals composing the breeding population of a particular species come to exploit different resources within or between habitats. The adaptive significance of such behavior is that different segments of a population essentially nonoverlap in terms of resource utilization, lowering the effects of intraspecific competition or allowing a population to become so scarce in terms of density per patch that the deleterious effects of interspecific competition can be avoided. The problem, however, is to recognize and detect those cases of habitat selection in which a genetic differentiation has occurred within the population so that different genotypes are exhibiting different preferences in the habitat(s). This is important because many cases of habitat selection may reflect short-term behavioral responses without genetic differentiation, possibly precluding the evolutionary response altogether. Thus not all cases of habitat selection involve shifts in genetic structure within populations. Maynard Smith[(111)] has shown that habitat selection can act to preserve genetic variation in populations. In *Drosophila* populations pronounced differences in allozyme gene frequencies and chromosome inversion frequencies often correspond to differential usage of the environment.

The inverse relationship between allozyme diversity and species diversity in species of *Drosophila* in Hawaii[(392)] suggests that areas of high species richness are relatively unpredictable environments so that genetic variation in local populations is unnecessary. Such data complement the notion that high amounts of some kinds of genetic polymorphisms in populations function to allow the species to occupy a range of microenvironments or exploit a range of resources, notably in areas where the environment fluctuates unpredictably and niches and resources are both changing in frequency and distribution. The role of environmental fluctuations on the genetic structure of populations has been discussed by Levins.[(3,381)] In groups of tropical butterflies exploiting a range of larval host plants locally, under some conditions polymorphisms may regulate the ecological flexibility to exploit a range of such resources, although under other circumstances high phenotypic flexibility per individual may be the optimal strategy.[(275)] The selective advantage to polymorphic individuals in tropical butterflies that participate in mimicry associations has been studied.[(393)]

The relevance of a discussion of the distributional patterns of insects in tropical habitats to the general subject of habitat selection is the potential application of the concept to explaining the high diversity of insect species in tropical habitats. The individuals forming the population of a particular species are subject to a variety of ecological factors, including competition, predation and parasitism, and subtle changes in the physical environment. Habitat preference largely becomes a matter of the species positioning itself in some optimal way within the existing environment so as to maintain a breeding population at or near the carrying capacity. Explaining the richness of insect species encountered in various sweep samples (e.g., those of Janzen[(72)]) is the task of explaining patterns of resource partitioning within and between habitats by insect species in each trophic level. Taxa capable of existing in more than one kind of habitat might exist, while others are habitat specialists. Specialization, or lack of it, is a function of the behavioral plasticity available among individuals in the population, genetic variation in the population for long-term responses, and the ability to use either the same resource among habitats or different resources in different habitats. Given the observed high levels of insect taxa in tropical habitats and over much of the tropics in general, the explanation for this diversity must be accounted for in terms of broad biogeographical phenomena promoting regional differences in species composition of faunas and also the niche-differentiating effects of local patterns of environmental heterogeneity such as within- and between-habitat resource diversity. The issue involves examining differentiation in ecological habits within a single population of a species and relating patterns to other species in the same habitats.

5.2. EXAMPLES OF HABITAT DIFFERENTIATION IN THE TROPICS

It is safe to say that the lowland tropical wet regions of the world probably contain the greatest diversity of habitats of any region, tropical or extratropical. Much of this diversity is the result of local instability in forest environments, creating light gaps, a variety of secondary habitats, and ephemeral habitats associated with the edges of rivers and streams. Connell[(65)] has suggested that the lowland tropical rain forest ecosystem is unstable in the sense that there is considerable replacement of plant species due to ecological processes such as tree falls, mortality of seeds and seedlings, and floodings. In many localities a series of disturbed and undisturbed habitats is expected to be found, some a result of natural perturbations in the ecosystem, others the result of man's activities such as agriculture and lumbering. Thus the lowland tropics is actually a mixture of many different kinds of habitats, some characterized by a wealth of plant and insect species, others by considerably lower biotic diversity. In these contexts I define *habitat* as a distinct assemblage of plant species occupying an area with more or less recognizable boundaries and probably occupied by a distinct assemblage of other organisms. The habitat may comprise several distinct microhabitats occupied by smaller subsets of the species characterizing the whole habitat.

Tropical habitats, particularly those in the nonseasonal or mildly seasonal lowlands and mid-elevations, are characterized by complex assemblages of different communities of plants and animals, reflecting the complex trophic organization of these ecosystems. The combination of relatively long stable environmental conditions and the dynamics of coevolved interactions between groups such as plants and insects have paved the way through evolutionary time for the evolution of complex communities. Many communities in tropical terrestrial habitats are relatively more complex than similar Temperate Zone communities in terms of number of species and the degree to which a portion of the community's membership are specialist species in terms of exploiting resources. In general, the species making up a tropical community may be more in equilibrium with their resources than species in a Temperate Zone community. In either the tropical or temperate region, different communities exhibit varying degrees of species complexity, specialization of member species, and degree of coevolved interactions. Yet on the average, the lowland and mid-elevational regions of the tropics contain more complex communities. The greater such complexity, in terms of numbers of species and ecological specialization, the greater the vulnerability of the community to changes in the environment in which it thrives. Relatively small changes in physical factors in a lowland or mid-elevation tropical habitat may have large, drastic changes on the community: the equilibrium between species and their resources can be shifted in such a way as to either promote the loss of one or more species from the community or result in a replacement of one or more species with invading species. Ecological specialization in some communities, and high species number, promotes conditions favoring an unstable equilibrium, and sustained perturbations, even if small in magnitude or intensity, may lower or shift diversity.[65] Such effects of environmental changes may impact less in many Temperate Zone communities, but because a spectrum of community complexity patterns is found at any latitude, this is not universal. Vulnerability to environmental perturbations will also vary with trophic status and the degree to which member species are specialists within the trophic level.

Habitats and their component microhabitats are both open systems, with energy flow within them and with other habitats. Thus epiphytic tank bromeliads are one kind of microhabitat within the primary tropical rain forest habitat, and there is certainly an exchange of energy with other components of that habitat, mediated by insects and other animals that may breed or develop within this microhabitat but also interact with other microhabitats. The death of such a bromeliad makes it a component of the ground litter microhabitat. Depending primarily upon the permanency of the habitat and the food chain complexity it contains, resident species will be specialists or generalists, or some mixture, for a given trophic level.

Richards[394] provides detailed descriptions of the floristic composition and structural organization of tropical forests, with an emphasis on the lowland tropical rain forest. Holdridge[395] and Holdridge *et al.*[396] discuss in detail the relationships between climate and other features of the physical environment as determinants of the flora of different tropical forests. It is easy to conceptualize the tropical forest as a single phenomenon with three distinct forms: rain forest, cloud forest, and semideciduous forest. Given these distinctions, tropical forests are most developed in the equatorial regions of Central and South America, Africa, and Indo-Malaysia. The vertical stratification of the rain forest is stressed, and the fact that groups of invertebrates associated with the soil in Temperate Zone forests are associated primarily with the vegetation in the tropical forest.

The ground fauna of the lowland tropical rain forest appears to be somewhat impoverished (Table 5.1). In semideciduous forests there are seasonal shifts in the density and species composition in some groups of invertebrates, with contractions of faunas in the dry season (Fig. 5.1).

Most of our knowledge of the distribution of insect species in tropical forests comes from studies of the faunas associated with the lower strata of primary forest or the low canopy faunas associated with relatively young secondary succession. Explorations and systematic surveys of the canopy faunas of primary forest will undoubtedly reveal not only new taxa, but also different types of ecological interactions among species. For example, in some tropical forests, substantial accumulation of debris is expected in the canopy or just beneath it, and these materials may function as food or domicile for various insects associated with the canopy. In the realm of risky speculation, insect groups less abundant in the ground litter and lower vegetational strata of primary-growth rain forest may reappear in great abundance in the canopy communities. Thus, rather than some insect taxa of the lowland tropics experiencing range expansions through horizontal movements, some may filter through the canopy via vertical movement, experience considerable ecological diversification there, undergo horizontal shifts along elevational gradients, and so forth. Such geographical shifts may involve a return to a ground-dwelling habit, especially if habitats at high elevations are less structurally complex.

Clearly, insect taxa also have distributed themselves across all the major vertical strata of a lowland tropical rain forest as well as among all habitats in a

TABLE 5.1
A Comparison of Mite and Collembola Densities in Temperate and Tropical Forest Soils[a]

Habitat	Mean density per square centimeter	Reference[b]
Temperate		
Beech mor	8.61	van der Drift (1951)
Spruce mor	14.6	Forsslund (1948)
Fen soil	3.19	Macfadyen (1952)
Hemlock mor	2.92	Wallwork (1957)
Tropical		
Rain forest (Brazil)	1.0–2.6	Beck (1967)
Sclerophyl forest (Chile)	1.0	

[a] From Wallwork.(786)
[b] References can be found in Wallwork.(786)

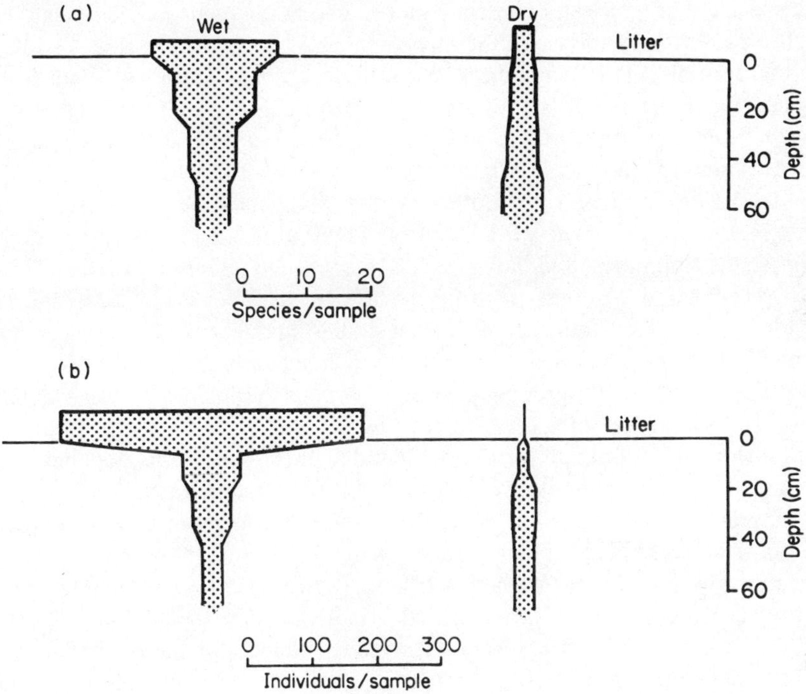

FIGURE 5.1. The vertical distribution patterns of cryptostigmatid mites in wet and dry seasons in El Salvador rain forest soil. (a) Number of species per sample; (b) number of individuals per sample. [From J. A. Wallwork, *The Distribution and Diversity of Soil Fauna*, Academic Press, London. Reprinted with permission.]

region. Ehrlich and Raven[6] discuss the patterns of coevolution between butterflies and their host plants, indicating that the great physical and chemical diversity of angiosperms in the tropics is a major determinant of the great diversity of butterflies found there. They argue that it has been primarily the evolution of secondary plant substances, especially in the tropics, that has driven the evolutionary radiation in many groups of phytophagous insects, including butterflies. In not all cases is oviposition preference in butterflies governed solely by chemical and mechanical stimuli from plants. Wolfson[397] found that the oviposition preferences of *Pieris rapae* were determined largely by leaf water content, with variations in leaf water content being the source of selectivity in ovipositing butterflies. Richards[394] has stressed the great diversity of leaf structure in tropical vegetation, pointing out considerable diversity in terms of leaf toughness, thickness, pilosity, etc. Some of these differences may reflect differences in leaf water content, and the flora of a lowland tropical forest is expected to provide a rich spectrum of leaf water content patterns among the plant species found there. Such a pattern only increases the environmental heterogeneity in terms of promoting specialized or generalized oviposition habits in tropical plant-feeding insects, including butterflies. Such complexity adds an additional layer of resource heterogeneity for these insects over the existing spectrum of chemical defenses among plant species in a habitat.

Insects, being small poikilotherms, are believed to be successful ecologically in terms of coping with physical features of the environment such as temperature, humidity, and moisture. But most of the selection pressures leading to diversification through evolutionary time have been the adaptations of phytophagous insects to the flora. Recent studies on the natural history of tropical butterflies support this view in the sense that not only are certain higher taxa associated with specific groups of plants, but considerable host plant specialization exists within certain genera and species of butterflies. A species, through the medium of the individuals in the population, and through the population as a functional unit in a community, adapts to the local environment by adapting to the physical environment, to potential or actual competitors, to a particular spectrum of resources (including food), and to the removal of genotypes from predation and parasitism. The joint phenotypic responses to these environmental factors molds the strategy of reproductive isolation in a species.

The view of Ehrlich and Raven[6] gives us a handle on the broad patterns of ecological relationships between phytophagous insects and their host plants. However, in trying to explain why a particular insect feeds on a particular host plant species it is necessary to examine the character states for the four classes of environmental effects mentioned here. The interplay of these factors and the constraints of genotype and phenotype frequencies within a population are what determine patterns of habitat selection in species of insects in the tropics and elsewhere. Where closely related species exhibit distinct patterns of host plant utilization within a habitat or occupy similar resources or patches in different habitats, we must examine these factors in determining the selection pressures for the observed character displacement. Spatial character displacement over relatively short distances in and among habitats within a region is one mechanism of accumulating a high number of species in a region. See Chapter 7 for a consideration of patterns of seasonality in tropical insects, a form of temporal character displacement.

Patterns of local character displacement among closely related species in tropical habitats may involve not only phenotypic traits such as host plant preference and shade tolerance, but also the ecological factors underlying population growth. A species of leaf hopper attacking trees in a forest understory may experience the environment very differently from an allied species attacking herbaceous plants in an adjacent disturbed habitat. The ease of nutrient retrieval may be different: the leaf hopper in the forest habitat allocates more energy to feeding and therefore has less energy for evolutionary increases in fecundity. These effects, coupled with the existence of larger host plant patches in the disturbed habitat, may result in a buildup of larger populations of the leaf hopper species in the disturbed habitat. Predation intensities may also vary, as well as the number and density of competing species on the same host plants. Thus evolutionary changes in the population growth parameters (see Chapter 6) in different directions in allied species in different habitats are phenotypic traits reflecting character displacement and habitat selection.

Some studies of aquatic insects in the tropics, such as the study by Hynes[(398)] of aquatic Heteroptera in Trinidad and Tobago, also indicate considerable niche diversification among species. But this particular survey includes both geographical and habitat collections, combining the effects of both factors in determining niche divergence within a habitat. Geographical differentiation within a particular group is expected, sometimes to the degree that selection pressures within a particular region will be markedly different than elsewhere in the range. The example of large, flightless weevils with plant growth on their elytra in the moss forests of Papua, New Guinea (Gressitt *et al.*[(399)]) exemplifies the effects of local adaptation to a set of environmental conditions not found elsewhere in the range of this subfamily of weevils. The phenotypic traits present on Papuan weevils indicate strong selection pressure favoring well-developed crypsis, probably as a result of the flightless conditions of the insects. Sedentary insects in tropical forests may be more vulnerable to predators than more mobile species because they are deprived of escape behaviors associated with movement. Silberglied and Aiello[(400)] studied the wettable body surfaces of some bark-inhabiting bugs on Barro Colorado Island in Panama; wettability conceals the bugs from predators by making the bugs match wet bark in the forests. These insects are relatively sedentary, spending most of their time on wet bark.

Jiron and Zeledon[(401)] report on the distribution of *Anastrepha* (tephritid) flies in various kinds of fruits in Costa Rica and demonstrated considerable overlap among three species of the flies for several fruits examined. The effect is regional: the species occur in different regions of the country but exploit some of the same fruits. Many species of drosophilid flies are commonly seen on rotting and fleshy fungal growths in tropical wet forests, including species that breed there and those whose adults only feed there. The occurrence of such food sources in tropical forests provide a microhabitat for considerable co-occurrence in drosophilids. Drosophilids are also found on rotting fruits in such habitats. Buruga and Olembo,[(402)] for example, found that large, fleshy fruits supported the greatest number of species of *Drosophila* in forest habitats in Uganda. Experiments in progress by Young in tropical wet forest in Costa Rica indicate that considerable habitat selection occurs in some groups of fruit-eating butterflies. In one study, a transect of rotting banana baits was set up in both primary forest and relatively open second growth and the numbers of Morphidae,

Satyridae, and Brassolidae were scored for as long as the baits lasted. Very little overlap between habitats was found, and the overall number of species of these butterflies in the second growth habitat was low compared with the number in the forest habitat. Three species of satyrids were seen at the second growth baits, as were a few individuals of *M. peleides*. In the forest habitat, three species of *Morpho* were found on the baits and at least three genera of satyrids and two genera of brassolids. Such data simply confirm something that most tropical biologists have known for some time, namely, that different groups of butterflies are associated with different habitats, primarily because their resources are not distributed evenly among all habitats in a region. In tropical communities with a high number of ecologically similar taxa, some pattern of resource partitioning is expected if some resources are in limited supply. Sometimes the partitioning is evident within a particular habitat or is more diffuse at the regional level. Thus Walter[(403)] discovered considerable evidence of resource partitioning among 28 species of dung beetles in Zaire, and these ecological differences were exhibited in terms of nest sites, flight patterns, and other phenotypic traits related to feeding and breeding. The observed patterns of differentiation were most evident on a regional basis where the species overlapped considerably.

In the central highlands of Costa Rica, some species of cicadas partition the local environment along rather distinctive boundaries. For example, *F. pronoe* emerges among coffee bushes adjacent to a forest-lined stream, whereas two others, *Q. gigas* and *Z. smaragdula* emerge on the trees along the stream, only a few meters away (Fig. 5.2). The mechanisms responsible for such patterns have not been determined, but similar surveys of cicada emergences in premontane tropical wet forest by Young[(71)] suggest that cicadas exhibit considerable divergence in host tree preferences; the nymphs may associate with the root crowns of different trees. Young was able to recognize those cicada species that emerged in secondary growth and those emerging in primary growth habitats in that region. Young also suggested that the creation of secondary habitats in the wet tropics may provide the opportunity for some cicada species, and probably many other insects, to invade new environments as they become available as a consequence of land-clearing activities of man and natural events such as major floodings of rivers.

Schuster[(404)] discovered that flower-visiting Orthoptera in the tropics often

FIGURE 5.2. Cicada habitats in the central highlands of Costa Rica. Above: strips of *Zygia latifolia* trees (Leguminosae) that border streams (to the left), with which cicada emergences are associated. Below: discarded nymphal skins of the large cicada *Q. gigas* clinging to a *Zygia* tree. In heavily disturbed habitats in Central America, cicada emergences are associated primarily with forest remnants as shown here, although some species have secondarily become associated with leguminous shade trees over crops such as coffee. The association is probably defined in terms of suitable oviposition sites (branches) and root crowns for suitable nymphal development.

exhibit considerable preferences for certain plant families. Working in the tropical wet regions of Peru and the Canal Zone, he concluded that orthopterans may seriously reduce the breeding capacity of the plants visited because the insects eat the anthers, probably as a source of protein. The study of Fraser Rowell[404a] shows that acridids (Orthoptera) in Costa Rican lowland tropical rain forest are specialist feeders on plants. Fraser Rowell suggests that great differences in secondary compounds between different plant species in such environments precludes the evolution of a generalist feeding strategy in the acridid species found in this habitat. I have observed that some orthopterans are opportunistic predators on insects at lights in Costa Rican wet forest regions, and one gets the impression that associations of orthopterans with flowers are not specialized associations but rather ones that occur opportunistically. Schuster[404] does make an interesting point about the activity possibly reducing the pollination of the attacked flowers by bees, since structures are destroyed and pollen is taken. A similar phenomenon with *Trigona* bees as the bandits, may be occurring in some cacao plantations in Central America. Young[405] found that *Trigona jaty* collects large amounts of pollen from cacao flowers but does not pollinate the flowers. The pollination of cacao is done primarily by ceratopogonid midges whose behavior of visiting the flowers is such that pollination can result.[406] Yet the high densities of *T. jaty* in some cacao plantations suggests that the bees are actually robbing pollen that would otherwise be available for reproduction. (See Chapter 9 for a more in-depth discussion of this complex matter.) An interesting further consideration concerns the kinds of habitats adjacent to cacao plantations that may alter the foraging habits of *T. jaty* on cacao. For example, I discovered a plantation bordered by dense flowering patches of the composite *Schistocarpha oppositifolia* in which very large numbers of *T. jaty* bees were foraging. On the adjacent cacao, the bees were virtually absent. The tentative conclusion of such descriptive studies is that the availability of alternative resources in one habitat will influence the foraging habits for another resource in an adjacent habitat. Our knowledge of insects in tropical habitats must include an attempt to understand how different sets of habitats, with differing resources, influence the patterns of exploitation of particular resources within each component habitat. Few studies are designed around such a goal.

Some kinds of agricultural habitats in the wet tropics may support a diversity of microhabitats, as suggested by Laessle's study[407] of the bromeliad flora of several cacao plantations in Trinidad. In that study, the cacao plantations were similar in design and the composition of the bromeliad flora and shade characteristics of habitats were studied for caco plantations and adjacent strips of forest habitat. The data obtained (Table 5.2) indicate considerable heterogeneity in the response patterns of different bromeliads to the three kinds of habitats: high-shade forest, low-shade forest, and very low shade cacao plantation. The author recognizes three different ecological groups of bromeliads on the basis of tol-

TABLE 5.2

The Density of Bromeliads per Five Trees of *Eyrthrina micropteryx* in Cacao Plantations in Four Distinct Rainfall Zones of Trinidad[a]

	Density				
Bromeliad	A	B	C	D	p values of indicated density changes[b]
Catopsis sessiliflora	2.1 ± 0.5	0.5 ± 0.5	0.6 ± 0.4	0.2 ± 0.2	C to F; $p < 0.05$
Vriesia amazonica	2.5 ± 0.5	0.0	0.1 ± 0.1	0.0	C to D, E, F; p negligible
Vriesia procera	7.8 ± 0.7	6.2 ± 1.6	7.3 ± 0.8	3.7 ± 1.8	No significant change
Guzmania monostachia	6.9 ± 0.8	8.8 ± 0.5	8.5 ± 0.4	7.1 ± 1.3	No significant change
Gravisia aquilega	8.0 ± 0.7	9.3 ± 1.0	8.5 ± 1.0	5.0 ± 2.2	No significant change
Wittmackia lingulata	(1.5 ± 1.0)	8.9 ± 4.3	6.9 ± 1.7	1.2 ± 1.2	E to F; $p < 0.05$
Aechmea nudicaulis	8.3 ± 0.8	13.0 ± 2.3	20.1 ± 2.3	8.3 ± 2.0	E to C or F; $p < 0.02$
Tillandsia fasciculata	5.6 ± 0.7	8.2 ± 1.2	7.5 ± 0.7	10.6 ± 1.3	C to F; $p < 0.05$
Tillandsia utriculata	0.0	0.0	2.1 ± 0.5	3.9 ± 0.8	C & D to E or F; p negligible
Tillandsia subimbricata	0.0	0.0	1.3 ± 0.6	6.4 ± 0.5	C & D to E or F; p negligible
Number of samples (of five trees)	12	6	10	6	

[a]From Laessle.[407]

[b]The values given are means of square-root transformations of actual frequencies per five trees.

TABLE 5.3
The Distribution and Abundance of Bromeliads in Two Forest Strips Differing in Light Intensity and in an Adjacent Cacao Plantation in Trinidad[a]

	Forest 1	Forest 2	Plantation
	Dark due to lianes	Well lit by river at side	Strongest illumination also drier
Exposure group			
C. sessiliflora	0	4	49
C. floribunda	0	8	3
V. procera	8	28	144
V. amazonica	3	5	5
Sun group			
T. fasciculata	0	1	61
G. monostachia	8	60	101
Aechmea Mertensii	1	5	0
Tillandsia bulbosa	8	10	2
G. aquilega	0	28	142
A. nudicaulis	6	79	201
Guzmania sanguinea	0	1	0
Hohenbergia stellata	0	4	0
Shade-tolerant group			
Tillandsia anceps	4	34	0
Tillandsia monadelpha	1	26	0
Vriesia albiflora	0	1	0
Vriesia simplex	1	10	0
Guzmania lingulata	48	78	0

[a] From Laessle.[407]

erance to shade (Table 5.3). The implications of such patterns for the insect species associated with the bromeliads are also apparent: tank bromeliads in shaded forest habitats will have larger communities of aquatic and arboreally cryptozoic insects than bromeliads associated with exposed or dry habitats in the same region. But the bromeliads found in cacao plantations may support damp-to-wet leaf litter communities as well as herbivores associated with leaves and flowers. Cacao plantations can support a high density of different bromeliads in shade trees such as *Erythrina* (Table 5.2). The differences in distribution of various bromeliads between different kinds of forest habitats and cacao plantations impose a pattern of environmental heterogeneity for insects and other organisms exploiting these microhabitats.

The studies of M. H. Robinson and B. Robinson[408] described another form of specialized microhabitat in tropical wet forests, the association of Drosophilidae with certain orb-weaving in forest understories. These are commensal associations in which the flies clean the mouthparts and anal areas of the spider and the flies take food from the spider as it is being eaten. Thus the spider and its web form a sort of specialized microhabitat for the flies. Young[408a] and some earlier workers, notably P. Calvert, described the association of certain bromeliad-breeding giant damselflies with the same orb-weaving spiders involved in the commensal associations with flies. Young observed that certain kleptoparasitic spiders in the webs of the orb weavers are taken as prey by these damselflies. Thus these damselflies essentially glean the web microhabitat of kleptoparasites.

Such studies imply that some types of resources are very limited in abundance, suggesting ecological constraints on species specialized to exploit them. Matthews[223,224] studied the nesting behavior of the social wasp *Microstigmus comes* in a tropical wet forest in Costa Rica and found that the wasp selectively nests on the undersides of leaves of certain palms in the forest understory. The wasps require scrapings of tissue from the undersides of these palm leaves to make their nests. Within a patch of approximately 75 palm plants, about 60% had no nests and the remainder one to five nests apiece. The data suggest that the wasp population is spatially distributed within a relatively small area of forest habitat (about 35 × 40 m) where a required nesting resource is located. Nesting site selection in some social insects may also depend upon the degree to which the species can defend itself against enemies.

In a very interesting study, Windsor[409] found that nest site selection in a docile tropical wasp *Mischocyttarus immarginatus* is determined largely by the locations of nests of another polybiine wasp, *Polybia occidentalis,* which is a fierce attacker. Such interspecific nesting relationships show that a docile associate nests immediately adjacent to an aggressive associate species, presumably as a means of protection against predators such as small vertebrates. In ant species with specialized nesting requirements, the distribution of nests in a habitat will be determined in part by the distribution of preferred nesting sites. For example, Young and Hermann[212] found that mature *Pentaclethra* trees in Costa Rican tropical wet forest tend to have a greater number of nests of the giant, aggressive ant *Paraponera clavata* associated with the roots of these trees. But that study did not include an experiment to demonstrate the apparent preference of this tree species over others as a nesting site in this ant. This ant is capable of various kinds of feeding, including expolitation of extrafloral nectaries,[211] and it would be interesting to determine if its nesting habits include some specialized relationship to the root systems of certain tree species in forest habitats. The ant is a subterraneous nester that forages primarily in the forest canopy, thus exploiting a broad range of microhabitats within the forest habitat.

Tropical forest and secondary-growth habitats may contain a wealth of microhabitats in the form of both nesting sites and food for the great many ant species found in these places. Jeanne[410] demonstrated that rates of predation by ants are greater in the tropics than in Temperate Zone areas and that predation rates in dissimilar tropical habitats, such as forest understory and open field, tend to be similar. He found more pronounced habitat differences in predation rates for ants in different Temperate Zone habitats than in tropical habitats (Table 5.4). The greater pools of ant species associated with both forest and secondary-growth habitats in the tropics result in high removal of prey under both conditions, whereas in the Temperate Zone, more of ant species are in the field habitat and therefore predation rates are higher. The availability of many kinds of prey and other food sources and a diversity of nesting sites account for the high species richness of ants in both tropical forest and field habitats. In Jeanne's study,[410] the number of ant species found in the tropical habitats was considerably higher than for the other regions examined (Table 5.5).

Ant species which depend upon the exudates of homopterous insects may be an example of a hypercontingent species as defined by Colwell.[411] A hypercontingent species depends directly upon a relationship between at least two other

TABLE 5.4
Degree of Habitat and Site Specificity among Foraging Ant Species[a]

Shared habitats or microhabitats	New Hampshire	Florida	Mexico	Costa Rica	Brazil
Forest microhabitats[b]					
1	17	57	58	59	57
2	17	13	20	16	15
3	8	10	16	2	17
4	25	20	4	18	7
5	33	0	2	5	4
Field microhabitats[b]					
1	38	30	46	56	46
2	31	40	46	18	27
3	31	40	8	26	27
Field and forest[c]	27	31	16	28	3
Canopy and forest understory[d]	100	50	36	44	44

[a] From Jeanne.[410]
[b] "Forest" and "Field" entries give the number of ant species collected from 1, 2, . . ., in microhabitats, as a percent of all species collected from that habitat at one locality.
[c] "Field and forest" entries give the number of species collected in both habitats, as a percent of the sum of all species collected in the two habitats at one locality.
[d] "Canopy and forest understory" entries give the number of species collected in the two set of microhabitats, as a percent of all species found in the canopy at one locality. ("Canopy" here does not include canopy trees at 15 m.)

TABLE 5.5
Ant Species Richness at Five Localities[a,b]

	New Hampshire	Florida	Mexico	Costa Rica	Brazil
Forest Habitat					
Ground	12(70)	16(72)	23(54)	26(71)	24(78)
Twigs	8(36)	11(42)	15(34)	16(58)	19(66)
Leaves	9(30)	12(49)	19(39)	14(60)	12(71)
Ground + twigs + leaves	12(136)	24(163)	40(127)	33(189)	35(215)
Trunks	5(33)	7(35)	18(44)	13(68)	15(77)
Canopy above 15 m	7(25)	12(42)	11(19)	16(37)	16(22)
Canopy at 15 m	—	—	—	14(29)	17(23)
Entire forest habitat	12(194)	30(240)	59(190)	50(323)	50(337)
Field Habitat					
Ground	8(45)	15(99)	12(67)	18(72)	12(71)
Twigs	11(55)	13(59)	13(53)	20(76)	18(69)
Leaves	12(58)	13(52)	14(53)	20(77)	17(47)
Ground + twigs + leaves	16(158)	20(210)	24(173)	34(225)	26(187)
Buildings	2(2)	7(39)	2(3)	13(60)	7(32)
Total	22(354)	40(489)	66(366)	68(608)	74(556)

[a] From Jeanne.[410]
[b] Data are numbers of species. Number of baits at which ants were collected at each microhabitat, habitat, or locality is given in parentheses.

species. Thus phytophagous Homoptera depend upon plants for food, and the ants obtain exudates from the bugs as a food source. As pointed out by Colwell, the hypercontingent strategy may involve a dependence for food, dispersal, defense, reproduction, or support. Thus the palatable insect species, such as many butterflies in the tropics (Fig. 5.3), some of which mimic highly unpalatable models, are examples of hypercontingent species, since survivorship of individuals in their populations depends on the interaction between host plant and larvae of the model species as the mechanism of unpalatability enforcing the relationship.

Colwell suggests that the interplay of various adaptive strategies in organisms accounts for part of the high tropical species diversity noted in some regions and habitats. He suggests a mixture of sequential specialists, namely, species that differentially exploit resources through time such as seasonal displacement (Chapter 7); interstitial species, which depend upon several different widely dispersed resources; grain specialists, which exploit specific resource patches of one or few kinds in a habitat; and hypercontingent species. Colwell developed these definitions of adaptive strategies from his studies of the interactions of hummingbird species with seasonally available flowers, the association of nectarivorous mites with flowers and hummingbirds, and the influence of a

a

FIGURE 5.3. The fourth-instar caterpillar of the Neotropical nymphalid butterfly *Hamadryas februa* actively feeding on a mature leaf of *Delachampia heteromorpha* (Euphorbiaceae) in the lowland tropical dry forest zone of northwestern Costa Rica (a); dorsal and ventral sides of the adult (b).

nectar thief on the system. The definitions are useful in examining a variety of interactions among species in different trophic levels within tropical habitats.

Tropical communities are believed to consist of species with a variety of different adaptive strategies and in which most of the adaptation is a response to biotic associations rather than physical environment. For example, one might hypothesize that some kinds of endoparasites of phytophagous insects are grain specialists, in that closely related species exploit different hosts and in proportion to their abundances in the habitat. And the abundance and distribution of the insect hosts are a function of the distributions of the various host plant species exploited. Some phytophagous insects in the same habitat, rather than being specialists on one or a few resource patches in the habitat, may be interstitial species in the sense that they feed on a variety of plant species dispersed in the habitat. The coexistence of two possibly competing endoparasites therefore depends, in this hypothetical example (and borrowing from Colwell[411]), upon the grain structure of the environment, determined in this case by the feeding patterns of the phytophagous insect species. The assumption is made, of course, that each species of endoparasite is a specialist on a particular insect host, and this may not be the case if most endoparasites of phytophagous insects in the tropics are generalists. Few data in either direction are available. In the first instance, specialist endoparasites would be considered hypercontingent species because their existence in the habitat depends upon the successful interaction of their hosts with their food plants.

As pointed out in Section 5.1, a major logistical problem with determining the degree of niche overlap among ecologically similar species in a habitat is the level of empirical resolution to detect differences. Several studies of resource partitioning in natural populations of microsympatric *Drosophila* species have emphasized this matter. Two or more closely related and ecologically similar insects may overlap tremendously in terms of diet, and the populations of each species may be large. The question of co-occurrence necessitates the examination of other components of comparative natural history to determine possible mechanisms of resource partitioning, assuming resources of limited supply. Thus two closely related chrysomelid beetles in a tropical field may coexist on the same host plant species with complete overlap of diet in terms of host plant selection. But resource partitioning may exist in terms of the spatial exploitation of the host plant population. Most individuals of one beetle species prefer to feed on host plant individuals with a certain degree of exposure, while a second beetle species prefers the opposite level of shade on the plant. Or each species may feed on different parts of the plant, in which case considerable nonoverlap of niches may exist at the level of secondary substance types and concentrations experienced by feeding beetles, and perhaps also in terms of nutrients received.

Other forms of spatial separation of the potentially competing species could include differing preferences for young versus older individual plants. If it can be demonstrated that the host plant is a limited resource to both species of beetles in that particular habitat, the observed patterns of resource partitioning can be assumed to result from competition between the species. We would also have to demonstrate absence of feeding on other plants and the degree of movement of beetles in other habitats, where the relationships of each species to the flora may be very different. Once the pattern of resource partitioning was determined, we would need to examine phenotypes of each species to determine whether morphological, behavioral, or physiological differences are correlating well with the observed spatial partitioning. The latter is not easy to do in most cases. Both species may respond to temporal changes in the habitat in the same way, adjusting to daily influences of thermal gradients and shade gradients, and doing so in such a way as to maintain an optimal spacing pattern of individuals within the population.

Good data are available on the overall species compositions of the insect faunas associated with various terrestrial habitats in tropical regions, including a few studies comparing faunas over altitudinal gradients. I am referring here primarily to the sweep sample studies of Janzen,[(72)] Janzen and Schoener,[(134)] and Janzen *et al.*.[(412)] There are also data on the specific host associations of some insect taxa such as seed-eating beetles and bugs, and larval host plants of Lepidoptera. We still need to determine the niche breadth of selected insect species in a habitat and to relate these measurements to competition, predation, and spatial and temporal availability of resources. Where very different kinds of habitats occur together in an area, it is also necessary to determine the amounts of species overlap among habitats, movement patterns of individuals among the habitats, and changes in the natural history of individual species due to differences in habitats for species that occur in more than one habitat. Such measurements should be incorporated into some of the major theories of maintenance of high species diversity in the tropics and the lack of high species diversity in some areas of the tropics and temperate zones. The plant–insect interface seems to be a good starting point, using the approach of comparative natural history.

5.3. INSECT COMMUNITIES IN TROPICAL HABITATS

The studies of Futuyma and Gould[(413)] indicate that few plant species in the Temperate Zone habitat studied have distinct assemblages of leaf-chewing insects associated with them. Many insects in such samples tend to be generalists on many plant species. Typically a Temperate Zone deciduous forest has considerably fewer species of plants than most lowland tropical forests, and a partial explanation for a lack of correspondence between leaf-chewing insects and host plant species is that the average insect species in such a habitat must generalize to feed on several different plant species in order to maintain a breeding population.

Although such a habitat does have a complement of specialist species,[413] ecological specialization in most phytophagous insects under such conditions may be more difficult to evolve than in a tropical forest habitat, where a greater number of plant species per family occurs and the balance might be partially tipped toward specialization as reflected in monophagy and limited polyphagy (defined here as the exploitation of two closely allied plant families). Specialization in tropical forests also depends upon the patch structure of each plant species relative to the population growth requirements for maintaining breeding populations.

Murdoch *et al.*[414] found that insect diversity in a Michigan old field correlated very well with plant diversity and plant structure. There is reason to expect similar correlations in tropical communities: both parameters of plant communities are higher in most habitats.

With the exception, perhaps, of some defogging studies of the Orthoptera associated with the canopy of Costa Rican and Amazonian tropical rain forests, the sweep-sample studies of Janzen[72] in Costa Rica and the Caribbean provide a picture of insect communities associated with different types of foliage in different tropical habitats. The reader is referred to Janzen,[72] Janzen and Schoener,[134] and Janzen *et al.*[412] for in-depth analyses of the extensive data obtained with the sweep-sample approach. The findings of these studies indicate the effects of seasonality, habitat type, elevation, and island effects on the composition of insect communities in the tropics. Some of the findings are discussed here.

A basic comparison in terms of distinct habitats is that between insects in secondary vegetation and in primary forest understory at a particular site. For example, in tropical wet forest in Costa Rica there is a drastic reduction in numbers of both bug species and individuals of each species for adults going from secondary to primary forest understory; the transition is far less noticeable for adult beetles.[72] The greater abundance of nonwoody stems in secondary habitats promotes a greater diversity of plant-sucking insects such as adult Homoptera, whereas leaf-chewing adult beetles have leaves in high abundances in both habitats. Janzen found little overlap in species composition between the two habitats. Interestingly, Janzen found the greatest diversity of insect species at intermediate-elevation habitats. He explains that there is a higher harvestable productivity in plant communities at intermediate elevations due to reduced plant maintenance costs on cool nights. Above intermediate elevations (near 3400 m), the physical factors are more rigorous and there is a drastic reduction in the number of species and individuals and responses to decreased diversity of flora. Lowland tropical areas with a long and severe dry season show a great decline in insects during the dry season, with many insects moving horizontally into moist forest refugia as the dry season advances. Areas with a mild dry season show increases in insects during the dry season, suggesting that the wet conditions of the wet season tend to suppress insect populations, and many species may exhibit greatest reproduction, survivorship from biotic factors, and flushes of population growth in the dry season.

Areas with pronounced dry seasons are characterized by insects passing the dry period in the adult stage and in reproductive diapause.[116] Because vegetative growth in most plants ceases during the dry season, insect populations decline or move to other habitats where the vegetation remains lush to some degree. Janzen[72] found reduced numbers of insect species and numbers of individuals per species on islands in the Caribbean and a greater percentage of predatory insects, principally ladybird beetles. The reduced floral diversity of these islands cannot support a phytophagous insect fauna comparable in complexity to those found in mainland tropical habitats; thus the proportion of predatory species increases. The greater ecological specialization in tropical herbivorous insects makes it more difficult for these species to establish and maintain populations in island habitats, much more so than for generalist predatory species.[415] The relatively

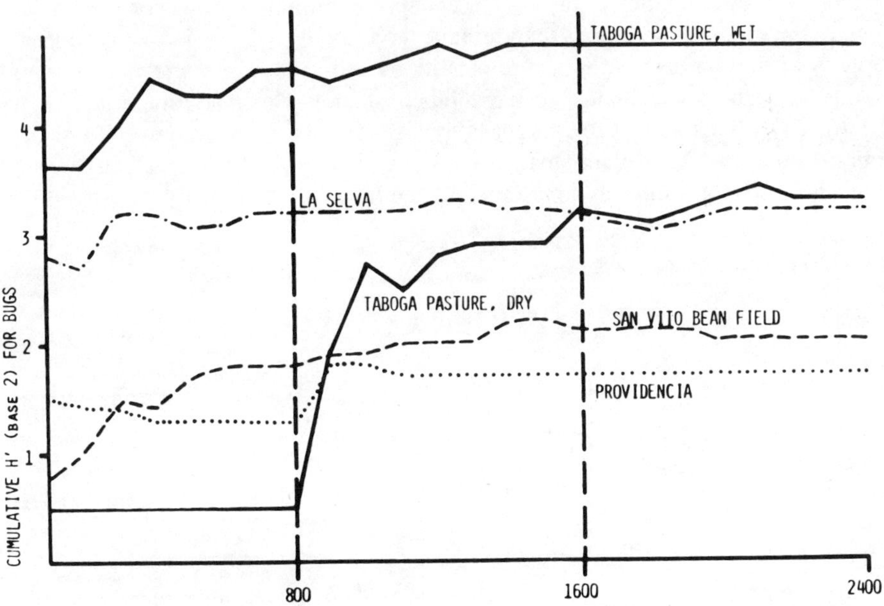

FIGURE 5.4. The cumulative species diversity index, H′, for species of adult bugs in 100-sweep increments for several localities in Costa Rica. See E. C. Pielou. The measurement of diversity in different types of biological collections, *J. Theor. Biol.* **13:**131–144, (1966), for a discussion of H′ and how it is calculated. [From D. H. Janzen, *Ecology* **54:**687–708 (1973). Copyright 1973 by the Ecological Society of America.]

greater flexibility in feeding habits may provide predatory insects with the resilience needed to survive unfavorable periods associated with colonization of islands[415] and may preadapt them to surviving dry seasons in such habitats.

The sweep-sample studies demonstrate that not all tropical terrestrial habitats will support the same kinds of insect communities associated with vegetation (Figs. 5.4 and 5.5). Trophic organization and insect diversity appear to be greatly influenced by the composition and structure of the vegetation in a habitat and the degree to which the harvestable productivity of the habitat changes with seasons. Insect species composition and population sizes within a particular habitat will also be influenced by the size distributions of adult insects present. Mean body size, and the variation of this statistic within an age class of a certain species in one habitat, influences how the population's membership can distribute itself upon the available resources and also affects interactions with individuals of other species. Schoener and Janzen[416] report that the range of size variation in insects is greatest in communities from tropical habitats. They found that spatial heterogeneity in moisture and fluctuations in the abundance of resources in time are two features of habitats that correlate well with observed differences in the size distributions of insects. Insect species from drier areas or from areas with longer growing seasons tend to have larger body sizes, whereas those from areas with more uniform moisture conditions and less distinct growing seasons (e.g., lowland tropical rain forest) exhibit considerably less variation in size (Fig. 5.6). Drier habitats promote larger body sizes in insects to circumvent dessication limits associated with small body sizes, and larger body sizes also result when growing seasons are long in some habitats; an evolutionary adjustment in

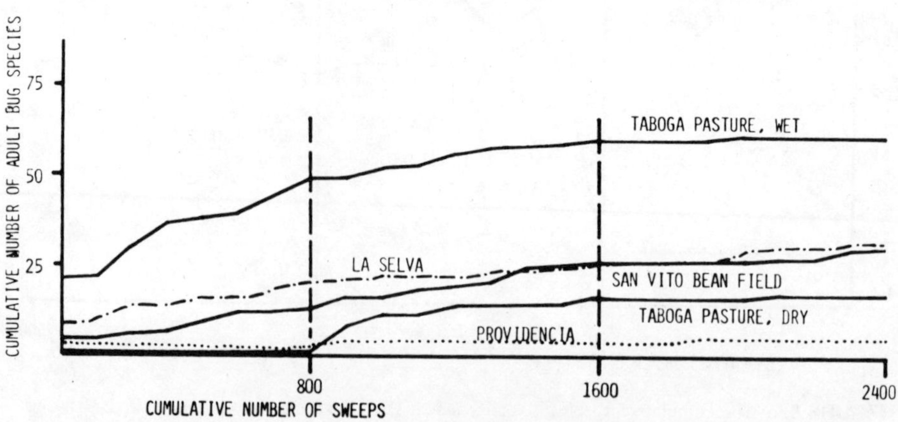

Figure 5.5. Cumulative number of adult bug species in 100-sweep increments for several localities in Costa Rica. [From D. H. Janzen, *Ecology* **54:**687–708. (1973). Copyright 1973 by the Ecological Society of America.]

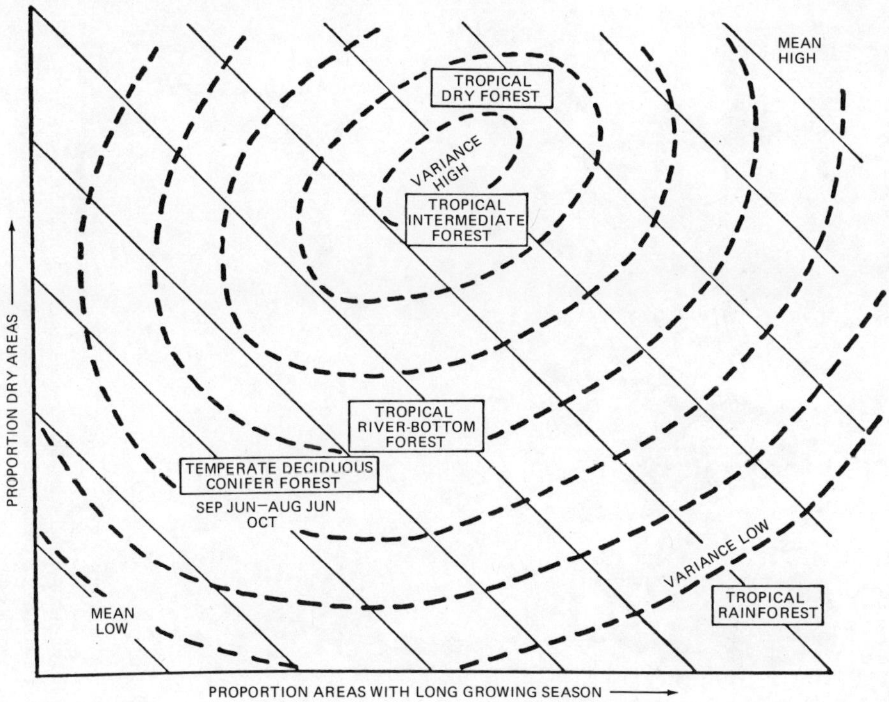

FIGURE 5.6. The effect of length of growing season and proportion of dry areas on the distributions of insect lengths. (———), Means; (- - -), standard deviations). [Reprinted from "Notes on environmental determinants of tropical versus temperate insect size patterns" by T. W. Schoener and D. H. Janzen, *American Naturalist* **102:**207–224 (1968). Reprinted by permission of the University of Chicago Press.]

developmental time can be positively correlated with the length of time habitat conditions are favorable for growth.

The greater variance in body size expected for insects in tropical rain forest regions can be examined from the standpoint of size variation within one group, such as a family. Take Neotropical cicadas as an example (Fig. 5.7). At both lowland and premontane wet forest sites in northeastern Costa Rica, five genera and 11 common species of cicadas represent a body length range of 10–55 mm with a mean of 23.5 ± 20.1 (S.D.). In the northwestern dry lowlands, four genera and 7 common species represent a body length range of 10–35 mm and a mean of 18.4 ± 12.4 (S.D.). There appears to be a greater range of body lengths for the common and characteristic cicadas in lowland tropical rain forest.

Of the mosaic of habitat types in the lowland tropical rain forest regions, some habitats are expected to comprise communities with many specialist

FIGURE 5.7. Shown here, from left to right, for both adults and final-instar nymphal skins, are *F. mannifera, F. sericans, F. amoena, F. pronoe, Z. smaragdina, Proarna sallei,* and *Proarna* sp. Interest centers on the determination of patterns of temporal and spatial utilization of the environment in the tropics, using these insects as a model system.

species, while others may contain major representation by generalist species. There is no reason to predict that all habitats in such a region will be dominated by specialist species, including phytophagous insects. For example, the edges of streams in the lowlands may favor generalist species of carabid beetles, which are then capable of becoming specialists in other habitats.[417] The stream edge habitat supports generalist carabids, which then can diversify into other habitats and undergo radiative adaptation, exhibiting both morphological and behavioral specializations to new selection pressures. Species of marginal habitats may be preadapted for such radiations into other habitats. Of interest is how the generalist species arrived in these marginal habitats in the first place. Stream-edge habitats are probably very old in the sense that they have existed along with the forests for long periods, giving time for the accumulation of generalists in them. These species might have been derived originally from the undisturbed forest faunas, responding to directional selection favoring invasion into these marginal habitats.

Young[71] has noted that some species of cicadas in lowland and premontane tropical wet forests in Costa Rica (and presumably elsewhere) occupy stream-edge habitats, and some of these species are also invading disturbed habitats resulting from the clearing away of forests for agriculture and lumbering. In these habitats, population densities of these cicadas may be very high, indicating successful colonization. The implication from such observations is that species adapted to marginal habitats such as stream edges in the lowland wet tropics also may be preadapted to colonize other kinds of disturbed habitats in addition to undergoing radiative adaptation to more undisturbed habitats such as primary forests. But apparently such invasions into other disturbed habitats do not necessarily result in ecological specialization and divergence of the kind expected for transitions to undisturbed habitats, as disucssed for carabids by Erwin.[417]

5.4. PLANT RESOURCES AND HABITAT SELECTION IN TROPICAL INSECTS

Plant species in different tropical habitats exhibit different growth profiles, different reproductive structures, and varying degrees of response to seasonality. Within a lowland tropical rain forest region, for example, the degree of species nonoverlap may be high between different habitats and considerable structural differences should exist among the plant communities in each habitat. Thus both the upper strata and the understories of primary forest habitats tend to be dominated by woody plants, trees, and vines of various sizes and patch sizes. Young, disturbed habitats less than 5 years old are dominated by herbaceous plant species, while various kinds of older secondary habitats are dominated by mixtures of herbaceous and woody trees and shrubs.[396]

Not only will differences among habitats in a region be expressed in terms of the collection of the chemical defense systems and mutualisms between ants and plants that affect the taxonomic composition of phytophagous insect communities in the tropics, but the spectrum of physical differences among plants will also influence habitat selection in these insects. Janzen,[72] for example, has suggested that the marked decrease in adult bugs in the forest understories of lowland wet forest is due to an absence of herbaceous stems, which represent a readily penetrable resource for these xylomycetophagous and phloeophagous insects. How the diversity of plant species in a particular tropical habitat will determine the diversity of phytophagous insects is expected to vary for different groups of insects, primarily on the basis of differences in feeding habits. The combined effects of less restrictive physical properties of plants such as woodiness, a large pool of taxonomic and defensive chemistry diversity, and a heterogeneous patch structure are expected to result in a high diversity of phytophagous insects in secondary habitats in the tropics. The harvestable productivity of secondary habitats is greater than that of forest understories, providing a larger food base for insects. Whether the canopies of lowland forests are also characterized by higher harvestable productivity similar to that of some secondary habitats remains to be analyzed, and the insect communities associated with canopies warrant measurement.

The data of Beaver[418] indicate that for the beetle families Scolytidae and Platypodidae the tropics contain a relatively higher number of polyphagous insect species than do Temperate Zone regions. The greatest increases are in two groups of these beetles, defined by the feeding habits: the xylophagous species, which live and feed on wood; and the xylomycetophagous species, which usually live in wood but cultivate and feed on ambrosia fungi growing in the galleries of these insects. The mechanism of specificity is therefore determined by the host specificity of the fungi in the case of xylomycetophagous species but can also be determined by the oviposition behavior of the beetle, which molds the pattern of host tree selection in both groups. The same study revealed that phloeophagous species, those that construct galleries and breed in the phloem and inner bark of trees, are more diverse in temperate zones, and this pattern is explained in that paper by the differences in host tree patch structures in temperate and tropical forests.

The hyperdispersion of tropical tree species selects for the generalist strategy since distances between individuals of a particular tree species (of right age, etc.) are too great for specialization. Yet patch sizes of tree species in temperate-zone forests, which have fewer tree species per unit area, are large and provide larger segments of biomass for several species to utilize. The gradient appears to be opposite what one would expect from the tree species-rich tropics, where selection should favor ecological specialization in phytophagous insects. But the key factor in such considerations is the patch structure and relative patches sizes for each tree species, relative to the dispersal abilities and reproductive cycles of the beetle species involved in these interactions. These beetles may not come into contact with the diversity of secondary substances generally present in leaf and meristem tissues of plants and are precluded from the same sorts of host plant selection mechanisms affecting the majority of phytophagous insects. Put differently, a forest patch of known size containing 10 tree species in a tropical region and forest patch of the same size but containing tree species in a Temperate Zone (and comparable numbers of individual trees in both patches) are not viewed differently by scolytid or platypodid beetle species present in each habitat.

The host tree selection mechanism is based on other signals in the trees rather than based on secondary substance chemistry associated with leaves and meristems, the kind of selection mechanism predicted. Although secondary substances might be involved, they are likely to be different from those of other vegetative structures, perhaps of kinds found in the phloem tissues of most tree species. Yet in considering the likelihood for specialization in phytophagous insects associated with leaves and meristems, it is necessary to assess dispersal and life table characteristics in order to determine the evolutionary advantage or disadvantage for specialization on a host plant species with a certain patch structure. Low population turnover coupled with low dispersal ability may act to favor the generalist strategy in many phytophagous insect species associated with tropical habitats characterized by a high diversity of plant species. Under such conditions the generalist strategy would include the physiological ability to feed on a variety of unrelated plant species in a small area of habitat whose boundaries for the insect species in question are set by the dispersal abilities of adults and the population growth parameters of the species in that particular habitat.

Clearly some studies of Temperate Zone insects such as butterflies have revealed few if any patterns of habitat selection based upon distributions of resources such as adult and larval host plants.[(419)] Rausher,[(420)] however, studying habitat selection in three species of *Aristolochia*-feeding butterflies in Mexico, found significant differences in the survival of eggs and larvae of these species between different habitats. Survival of eggs and larvae was higher in shady areas than in exposed areas (Figs. 5.8 and 5.9), and selection should favor oviposition in shady habitats. But Rausher found that two of the species actually preferred to lay eggs in exposed habitats (Fig. 5.10), presumably locked into host plant cues having little or no relation to survival of eggs and larvae.

Breedlove and Ehrlich,[(327)] studying a Temperate Zone butterfly population, concluded that in some cases the intensity of predation on the plant species in a region may be so high as to be a source of selection regulating the distribution of the plant populations among habitats. But the pattern may be heterogeneous (Dolinger *et al.*[(421)]) in that if the butterfly is not "tracking" the plant populations, there will be considerable variation in the phenotypic commitment within different plant populations in the expression of secondary substances used in

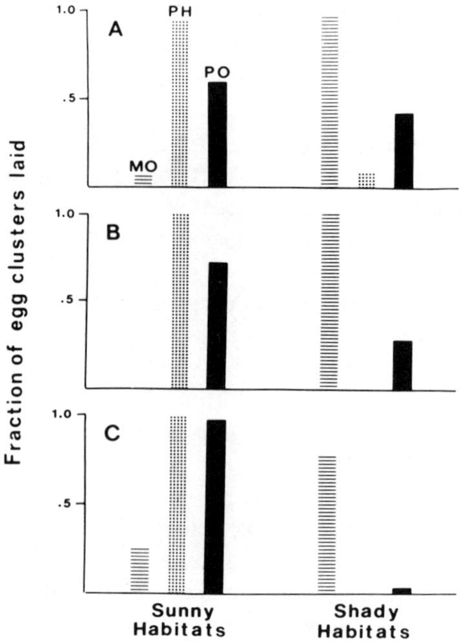

FIGURE 5.8. The distribution of egg clusters in sunny and shady habitats for three Neotropical papilionid butterflies, *Parides montezuma* (MO) (shaded bars), *Battus philenor* (PH) (stippled bars), and *Battus polydamas* (PO) (black bars) at two localities, El Encino (A, C; two different study periods) and Pico de Oro (B) in Mexico. There is a definite preference for *B. polydamas* and *B. philenor* to oviposit in sunny habitats and for *P. montezuma* to oviposit in shaded habitats. [From M. D. Rauscher, *Ecology* **60:**503–511 (1979). Copyright 1979 by the Ecological Society of America.]

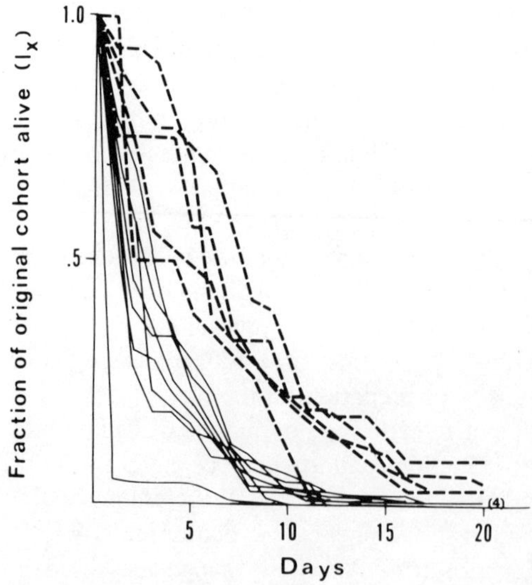

FIGURE 5.9. Patterns of egg and larval survivorship for *P. montezuma, B. philenor*, and *B. polydamas* for all combinations of these species and habitat patches studied. Solid lines represent curves for sunny habitats and dashed lines for shady habitats. Day zero is the day the egg cluster was discovered in the wild. The "(4)" in the figure indicates the point of convergence for all four curves, thereafter represented as one line. [From M. D. Rauscher, *Ecology* **60:**503–511 (1979). Copyright 1979 by the Ecological Society of America.]

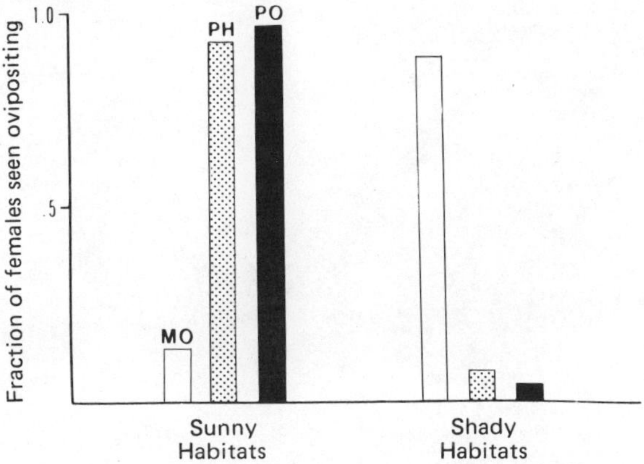

FIGURE 5.10. The proportions of females of *P. montezuma* (MO), *B. philenor* (PH), and *B. polydamas* (PO) searching for oviposition sites in sunny and shady habitats over all observation periods at two localities in Mexico. [From M. D. Rauscher, *Ecology* **60:**503–511 (1979). Copyright 1979 by the Ecological Society of America.]

herbivore defense. Thus populations where the butterfly is absent or scarce have low alkaloid profiles, and where abundant, high profiles.[(421)] The plant species might be able to occupy different habitats and the butterfly only a subset of these.

Another factor to consider is the availability of alternative host plants in a habitat. Chew,[(422)] working with montane *Pieris* butterfly species in the Northern Temperate Zone, found imprecise correlation between oviposition behavior and the suitability of host plants for larval growth. She concludes that the patchy distribution of some crucifer species used by the butterflies promotes the use of less suitable crucifers. Habitat hereogeneity, in the form of both suitable and less suitable host plant species having patchy distributions within the habitats studied, promotes the exploitation of all the host plants by the average female in each species. It is almost as if the environment is providing too much information for the female butterfly to process per unit time during the oviposition periods.

Young[(313)] found some evidence of habitat separation in heliconine butterfly species in Costa Rica. For example, although *Eueides lybia* and *Eueides aliphera* are sympatric and utilize the same larval host plant, *Passiflora vitifolia,* in the premontane tropical wet forest region of northeastern Costa Rica, the former species oviposits primarily along shaded forest edges while the latter does so primarily in open secondary habitats. In this case the larval host plant occurs in more than one kind of habitat, while the two butterfly species, probably potential competitors, have different habitat preferences, presumably mediated by tolerance to sunlight and related factors. Other patterns for heliconiines (Fig. 5.11) have also been noted utilizing other *Passiflora* species.[(313)]

FIGURE 5.11. The heliconiine butterflies *Dryas iulia* (above) and *Eueides aliphera* (below), which occur together regionally in Central American rain forests but utilize different larval food plants (Passifloraceae).

Slater[423,424] analyzed the vertical distribution of seed-eating lygaeid bugs in South Africa and the West Indies and found four ecologically distinct groups based on feeding behavior: (1) arboreal seed predators, (2) obligatory terrestrial seed predators, (3) frequent facultative seed predators, and (4) accidental terrestrial seed predators. He noted that most of the species in these groupings belong to one subfamily. The evolutionary divergence of these bugs into these ecological groups may have resulted from competition among the bugs for the same kind of resources, and the groups reflect differing amounts of ecological specialization, with accidental terrestrial seed predators being the most flexible group in terms of utilizing resources other than fig seeds. An important component of such interactions involving many insect species is the area of habitat available for the species to thrive in and the amount of time available for ecological divergence.

Janzen[375] has shown that the dissolution of mutualism between *Azteca* ants and *Cecropia* trees in Puerto Rico, compared with the well-developed mutualism present in many mainland Central American localities, is an adaptive response to reduced competition among plant species in *Cecropia* habitats and reduced herbivore pressure on the trees on islands. Some studies, (e.g., Lawton[425]) indicate that vegetation complexity, in terms of plant species composition and the structural diversity of plants present in the habitat, is a good indicator of the diversity of phytophagous insects present. Habitats with high structural diversity and species composition may accumulate a greater diversity of insects in certain trophic classes.

Young[215] found that many species of Neotropical cicadas will form dense aggregations of adults in certain trees in disturbed habitats, while such aggregations are far less numerous in forest habitats (Table 5.6). It was suggested that cicadas in disturbed habitats select on such trees for optimal mating sites and feeding and that the absence of aggregates in undisturbed forest is due to the patchy distribution of each tree species found there, making it difficult so to speak for cicadas to discover and congregate in those trees that optimize feeding and mating. Thus certain kinds of behavior in insects will also influence how populations are superimposed upon available resources, again with the patch structure of the vegetation in each habitat being a major determining factor.

5.5. DISTURBED AND UNDISTURBED HABITATS: SOME GENERAL EFFECTS

Temporary or rapidly changing terrestrial habitats in the tropics will present very different types of selection pressures from more stable habitats. For example, wide oscillations in the abundance of resources for insects might result in large fluctuations in abundance as a consequence of such temporal environmental heterogeneity. The study of Irmler and Furch[426] on cockroach populations in an Amazonian inundation forest illustrates the dependence of scavenger insects occupying such changing habitats on more than one kind of food source, namely, decomposing leaves and decomposing animal tissues. Both kinds of resources are necessary for juvenile roaches to receive the correct balance of nutrients, but both are expected to fluctuate in abundance with flooding cycles.

In some temporary habitats such as a crop field, large populations of relatively few species of phytophagous insects might be exploited by several species of parasitic insects. The insects feeding on the crops are derived from the local fauna associated with either other kinds of disturbed habitats or undisturbed habitats. The degree of parasitism present in populations of a particular species will depend in part upon the kind of habitat where the population occurs. For example, the degree of parasitism within a population in an undisturbed habitat may be considerably lower since a parasite species such as a tachinid fly is expected to be a generalist on many different host species within relatively small patches of forest. The biomass per host species is low per unit area of habitat because each host plant species is hyperdispersed; the result is a great diversity of potential host insects accumulating over small patches of forest. The greater the disruption of the habitat,

TABLE 5.6

Abundances of Adult Cicadas on Isolated and Aggregated Individuals of Various Tree Species in Costa Rican Tropical Forests[a,b]

Tree	Season of peak adult activity	Number of single trees	Number of days of observation	Total number of cicadas	Mean number of cicadas ($\bar{X} \pm$ S.D.)	Number of tree aggregates observed	Range of number of trees in aggregates	Number of days of observation	Total number of cicadas	Mean number of cicadas ($\bar{X} \pm$ S.D.)
					F. sericans					
Pourouma aspera (Moraceae)	Dry	37	13	103	2 ± 2.4	12	5–20	10	650	51 ± 10.5
Goethalsia meiantha (Tiliaceae)	Dry	10	5	18	2 ± 1.6	1	54	23	735	30 ± 8.6
					Z. smaragdina					
	Wet	10	8	27	3 ± 0.6	1	54	52	1483	31 ± 11.3
Pentaclethra macroloba (Leguminosae)	Wet	32	27	138	5 ± 7.4	3	6–11	20	753	35 ± 9.6
					F. pronoe					
Vernonia patens ([illegible])	Dry	12	10	23	2 ± 3.2	4	5–30	10	420	38 ± 11.4

					Q. gigas					
	Dry	12	10	11	1 ± 0.5	4	5–30	10	184	17 ± 8.3
					F. pronoe					
Gliridicia sp. (Leguminosae)	Dry	19	8	15	1 ± 1.4	1	35	8	512	60 ± 7.5
					Diceroprocta sp.					
Tamarindus indica (Leguminosae)	Wet	10	6	10	1 ± 0.4	1	28	5	105	19 ± 5.3
					Q. gigas					
Zygia latifolia (Leguminosae)	Dry	4	14	0	0	1	22	17	2530	124 ± 15.6
					Z. smaragdula					
	Wet	4	12	0	0	1	22	15	568	37 ± 10.2

[a] From Young.[215]

[b] With the exception of the last two tree species in this table, all observations were made at "Finca La Tirimbina," near La Virgen de Sarapiqui, Heredia Province, in the premontane tropical wet forest zone. The observations for *T. indica* were made at Playas del Coco, Guanacaste Province, and for *Z. latifolia* at San Rafael de Ojo de Agua, San Jose Province.

going from advanced secondary to monoculture or simple polyculture (two to three) crop arrangements, the greater the number of parasite species the habitat can support per unit area. In disturbed habitats, parasites are therefore expected to be specialists, since the biomass (harvestable productivity) of the few host insect species present is large enough to support breeding populations of several parasite species. There might be considerable niche overlap among parasite species under such conditions since the resource base is large enough to be exploited by several parasites. This argument is developed for spatial changes in the environment, in that different kinds of habitats will have different species composition and abundance of phytophagous insects and such variations will also be manifested in the diversity and abundances of parasitic insects attacking these herbivores. Miller[427] developed a similar argument based upon temporal changes in habitats in a Temperate Zone environment.

Depending upon the trophic level considered, highly disturbed habitats in the tropics may contain specialist or generalist insect species. As discussed above, the lower abundance of some plant groups and larger patch sizes per plant species expected in some disturbed habitats may select for specialist feeding in many phytophagous insects, assuming the ability to locate suitable patches and the ability to adjust evolutionarily population growth parameters (see Chapter 6) to prevent overexploitation of harvestable productivity while maintaining breeding populations. But if other kinds of resources are patchy and patch sizes are small, the generalist strategy may be optimal for some insects. For example, predatory insects and opportunistically omnivorous insects such as carabid beetles may be generalists in some disturbed habitats. In terms of prey actively sought by predatory insects, the patch structure of these resources is defined largely by size ranges of the prey and their distributions in the habitat. It is therefore difficult to generalize about the relative frequencies of generalist and specialist insects in tropical habitats without considering trophic level and distribution and abundance of preferred resources.

The degree of specialization, of course, will set the limits on the amount of flexibility there is among individuals in a population of a particular species for the occupancy of more than one kind of habitat. The foraging habits of many social Hymenoptera probably involve the retrieval of prey from more than one kind of habitat. It was discussed in Section 5.2 that the great diversity of ant species in the tropics is due in part to the exploitation of arboreal nesting sites, not to mention the great horizontal diversity of available microhabitats in some habitats. Ant species that have entered into specialized mutualisms with other orgnisms are expected to have definite patterns of habitat selection, while ant species with flexible nesting and foraging requirements may occupy more than one kind of habitat in an area. Leaf-cutting ants may nest in one kind of habitat, forage there, and also forage in adjacent habitats. The extreme in habitat flexibility is attained in some species of army ants, whose raiding columns and bivouac locations can be found in different habitats. The degree of raiding in army ant colonies varies directly with the amount of food intake, but emigration rate does not appear to fluctuate under different food intake regimes. The food-gathering component of emigration behavior in army ants may not be the major functional role of such movement patterns in these insects.

Foraging leaf-cutter ants retrieve leaf material from many tree species over a large area encompassing more than one habitat type (Table 5.7). Foraging in leaf-cutter ants also has a strong diurnal component (Fig. 5.12), workers leave nests late in the day and return later at night with booty. Young and Hermann[(212)] found a similar foraging pattern for the giant ponerine ant, *Paraponera clavata,* involving both prey and various plant materials as booty. The marked vertical stratification of the tropical wet forest habitat may result in different resources being spatially separated for a particular species. A good example is *P. clavata,*

TABLE 5.7
The Distribution and Abundance of Plant Species as Indicated by Leaf Sampling, Encompassing an Areas of about 1200 m² around the Leaf-Cutter Ant Nest Studied[a]

Species	Number recorded	Percent frequency
Marlierea montana (Aubl.) Amsh.	87	21.2
Terminalia amazonica (J. F. Gmell) Exell.	26	6.3
Eperua falcata Aubl.	16	3.9
Pithecellobium eperuetorum Sandw.	16	3.9
Ocotea schomburgkiana (Nees) Bth.	13	3.2
Calycolpus glaber (Bth.) Berg.	10	2.4
Ecclinusa psilophylla Sandw.	10	2.4
Licania buxifolia Sandw.	10	2.4
Chondodendron candicans (Rich) Sandw.	9	2.2
Catostemma fragrans Bth.	8	1.9
Ischnosiphon gracilis Koern	8	1.9
Emmotum fagifolium Desv.	7	1.7
Cassia pteridophylla Sandw.	6	1.5
Eperua grandiflora (Aubl.) Bth.	6	1.5
Anaxocorea petiolata R. E. Fr.	5	1.2
Myrcia silvatica (Mey.) D.C.	5	1.2
Tovomita cephalostigma Vesque	5	1.2
Calycolpus pyrifer Riley	4	1.0
Dugetia inconspicua Saget.	4	1.0
Pouteria filipes Eyma.	4	1.0
Tapirira guianensis Aubl.	4	1.0
Simaba cedron Planch.	3	0.7

(*continued*)

TABLE 5.7 (*Continued*)

Species	Number recorded	Percent frequency
Talisia squarrosa Radlk.	3	0.7
Vismia guianensis DC.	3	0.7
Alexa leiopetala Sandw.	2	0.5
Aspidosperma excelsum Bth.	2	0.5
Casearia densiflora Bth.	2	0.5
Catostemma altsoni Sandw.	2	0.5
Connarus lambertii (D.C.) Sagot.	2	0.5
Duroia eriopila L.F.	2	0.5
Eschweilera corrugata (Poit.) Miers.	2	0.5
Matayba opaca Radlk.	2	0.5
Neea constricta Spruce ex Schmidt	2	0.5
Oxandra asbecki (Pulle) R.E.Fr.	2	0.5
Ryania speciosa var. *subuliflora* (Sandw.) Monachino.	2	0.5
Simaba multiflora A. Juss.	2	0.5
Tabebuia capitata (Bur. & K. Schum) Sandw.	2	0.5
Tachigalia rusbyi Harms.	2	0.5
Unonopsis guatterioides (A.DC.) R.E.Fr.	2	0.5
Xylopia benthamii R.E.Fr.	2	0.5
Byrsonima incarnata Sand.	1	0.2
Clusia fockeana Mig.	1	0.2
Cupania scrobiculata L.C.Rich	1	0.2
Cybianthus brownii Gleas.	1	0.2
Eugenia essequiboensis Sandw.	1	0.2
Guatteria schomburgkiana Mart.	1	0.2
Himatanthus bracteatus (D.C.) Woodson	1	0.2
Hirtella davisii Sandw.	1	0.2
Maprounea guianensis Aubl.	1	0.2
Marlierea schomburgkiana Berg.	1	0.2
Miconia guianensis (Aubl.) Cogn.	1	0.2
Nectandra grandis (Mez.) Kosterm.	1	0.2
Pouteria cladantha Sandw.	1	0.2
Pouteria egregia Sandw.	1	0.2
Pouteria engleri Eyma	1	0.2
Pouteria jenmani (Pittier) Sandw.	1	0.2
Rinorea flavescens (Aubl.) Kze.	1	0.2
Smilax schomburgkiana Kth.	1	0.2
Tapirira marchandii Engl.	1	0.2
Toulicia pulvinata Radlk.	1	0.2
Weigeltia surinamensis (Spr. f.) Mez.	1	0.2
Lianes (unidentified)	51	12.4
Unidentified	34	8.3
61 species TOTAL	410	

[a] From Cherrett.[581]

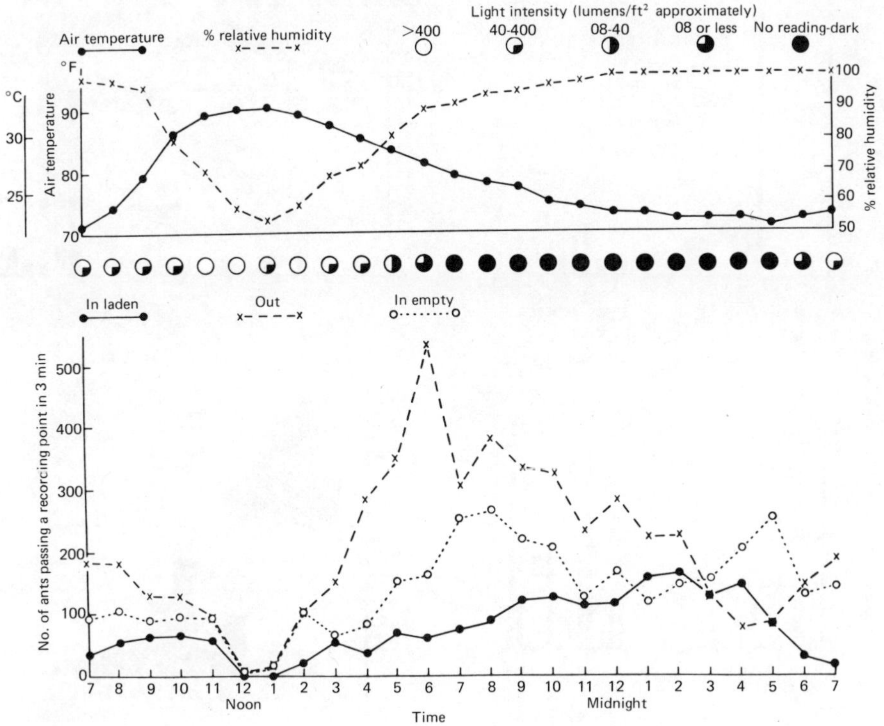

FIGURE 5.12. The daily cycle of foraging activity in a Neotropical leaf-cutter ant, *Atta cephalotes*, and associated climatic data. [From J. M. Cherrett, *J. Anim. Ecol.* **37:**387–403 (1968). Copyright 1968 by the British Ecological Society.]

a species that nests beneath certain species of canopy-size trees but which concentrates most of its foraging in the canopy and does so with a marked diurnal cycle of activity.[(212)] In a study of the nesting and foraging of the Neotropical ant *Zacryptocerus varians,* Wilson[(428)] hypothesized that arboreal nesting resulted from competition with ground-dwelling ants and scavenging resulted from competition with other arboreal predaceous ants. Wilson was able to associate other behavioral (and specialized) traits with these aspects of habitat utilization. Gotwald[(429)] showed that the diet composition in three species of African driver ants was determined primarily by habitat, with the major clases of prey being insects, arachnids, and earthworms. The relative abundance and accessibility of particular prey types also determined diets[(430,431)] (Fig. 5.13).

Carroll and Janzen[(197)] report that the abundance of soft wood stems is a limiting factor in the abundance of arboreal nesting ant species in tropical deciduous forests. They suggest that the abundance of ant nests in tropical forests

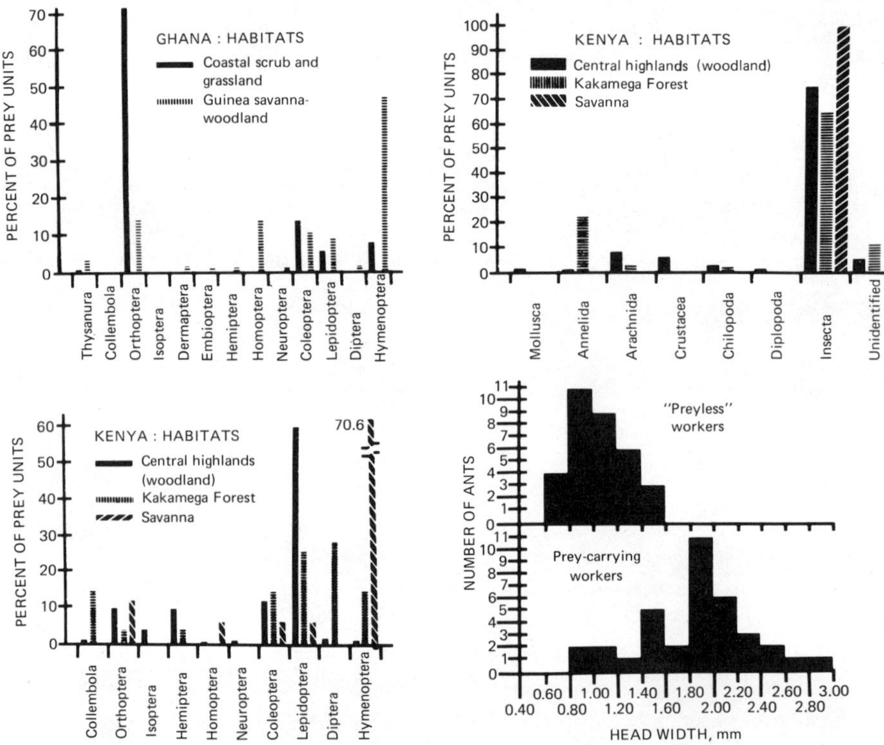

FIGURE 5.13. Taxonomic composition of prey taken by African driver ants, *Dorylus* (*Anomma*) species in Ghana and Kenya habitats, and the frequency distributions of head widths of "preyless" workers and prey-carrying workers of *D.* (*A.*) *nigricans* var. *molesta*. [From W. H. Gotwald, Jr., *Ann. Entomol. Soc. Am.* **67:**877–886 (1974). Copyright 1974 by the Entomological Society of America.]

will vary directly with the harvestable productivity of the forest patch available for colonization. Harvestable productivity will be a function of the kinds and distributions of potential resources in the patch and the amount of competition for the resources with other ant species. Carroll and Janzen(197) suggest that small colony size in tropical ants may result not only as an adaptation for nesting in small cavities but also as a niche contraction resulting in food specialization under selection pressure from competitors. Competition for food is believed to be a major ecological factor determining the local ant diversity in tropical forests. Studies of ant nest distributions in Temperate Zone grasslands (e.g., Yasuno(432)) indicate that intraspecific competition for nesting sites accounts for considerable spacing in a rather uniform habitat. Similar patterns may exist for ant species in tropical grasslands if colonies are very large. But the more heterogeneous nature of most other terrestrial habitats in the tropics probably reduces the possible role of intraspecific competition for both nesting sites and food to nothing.

Other social Hymenoptera may exhibit specific habitat associations based largely upon the kinds of food found there. Wille[21] discusses the specialized foraging behavior of certain species of *Trigona* bees for gathering pollen from *Cassia* trees. The distribution and abundance of *Cassia* trees among habitats at a locality will contribute a major portion of the grain structure of the food environment for these bees. Most species of *Trigona* are fairly opportunistic foragers at many different plant species in tropical forests, and collectively these bees probably acocunt for a very large percentage of the pollination in lowland tropical rain forests such as found on the Osa Peninsula of Costa Rica (A. Wille, personal communication). *Trigona* bees probably fit Colwell's definition of interstitial species in the sense that they forage on a variety of plant species in habitats where patch sizes are relatively small and each plant species is highly dispersed in the habitat. Many species of *Trigona* bees collect pollen from tubular anthers of many different plant species by biting through the anther and releasing the pollen grains. The grains are picked up with the proboscis, then gathered on the thoracic sternum, from which it is transferred to the corbicula. A. M. Young[677a] has found that at least two species of *Trigona* exhibit this form of pollen-collecting for the introduced ornamental vine *Thunbergia grandiflora* (Acanthaceae) in Costa Rica. The bees also rob nectar from this vine by biting through the base of the tubular corollas before hummingbirds visit near dusk, an activity that probably reduces pollination frequency considerably. In Costa Rica the vines are propogated only by cutting, as there is no fruit set.

Different species or groups of species of *Trigona* bees exhibit various forms of pollen-collecting behavior, and some species are capable of changing this behavior for the exploitation of different flowers with different anther structures. It is perhaps this very flexibility in feeding that has resulted in the evolution of aggressive behavior among species of *Trigona* occupying the same habitats. Johnson and Hubbell[433] found that several species of these bees exhibit both intraspecific and interspecific aggressive behavior toward flowers and artificial baits that result in considerable lowering of the amount of time bees spend on flowers. Johnson and Hubbell[433] found that aggression increases directly with increases in the sugar concentrations of artificial baits. Such behavior reduces the amount of nectar or pollen that bees gather on individual visits to flowers in the wild, and results in a spatial partitioning of the habitat so as to reduce aggressive encounters. The ability of many species to exploit the same species of flowers promotes the aggressive-behavior syndromes as a means of enforcing spatial separation of foraging areas among different species or different colonies of a single species in a habitat. Such aggression is expected to increase in habitats where many bee species occur and where the diversity of flowers is relatively low through time, or where the few flowering species available at any one time have small patch structures and therefore low, harvestable productivity for bees.

Because pollen is presumably rich in a variety of nutrients whose concentrations and types vary considerably among different plant taxa in the tropics, pollen itself is a critical resource for many insects. As studies of flower-visiting insects in the tropics continue, elucidating the patterns of exploitation of pollen by many insects other than bees will be possible. Gilbert[31] has determined that some species of *Heliconius* butterflies feed on pollen, which in turn acts as a source of amino acids and proteins used in the production of eggs by the butterflies. That female ceratopogonid midges visit cacao flowers suggests that these insects also use pollen as a source of amino acids for eggs. We must also examine the kinds of selection pressures operating in such species to promote the expression of pollen feeding and the accompanying physiological mechanisms involved in the assimilation of pollen amino acids into egg proteins in these insects. Gilbert,[31] for example, argues that pollen feeding evolved in some *Heliconius* to counteract the heavy predation on immature stages in populations, the net result being that fecundity is increased evolutionarily by such a mechanism. H. G. Baker and I. Baker[216,217] have emphasized the coevolution of pollinators and plants on the basis of pollen and nectar nutrient concentrations, and it is necessary to examine such relationships for a variety of pollinating insects, including bees, lepidopterans, beetles, and flies. Specialized forms of pollen feeding may involve insects utilizing only certain plant species, and such interactions may result in considerable habitat selection. If a particular pollen source is found only in a certain kind of habitat, the insect exhibits a preference for that habitat. One might expect that specialized pollen-feeding associations would be most frequent in rather stable habitats where such resources may be most predictable over long periods of time. The strategies of flower foraging developed by Janzen[40] are the kind that might involve specialized pollen feeders. Less stable or changing habitats may select for flower visitors that are not specialists on certain species. Temporal and spatial predictability of flowers for each plant species seems to be a requirement for the evolution of specialized flower-visiting strategies in insects, and some habitats might be better for this than others.

CHAPTER 6

POPULATION RESPONSES TO THE ENVIRONMENT IN TROPICAL INSECTS

The association of an insect species with a particular habitat means that the species maintains a breeding population in that habitat. We therefore need to distinguish between resident and nonresident species associated with different habitats. The great differences in the taxonomic composition and structure of vegetation in disturbed and undisturbed habitats in tropical regions, discussed in detail by Holdridge *et al.*,[(396)] indicate that the composition of local insect faunas may differ greatly between habitats in the same region, a phenomenon well documented by the studies of Janzen and Schoener[(134)] and Janzen.[(72)] Thus different collections or assemblages of insects and other vegetation-dwelling arthropods are expected for different kinds of terrestrial habitats in tropical regions.

6.1. ADAPTATION TO HABITATS

In order to understand how different habitats may support species-rich assemblages of insects in the tropics, and the differences in species abundances among different habitats in a tropical region, it is necessary to consider how the breeding populations of individual species "fit in" to the habitat. Natural selection determines various kinds of so-called life history patterns of organisms as a function of varying environments. Life history refers to the age-specific fecundity and mortality schedules and other parameters that go into the formulation of the growth rate of populations. Natural selection is expected to mold these ecological parameters within populations in such a manner as to increase the likelihood that the species will survive in a particular habitat or environment. Because the phenotype is a compromise, there is always a tradeoff in the allocations of resources to different traits such as vegetative maintenance, reproduction, feeding, and defense against biotic mortality factors. The patterns of resource allocation among these and other phenotypic traits affecting population growth will vary in different kinds of environments with different forms of selection pressure.[434] Natural selection favors increases in fecundity, defined here either as the age-specific schedule of births (eggs) or the total number of births in an individual's lifetime, adult survival (from biotic and abiotic environmental factors), and juvenile survival (egg to age of first reproduction for most species). But there might be compensatory tradeoffs between these ecological parameters, with an increase in one through evolutionary time being accompanied by a decrease in another during the same period. Again, the phenotype is a compromise in terms of total resources available for various functions related directly to fitness.

Shifts in the environment may change selection pressures, which in turn alter the resource allocation profiles for individual genotypes in the population. Selection pressures affecting fitness parameters associated with population maintenance and growth will include fluctuations in food availability and accessibility, fluctuations in the physical environment, shifts in the abundance and density of competing organisms, and shifts in the intensity of predation and parasitism. Shifts in genotype frequency resulting from emigrations and large fluctuations in population size will at the same time alter the degree of genetic and phenotypic resiliency in the population to respond to these selection pressures. In both the lowland seasonal and nonseasonal tropics, changes in food availability, abundance of competing species, and changes in the intensity of predation or parasitism are probably the major classes of selection pressures affecting the population structures and dynamics of insect species in disturbed and undisturbed habitats. At the high altitude (above mid-elevation) regions of the tropics, fluctuations in the physical environment, with manifold effects on resource availability, may be the major form of selection pressure. Changes in fecundity, developmental time, and longevity in response to severe fluctuations in the physical environment, including long-term unfavorable growing conditions, may be very different from the changes needed to maintain a population due to shifts in competition or predation. I use the term *maintain* in this context to mean "keep growing or remaining stable in size in a particular habitat."

Lewontin[87] discussed the differences in various kinds of tradeoffs arising from shifts in certain populational parameters over others for species exhibiting colonizing episodes. Thus a tradeoff between adult survival and increased fecundity may depend on the gain in fecundity per unit of adult survival lost, coupled with the direction of selection. Lewontin[87] showed, using data from *Drosophila,* that about a 10% increase in developmental time is the same as a 100% increase in fecundity, implying that selection pressures would favor the former phenotypic change in increasing the growth rate of the population in a new environment. Selection is expected to favor earlier reproduction for a species invading a new habitat, but eventually population growth rate might decrease as the species becomes integrated into the community, and the shift might then be toward postponing reproduction. MacArthur and Wilson[341] pointed out that the relative value (increase in fitness) of decreasing developmental time and increasing fecundity depended upon the net reproductive rate (R_0), since that latter parameter is an estimate of the total average fecundity of individuals in the population. For example, if R_0 is relatively low within a certain population, the advantage for increasing fecundity would not be as great as for a higher value of R_0 in another population, or within the same population at different time. But in some stable populations increases in adult survival are the optimal solution to increasing fitness.

Increasing fecundity or decreasing developmental time appear to be most adaptive in populations in the early phases of population growth. Thus, depending upon the stability of the population, selection will act to increase fitness by affecting different parameters underlying population growth. The avenue of response embarked upon by the species will be a function of the type of habitat occupied. We need to distinguish between relatively stable and unstable tropical habitats and the unique assemblages of insect species found in each of them. It is not at all clear whether primary forests in the lowland tropical regions are stable from the standpoint of turnover in floras at the local (habitat) level. The data seem to indicate the opposite.[65,435] Yet certain kinds of secondary tropical habitats may be incredibly stable, with little or no turnover in plant species composition.

From the viewpoint of phytophagous insects and the insects that feed on them, shifts in plant species in some tropical habitats are a form of selection pressure perhaps favoring colonization to more suitable areas. The total insect faunas may change under such conditions, but so may the population dynamics of individual component species. Some tropical insects may in fact consist of shifting mosaics of plant species and associated insect species (and their parasitoids), while others may consist of stable elements. Shifting conditions are expected to result in shifts in the direction of selection affecting the parameters of population growth.

6.2. DEMOGRAPHY AND FITNESS IN POPULATIONS: IMPLICATIONS FOR TROPICAL INSECTS

The basic theory of individual selection is Fisher's Fundamental Theorem of Natural Selection, which states that the rate of change of fitness within a population is equal to the genetic variance in fitness. The rate of evolution is therefore limited by the variation in fitness of the units being selected. The rapidity of response to selection within a population of a species depends upon the heritability of differences in fitness between the units selected. The rate of evolution depends upon the rate of reproduction and mortality. Fitness, defined for the individual in a population, has units of time, as seen from Fisher's Theorem:

$$\frac{d\bar{W}}{dt} = \text{Var}\ (\bar{W}), \text{ so that } dt\left(\frac{d\bar{W}}{W^2}\right) \text{ and } t\left(\frac{1}{W}\right)$$

where W is the mean fitness and Var (W) is the variance in fitness.

A species with a very long life cycle, such as some cicadas, cannot evolve as fast as one with a short life cycle, and the intensity of selection for a unit is directly related to the cycle time for the unit. Thus populational selection must be generally weaker than individual selection since births and deaths of whole populations encompass many individual lifetimes. But it is possible that populations and communities are also units of selection. The interactions of individuals within a population of a single species, and the interactions among individuals of different species and with the physical environment, result in forms of selection pressure not found at the level of the individual. In the tropics, where biotic interactions, often highly specialized, are common in communities with large numbers of genera and species, selection at the population and community levels of ecological organization might be particularly important in determining the structure of communities associated with different habitats in a region. The present discussion, however, is limited to a consideration of life history traits in populations.

As discussed in Chapter 2, natural selection and other factors such as drift, mutation, and recombination can alter the genetic composition of a population. The frequency distribution of genotypes is changed to maximize the average fitness in the population, as the result of individual selection. Each genotype in the population has a fitness value that contributes to the overall growth rate of the population. The measure of heritability provides an estimate of the amount of phenotypic variation in a population that arises from genetic variation and from the distribution of genotypes and their fitnesses. Phenotypic variance is the sum of genetic variance (V_G) and environmental variance (V_E). Phenotypic variance is measured for a given trait in the population. The measure of heritability is given by

$$h^2 = \frac{V_G}{V_P} = \frac{V_G}{V_G + V_E}$$

and it varies from 1.0 to 0.0. An h^2 value of 1.0 results when all of the variation in the population for a given trait is caused by differences among genotypes and no differences result from environmental effects. Clearly, most h^2 values will reflect a mixture of both effects for many traits. The genetic variance in turn result from both additive and nonadditive genetic effects, the latter including dominance and epistasis. Populations are characterized by population growth parameters that may have a strong genetic component:

l_x the age-specific survivorship schedule of individuals in the population, from birth until death.

m_x the age-specific fecundity schedule of individuals in the population, from birth until death.

r the intrinsic rate of increase of the population, the growth rate itself, computed from the l_x, m_x tables, initial size of the population, and developmental time.

R_0 the net reproductive rate of the population, the total number of offspring produced by the average female during her lifetime.

T the generation time of the population.

The size, N, of the population is then described by the following formulation, which brings together some of the above ecological parameters:

$$1 = \sum_{x=0}^{\infty} e^{-rx} l_x m_x$$

It is important to realize that these parameters are determined in part by the distribution of genotypes in the population and that an optimal strategy, based on the assumption of individual selection, is to maximize the average fitness of the population. This should be reflected in the maximization of growth rate of the population. The differential fitness of genotypes within a population and the variations in environments account for differences in population growth rates. Lewontin and White[(54)] examined these considerations for populations of the Australian grasshopper *Moraba*. Different populations of the same species will maximize fitness in different ways, depending upon the initial genetic composition of the population and differential selection pressures affecting components of fitness such as population growth parameters.

The evolutionary responses made by species to the environment occur by the following mechanisms: (1) adaptive responses within the population; (2) differences in adaptive responses between populations of the same species; (3) differences in adaptive responses among closely related species in the same habitat or region, or in different habitats or regions; (4) adaptive responses arising from interactions among unrelated species; (5) responses to changes in the physical features of the environment. Depending upon the kind and direction of selection pressures in the environment, a population will either increase, decrease, or remain the same. As the result of both laboratory and field censuses of animal populations, populations exhibit different kinds of census histories, reflecting different causal mechanisms (Fig. 6.1). No single population is expected to be governed by one of these mechanisms; rather the interplay of different factors usually determine the census history of the population. But census histories are expected to vary considerably with habitat, region, and period for a given species.

The expression $dN/dt = rN$ describes optimal growth of a population with

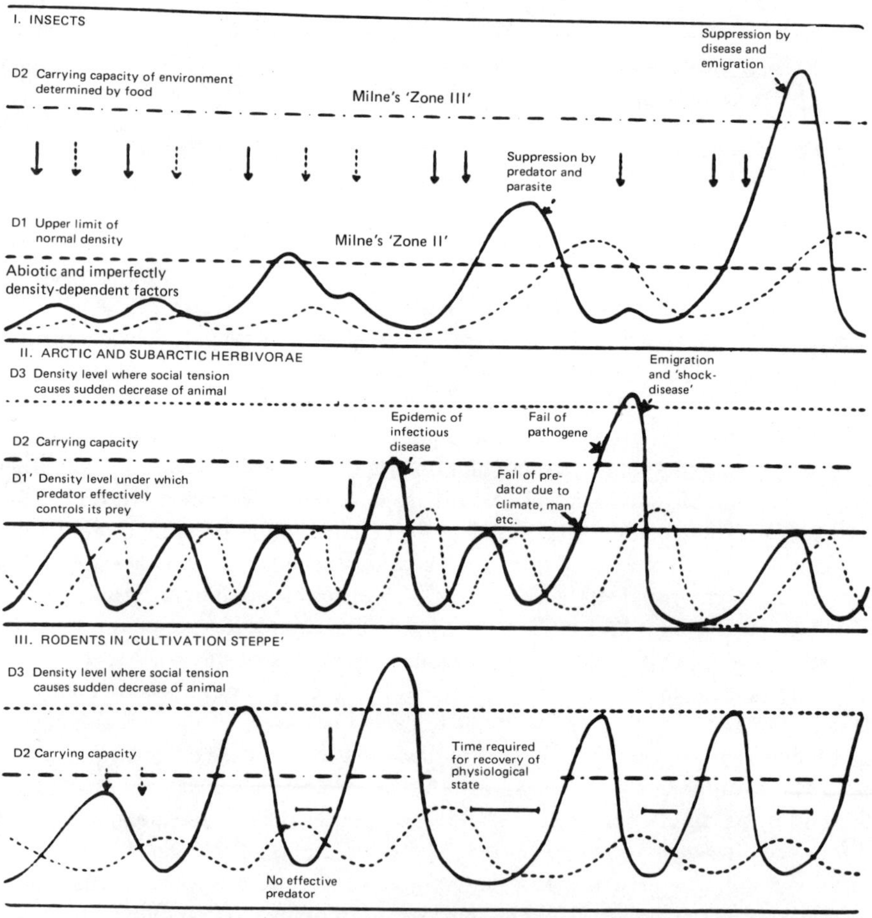

FIGURE 6.1. A summary of the various mechanisms of census history fluctuations in three distinct types of animal populations, including insects. Solid lines show changes in population density, and broken lines indicate census histories of predators (natural enemies). The solid arrows indicate climatic changes favoring increase in the animal (host) populations, and the dashed arrows show the effects of changes decreasing population densities. [From Y. Ito, *Bull. Natl. Inst. Agric. Sci. Japan Ser. C,* No. 13 (1961).]

unlimited resources. Yet we know that few if any populations can grow for any appreciable length of time under such ideal conditions. If we were to examine one parameter of population growth, such as the age-specific survivorship schedule l_x, various kinds of coactions (defined here as interactions between two or more unrelated organisms) alter the l_x schedule from what has been an ideal or "physiological" condition, reflecting the intrinsic capacity of the species to survive under ideal environmental conditions (in the absence of coactions and deleterious effects of the physical environment). The five major kinds of coactions, namely, plant–herbivore, prey–predator, host–parasite, host–pathogen, and mutualist–mutualist, endow the population with a pattern of "ecological longevity" considerably lower than that resulting from the physiological l_x schedule (Fig. 6.2).

It is important to realize that the length of time a population of a particular species has occupied a certain habitat will determine to a large degree the level of adaptation to the environment. For example, a species in the early stages of arrival into a new habitat may still possess ecological traits that adapted the species to its previous environment. Certain phenotypic traits, evolved in response to the previous environment, can influence the trajectory of adaptation in the new environment. An herbivorous insect that feeds on certain Solanaceae in

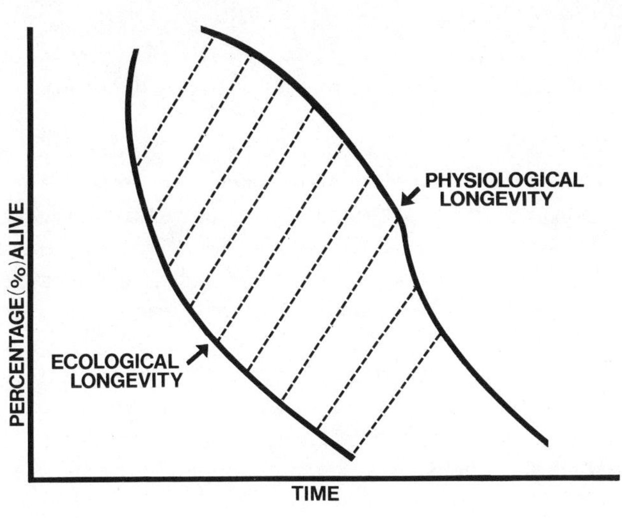

FIGURE 6.2. A hypothetical representation of the effects of ecological processes on the survivorship curve of an insect species in a particular habitat. Individuals in such a population are believed to be genetically and physiologically programmed for some optimal pattern of survival (upper curve), but various ecological interactions (e.g., coactions) in nature actually lower survival in different age classes (lower curve). The area between the two curves approximates the actual losses in harvestable biomass from the population as the result of ecological processes (e.g., density-related intraspecific competition, predation, etc.).

the present environment might in fact have developed a feeding relationship with these plants in the previous environment. Thus the coevolved association is uncharacteristic of the present environment.[436] Similarly, parameter character states determining the rate of population growth in the present environment may still be those carried over from the previous environment.

Species involved in colonizing episodes may have genotypes in the previous environment that allowed the colonizing episode to occur, and some of these traits may still be present in the new habitat. In the new habitat the species might not have occupied all resource patches for the spectrum of resources it is capable of exploiting. A reason for this could be insufficient time to fulfill population increase to the point where all resource patches are occupied. When this occurs, the population will no longer be able to increase, and strong selection might favor an alteration in the parameters underlying population growth. There might be an evolutionary reduction in r. Other adjustments in r may result from resources being in short supply, or being inaccessible as the result of competition being intense, or being different in some way that precludes their penetration by a newly arrived species. For example, the chemical defenses of the host plants may be sufficiently different from the phenotypes of the same host plant species in the previous environment so as to block attack by the herbivorous insect associated with plant in the previous environment. In the lowland tropics, insect populations are believed to be regulated largely by various biotic factors, most notably (1) limited supplies of resources, (2) intraspecific competition resulting from (1), and (3) intense predation and parasitism that keeps populations below the carrying capacities.

The ecological processes under (1) and (2) above are expected to be density dependent. Insect species in high-altitude regions are expected to be largely regulated by density-independent factors such as large fluctuations in the physical environment. But to some extent all populations will be affected by both classes of factors in the environment. For example, most insect species in the temperate zones are known to experience rather high mortality in the juvenile stages (Fig. 6.3), so that survivorship curves for many insects reflect generally low recruitment potentials for adults. Similar patterns are expected for the tropics, although far fewer data are available. The high density and diversity of predatory and parasitic insects and other arthropods in many tropical habitats, particularly in lowland regions, suggest that mortality on juvenile stages is probably very high in many species of insects.

Coactions such as prey–predator and host–parasite are expected to be frequent in the tropical lowlands, where the abundance and diversity of these insects are high. Thus mortality and natality, in the forms of the l_x and m_x schedules for new cohorts in the population of a species in a particular habitat, will determine to a large degree the size and density of the population in that habitat. But dispersal, the shift in population numbers and distribution through movement of

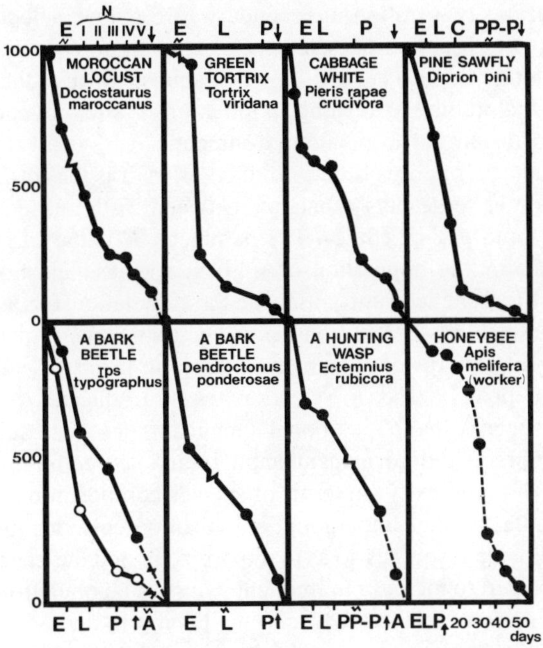

FIGURE 6.3. Examples of survivorship curves determined empirically for a variety of insects, with analysis by stages of the life cycle. In the case of *Ips* bark beetle, data are given for two different populations. The letters on the abscissa refer to immature stages (E=egg, N-nymphal instars, L=larva, PP=prepupa, P=pupa, and C=cocoon); the ordinate is in number of individuals. [Redrawn and modified from Y. Ito, *Bull. Natl. Inst. Agric. Sci. Japan, Ser. C,* No. 14 (1961).]

the population's membership, will also contribute to the size of the population at any one point in time. Populations may select for dispersal-prone genotypes in response to increased population densities, changes in the habitat resulting in reductions of resources, and the sudden availability of new habitats suitable for occupancy by the species. Such dispersal functions as a flush mechanism that promotes escape in space to other suitable habitats, either permanently or temporarily (as in the case of some butterflies and dragonflies). Dispersion refers to the distribution of individuals in the population at any one point in time, and some forms of dispersal concern adjustments in the pattern of dispersion within that population rather than a shift of the population, or a portion of it, to a new habitat.

Terrestrial tropical habitats with many plant species and small average patch size for each species will result in specialized phytophagous insect species also having patchy distributions, which is a pattern of dispersion resulting from within-habitat dispersal to certain plant patches. The sudden appearance of a forest light gap, or regeneration of secondary forest, may select for dispersal to new habitats in two directions: insect species arriving to colonize new resources in the new habitat, or species exiting in response to a reduction in suitable resources for maintaining breeding populations. Such events point to the dynamic state of affairs of tropical environments.

The parameter *r* requires little introduction, as it is the intrinsic growth rate of a population and represents a balance between birth and death rates (determined by the l_x and m_x schedule). The parameter *K* is the carrying capacity of the environment for the population of a given species, and both *r* and *K* are determined by the genetic composition of the population for a particular environment. Different kinds of environments, including many in the tropics, will select for different patterns of resources allocated to levels of *r* and *K* in populations. The concept of *r* and *K* selection was introduced by MacArthur and Wilson[(341)] to specify the populational characteristics best suited for species occupying temporary and permanent habitats and for communities varying in different levels of complexity (in terms of species composition and trophic structure). For example, species undergoing colonization episodes into new environments should allocate resources so as to maximize *r,* and the genotypes with high *r* values are expected to increase in frequency in such a population. High *r* allows a population to grow quickly and swamp the habitat in the face of certain levels of environment uncertainty associated with new habitats. Environmental uncertainty usually results in considerable density-independent mortality of individuals in the population. The tradeoff for high *K* and therefore low *r* adjusts the species to occupying a position in a community at the time the species becomes integrated into the community. Permanent and somewhat stable communities are generally associated with permanent environments. In a temporal sense a colonizing species passes through a temporary phase of population growth in the new habitat, but to avoid extinction the species eventually becomes integrated into the existing communities in the environment.

The "*r* strategy"[(92)] is to also make use of temporary habitats such as a weedy mud flat in a slow-moving stream. Both in this instance and in colonizing episodes involving the invasion of permanent habitats, genotypes with high *r* capacities are favored by selection. There is less selection for genotypes with good competitive ability, for both intraspecific and interspecific competition. Competitive ability is conferred primarily by genotypes in the "*K* strategy."[(92)] This strategy allows the species to adjust to a portion of the habitat and to become a member of the community. By relaxing selection for high *r,* energy is freed to invoke mechanisms that relate to competition and escape from predation and parasitism. The result is that the population approaches the carrying capacity of the habitat, whereas the population experiencing selection for high *r* tends to remain well below the carrying capacity if both abiotic and some forms of biotic mortality are intense, which is expected in colonizing episodes and the exploitation of ephemeral habitats.

For plant-feeding tropical insects, the *r* strategy is predicted to occur as new habitats are created by natural catastrophes such as mudslides, floodings, tree falls in forests, and volcanic eruptions, and biotic changes in existing habitats select for colonizing episodes in some resident species. Some tropical habitats may be characterized by a flux of existing species being replaced by invading species, even though the numbers leaving and arriving may not be the same. Such effects may cause herbivore guilds associated with some plant species in a particular habitat to shift in composition.

Intense *K* selection may lead to considerable ecological specialization in terms of an insect species exploiting certain host plants, but some residual flexibility is expected in order to buffer against selection resulting in niche shifts of potentially competing species. One might expect essentially *K*-selected populations to be a mixture of predominantly *K*-adjusted genotypes but also some *r*-adjusted genotypes.

If a species of lepidopteran is a specialist on a few host plant species in a particular habitat, and its population is regulated largely by density-dependent factors such as predation or parasitism, the population appears to be *K*-selected. If, however, the intensity of predation or parasitism shifts away from this insect species, the butterfly population may increase in size and density as selection increasingly favors the *r*-selected genotypes through successive generations. These prevailing genotypes will be most adapted to density-independent factors, and as the population increases, the specialist feeding habit results in a rapid depletion of the host plant population and either the butterfly goes extinct locally (the population crashes) or the species experiences a colonizing episode to another habitat or region. Other consequences may also result under such a scenario. As the butterfly becomes very abundant in the original habitat, it may become a favorite prey for other predators, perhaps in a different stage of the life cycle, and a new form of density-dependent regulation may result.

If the butterfly undergoes a colonizing episode to another habitat, the emptied niche left behind may be filled by a competing species in that community, or a new species may arrive from another habitat through colonization. The result is that the community may continue to have the same number of species, or one less. Multiplying such effects for many species in tropical communities illustrates that such assemblages may be continually changing in composition.

From this discussion, it is evident that the basic issue in the population biology of insects in the tropics distills down to the analysis of how resources and energy are allocated to different vegetative and reproductive functions in the individuals making up a breeding population of a species. The problem is complex because the community in which the species is found is continually changing. On the one hand, selection within the population will favor the evolutionary maximization of fitness either through increasing the capacity for the population to increase (*r* selection) or through adjusting the population growth parameters to maximize adjustments to competitors (K selection). These are two ways of optimizing the use of the environment. Borrowing from researchers such as Lewontin,[87] I suggest the two different joint l_x and m_x functions shown in Fig. 6.4. For insects in terrestrial tropical communities, the local faunsitic composition will include species with either strategy. Which one is used will depend largely upon the degree of genetic and physiological specialization on a set of resources, the intensity of biotic mortality factors, the intensity of competition, and the intensity of abiotic mortality factors. If most parasitic flies and wasps are generalists in the tropics, their impact on phytophagous insect communities may be patchy; certain species are repeatedly attacked while others escape repeatedly. This is expected to be the case especially if the parasite uses the host's food plant as a means of locating the host.[437]

Thus some phytophagous insect species may be more *K* selected than others in the same habitat: species escaping from such predation will build up larger populations and therefore be more likely to compete with other species for the same resources. Intense predation favors the *r*-selected strategy, since populations are cropped severely, precluding the activation of density-dependent and competitive effects. We clearly need to explore the comparative natural history of insect species exploiting particular plant species in a habitat and to analyze life table parameters and mortality factors. We must also examine to what extent the

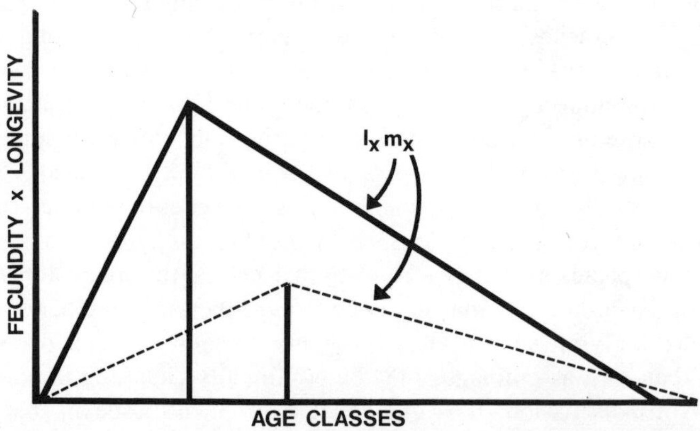

FIGURE 6.4. Reproductive effort as a function of fecundity (age specific m_x curves) weighted by age specific survivorship schedules (l_x). The function assumes a triangular shape over all age classes in the population considered. Some environments select for $l_x m_x$ curves to be skewed toward maximal reproduction in young age classes (upper graph), while others select for a postponement of reproduction until later (lower graph).

population of each host plant species is utilized and to what degree by each member of the herbivore guild found on each plant species.

When reproductive effort is spread out over many age classes of genotypes within a population (Fig. 6.4), natural selection acts to adjust the reproductive effort at each age in such a manner as to maximize the overall fitness of the population. Colonizing episodes for an insect species into a new environment are believed to be characterized by genotypes maximizing rate of development to the adult stage, and higher levels of egg production in younger age-classes within the adult stage. More predictable environments, with species well integrated into stable communities, may allocate less energy to early breeding and disperse reproduction over a greater number of age-classes. Of course, one might expect selection to favor high reproductive efforts at all age classes capable of reproducing. But such a pattern of resource allocation may also diminish other components of fitness in other stages of an insect's life cycle. Natural selection functions to balance these cost–benefit effects in the attempt to maximize overall fitness. Gadgil and Bossert(92) discuss the ramifications of these considerations in terms of when an organism should maximize reproduction at one point or else spread it out.

Natural history studies are needed in the tropics to determine the factors affecting the recruitment of reproductively competent adults to the population of a species. Populations continually change in space and time, and shifts in genotype frequency are expected as the result of selection. Different forms of selection pressure affect different stages in the complex life cycles of insects. We know little about the impact of mortality factors, resource quality, and impact of abiotic mortality factors on the population dynamics of any tropical insect. Some census histories are discussed in a later section of this chapter. For the present, however, it is necessary to realize that different stages in the life cycle may be affected by predominantly density-independent or density-dependent factors. This is true for insect populations in temperate regions as well. For example, Ito *et al.*(438) found that fluctuations in the population densities of the moth *Hyphantria cunea* in Japan were the result of the interplay of both density-independent and density-dependent factors affecting eggs, larvae, pupae, and adults (Fig. 6.3).

In the insect communities of the lowland tropical regions, population outbreaks tend to be rather scarce. From studies in the Southern Temperate Zone, Andrewartha and Birch(439) concluded that population outbreaks were generally caused by changes in prevailing weather conditions, having little to do with density-dependent effects of predators, parasites, or competitors. In the lowland tropics, the annual patterns of rainfall and temperature do tend to be similar over long periods, suggesting that density-independent regulation of insect populations may not be prevalent. A very different picture might emerge from long-term studies in high-altitude regions of the tropics. But cooler temperatures at high altitudes generally reduce insect diversity too. Population regulation of insects in the lowlands, if it exists, is expected to result primarily from (1) restrictions in accessibility of resources either because they are very patchy or because competitors mobilize portions of them (i.e., resource limitation), or (2) intense predation and parasitism. Both effects may be strongly density dependent. The predation–parasitism model would involve host species being r-selected, since their populations would buffer the effects of predators and parasites through increased reproduction. Such an effect is brought about most efficiently by an evolutionary reduction in developmental time or secondarily so by increasing fecundity.[87] The assumption is made here that the predation or parasitism is indeed intense to the point that increasing the host populations increases the mortality.

Evolutionarily increasing r may result in a satiation of predators and parasites so that a breeding population of the host species can be maintained in that habitat. If, however, predation or parasitism is low or weak in intensity, its effect may be density independent and have little regulatory impact on the insect population. Under such conditions, other forms of density-dependent regulation such as resource limitation and competition may take over in regulating the population. The latter situation, of course, promotes the K-selection strategy. Few data exist on the impact of predator and parasites on insect populations in the tropics. The concept of biotic control of host populations in the tropics, as opposed to control from resource limitation and competition, comes from the studies on seeds and seed predators.[29,65,435] In lowland seasonal regions of the tropics, the dry season may be a time of considerably density-independent mortality on insect populations, especially if the beginning of the dry season, and first few months, is an unpredictable event in terms of the synchronization of life cycles. The sweep-sample studies of Janzen and Schoener[134] indicate that many insect species drop out of the field communities in lowland Guanacaste Province, Costa Rica, during the dry season. Such an effect may result in considerable mortality for the adults of many small insects and cause many small-bodied parasitic flies and wasps to drop out of the communities for this period. The latter could relax predation and parasitism on larger insects that remain active in some habitats during the dry season.

6.3. FURTHER NOTES ON POPULATION REGULATION

We can generalize about the ecological genetics of populations by echoing Cole,[434] who predicted that decreasing the age of first reproduction of individuals in a population results in increased growth rate, r, of the population. Or we can be cognizant that populations are affected positively or adversely by many factors in the environment. Cole's early formulations concerned the statistic r_{max}, which is of course the familiar intrinsic rate of natural increase or the population growth rate when all conditions of the environment are ideal. Under such idealized conditions fitness is maximized through the realization of r_{max}. Yet we know that populations are characterized by a realized optimal fitness, which is generally lower than r_{max}, if we equate fitness with r. In nature, populations are regulated by various factors, including resource limitation and predation.[440] In fact, the interplay of factors such as limited food supply and predation may effectively regulate populations.[440]

In a lowland tropical field or secondary habitat, many dicotyledenous plant species often show signs of heavy damage by a variety of herbivores, although seldom is a plant completely defoliated by a single species of herbivorous insect. Such observations imply that a plant species may be attacked by several unrelated insects, although densities of the individual species on each host plant individual are low. Presumably the population of any one insect species is distributed among many individuals of that plant species or, in the case of polyphagy, among individuals of different plant species in the same habitat.

The usual absence of complete defoliation by a single species is the result of the dispersion of the insect population among different resource (plant) patches so as to lower the density per patch and predation or parasitism on the insect population that further lowers the abundance of individuals in each patch. Such a population system is thought to be r selected in the sense that the insect population is regulated primarily by biotic mortality. Depending in part upon the egg placement habits of the insect species, there may also be some form of intrinsic control of the population, such as cannibalism among young larvae on each other or unhatched eggs in a species that deposits eggs in large clusters. Although female *Mechanitis isthmia* butterflies carefully place large clusters of eggs on the larval host plants, the resulting clutches of first-instar larvae are usually reduced by as much as 20%, even when egg predators other than *Mechanitis* larvae are prevented from reaching them (A. M. Young, personal observation). In laboratory populations of some *Tribolium* flour beetles, egg cannibalism by adults is sometimes a form of regulation for the populations.[441] However, in other tropical insects, such as the butterfly *Perrhybris lypera,* egg cannibalism is absent even though the eggs are laid in tight clusters.[77,291] If there is strong selection for egg clustering in response to some phenotypic feature of the host plant requiring coordinated group feeding, there should also be selection for ensuring that potential losses in fitness arising from cannibalistic activity of early hatchlings or other life stages present on the plant are minimized. Such a form of population self-regulation is probably uncommon in insect populations, despite the many examples of egg clustering and communal larval (or nymphal) activities in the tropics. Given the high densities of predatory and parasitic insects in the tropics, such as ants, wasps, and tachinid flies, the early stages in particular of many insects are probably cropped down by these predators and parasites; such effects are a form of population regulation. At high altitudes many small predatory insects drop out of the communities, and scavenger, herbivore, and omnivore assemblages of such regions are probably regulated by limitations in the food supply and by the rigors of the physical environment.

Comprehensive theories of population dynamics involve both density-dependent and density-independent factors in the regulation of insect populations. Some emphasize density dependence in the regulation of populations, and still others (e.g., Milne[442]; Andrewartha and Birch[439]) stress that insect populations were regulated primarily by density-independent factors such as weather and climate. The favorable physical conditions of the lowland tropical rain forests may preclude density-independent regulation for insects in such regions. Density-dependent factors include intraspecific competition, predators, parasites, and pathogens. While many biotic factors may affect insect populations in a density-dependent fashion, others, such as generalized predators and parasites with many alternative hosts, may be density independent. Density-dependent factors provide a form of natural control on some insect populations in the sense of being extrinsic mechanisms of population regulation. In the lowland tropics, the interplay of several different kinds of biotic factors on insect populations, including both density-dependent and density-independent factors such as intra- and interspecific competition, predation, or parasitism, probably keep many insect species well below the carrying capacity of the habitat. Such effects are examples of intrinsic (intraspecific) and extrinsic natural control of insect populations. These biotic factors are responsible for removing individuals from a population in various stages of the life cycle, generating rather concave survivorship curves as found in the populations of many different insects (Fig. 6.3). Because biotic control is expected to operate on many insect species in different trophic levels in the tropics, outbreaks of insects are not expected to be frequent in relatively undisturbed to lightly disturbed habitats. In this context, population outbreaks occur either when the effectiveness of the existing biotic control factors breaks down or when insect species occupy physically controlled environments[104] where biotic control mechanisms are not prevalent. Much of the basic research in tropical biology should focus upon the natural history of control mechanisms in insect populations as related to the kind of habitat and degree of habitat complexity, community complexity, and fluctuation patterns of abiotic factors such as temperature. We need to know something about the range of hosts, for example, of the average species of braconid wasp or tachinid fly in different habitats. We also need information on the relative roles of different biotic factors in the regulation of an insect population in different habitats. Such data are lacking for tropical insects. When we have information on the composition of a community found in a particular habitat in the tropics, we will need also to gather data on the following parameters:

1. Resource abundance and accessibility and the degree to which resources are a limiting factor for a particular insect species being scrutinized.

2. The degree to which the population of that insect is affected by predators or parasites (and pathogens).
3. The degree to which effects of overcrowding induce mortality or dispersal to new habitats.
4. The degree to which competition among ecologically similar organisms lowers the carrying capacity of the insect population in that habitat.
5. To what extent mortality is induced by fluctuations in parameters of the physical environment, and over what time frame.

When discussing the impact of density-dependent and density-independent mechanisms or factors on the population dynamics of an individual insect species in a particular habitat, it is important to realize that any single individual in the population can be subjected to both kinds of mechanisms and that "population regulation" is essentially the combined effects of both mechanisms. In many terrestrial tropical habitats, insects will be subject to subtle differences in environmental factors affecting survival and reproduction, some of which may be density-dependent factors and others density independent. Thus the collection of individuals of a grasshopper species occupying a premontane tropical rain forest light gap will be differentially influenced by a wide range of such subtle factors so that understanding survival and reproduction become a matter of individual-to-individual variation in responses to such factors. Collectively, the sum of such differences in response will determine the tenacity of that population for that light gap. Conceptually it is easy to examine issues of population regulation in terms of density dependence and density independence. But the reality of tropical insects in their habitats is that a wide range of often subtle environmental factors creates a complex system of population regulation, a system that will vary within a species over different communities in the same region, among regions (geographical), and between seasonal and nonseasonal environments.

For lowland tropical regions, it is necessary to assess the role of periodic catastrophes in the same way that Ito[(443)] analyzed typhoon effects on populations of the beetle *Ips typographus* on the subsequent potential for population regulation in the region for that species and the likelihood of periods of overpopulation as part of a general return to equilibrium conditions in the community. It is easy to imagine and determine that fluctuations in insect populations come from a variety of sources, even in tropical regions (Fig. 6.5), and we must not generalize about such effects for all species in all regions. Insect populations in the tropical regions of the world are believed to exhibit diverse forms of census histories reflecting mechanisms of food limitation, predator limitation, competition limitation, and limitations from the physical environment. We do not necessarily expect one of A. M. Shapiro's high-altitude pierid butterflies in Ecuador to possess the same kinds and intensities of regulatory agents as a related species in the lowlands on the eastern slopes of the Andes below 200 m.[(760)] The key, it seems, is to determine the relative role of various types of environmental factors

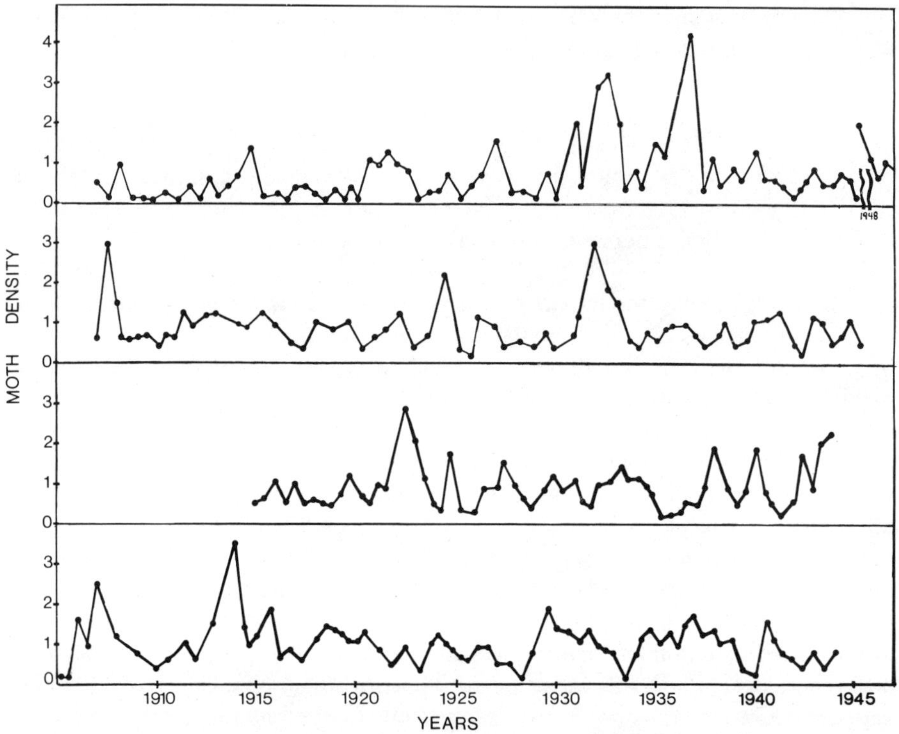

FIGURE 6.5. Insect population census history as related to fluctuations in the physical environment. The pattern of census history depicted is for the paddy borer moth, *Schoenobius incertellus*, in Japan. Redrawn and modified from Y. Ito, *Bull. Nat. Inst. Agric. Sci. Japan Ser. C*, No. 13 (1961).

(biotic and abiotic) on insect populations in different kinds of regions in the tropics and then to relate these effects to community structure, habitats, and patterns of fluctuation in physical parameters. We also need to consider the effects of trophic status on the type of regulation, if any, operative on the population of a particular insect species occupying a particular habitat.

This is really a discussion of the differences between *r* and *K* selection: species in physically controlled environments are believed to be ecological generalists exhibiting large populations and with few if any competing species and a lack of biotic control from predators and parasites. This is an example of *r* selection of the type expected for most high altitude regions in the tropics and tropical deserts. *K*-selected species have smaller populations, more specialized patterns of resource utilization, biotic control, and perhaps competition from other species. *K*-selected species are believed to be more prevalent in some kinds of tropical wet lowland environments. But one must keep in mind that we are referring to relative effects, and that at any elevation in the tropics we expect to find mixtures of *r*- and *K*-selected species, depending upon the habitat. Given the diversity and patchiness of plant species in some lowland tropical terrestrial habitats, one might expect that many phytophagous insect species are *K*-selected in the sense of being specialists on certain plant species, with low average food niche breadth. The comparatively benign physical features of the lowland tropical wet and dry forests (with the exception of certain coastal habitats) contribute to the accumulation of plant species in some habitats, which in turn promote ecological diversification among phytophagous insects. *K* selection is also manifested in terms of competition between these insects, as evidenced indirectly by the accumulation of species-rich insect guilds on some plant species, with low numbers of individuals per species, and in some instances with each component species specializing on a different part of the plant species.[(93)] Yet some of these species may be kept scarce by predation and parasitism. But tropical predators and parasites are believed to be generalists with some tendency toward *r* selcction attributes in order to take advantage of diverse resources per unit area of the habitat. I am referring here primarily to tachinid flies and various hymenopterous groups such as the Braconidae. If each host patch is small, a predator and parasite both need to visit many patches within the habitat, and selection favors the ability to exploit patches of different host species, within some limits. Those limits are set largely by the searching and feeding physiology of the predators and parasites. The obvious implication for such considerations is that some of these insects become ideal candidates for biological control programs, even in Temperate Zone countries.[(48)]

While lowland tropical regions may be relatively benign in terms of temperature fluctuations and excessive dryness throughout the year (i.e., nondesert), some tropical butterflies such as *Parides* are killed as adults (Fig. 6.6) by severe rainfall.[(444)] Such effects produce considerable fluctuations in the abundance of adult butterflies during the rainy season (Fig. 6.7). In some cases the resulting survivorship of adults is comparable to that of many Temperate Zone species (Table 6.1). Adult butterflies sometimes shift their patterns of dispersion markedly in response to seasonality, increasing their population densities and accessibility to some vertebrate predators. In some instances highly palatable butterfly species are involved, and avian predators expertly remove the more palatable abdominal contents and discard the exoskeletons.[(205)] Such predation is density dependent.

Predation may be high in tropical butterflies that are supposedly palatable but have other means of defense such as disruptive wing coloration patterns[(445)] and in which the disruptive coloration does not reduce predation. Butterflies in lowland tropical regions may be subject to certain kinds of predation from vertebrates[(186,188,189,207)] while both feeding and flying. We need to reexamine some of the classic concepts of protective coloration and mimicry with more data on

FIGURE 6.6. Adult female of the Neotropical butterfly *Parides iphidamas,* resting on a melastomaceous plant at the edge of primary growth tropical rain forest in Costa Rica.

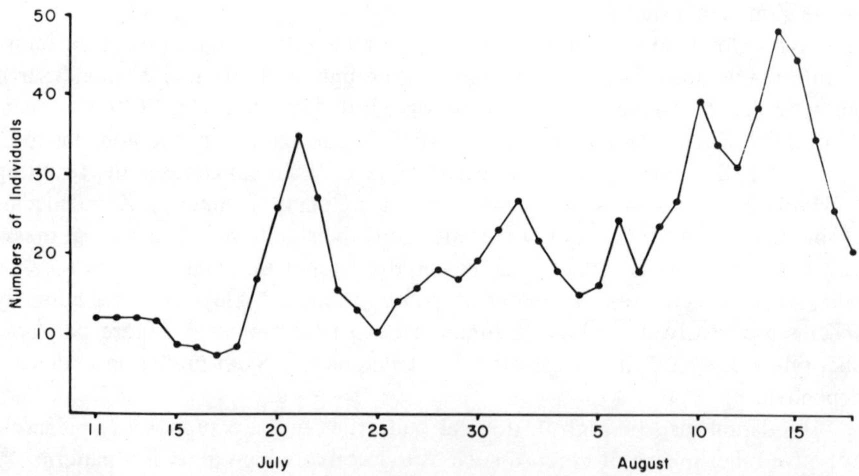

FIGURE 6.7. Changes in adult population size for the Neotropical butterflies *Parides neophilus* and *P. anchises* in Trinidad during the 1965 wet season. Daily population size (both species) is smoothed by application of three-point moving average. [From L. M. Cook, K. Frank, and L. P. Brower, *Biotropica* **3:**17–20 (1971).]

TABLE 6.1
Estimates of Daily Survival Rate of *Parides* on Trinidad Compared with That of British Lepidoptera[a]

Species		Survival rate	Reference[b]
Parides spp.		0.84	This paper
Polyommatus icarus (von Rottenburg)		0.83	Dowdeswell, Fisher, and Ford (1940)
P. icarus		0.74	Parr, Gaskell, and George (1968)
Parage megera (Linnaeus)		0.79	Parr, Gaskell, and George (1968)
Maniola jurtina		0.89	Dowdeswell, Fisher, and Ford (1949)
Coenonympha tullia (Müller)		0.70	Turner (1963)
Lysandra bellargus (von Rottenburg)	June	0.79, 0.92	Davis, Frazer, and Tynan (1958)
	September	0.91	
Lysandra coridon (Poda)		0.84	Davis, Frazer, and Tynan (1958)
Panaxia dominula		0.84	Fisher and Ford (1947)
P. dominula		0.78	Sheppard (1951)
P. dominula		0.62	Cook and Kettlewell (1960)
Zygaena filipendulae		0.71	Manly and Parr (1968)

[a] From Cook *et al.*[(444)]
[b] References can be found in Cook *et al.*[(444)]

actual surivorship in natural populations. As with populations of other insects, tropical butterfly populations may be regulated by a variety of biotic control agents.[(221,222)]

Whether or not such patterns of mortality behave in a compensatory manner through time in a fashion similar to that described by Murdoch for Temperate Zone Carabidae,[(446)] long-term studies in the tropics are warranted. We also need to recognize cases in which shifts in the breadth of food resources utilized in a habitat by a single insect species result from changes in population density: under conditions of low density, food niche breadth might be lower than for the same species and population at higher densities. There might be some cases of "facultative broad-base or flexible monophagy or even polyphagy" when population density becomes unusually high. Such effects will result from changes in the habitat affecting competition and biotic control agents: periods of relaxed control or competition may be the prelude to periods of increased population density and facultative shifts in feeding spectra within a species. Some habitats may exhibit more changes through contemporary time than others in the tropics. If the habitat changes in such a way as to relax some forms of biotic control on some insect species, and if the response to selection is rapid by the insects, there could be a shift from K selection to r selection in terms of the population dynamics.

Given sufficient generations, directional selection may result in an adaptive intrapopulational shift toward genotypes capable of increasing population growth rate. In some tropical habitats such shifts may occur regularly for insect species, depending upon the tenacity of the species for that habitat and the degree to which gene flow from populations in nearby habitats delays or promotes such responses. The resulting "outbreak" is successful if the increasing population of the insect species is able to find and utilize suitable resources. If the habitat contains large patches of the single preferred resource or consists of a mosaic of several suitable alternative resources, the species will colonize the resource base to a degree not realized under K selection. It may be easier in terms of energetic considerations for the species to exploit a single suitable resource if differences among resource types are sufficiently different to reduce fitness additively. Outbreaks of phytophagous insects are known from single-species plantings in the Temperate Zone,[(447)] and some regions of the tropics may have regular or occasional "mini-sagas" of such effects as habitats change in cyclic or acyclic manners.

Depending upon the degree to which scavenger insects and other arthropods are resource specialists, occupants of benign or harsh regions, and regulated by biotic control agents, these insects will also respond to habitat changes in a manner implied above for phytophagous insects in particular. Generalist predators and parasites of phytophagous and scavenger insect species may respond less to such perturbations in the environment since they are already generalized to some degree. But the direction of selection pressures may be sufficiently different so that some portion of the resources of endoparasitic forms and predators are shifted in a significant manner: if some phytophagous species drop out of the community as a result of a habitat change, the food base for predators and parasites may contract, changing the population dynamics of these species. Such species may increase exploitation on some or all remaining species suitable for exploitation or experience a reduction in population growth since fewer resources are available. The generalist tendencies of these species, however, probably precludes their dropping out of the community: they are tenacious enough to adjust to the change accordingly. Such predictions warrant long-term studies for verification.

The previous discussion stresses the great heterogeneity of the tropics in terms of how component species of specified communities respond to perturbations in the environment. Relatively benign habitats and regions of the tropics, that is, those where physical conditions are not severe, may be more vulnerable to such perturbations (floods, severe periods of rainfall, mudslides, etc.) than other habitats where the environment is generally harsh most or all of the year (high elevations, deserts, certain coastal formations, etc.). Organisms occupying the latter regions are already evolutionarily adapted to conditions favoring *r* selection in response to abiotic mortality factors and an absence of density-related biotic control agents. Any conclusions about the kind of regulation of insect populations in the tropics must therefore be qualified with a consideration of the regional and habitat effects on individual species associated with each trophic level. We anticipate a spectrum of different habitats at various intervals along elevational gradients in the tropics; there is also the effect of increasing or decreasing elevation on the physical environment, in turn affecting organisms found there, and the ability of individual species to move in and out of communities at different elevations. Physically controlled communities are composed of generalist species, and there is room for the addition of additional species through the effects of migration and range expansions. Biologically accommodated communities associated with some lowland tropical regions may be saturated with species, many of which are specialists, and the addition of new species is a gradual process made possible by the increase in harvestable biomass of such environments (see Chapter 1).

6.4. SOME EXAMPLES OF CENSUS HISTORIES IN TROPICAL INSECTS

The modes of adaptation to the environment are largely determined by the properties of breeding populations of a particular species. The pioneering studies of Sewall Wright elucidated rigorously the effects of population size on the evolution of adaptations in populations.[448,449] When populations are small on a relative scale, certain events affecting gene and genotype frequencies, such as drift, may override the effects of selection and gene flow as major determinants of evolutionary history within species. On the other hand, large populations may be characterized by considerable genetic variation of various kinds,[1] providing greater opportunities for new adaptations to appear and stabilize. Populations of individual species may pass through evolutionary filters of various kinds, characterized by periodic contractions and subsequent expansions, thereby clouding the effects of what might have been a direct selective pathway to optimal use of the environment. Initial gene frequencies may vary considerably in different populations of the same species in a region and the differential effects of selection in different areas within the region result in different genomes being stabilized,[54] even if only temporarily. The rate of stabilization of characters through stabilizing selection largely depends upon the amount of genetic variation in the population and the size of the population. Different populations of a species in more or less continuous contact may result in a postponement of local adaptations being expressed within each population because the effects of local selection are sometimes swamped by shifts in genetic composition arising from gene flow.

Any attempt to understand the interplay of regional and habitat diversity on the development and maintenance of faunas in the tropics must also take into account the discreteness and sizes of breeding populations for a species. The problem for the tropical biogeographer is to confidently recognize the boundaries of populations in making analyses of historical versus contemporary selection pressures acting on species; the tropical population biologist requires the same data for understanding how different species fit into a community and where the boundaries of communities occur. Yet one major obstacle remains the lack of long-term studies on the population dynamics of any tropical organism[42] for periods of at least 10 consecutive years. Some short-term studies of tropical butterflies, (e.g., on such as that for *M. peleides* by Young and Thomason[190]) provide information on the effects of habitat boundaries on the size of an insect population and the dynamics of dispersal within it. In that study it was possible to estimate population size from various methods, something greatly facilitated by the confidence in actual boundaries of the butterfly population (Table 6.2). Daily survival rate was calculated from the same capture-mark-release data from the expression

$$(1 - S)\,\frac{1 - R}{{}_1N_1 + N_R}$$

where S is the survival rate per day, N the estimated daily population size, and R the last day of collecting.[450] Such studies also reveal interesting data on sex

TABLE 6.2
Comparative[a] Daily Population Size Estimates and Survival Rates for *M. peleides* in the Barranca Study Area during the 1973 Dry Season[b]

	Population size estimates			
Date	Fisher and Ford (1947)	Jolly (1965)	Manly and Parr (1968)	Survival rates [Fisher and Ford (1947)]
23 January	—	—	—	0.56
24	99	84	84	0.60
25	72	72	73	0.63
26	105	103	105	0.67
27	79	66	64	0.71
28	78	73	75	0.75
29	81	97	93	0.79
30	82	80	78	0.84
31	68	66	66	0.89
1 February	55	58	58	0.94
2	61	—	—	

[a]Population size estimates are calculated by three methods.
[b]From Young and Thomason.[190]

ratios, which in the case of *Morpho* butterflies are generally skewed toward males, are not due to aberrant sex ratios at birth, but rather result from different behaviors of the sexes biasing the probabilities of recapture toward one sex or the other.

Although the study of populations may be more challenging in areas where habitats are not as restricting as in this study, Ehrlich and Gilbert[451] found convincing evidence that a population of the tropical butterfly *Heliconius ethilla* remained stable in size for more than 2 years (Fig. 6.8). That population was composed of two colonies with virtually no exchange of butterflies between them. The stability of the population is explained in terms of density-related forms of predation on early stages and the continual supply of adult resources found in the habitat. *Heliconius* butterflies are relatively long-lived as adults,[451,452] another factor contributing to the observed stability of populations. In some species of *Heliconius* such as *H. charitonius,* communal roosting (Figs. 6.9, 6.10) is a major mechanism contributing to the spatial structuring of the adult butterfly population.[452–455] Such behavior contributes to observed patterns of population stability and cohesiveness within the habitat in *Heliconius*.[452] Adult butterflies form conspicuous aggregations on the same clump of dead vines or branches (Fig. 6.9) month after month.[452–454] Capture-mark-release studies of species such as *H. charitonius* reveal that the same individuals return to a roost site over long periods (Fig. 6.10). Longevity in *Heliconius* is also promoted by specialized feeding on pollen that is used as a protein source for eggs by the butterflies.[31]

Another feature of such populations is a general tendency for age classes to

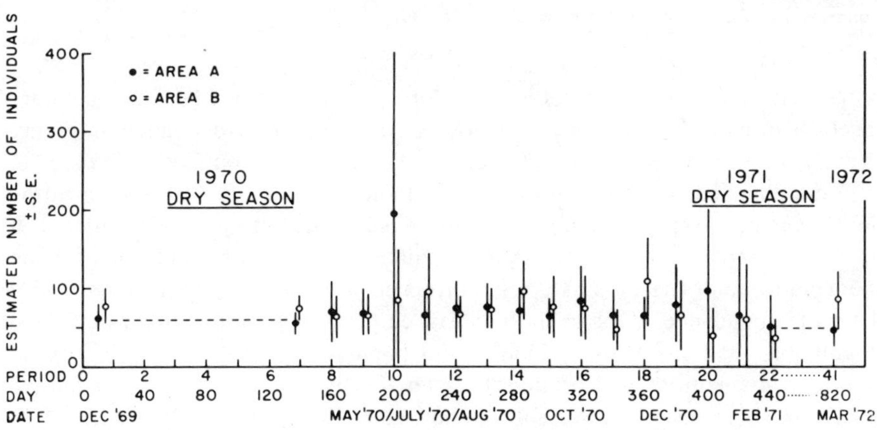

FIGURE 6.8. Estimates of adult population size for the Neotropical heliconiine butterfly *H. ethilla* at St. Andrew's Trace, Trinidad, for different areas (A and B) during two successive dry seasons. [From P. R. Ehrlich and L. E. Gilbert, *Biotropica* **5**:69–82. (1973).]

FIGURE 6.9. H. Beth Braker looking at a roosting site for the Neotropical heliconiine butterfly *H. charitonius* along a shady footpath near the edge of mixed primary and advanced secondary premontane tropical wet forest in northeastern Costa Rica. Several species of *Heliconius* form nightly communal roosts in clumps of dead vines such as the ones shown here. Communal roosting is a major component of the population dynamics of these butterflies.

remain rather similar in distribution over long periods (Fig. 6.11). The regulatory mechanism in such an insect population consists of density-dependent mortality in early stages coupled with high longevity. Predation rates on *Heliconius* are not well documented, although one study[(339)] indicated that survival is high in the adult stage due to specific types of defensive behavior. In tropical butterflies such as *Acraea encedon* in Uganda, some populations are predominantly female and the abundance of males is inversely correlated with overall population size.[(456)] Thus the abundance of males in this instance plays a self-regulating role for the population: mating frequency increases as population size diminishes. Tropical mosquito populations are usually characterized by considerable fluctuations in relatively short periods, the result of short adult life-spans (4 days or less).[(457)]

Fluctuations in the abundance of an insect species in an area are expected if a predator or parasite of that species interacts with it in a patchy manner, as

FIGURE 6.10. A communal roost of the butterfly *H. charitonius* in Costa Rica. Note marked butterflies. Marking studies reveal considerable individual fidelity to roosts in these insects, a factor contributing to population cohesiveness.

expected for generalists exploiting several different resources in a habitat. If the predator or parasite is a specialist on a few host types, host populations may not exhibit fluctuations through time since the intensity of the mortality generated may be more or less constant through successive generations. For insects with effective defensive mechanisms, particularly in the adult stage, adult survival may be high and adult population size stays rather constant. Strongly seasonal environments in the tropics add another dimension, in that periods of uneven abundances may result from effects of shifting physical conditions on biotic control agents such as small-bodied parasitic wasps. For example, a large-bodied lepidopteran may continue to breed during the dry season if larval host plants

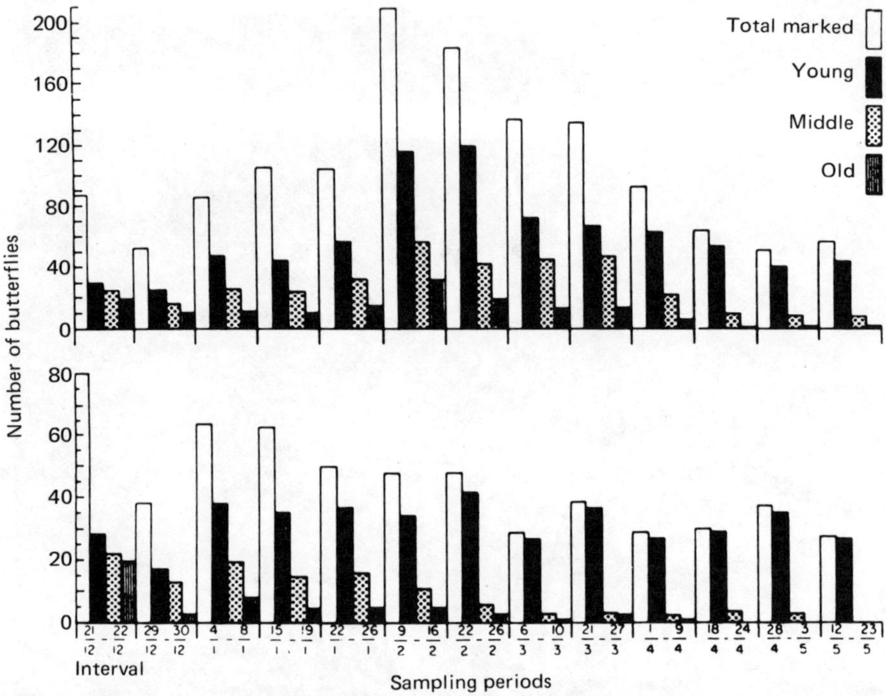

FIGURE 6.11. Number of adult butterflies, and abundances within three approximate age classes, for *H. charitonius* in Costa Rican mountain rain forest (above), and the numbers of newly marked butterflies (below). Units on the abscissa are days above and months of the year below. [From L. M. Cook, E. W. Thomason, and A. M. Young, *J. Anim. Ecol.* **45:**851–863 (1976). Copyright 1976 by the British Ecological Society.]

remain lush (as is expected for some riparian forest habitats), but the relatively small-bodied endoparasitic braconids that parasitize the larvae of such a species may disappear temporarily in the dry season, relaxing one type of biotic control on the herbivore population. Regulatory interactions of this kind may return during the following rainy season. At the same time, other forms of biotic control, such as increased predation on adult butterflies in dry season forest refugia, may occur when populations of individual species become compressed into relatively small areas of moist habitats. We therefore need to examine census histories of insect populations in seasonal and nonseasonal tropical regions over several cycles of seasons to determine the patterns of fluctuations in adult recruitment and the kinds of environmental regulatory effects operative at different times. Shifts in the kinds of resources available to the target insect species being studied must also be considered, especially for species that are herbivores and pollinators.

6.5. POPULATION RESPONSES TO ENVIRONMENTAL HETEROGENEITY IN TROPICAL INSECTS

The challenge to any species, under the assumption of individual selection, is to maximize fitness. The evolutionary maximization of fitness in populations is reflected in the tendency for such populations to increase. A useful conceptual framework for population increase is provided by the familiar formulation

$$1 = \sum_{x=0}^{\infty} e^{-rx} l_x m_x$$

with parameters as defined earlier. Selection should favor the maintenance of an optimal value of r or the increase in r. If environments were constant and full of infinitely abundant resources, the populations of individual species would realize r_{max}. But environments are heterogeneous to varying degrees and in different ways, creating an evolutionary challenge for species to maximize r as much as possible given prevailing environmental conditions. The response of selection for a given species will be a function of the generation time, generally expressed as

$$T = \frac{\Sigma l(x) m_{(x)}(x)}{\Sigma l(x) m_{(x)}}$$

where T is the mean generation time in a particular population. The rate of increase of the population will therefore be

$$r = \frac{\log_e R_o}{T}$$

where R_o is the net reproductive rate of the sum of reproductive expectations ($R_o = \Sigma l_x m_x$). These formulations are based upon the assumption of a stable age distribution in a population, a condition which virtually never exists. Populations generally fluctuate to some degree in terms of density, so that the distribution of age classes is not stable from generation to generation. Reproductive expectations, the product of each l_x and m_x value for females, are changing in different age classes, so that fitness is determined by an additive effect summed over all reproductively active age classes. Periods of colonizing episodes in species favor age classes with high fecundity in early adult age classes so that the production of offspring is maximized.[87] Populations experiencing K selection will be characterized by fecundity being distributed over older age classes (see also Fig. 6.4).

One type of environmental heterogeneity affecting populations is the transition from favorable to unfavorable conditions, a situation that promotes colonizing episodes in space: the species migrates to a new environment. Periodic temporal fluctuations of this sort may occur in very seasonal lowland tropical regions where species migrate into higher elevation areas where conditions have generally remained benign (in terms of moisture and responses of resources to moisture gradients). The studies of Dingle[140–142] have indicated that *Oncopeltus* milkweed bugs exhibit survivorship curves that correspond to different environmental character states such as light and temperature regimes (Fig. 6.12).

Temperature change may alter the survivorship curve even when day–night cycles remain 12 : 12 hours: the survivorship curve for 27°C indicates considerable mortality in early adult age classes, whereas 23°C conditions promote longevity and continued fecundity (Fig. 6.12). The implication is that hot tropical conditions, such as those of the dry season in some lowland areas, may promote a form of *r* selection in which bugs are programmed to reproduce early in adult life. Such a population structure may be an effective adaptation to bugs locating and colonizing patches of suitable habitat along an elevational gradient or horizontally into moist refugia as the dry season advances. More temporally favorable lowland tropical regions, such as those of tropical rain forest, would tend to select for extended reproduction to match a survivorship curve similar to that of *Oncopeltus* for the 23°C and 12 : 12 cycle (essentially tropical). Relative to

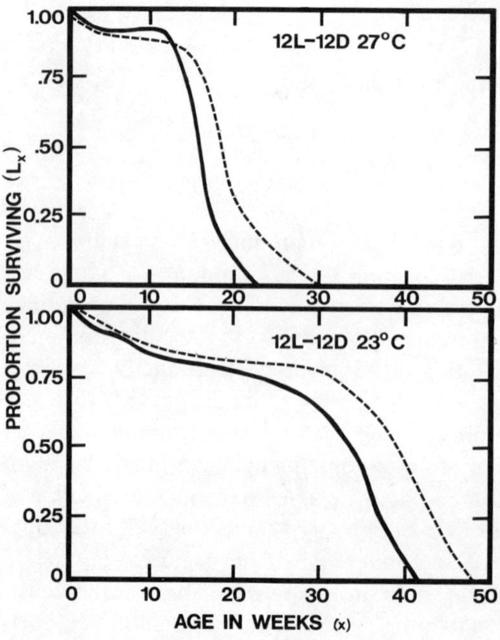

FIGURE 6.12. Survivorship curves for the milkweed bug, *Oncopeltus,* under the same light regimes but different temperature regimes, in the laboratory. Female survivorship is indicated by solid lines, males by dashed lines. x is a unit of age and L_x is the same as L_x in the text. [Modified and redrawn from "Life history and population consequences of density, photoperiod, and temperature in a migrant insect, the milkweed bug, *Oncopeltus,* by H. Dingle, *American Naturalist* **102:**149–163 (1968) by permission of the University of Chicago Press.]

TABLE 6.3
Rates of Increase in Laboratory Cultures of *O. fasciatus* under Different Environmental Conditions[a]

Photoperiod	Temperature (°C)	Initial density	Increase per day (r)	Doubling time (days)
16L–8D	27	10 pairs/box	0.0861	8.42
16L–8D	27	20 pairs/box	0.0810	8.90
16L–8D	23	10 pairs/box	0.0736	9.75
12L–12D	27	20 pairs/box	0.0593	12.03
16L–8D	23	20 pairs/box	0.0499	14.24
12L–12D	23	20 pairs/box	0.0369	19.13

[a]From Dingle.[140]

Temperate Zone conditions, however, tropical regions tend to select for lower population growth rates (Table 6.3). In tropical conditions, it is the age to first reproduction that is extended (Table 6.4), thus lowering r. In some *Oncopeltus* populations it is the joint effect of earlier age of first reproduction and higher levels of fecundity that promote Temperate Zone populations to be larger.[458] The timing of migration to favorable habitats in insects such as *Oncopeltus* will be prior to reproduction, and these individuals suppress feeding and reproduction for the sake of maximizing energy for flight.[141,142] These individuals possess high reproductive values (V_x), since the expectation for breeding is high once the colonization or migration is completed. The reproductive value is the expected contribution of an individual of specified age to future growth of the popula-

TABLE 6.4
Development Time from Birth to Adult and Age at First Reproduction for *O. fasciatus* in Laboratory Culture[a]

Photoperiod	Temperature (°C)	Development time (days) from birth to first adult	Age (in days) at first reproduction
16L–8D	27	27	38 (10 pairs/box)
			47 (20 pairs/box)
16L–8D	23	36	46 (10 pairs/box)
			63 (20 pairs/box)
12L–12D	27	27	61[b] (20 pairs/box)
12L–12D	23	36	95 (20 pairs/box)

[a]From Dingle.[140]
[b]One female began egg laying at 48 days. All others began at 61 to 65 days.

tion.[459] Reproductive value is defined relative to the value of this statistic at birth (v_o) for the species:

$$\frac{V_x}{V_o} = \frac{e^{rx}}{x} \left(\int^{\infty} l_t m_t e^{-rt} dt\right).$$

Integration is from x to X, so that only future births are considered.

The studies of Dingle, Landahl, and Root primarily concerned laboratory populations of milkweed bugs, and there is some difficulty in assuming that similar survivorship and fecundity curves occur in natural populations. That the tropical form begins reproduction later in adult life and has a lower m_x curve in the laboratory may not hold in the field, where survivorship will also be a function of biotic control agents affecting adults, a factor ruled out in the design of laboratory studies. Nonetheless, such laboratory studies do measure relative differences between forms or species under controlled conditions of abiotic factors and food supplies, and these studies provide estimates of the physiological capacities of forms or species to build up populations. One might attempt to infer that l_x and m_x curves of the kind found in the tropical form by Landahl and Root result in gradual, low recruitment of adult bugs in the field and that predation rates on adults are low. Field studies are needed to confirm such inferences.

Colonizing species should maximize reproductive value in order to maximize the likelihood for success in a new environment. In the lowland tropics, I envision the following situation promoting a genetic structure of an insect population that includes an ability to undergo occasional or periodic colonizations of the kind discussed above for milkweed bugs.

1. *Annual severe dry seasons, leaving behind a mosiac of relatively small moist refugia-type habitats that some insects can exploit, by horizontal movement into them, until the rainy season*. In addition to this horizontal movement, some vertical movement is expected in which adult insects move into moist forests at higher elevations during the dry season and recolonize the lowlands the following rainy season. This phenomenon is best described as an ecological contraction-and-expansion cycle that selects for some colonization in species unable to maintain the population structures established in the rainy season during the severe dry season.

2. *Periodic floodings of some stream or river-edge flood-plain forest habitats*. Rather than the environmental stress being water-stress effects on insects or their resources, as is the case for dry-season effects, the stress here is the alteration of plant communities so that associated insect communities are also affected adversely, to the extent that individuals may make long- or short-distance movements to colonize unaffected areas. The success of such events will be relatively less frequent than for effects of seasonality, since the former catastrophes are probably irregular events, and thereby making it difficult for selection to adapt the species to them. Success will largely be defined by the intensity of

the perturbation and the amount of phenotypic flexibility or genetic variation in the population of individual species.

3. *Pronounced perturbations in the structure and composition of habitats resulting from natural and human-induced activities.* Again, predictability of such events is low and local extinctions may be high if species are unable to colonize new habitats.

In all three categories of tropical environmental heterogeneity, success in colonization will also be determined by the interaction of a colonizing species with species already established in the communities associated with the environments being colonized. If lowland tropical habitats on the mainlands tend to be saturated with species,[15] competition may preclude successful establishment of new species. But if harvestable productivity is sufficiently high, resources might be sufficient for new species to be added to some communities under these conditions. Tropical mountain tops may be relatively easier to colonize if the arriving species can adjust to the stressful effects of the physical environment.

The above considerations deserve comment with respect to trophic organization in tropical terrestrial communities. Some river-edge tropical insects in Amazonian forests are efficient mobilizers of detritus from the ground litter,[460] and such forms are expected to adapt well to new areas where litter is present. Other generalists, such as endoparasitic flies and wasps and some predators on insects, are also expected to do well, assuming they can migrate successfully. Perhaps the greatest opportunities for local extinctions during such periods are to be found in the phytophagous insects, especially in communities where species are specialists on a few kinds of host plants.

Ecological escape from an unfavorable environment by an insect species may not necessarily be an ''all or none'' strategy in which a major portion of the population leaves and the rest of the membership goes extinct. Particularly in the lowland tropics, various kinds of disturbances may alter only portions of a habitat, and some portion of the insect communities found there may survive without leaving. The effect of disturbance may often be patchy, and if so, selection may be for a mixed strategy within a species: genetic polymorphisms may permit some morphs to survive in the altered habitat while other morphs are prone to dispersal to new habitats. Populations in some unpredictable environments are predicted to be polymorphic for parameters underlying population growth.[461] Polymorphic insect populations may have a higher average fitness because they endow a mixed strategy in terms of how temporal environmental heterogeneity is responded to within a population. As discussed above, various kinds of temporally fluctuating habitats or environments can exist in the tropics, with the major differences among these being the regularity and intensity of fluctuation upon insect communities. The two morphs predicted under such conditions are morphs adapted for early reproduction (r selection) and morphs adapted for late reproduction (K selection). Each kind of morph may increase or decrease in the population according to the size of the population at any one time.[462]

The lowland tropical dry season of dry forest regions in particular represents a major, regularly occurring environmental fluctuation to which resident species may respond by a mixed strategy. Some butterfly species in such regions have wet- and dry-season morphs.[317] Within the dry season, a period of environmental stress on many insects and other organisms, there may be a mixed strategy within a species, in that one morph adapts to the dry conditions while another migrates to less stressful habitats. Different habitats provide cues favoring the timing of reproduction within a species at a certain time and with a certain pattern of pulsation through time. Peak reproductive activity should occur at times when the chances of survival of juveniles are highest or when predators are most readily satiated.[363] But the timing of reproduction, and the expression of specific kinds of l_x and m_x trajectories within populations, is expected to be fluid for most environments in the sense that every species requires some degree of genetic variation and phenotypic flexibility in terms of the potential for population growth. Even in the lowland tropics, where physical conditions are generally held to be benign relative to very high altitude tropical regions, there will be some degree of temporal uncertainty in most habitats. This uncertainty or heterogeneity arises from occasional to regular fluctuations in the environment, and shifting distributions of species resulting from colonizations, range expansions, and shifts in harvestable productivity. These effects may be somewhat patchy in space, since some communities may be more sensitive to such disturbances than others. It is therefore the interaction of temporal variation of the environment, and the resulting spatial patchiness of its effect, that selects for some degree of adaptive flexibility in species. Insects are expected to exhibit response patterns to environmental heterogeneity because of their relatively short generation times, and these organisms therefore provide suitable empirical models for investigations on the dynamics of communities in the tropics.

Temporal heterogeneity is also evident in the tropics in terms of how seasons affect the patterns of plant growth and reproduction in different regions. The resulting patterns, in turn, determine resource distributions for animals including insects. Levels of harvestable productivity in tropical communities change with seasons, and some plant structures (i.e., vegetative biomass) are replaced with others (i.e., reproductive biomass: flowers and fruits) during a season in some regions. But regional effects also come into play: the tropics are variable in terms of spatial heterogeneity of the environment.

Spatial heterogeneity in tropical regions can be classified as follows:

1. Among regions, differences exist in terms of climate, topography, and the kinds of organisms capable of being residents.

2. Within a region there may exist several distinct or similar habitats, each with a more or less distinct flora and fauna, perhaps with some degree of restriction in gene flow among the habitats.

3. Within a habitat the distributions of organisms will be patchy to some degree; each species may have a characteristic kind of patchy distribution, and the sum of these distributions within a trophic level defines the pattern of community structure associated with that habitat. Patchiness of this sort is believed to vary among some trophic levels: plant species distributions and their associated guilds of phytophagous insects are expected to be patchy in species-rich habitats, while generalist predators, parasites, and scavengers may be less patchy since much of their foraging and feeding is opportunistic. Omnivores may present a mixture of degrees of patchiness in distribution because periods of specialized feeding on certain plants may be intermingled with opportunistic and generalized feeding on other insects. A good example may be some tropical orthopterans.

The spatial heterogeneity of vegetation depends upon many factors, including seed dispersal mechanisms, optimal seed germination and seedling growth conditions, and microclimate.[(463,464)] The bulk of the insect biomass found in a habitat depends in some way upon the vegetation found there, either directly (feeding on leaves, stems, roots, flowers, seeds, and fruits) or indirectly (feeding on fallen and drying or dead plant parts and products). The spatial distribution of a plant species within a habitat is molded by the interaction of both physical factors and biotic interactions among individuals of the species and among species.[(463)] Such factors also underlie the mechanisms of habitat selection in plants. In fact, an optimal habitat is sometimes referred to as an optimal patch,[(463)] although patchiness can also refer to the distribution of resources within a habitat. Wiens[(463)] distinguishes between true habitat selection and habitat correlation. True habitat selection occurs when individuals of a species have a choice between two or more habitats and choose the optimal one. Habitat correlation refers to an enforced distribution among habitats resulting from some environmental factor affecting the species. The response to habitat is determined for a species largely by the patch structure of the region.[(465)] Habitat suitability is also a function of population densities of species: as density increases, habitat suitability declines[(463)] (Fig. 6.13). Under low population density for a species, habitat selection will be most effective as a mechanism to maximize fitness within the population. But comparisons in habitat quality for a species should be made only when populations among different habitats are examined at their equilibrium densities.[(464)] Niche breadth of a species within a habitat is often density dependent.[(49)]

These considerations may also apply to insects because these organisms are closely associated with vegetation as a resource. In many tropical habitats, especially in the lowlands, true habitat selection may occur since plant species do not reach very high densities. This is not universally true as habitats exists where densities of a few plant species are very high. But in habitats with high species richness of plants, the observed distributions of species may reflect true habitat selection, with minimal or no clouding effects of density (Fig. 6.13). The same reasoning may follow for phytophagous insects. Densities of individual insect

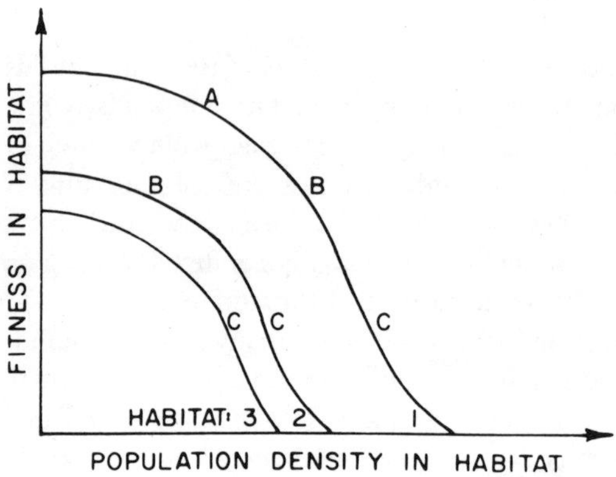

FIGURE 6.13. The Fretwell–Lucas model of habitat selection and distribution. Habitat 1 has the highest quality. When total population size is low (A), all members of the population should reside in it. But as density increases following colonization, the quality of this habitat declines to a point where fitness prospects between habitats 1 and 2 become equal, resulting in the individual selecting both (B). As density increases further, other habitats (C) will also be colonized. [Reproduced, with permission, from J. A. Wiens, *Ann. Rev. Ecol. Syst.* **7**:81–120 (1976). Copyright 1976 by Annual Reviews, Inc.]

species may not attain high levels in some habitats, and the distribution of insect species under such conditions will reflect optimal use of resources.

The actual distribution of individuals over the resources within a habitat or among habitats will depend also upon the "grain-response pattern".[463] Levins[3] distinguished between fine- and coarse-grained environments in terms of the manner in which organisms encounter different environments or different resource patches. An environment is coarse grained if the organism spends all of its time in one patch, whereas an environment is fine grained if the organism passes through many patches.[3,381] Individuals of a species adapted to exploiting resources in a fine-grained manner are predicted to spread more or less uniformly over resource patches as population density increases, whereas an initially clumped coarse-grained specialist at low population density may spread out over other resources as population density increases and suitable resources become available (Fig. 6.14). Thus an insect species specializing on one kind of legume herb in an advanced secondary habitat in the lowland wet tropics may be clumped on patches of that resource when density is low. But as density increases, and if the insect is genetically or physiologically flexible enough to exploit allied legumes in the same area, the population may spread out over the other patch types in the habitat. The key factor is determining the degree to which insect populations in fact increase to densities in the tropics to bring about such predicted changes in distribution. Increased niche breadth may not always be a reflection of increased population density, since some insects may be adapted to feeding on a broad range of plant species regardless of density considerations.

An interesting problem to examine is the transition of an insect species from an association with a natural habitat to outbreak conditions associated with ag-

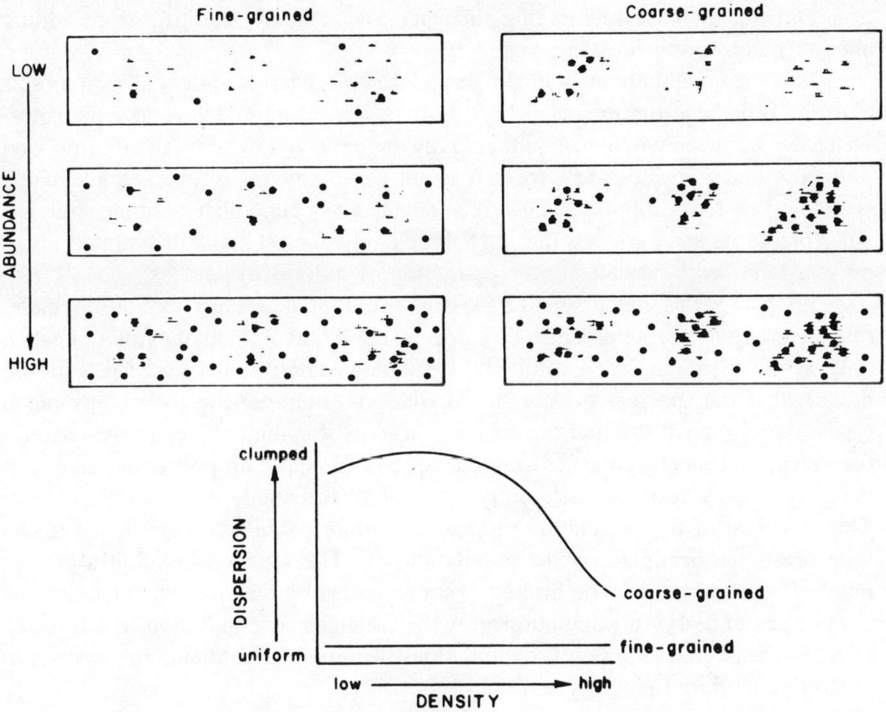

FIGURE 6.14. A model of habitat selection in fine- and coarse-grained environments. The fine-grained population is expected to retain a uniform dispersion pattern over a broad range of population densities, whereas the coarse-grained population initially may have clumped distribution (due to colonization of preferred patches) but later distribution that switches to a uniform pattern as density increases (occupancy of suboptimal patches are imposed upon the expanding population). [Reproduced, with permission, from J. A. Wiens, *Ann. Rev. Ecol. Syst.* **7**:81–120 (1976). Copyright 1976 by Annual Reviews, Inc.]

roecosystems: niche breadth may actually decrease as resources increase (super patch) and population density also increases. Wiens[(463)] summarizes a variety of investigations suggesting that predators forage so as to maximize patch profitability rather than necessarily selecting individual prey types. This is somewhat a restatement of the predicted generalist feeding habitats of predators and parasites in the lowland tropics. Predators lock into a patch where several different kinds of suitable prey may occur, and remain there until it is profitable to discover a new patch with similar properties. Habitats with many patches of suitable prey will promote considerable dispersal within predators and contact with many patches of prey.[(466)] The number of different patch types to be exploited by a predator or parasite species will be a function of interpatch travel time and the increase in foraging time per prey capture resulting from adding another patch to the foraging route.[467]

Many tropical habitats will support a large biomass of insects in various size classes, and these organisms in turn will be prey for predators and parasites. Distances between patches of suitable prey are expected to be relatively low, and predators and parasites are expected to move among many patches and patch types during foraging. Such an effect results in a generalist feeding strategy, since predators and parasites in such habitats may exploit many different kinds of prey or hosts without causing the extinction of individual patches.

Wiens[463] has discussed the possible role of a species exploiting many resources or patch types as a means of decreasing the likelihood of extinction when only a few resources are exploited. Population stability, it is argued, may be increased if the species spreads the population's membership over many patch types, thereby ensuring that the species survives if some patches prove unsuitable. Each patch type presents a unique sphere of selection pressures, and each patch type may respond differently to local disturbances in the environment. Thus survival of the population increases as more patch types are added to the total resources occupied by the population.[463] The ability to exploit a greater number of resources will be limited to some degree by the character states of the parameters underlying population growth. Changes in population density may result in the degree to which subpopulations become differentiated in response to exploiting different patch types.[112]

If a species in the tropics has a wide geographical distribution and gene flow is restricted between many of the populations, there may be considerable evolutionary divergence in response to this broad-scale spatial heterogeneity (regional). Relaxation of gene flow may cloud such effects, however. The expansions and contractions of forests in the Pleistocene in South America have resulted in mixed distributions of completely and partially differentiated species of butterflies and other organisms.[468,469] Differences among populations, when considering each population in terms of a "populational phenotype,"[112] result from selection and drift, processes dependent upon population size and density. The geographical complexity of tropical regions[16] provides the opportunity for considerable divergence in a species in response to regional spatial heterogeneity. Thus the phenotype of the very widespread butterfly *M. peleides* in Central and South America varies greatly over large regions (A. M. Young, unpublished data).

What relationship, if any, such variation has to differences in the environments from region to region is not known. Phenotypic variation in geographically widespread species like *M. peleides* may be related to differentiation of populations in response to regional environmental effects. The degree to which different populations of the butterfly are continuous among regions is not known. (I use the term *region* to refer to large land-mass areas such as northern Central

America, southern Central America, northeastern South America, etc. Within a region there may be considerable spatial heterogeneity through mountains and different climatic conditions over large areas, or zones. Thus Costa Rica is a region within southern Central America in which there is considerable geographical complexity and elaboration of subregions or zones.) *M. peleides* exhibits at least three distinct phenotypic patterns in Costa Rica, corresponding roughly to major climatic zones of the country. A morph found in some parts of the central mountains in Costa Rica has deep blue, almost purple wings; an Atlantic watershed morph has bright blue wings with a black band, and a western coastal form has almost completely brown wings (A. M. Young and P. DeVries, unpublished data). The boundaries of these forms have not been studied in Costa Rica or elsewhere in Central America. How many distinct populations there are of the species within each subregion is not known. Yet an understanding of the ecology of this insect in Central America necessitates determining distributional patterns and pinpointing areas of restricted or absent gene flow. Similar analyses should be performed for other groups of tropical insects as a means of elucidating the relationship between regional spatial heterogeneity and changes in phenotypic means for distinct populations. Special attention should be given species with broad distributions across elevational gradients in the tropics, since environments along these gradients, in terms of habitat type and number of habitat types, are expected to change in response to transitions from biologically accommodated communities in the lowlands to relatively physically controlled communities[(104)] in the high altitudes.

Both spatial and temporal patterns of environmental heterogeneity provide selection favoring the elaboration and maintenance of genetic variation and phenotypic flexibility within populations of insects. The amount of genetic variation, for example, is expected to be greatest for insect species requiring the qualitatively greater shifts in resource utilization patterns in response to environmental fluctuations in space or time. Some tropical specialist species may be coarse grained in distribution in response to patchy distributions of resource patches. Such species will experience different selection pressures in different resource patches if the patches are small; effects of drift may also influence the direction of adaptation in small patches.[(463)] Populations may be characterized by a "populational mean" for certain characters, but we need to consider the character states unique to subpopulations or demes as well, as emphasized in the recent studies of McCauley and Wade[(112)] with laboratory populations of *Tribolium*. Genetic polymorphism is expected under such conditions, with different morphs associated with different patch types. For fine-grained species the environment is encountered as an average of conditions by each individual. Such conditions favor high phenotypic flexibility within each individual rather than a mixture of specialized morphs.[(463)]

Young[(275)] examined such effects for the relationship between patch structure of host plant resources and the population structure of tropical butterflies. Of considerable interest is the possibility that fine-grained phytophagous insects possess a diversity of enzyme forms to degrade plant toxins associated with different plant species. Different levels of fine-grained responses in plant-feeding tropical insects may exist: (1) the insect is polyphagous and requires a diverse set of enzymes to handle diverse sets of plant secondary substances; (2) the insect possesses a broad monophagy in which it feeds on several closely related plants or plants with similar secondary substance profiles, requiring a limited number of enzymes or perhaps one all-purpose enzyme. The first situation might be genetically controlled by a heterozygous system of alleles, while the latter requires fewer alleles or homozygosity. But the alternative explanation, namely, that coarse-grained adaptations imply homozygosity and fine-grained adaptations require heterozygosity, has also been considered.[(463)] If considerable diversification has occurred among the patches or patch types exploited by a coarse-grained insect species, then genetic polymorphism may be the optimal strategy of adaptation. The key factors appear to be the degree to which the resource types are different and the capacity for certain loci to handle each resource type.

It is clear, however, that phytophagous insect species in particular, encounter considerable spatial heterogeneity within tropical habitats, as a function to the plant species richness of these habitats. The overall number of plant species found in the habitat and the spatial distribution of individuals of each plant species in that habitat are major factors promoting the evolution of feeding strategies in tropical insects. If an insect species is locked in as a specialist on a certain plant species, and if that plant species is too patchy within one habitat but not in another, the insect will select the latter habitat, based upon a premise of degree of relative profitability. A complete understanding of the observed patterns of insect species found in a particular tropical habitat will involve the construction of testable hypotheses related to the following issues:

1. The structure and dynamics of the plant community found in the habitat.
2. The patch structure of the vegetation within the habitat, in terms of the patchiness of individual plant species.
3. The degree to which both plants and insects are regulated in abundance by fluctuations in physical conditions of the local environment.
4. The niche breadth of each insect species and changes in niche breadth with respect to changes in population density of the insect species, and similar changes in potentially or actually competing species.
5. Changes in the patch structure of the vegetation over time and possible results of shifting the degree of food limitations for insects.
6. The frequency of local colonizations and extinctions of individual species, for both insects and for plants.

7. Assays of the potential feeding capacities of individual insect species on alternative resources in the habitat.
8. The proximity of other similar habitats as candidates for colonizing episodes of insect species from the habitat studied.
9. The dispersal abilities of insect species.
10. The degree to which insect species are regulated by biotic control agents, and an examination of the natural history of these control agents.

Although much of what is advocated here is strictly natural history in approach, included is the potential for some interesting simple experiments.

The above discussion strongly implies the need to look at the natural history of individual insect species in determining why a species is associated with a particular habitat or community. Mentioned, however, was the shaping of the distribution of a species in a habitat by interactions among closely related species. Tropical lowland faunas in particular are full of taxonomically closely related species of insects, as well as taxonomically unrelated species that may be ecologically similar. Many workers have emphasized that many closely related insect species (i.e., many congeneric species and many allied genera within a tribe or subfamily) co-occur in some regions of the tropics through divergence in habitat preferences or modes of exploiting common resources without conflict. The degree to which each species is a specialist in the habitat, if it co-occurs with other species in the habitat, is also pertinent.

Three species of African cotton stainer bugs *Dysdercus* illustrate different patterns of reproduction and migration that correlate with their differing patterns of exploiting resources. All three species have the same basic life history pattern: a short period of intense flight soon after the molt to the adult stage is followed by feeding, wing muscle degeneration, and maturation of ovaries.[470] Furthermore, when all three species are deprived of food, wing muscles do not break down and eggs do not mature: flight continues until food is encountered. The ecological similarities among these three species of *Dysdercus* stop here, as each species has evolutionarily modified this basic life-history pattern to meet its own ecological requirements. *Dysdercus fasciatus* is the most opportunistic species in that it feeds on abundant fruit supplied at certain times of the year, reproduces early in adult life, and has the highest fecundity of the three species. It is a colonizing species characterized by large adult populations at certain times of the year. At the opposite end of the ecological spectrum is *D. superstitiosus,* which feeds on a greater variety of fruits than *D. fasciatus,* reproduces much later than *D. fasciatus,* and has a much lower fecundity (R_0). It is a strongly K-selected species that thrives in a more certain environment, being genetically and physiologically flexible enough to feed on many different resources; it is a fine-grained species, while *D. fasciatus* is a coarse-grained species with respect to food supplies. The third species, *D. nigrofasciatus,* represents an intermediate or compromise strategy between the other two species in that the age of first reproduction and total fecundity are intermediate between those of the other species, and diet breadth is also intermediate. The result is that each species possesses a distinct population structure and is integrated into communities to varying degrees. The differences among the species may contribute to their ability to co-occur in the same region of Africa but not necessarily imply that selection favoring such ecological divergence came from interspecific competition or competition with other ecologically similar species.

We need to be able to historically distinguish between true habitat selection and habitat correlation[463] in attempting to understand the present-day life histories and distribution of these insects. Whenever possible, it is also advantageous to consider the previous distributions of the species as well as learn whatever we can of the past habitats occupied by the species. Present-day selection on a species in a particular region may have little to do with the kind of selection previously operative on the population. Tropical communities are characterized by the great numbers of closely related taxa of insects and other organisms. The observed distributions may have resulted from past biogeographic events, past selection pressures and habitats, and the forces of contemporary selection.

The incidence in tropical communities of the occurrence of aggregations of conspicuously colored insects on plants is higher than in comparable Temperate Zone communities. Many aposematically colored tropical insects are specialist feeders on plants rich in toxic secondary substances,[343] and aggregative behavior may reinforce distasteful properties to potential visually hunting predators such as lizards and frogs. The conspicuous aggregations of some phytophagous hemipterans on different food plants[183,355] in the tropics may also represent coarse-grained exploitation of a patchy resource. Aggregating behavior may evolve to keep the membership of a subpopulation together in a resource patch while at the same time providing for effective defense against predators. I am not advocating that all patchily distributed insect species on plants have effective means of deterring predators of the kind discussed here. If the insect is associated with a food plant type rich in toxic secondary substances, and if the food plant has a patchy distribution, the juvenile stages in particular of the insect may develop aggregative behavior to deter predators. Perhaps such behavior was made possible by the insect's being a coarse-grained specialist on that plant: the coarse-grained response kept the juvenile stages together on the same plant or in the same small plant patch and provided a preadaptation to the evolution of true aggregative behavior of the kind noted for hemipterans such as *P. blandus* on some Cucurbitaceae.[355] Adults tend to disperse, facilitated perhaps by a shift to feeding on seeds and fruits.

Cluster oviposition might have also evolved as a response to a coarse-grained resource for a specialist insect: cluster oviposition is often associated with highly coordinated aggregational behavior in phytophagous insects. In the past evolutionary history of the insect species, the original host plants might have been very patchy in distribution and the coarse-grained strategy prevailed. Such conditions provided the opportunity for increasing population or subpopulation densities of juveniles on the host plant patch or individual plant, and this was followed by the development of adjustments in the phenotypes so as to maximize survival under crowded conditions, namely, through the evolution of cluster oviposition and aggregated juveniles (and in some cases adults). If the plant species involved was very patchy and the subpopulation was associated with a large patch, some forms of aggregational behavior could be carried over into the adult stage, such as communal roosting in some tropical butterflies. A key factor is the unpalatability of insect to predators and the reinforcement of unpalatability by aggregates on the food plant. Such an explanation for cluster oviposition in butterflies, for example, is very different from others.[265] Even though today the insect exhibiting such traits might be exploiting fine-grained resources,the original pattern of exploitation, perhaps no longer present, would have been coarse grained under this hypothesis. Some aspects of this hypothesis are testable in the tropics today.

6.6. COMPETITION AS A SELECTIVE FORCE IN POPULATION RESPONSES

Because tropical floras and faunas tend to be markedly diverse and rich in terms of taxa, it is tempting to ascribe such diversity, at least in part, to the effects of interspecific competition. Although some forms of character displacement result from competition among species, character displacement or divergence also results from an overabundance of resources rather than from restrictions of resources.[471] If species populations are kept well below the carrying capacities of the environment, and if a suitable resource becomes more abundant, one or some species may diverge to exploit the abundant resource, even though the species may not have been competing. Hespenheide's reasoning[471] is well taken for the tropics, where there can be shifts in harvestable productivity of habitats, increasing or decreasing resources for species. If harvestable productivity is increased, a species previously kept well below its carrying capacity may respond by shifting to exploit the abundant resources. This shift may involve character displacement even though the species was not in competition with other species. Individual selection predicts that populations will increase in size in response to individuals increasing their own fitness through increased reproduction. One way to increase reproduction is to mobilize more energy from an abundant food supply. When two or more ecologically similar species in the same habitat are kept below the carrying capacity of the environment, from predation or parasitism, competition cannot be taking place since resources, even if limited in supply, are not fully utilized by the species. When species exploit patchy resources and are fine-grained, subpopulation densities associated with individual resource patches remain low, so that competition with other species is unlikely. If two or more species are coarse-grained specialists on the same resource, competition may occur if the populations are not kept down by predation, parasitism, or disease.

Insects associated with terrestrial habitats in tropical lowlands probably exhibit various forms of limitation on their populations. Both coarse- and fine-grained species may be limited in abundance by biotic mortality factors such as parasites, in which case populations are below K. Food resources tend not to limit in this case. Competition is unlikely. The fine-grained species probably experience fewer episodes of resource limitation if biotic controls are not operative, because their populations are distributed over several resource types.

Coarse-grained species stand a better chance of entering into competitive interactions with other species, especially if all participants are specialists on the same resources. Resource limitation may result for such species if biotic control is low or absent. Phytophagous insects may be phased in and out of such situations if they are specialists on young or old plant parts. Vegetative growth patterns of plants may not be continuous throughout the year, especially if a dry season occurs in the region, so fresh meristem may be a limiting resource for some insects specializing on both a particular plant species and the new growth of individuals. Similar patterns may occur for insects exploiting flowers and fruits, structures which do not necessarily occur evenly throughout the year. Thus insect populations may fluctuate considerably if resources are fluctuating in abundance and if the species in question tend to be specialists. If many plant species respond to the environment in the same direction, even generalist species may be similarly affected. Detailed natural history studies of each plant species and its associated insects are needed over a several-year period to determine such patterns of abundance, and it is necessary to determine the full spectrum of resources utilized by each of the insects in the guild.

Individual insect species may be temporarily phased in and out of competitive interactions with other insect species, even those feeding on other structures of the food plant,[93] as a function of fluctuations in the physical environment affecting vegetative and reproductive cycles in plants. Communities associated with high-altitude areas of the tropics may exhibit seasonal-type fluctuations in the abundance of some resources (such as new meristem for pierid butterflies[472]), but since the overall diversity of insects within a community tends to be relatively low, competitive interactions are far less likely. Ecological divergence and speciation in high-altitude insects may be relatively infrequent once these species colonize such regions, because the environment tends to be harsh even though perhaps predictable in terms of fluctuations. Harvestable productivity is kept low by the harshness of the physical environment. Unusual topographic local effects may create localized and patchy habitats of a particular kind, and these ''islands'' may serve as sources of speciation in some insects at high altitudes; other than these, the environment is rather homogeneous and harsh. Interspecific competition or resource limitation effects on insect populations are less likely at these altitudes, even though populations may be controlled by harsh climatic conditions resulting in some density-independent mortality and suppression of breeding activity at certain times of the year. Clearly, long-term studies at high altitude regions of the tropics are needed to examine these ideas, as the data base is lacking.

Although there are cases of competition among parasitic ichneumonids for the same host,[473] the predicted generalist feeding strategies of endoparasitic insects in the lowland tropics as discussed in this book probably reducc the intensity of competition among species for any single host species. Clearly many instances of multiparasitism are likely in tropical insects, but parasitoids may be largely fine grained with low investment per host patch. The same may be the case with insects and other arthropods predatory on other insects. However, some predators, such as ambush bugs requiring a certain kind of flower to sit in, may experience resource limitation if flowers are not always available in the habitat. In the case of sit-and-wait predators such as ambush bugs, perching sites such as certain species of flowers may be a limiting factor in their populations, with annual periods of abundance of these insects corresponding to periods of flowering. If there is a temporal succession of different but suitable species of flowers for these bugs throughout the year, populations may stabilize. Highly mobile predators such as asilid flies and various solitary or subsocial hymenopterans may show less responses. Insects which visit certain flowers for nectar or pollen may be capable of facultative shifts to other flowers as species composition of blooms changes throughout the year. Insect species incapable of such facultative switches within and among adjacent habitats may exhibit large fluctuations in population sizes at different times of the year. Fluctuations in the physical environment may also limit populations of pollinators in some instances. Although cacao (*Theobroma cacao*) trees flower throughout the year with some fluctuations, regions in Africa with a strong dry season adversely affect populations of the pollinating midges (Ceratopogonidae), presumably as the result of dessication effects on these small insects. Pollinator activity declines, and fruit set is even lower than usual. Small-bodied pollinating insects in seasonal lowland tropical regions may exhibit considerable fluctuations in abundance between dry and rainy seasons, and fruit set in preferred plant species may also reflect these events.

One of the problems in attempting to explain the distribution of closely related species in tropical communities is the lack of information of the history of species in previous environments. Closely related species in the same region may exhibit various forms of allopatry or microallopatry, but it cannot be concluded that such distributions result from competition. Biogeographical events and opportunistic responses of species to new and abundant resources[471] may result in shifts in distribution having little or nothing to do with interspecific competition. But Janzen[93] has called attention to a much more subtle kind of competition that may be prevalent among species in tropical communities. Janzen argued that unrelated insects utilizing different portions of the same plants within a population of a plant species may be in competition with one another. Insect species A may be attacking the leaves of plant individual A while insect species B is attacking the roots of individual B of that plant species, and since each insect species takes energy away from the plant species, competition is taking place among the species. The combined effects of all such species attacking the plant population, from the viewpoint of any one species in the interaction, is increasing resource limitation because each species will have less energy available through successive generations. The studies of Rockwood[474] demonstrated that increasing herbivore attacks on a plant species lowers the fitness of the plant through reduced production of fruits and seeds. Janzen's concept represents a "super-herbivore effect" arising from many different herbivores removing energy (biomass) from the plant, which in turn reduces its fitness. As a result it becomes less attractive a resource for each predatory insect species.

Andrews[475] presented some interesting data on the arthropod communities associated with cacao plantations in Costa Rica and Dominica that illustrate another aspect of community structure. She found a greater range of arthropod species and numbers per species in Costa Rica for the same number of sweeps taken in the two localities (Figs. 6.15 and 6.16). The Costa Rican (La Lola) plantations seem to support a larger community of arthropods than the Dominica plantations, with ants and diperans forming the bulk of species in both localities (Table 6.5). Yet the Costa Rican samples had greater percentages of other insect orders such as beetles and lepidopterans (Table 6.5), suggesting a greater degree of complexity in community structure assuming that these species are residents of cacao plantations. One is tempted to suggest that the likelihood of competition is greater in the Costa Rican cacao plantations, especially for a group such as ants. Yet the data may reflect differences in habitat complexity within cacao plantations having little to do with competition. Mainland tropical communities are expected to have a greater number of insect species than island communities, even though trophic structure may be different.[415] The great diversity of potential nesting and feeding sites for ants in cacao plantations may be the consortium of environmental features promoting ant diversity.[476]

Carroll and Janzen[196] discuss the relationship between ant diversity and availability of nesting sites in the tropics. Ecological diversification may have

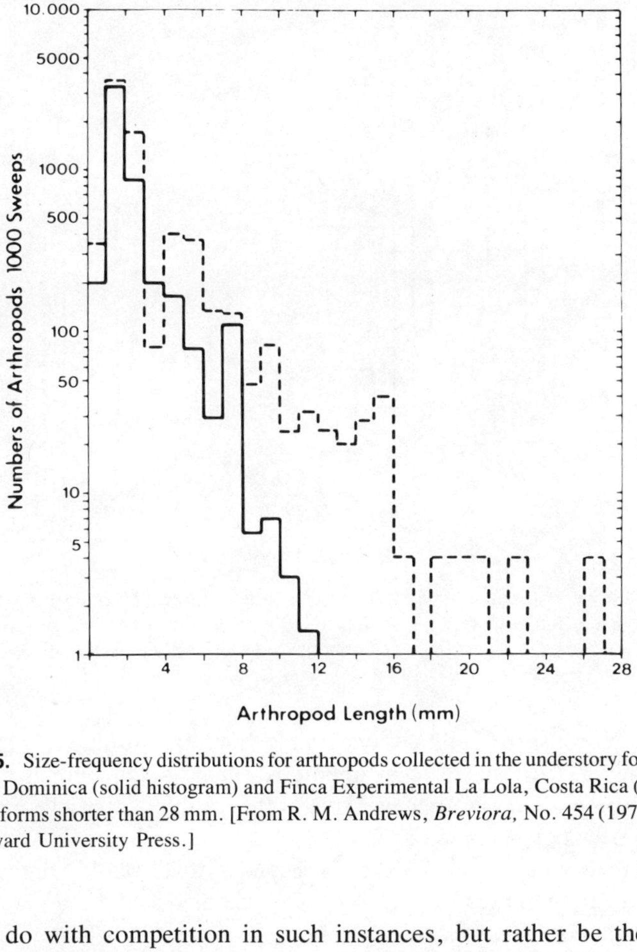

FIGURE 6.15. Size-frequency distributions for arthropods collected in the understory foliage in cacao plantations in Dominica (solid histogram) and Finca Experimental La Lola, Costa Rica (dashed histogram), for all forms shorter than 28 mm. [From R. M. Andrews, *Breviora,* No. 454 (1979). Copyright 1979 by Harvard University Press.]

nothing to do with competition in such instances, but rather be the result of different species encountering different opportunities for fitting into the prevailing environment. In the case of cacao plantations, resident insect species bring to these agroecosystems the host of ecological traits evolved in natural habitats before the plantations were established. Depending upon the type of shade cover, number of tree species providing shade, and other factors, cacao plantations vary considerably in complexity and as environmental mosaics for insects.

Competition may in some instances play a prominent role in determining the diversity of a particular group of insects, as illustrated by Johnson and Hubbell's study of stingless bees.[433] Niche shifts involving insect species making transitions to secondary habitats from forest habitats in the tropics may be another example. Young and Moffett[221,222] suggested as a working hypothesis that the ithomiid butterfly *M. isthmia* may have moved onto tough and hairy solanums in

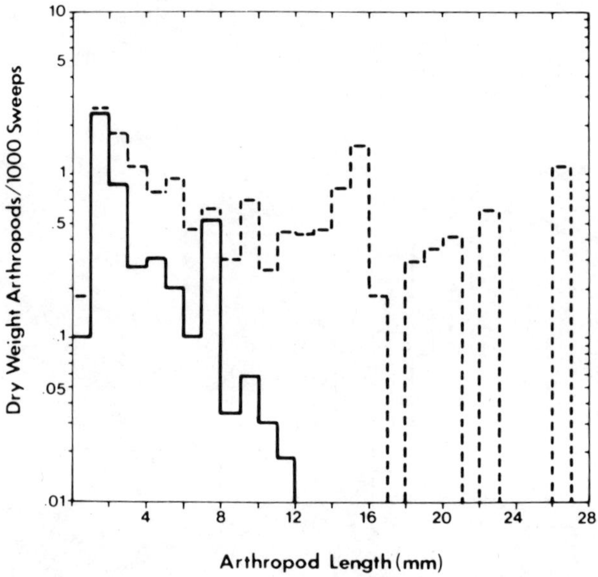

FIGURE 6.16. The distribution of dry weights for different size classes of arthropods censused from cacao plantation foliage in Dominica and Costa Rica (solid and dashed histograms, respectively). [From R. M. Andrews, *Breviora,* No. 454 (1979). Copyright 1979 by Harvard University Press.]

pastures in response to competition among ithomiids for thin, papery-leaved solanums in forest habitats. Most ithomiids exploit such solanums, and cases exist where more than one species exploits the same plant species.[290,302,304,310] Furthermore, woody forest trees and shrubs may be less attractive to insects as host plants than some related species in secondary habitats.[477] Forest solanums are small trees or shrubs, while pasture solanums tend to be spine-covered herbaceous plants. *Mechanitis* might be one of the few ithomiids capable of making the transition from a less suitable resource base to a more suitable one (assuming the results of Scriber and Feeny[477] for Temperate Zone plants) even though the defenses of the pasture solanums may be more difficult to overcome initially. If there was ecological convergence by ithomiids upon the more suitable of forest solanums in response to many others being unsuitable,[477] competition may have resulted among some or all of these species. Forest ithomiids tend to lay few eggs, and larval densities per species are generally low (A. M. Young, unpublished data), suggesting an adjustment in population growth parameters to reduce competitive interactions. These species may be K strategists in the sense of allocating, by necessity, considerable energy to overcoming the nonnutritive properties of the host plants, thereby reducing reproduction. Such effects lower the carrying capacity for each species and reduce the likelihood of competition.

TABLE 6.5
The Composition of Arthropod Collections Made by Sweeping Understory Vegetation on the Dominica and Costa Rica Cacao Sites.[a,b]

	Dominica		Costa Rica	
	Cacao foliage	Forest foliage	Cacao foliage	Forest foliage
Lepidoptera larvae	0.2	0.0	1.9	0.3
Lepidoptera adults	0.3	2.7	5.8	0.7
Orthoptera	1.9	0.5	2.3	4.1
Blattidae	0.1	0.0	0.0	1.1
Hymenoptera: Formicidae	40.8	52.5	10.3	21.4
Hymenoptera: other	10.7	3.1	18.5	12.1
Homoptera	5.5	5.8	10.6	6.4
Hemiptera	1.5	0.0	1.8	0.7
Coleoptera adults	1.2	2.3	13.3	11.2
Diptera adults	32.3	20.6	29.3	21.7
Holometabolous larvae (except Lepidoptera)	0.1	0.0	0.1	0.1
Araneida	4.5	11.8	5.5	6.3
Isopoda	0.0	0.0	0.1	2.4
Isoptera	0.3	0.0	0.0	9.6
Miscellaneous	0.5	0.7	0.5	1.9
Sweeps	700	450	250	350
Total arthropods	3552	3180	1952	1219
Number of 5 mm/1000 sweeps	243	44	972	526

[a] From Andrews.(475)
[b] Values are percentages of total numbers in each collection and do not include collembola and mites.

In such a system, in which one or a few resource types in the forest understory are more or less saturated with butterfly species, there will be opportunistic selection for some species to migrate to new habitats as they appear and where suitable host plants occur. The creation of pastures in the lowland tropics of Central and South America may have provided the opportunity for a few ithomiids such as various species of *Mechanitis* to invade these habitats and overcome an ecological barrier. Such transitions may represent population responses to competition in the original habitat, although all of this is speculation. Indirect evidence is provided, however, by the observations on the natural-history and host plant exploitation patterns of ithomiids today.(290,298,301,302,304,310)

The transition might have been done by a pre-*Mechanitis* form, and the genus may have subsequently undergone its major adaptive radiation in tropical pastures.(478) Other butterflies and tropical cicadas may also be experiencing niche shifts, some of which may result from competition.(71,138) But such suggestions require confirmation with long-term field studies and analysis of the extent to which resources of individual species are limiting in different adjacent habitats. It is necessary to differentiate between noncompetitive responses to new resources or changes in the abundance and accessibility of original resources and to actual competitive interactions when examining shifts in insect species to new resources within a habitat or to new habitats. Regardless of origin, such shifts may involve drastic changes in population structure and dynamics. But a shift arising from a competitive syndrome for a species may be more challenging since the species was probably strongly K-selected in the original habitat. Species with populations characterized by a mixture of r- and K-selected genotypes may have an easier time making such transitions, and such species would be primarily those responding to new resources without coming from a sphere of strongly competitive influences.

6.7. SPATIAL ENVIRONMENTAL HETEROGENEITY AND DEMES IN THE TROPICS

The lowland tropical wet or dry forests consist of mosaics of different habitats, each possessing some dynamic pattern of plant species and insects associated with them. Each habitat has a characteristic distribution of plant taxa, and those with many species with patchy distributions can be termed species rich. Species-rich tropical habitats may include both primary and secondary growth. Janzen[29] has pointed out the patchy distributions of tree species in tropical forests, showing that the observed patchiness could be explained by specialized forms of seed and seedling predation by insects and other animals, coupled with the effects of long-distance pollinators promoting outcrossing.

In habitats where plant species are abundant and patchily distributed, some insect species may have their populations subdivided into demes. A deme of a phytophagous insect species may be associated with one or few resources patches (e.g., host plants), with the demes being separated by the patch structure of the habitat. An insect species may be associated with resources in several adjacent habitats, in which case the deme structure is distributed across those habitats. Wright[448] pointed out that the subdivision of a population into small demes is an effective means to bring about rapid evolutionary divergence if gene flow among the demes is at least partially restricted. In the lowland tropics in particular, whole guilds of insects may be associated with the patchily distributed individuals or clumps of a host plant, and under such conditions one might speak of a form of community evolution: multispecies superdemes may diverge considerably in response to localized selection pressures affecting all members of a guild. Such effects may be more pronounced in the lowland tropics, where many plant species may have patchy distributions over one or more habitats in the region.

There is some evidence of differentiation among demes for some Temperate Zone species of flies[391] in which species are specialists on certain resources. In some tropical habitats and regions we expect to encounter many specialist insect forms, particularly in the case of phytophagous insects: the rich diversity of floras in some regions promotes ecological diversification in phytophagous insects associated with them, even though in some cases a breeding population of a species becomes partially subdivided as a consequence of host plant individuals or clumps being highly dispersed. Such conditions may reduce some gene flow within the population. Ecologically enforced partial isolation of this kind for some insects may result in various levels of drift effects depending upon deme sizes, at the same time buffering against environmental perturbations.[448]

The extinction of demes can be high if environmental perburbations are intense and frequent. The spatial distribution of partially isolated demes for a species promotes and maintains some forms of genetic polymorphisms as a result of disruptive selection. Thus if we discovered an insect species with a distribution on a plant population associated with several habitats in a lowland tropical region, we would expect to find some degree of isolation among subgroups, as is known for some species of *Heliconius* butterflies[479] and probably also occurring in other specialized insect groups. The constancy in membership in some communal roosts of these butterflies may provide some indirect support for the deme concept as applied to tropical insects.[190,479] Yet sometimes whole roosts of these butterflies are disturbed by environmental factors to the point that they disappear from an area.[455] Demes in phytophagous insects and in other specialized insect species in the tropics may represent an enforced diversification in response to the patchy spatial distribution of critical resources necessary to maintain a breeding population. But such demes may be vulnerable to frequent extinctions in habitats where unpredictable environmental perturbations occur. Yet there is also evidence that some demes of tropical butterflies are very stable over long periods.[451]

In strongly seasonal tropical environments, demes may expand and contract periodically and according to season. Thus we might envision that a population of the common nymphalid butterfly *Anartia fatima* in the northwestern lowlands of Costa Rica is distributed rather evenly over large pastures and young secondary-growth areas during the rainy season. These butterflies exploit a variety of herbaceous Acanthaceae as larval host plants,[322] and these resources tend to be abundantly distributed over large areas of secondary growth and pastures in many regions of Costa Rica. Population density of adult butterflies may be quite high in an area. But during the 6-month dry season of this region, food plant patches disappear or shrink in size owing to cessation of vegetative growth and initiation of leaf drop. Such a drastic change in the host plant population results in a contraction of the adult population into many small pockets associated with edges of moist forest patches and similar areas. Late into the dry season the butterflies enter into reproductive diapause. There might be some breeding in these subpopulations or demes early in the dry season, and there might be some gene flow among them. But late into this season the demes cease breeding and gene flow is cut off. In the following rainy season the populations again expand, but the genetic structure of the population may have been changed by "sampling error" arising from deme formation in the dry season. Demes of different size may be subject to different kinds of mortality during the long dry season, enhancing shifts in gene frequencies by the beginning of the rainy season. Similar effects may occur in other insects during severe tropical dry seasons in the New and Old World tropics.

Some species of Lepidoptera in the Temperate Zone with broad geographical ranges often exhibit populations with varying amounts of genetic variation as measured by allozyme patterns.[480] At the same time, there may also be some evidence of substructuring of individual populations as reflected by a degree of intrapopulational variation in allozymes.[480] Depending upon local population sizes, the observed patterns of allozyme variation result from differential effects of natural selection and genetic drift. In the moth *Yponomeuta* studied by Menken,[480] the gregarious habits of the larvae are a factor in promoting the substructuring of breeding populations. Gregarious larvae or nymphs are found in many species of Temperate Zone and tropical insects, and together with the fractionating effects of spatially patchy plant species may result in considerable substructuring of breeding populations (subdivision of demes). On the average, many herbaceous plant groups have very patchy distributions in some tropical habitats, as do tree species,[29] and therefore a greater incidence of substructuring of populations might occur in such environments. Alternatively, some insect species may exhibit considerable phenotypic flexibility of a nongenetic form in exploiting different microenvironments along a gradient of resource patchiness,[481] and in such cases drift may be the major component governing shifts in genetic composition within subpopulations.

The dry season (see Chapter 7) is a periodic form of environmental perturbation affecting the abundance and distribution of plants in some habitats. Such patterns of temporal heterogeneity create changes in the spatial heterogeneity of some resources for insects, which in turn may result in changes in the genetic structure of insect populations. Such populations may be subject to various forms of directional selection and random genetic changes (drift) that shift in direction and intensity between seasons according to variables such as initial population size, genetic variation in the population, dispersal, and the extent to which resources such as food plants are affected in abundances (population sizes, densities) and distribution (patchiness). In the Temperate Zones some insects exhibit polymorphisms that correspond to the exploitation of different kinds of resources over sizable geographical regions. Halkka[482] showed that different colored morphs of a spittle bug *Philaenus* are found in different kinds of host plant patches, and that gene flow is somewhat restricted among these demes. The morph found in each deme is adapted to feed on a particular combination of host plants ("minimeadows"). Dissimilar degrees of dispersal ability between adults and nymphs preserve the polymorphism in response to ecological dissimilarities among widely scattered host plant patches, but speciation is prevented by some exchange of adults. In the tropics we might expect to encounter many examples of such effects since many plant species have patchy spatial distributions in some habitats. Studies of groups such as the Homoptera, Hemiptera, and Orthoptera may reveal some interesting patterns of deme structure and quasi-isolation among demes and how such patterns vary between seasonal and relatively nonseasonal regions. There is probably a greater occurrence of effects similar to those described by Halkka in the Temperate Zone for many lowland tropical areas and involving a greater number of species of insects associated with plants.

Whenever individuals in the population of a given species enter into a period of colonization of new habitats or regions, the species often encounters a change in the spatial heterogeneity and the way the species responds to such a shift in the environment is also different. It is important to bear in mind that colonization is a relative condition of one population compared with another of the same or closely allied species. If a population of a particular species is highly subdivided into several distinct groups or demes, the dispersal among these units is low or absent, one may speak of colonization episodes associated with particular demes rather than with the entire population in the habitat or region (collection of habitats). For a given insect species, we can envision two general types of ecological situations in the tropics that promote colonizing episodes.

First, man-induced changes in habitats may result in a mixture of extinctions and colonizations among the insect species affected by the disturbance. If populations are small and fragile, if genetic variation for dispersal-prone genotypes is low, and if feeding flexibility is low, we would expect the species to experience a localized extinction at a rate directly proportional to the size and age distribution of the breeding population. If these ecological traits are of the opposite character states, colonization is more probable. But the chances of success in colonization are low.[87] Second, recurring localized catastrophes brought about by natural conditions may lead to colonization. In the tropics, these may be classified as landslides, floodings, volcanic eruptions, and hurricanes. These agents function in a density-independent fashion to alter large portions of vegetation cover in some regions. Schwerdtfeger,[45] for example, analyzes the properties of hurricanes affecting northern Central America and the Caribbean archipelagos. Localized landslides may alter the vegetation cover in a particular habitat, making the area potentially available for colonization by other organisms. For example, in the central highlands of Costa Rica, frequent small mud slides provide bare patches of earth that are colonized by seeds of *Machaerium seemani,* a leguminous vine, within a few months (A. M. Young,

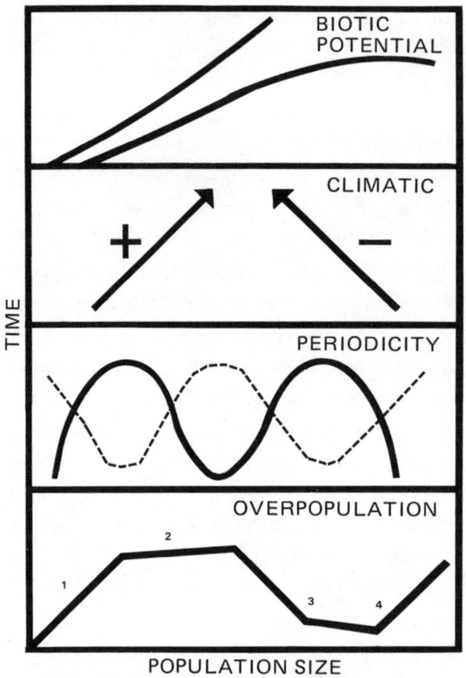

FIGURE 6.17. Four general theories of the regulation of population size. Broken line in periodicity section: cycles of abundance resulting from population's being exploited by a predator or parasite.

unpublished data). These young vines in turn are colonized by the butterfly *M. peleides* (Fig. 4.2), a species adapted to a variety of secondary habitats.[311,316] In more stable secondary habitats in the region, the butterfly oviposits primarily on mature leaves of *Mucuna urens* and several other adult-size leguminous woody vines and trees.[316] The species appears to opportunistically exploit young seedlings of *Machaerium* as they appear in patches following landslides and the construction of new road cuts. Such colonization is somewhat different from full-scale colonizing episodes since the species is probably only occasionally trying to exploit new habitats and resources, while the major bulk of the population continues to thrive in more stable conditions. Yet we may expect that some insect species are in fact experiencing colonizing episodes in response to changes in the availability of optimal resources in existing habitats. As habitat quality declines for a particular species, the likelihood of colonization increases, i.e., migration to another habitat. Colonizations shift the species richness of insects in some tropical habitats in response to changes in the distribution and abundance of resources. Comparisons of species experiencing colonizing episodes with allied species not experiencing colonizing episodes under the same or similar conditions are needed to elucidate the ecological and genetic traits fostering colonization.[483]

It is the interplay of intrinsic and extrinsic properties of insect populations that results in the observed pattern of abundance within a breeding population of a species and the temporal and spatial changes in the population. In this chapter an attempt was made to elucidate some of the mechanisms that regulate insect populations, even though such mechanisms will vary considerably among species, habitat, and geographical region. The array of genotypes in a population during a single generation endows the species with a "biotic potential" or the ability to increase population size. The biotic potential usually results in a leveling off of abundance (Fig. 6.17) as the result of density-related factors curtailing the biotic potential. Regionally, various kinds of climatic factors promote the increase or decrease of the population, and the interaction of the species with its resources and predators or parasites may generate cycles of abundance over time (Fig. 6.17). When the population size is largely a function of prevailing climatic factors and their fluctuations, the population will exhibit periods of both exceptionally low and exceptionally high abundances at different times (Fig. 6.17). The analysis of the census histories of tropical insect populations should take into account the varied sources of regulatory factors affecting any one population of a species.

CHAPTER 7

EFFECTS OF SEASONALITY ON INSECT POPULATIONS IN THE TROPICS

7.1. TEMPERATE ZONE AND TROPICAL SEASONALITY

The entomological and natural history literature is replete with examples of the effects of Temperate Zone cycles of annual seasonality on insect populations. In the Temperate Zone regions there is a succession of four major seasons at most latitudes, although the duration and intensity of each season vary considerably. The growing season or northern summer alternates with the winter season, a period of little or no growth in plant communities and associated animal communities. The transitional seasons of spring and fall are periods of buildup and decline, respectively, of plant and insect populations at northern latitudes. As discussed recently by Lawton,[425] there is considerable interest in how the dynamics of Temperate Zone plant communities influence the diversity of phytophagous insects throughout the growing season.

Seasonality in the temperate zones influences plant species in at least two ways significant to insect populations associated with them: (1) herbaceous plant species with few or no effective chemical defense systems build up large quantities of new biomass each year, a major portion of which is harvestable by insects; and (2) forest trees have most of their biomass already in place each growing season but show seasonal changes in the production of chemical defenses such as tannins.[258] In regions of the world where abiotic factors fluctuate considerably and unpredictably, phenotypic differences among individuals within a population, as related to feeding, disease resistance, and other factors, become of paramount importance in establishing and maintaining populations. For different reasons, a spectrum of phenotypic variation within populations of tropical insects provides the flexibility for responding to shifts in biotic factors affecting average fitness. Lawton,[425] citing other studies, presents data showing that herbaceous species exhibit a gradual buildup of phytophagous insect populations throughout the growing season (data from Price[484] and others) and that insect species on forest trees exhibit a rapid increase in diversity at first, followed by a period of decline and some subsequent recovery (data from Feeny[258]). These effects are stylized in Fig. 7.1. A greater number of insect species are found on trees and woody vines than on herbs in both the temperate zones and the tropics, because trees and woody vines possess a greater biomass per species and greater structural complexity. Thus trees and woody vines have more "microhabitat space" for insects than herbs. This difference between woody plants and herbs for building up insect communities is most pronounced in seasonal environments since populations of individual insect species have a relatively shorter annual time period (compared with the nonseasonal and benign tropics) to build up and maintain breeding populations prior to the arrival of the unfavorable season.

As shown in Fig. 7.1, some interesting similarities in the seasonal distribu-

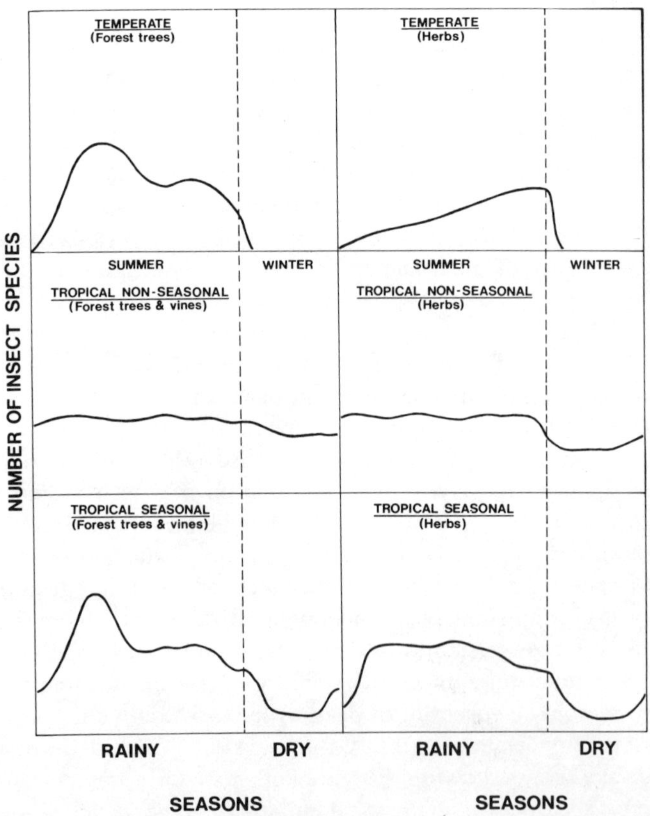

FIGURE 7.1. Hypothetical model for the effects of seasonality, or lack of it, on the species richness of herbivorous insects associated with forest trees and herbs.

tion of insect species are predicted for both Temperate Zone and strongly seasonal lowland tropical regions: the tropical dry season overall is a period of reduced abundance of insect species, whereas the northern winter generally results in a complete cessation of insect activity on vegetation at most latitudes. In the tropics a generally higher harvestable productivity, is expected particularly in herbeceous secondary growth plant species, resulting in a greater accumulation of insect species as the rainy season advances (Fig. 7.1). This effect may even be greater in the nonseasonal tropics for herbaceous plants, since vegetative growth continues to some degree throughout most of the year (Fig. 7.1). Insect faunas of woody vines and forest trees may be relatively less rich in both the seasonal and nonseasonal tropics, if annual vegetative growth is less relative to herbaceous plants in the same region and if trees and vines elaborate large concentrations of defensive compounds in a manner similar to that found in Temperate Zone oaks.[258]

The predicted depression of overall insect species abundance during the tropical dry season in both the seasonal and nonseasonal tropics (in the latter there is usually a short and erratic *veranillo*) does not imply that all species decline: there is some replacement of species in that some become more active than others in the dry season, but in most cases the net result is a decline. Excluded from this reasoning, of course, is the concentration of many species into moist forest refugia during the dry season in strongly seasonal tropical regions. Janzen and Schoener[134] reported that many small-bodied insects become less abundant in the dry season in lowland Guanacaste Province, Costa Rica. Large-bodied bee species, however, become more active during the dry season of this region since many plant species are flowering at this time.[116,118] The depression in insect species during the tropical dry season at some localities is probably heavily skewed toward those with small body size distributions. Comparing the seasonal and nonseasonal tropical regions, fluctuations in species abundance may be more noticeable in the strongly seasonal regions since the populations of individual insect species are believed to be synchronized with the rainy season (Figs. 7.1 and 7.2). Such oscillations are expected to be less noticeable in the nonseasonal tropics since many overlapping generations may occur throughout the year. Several overlapping generations are expected for the seasonal tropics during the growing season, and because defensive systems against herbivores are observed to be well developed in both trees (and woody vines) and herbs, populations sizes are expected to be similar for both kinds of plant growth forms. The nonseasonal tropics provides a greater opportunity for predators and parasites of herbivores to regulate their host (prey) populations, thereby keeping the populations of these insects well below the carrying capacity of the habitat(s) for the average herbivore species.

It must be kept in mind that this is an attempt at a general synthesis, and that the nonseasonal tropics will contain many insect species with marked cycles of

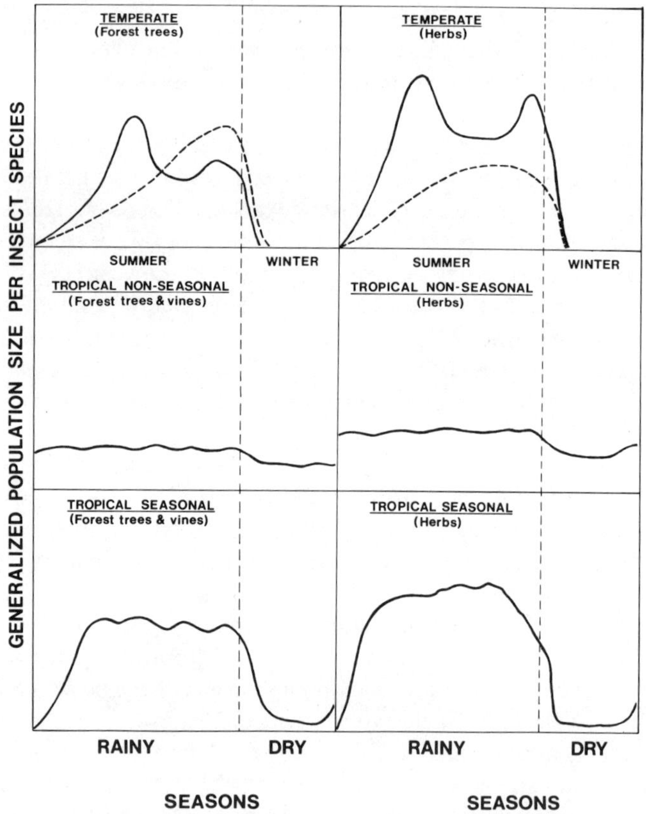

FIGURE 7.2. Hypothetical model for the effect of plant growth form and seasonality on the population census histories for plant-feeding insects in the tropics andTemperate Zone. On a per species basis, population size is probably greater during the growing season in seasonal environments, and in theTemperate Zone a species may be multivoltine (solid lines) or univoltine (dashed line) depending upon the developmental time and absolute length of the growing season.

species abundance and population size fluctuations. Critical factors determining such cycles include the mean generation time of the population for a species the degree to which its food plants are affected by slight or strong seasonal effects throughout the year, and the degree to which the predators and parasitoids of the insect species are also affected by seasonal changes in climate.

We need to examine the population census histories of individual insect species over several years at a locality to determine the extent that populations fluctuate and how fluctuations correlate with changes in the local environment. Such data are generally lacking for most tropical insects. Tropical populations in general are believed, in most species, to have many overlapping generations, although Temperate Zone species may be univoltine or multivoltine (Fig. 7.2). Tropical insect populations, especially those associated with herbaceous vegetation, are predicted to build up more rapidly than comparable Temperate Zone populations because the first rains of the rainy season result in a sudden and massive flush of new growth in many plant species[116] and the rate of increase in new harvestable biomass from plants is generally higher. Temperate Zone plant communities are marked by a more gradual buildup in harvestable biomass, particularly in herbaceous species.

In the strongly seasonal tropics, the number of active insect species may surge at the beginning of the rainy season, followed by a gradual decline and quasi-equilibrium near the end of the rainy season (Fig. 7.1) resulting from a reduction in harvestable biomass, maturation of surviving foliage, and the bringing into synchrony of biotic control agents on phytophagous insect populations. In tropical regions insect communities on vegetation are predicted to come into some form of equilibrium with resources as food quality diminishes in the latter half of the rainy season, and the pattern will be similar to that presented by Lawton[425] (his Figure 7.10) for Temperate Zone communities.

Not all insect orders will respond to seasonality in the same manner, since feeding habits differ greatly. Janzen[72] suggested that plant-sucking bugs exhibit fewer gradients in species composition among habitats in the tropics since they probably come into contact for fewer defensive compounds than chewing insects such as Lepidoptera or Coleoptera. As host plants dry up in open fields in the dry season, homopterans may move into moist forest pockets, undergo reductions in population size, and feed on plant species very different from their original hosts. Homoptera are generally less abundant in forest understories in the tropics since there is less biomass of soft meristems for feeding.[72] As with Temperate Zone faunas, orders such as Lepidoptera and Coleoptera[425] are expected to be the most abundant groups in both forest trees[413] and herbaceous plant communities in the tropics.

The greatest number of discontinuities in species within these orders are predicted in moving from one kind of habitat to another in the same region, and by virtue of their feeding habits Coleoptera and Lepidoptera are specialists on a narrow range of plant species. Thus the spatial rearrangement of the lepidopterous and coleopterous (and others such as Orthoptera) subsets of the local phytophagous insect communities in the tropics will be most drastic in strongly seasonal regions since both trees and herbaceous species may respond considerably to seasonality.[116] Both temperate and tropical communities of bracken-exploiting insects are overwhelmingly dominated by chewing insects, notably Lepidoptera and Coleoptera.[425] The influence of the tropical dry season is probably greatest on chewing insects if these insects exhibit host plant selectivity on the basis of chemical and other defense properties. In temperate forests, disturbed forest patches where the canopy is thinned have higher rates of removal of fleshy fruits and appear to be bimodally synchronized with the end points of the growing season, periods when damage to fruits by insects and other invertebrates is minimal.[485] Light gap patches in mildly seasonal or nonseasonal tropical forests tend to have a high number of bird species as well as insects, and because of these conditions, both fruit damage and fruit removal are expected to be high throughout most of the year, especially since a succession of fleshy fruit types might be available over the year.[101]

The above considerations suggest that the tropical dry season, where and when severe and long in duration, disrupts insect populations for many species by reducing the population and reducing the overall species richness. If the number of species remained about the same during such a dry season, and only population sizes of individual species declined, species diversity would of course increase, as calculated from the well-known formulation

$$\log H = \Sigma\, p_i \log_{p_i}$$

where H is the index of diversity, and p_i is the proportion of individuals in the *ith* species.[3] Yet biologically a species may be replaced with others, and the observed net changes in community structure with regard to insects would be due to factors such as physiological constraints of small body sizes and constraints imposed by specialized feeding habits and requirements.

Examining other levels in the trophic organization of ecosystems indicates that changes in the species composition and population sizes of various phytophagous insects will influence the distribution and abundance of predators and parasitoids on these forms. Again, there might be a replacement of species from season to season in some cases: parasitoids of large-bodied insects may become more active in the dry season if they can remain in a moist enough microhabitat for sufficient periods to allow breeding. Then again, parasitoids and predators may shift to moist forest refugia as a means of "tracking" their hosts. The average amount of time a tachinid fly can spend searching for a succulent lepidopteran larva in an overgrown pasture in a strongly seasonal tropical region is probably far less in the dry season that it is during the rainy season. The same is true for braconid wasps and other forms.

Communities of scavenging insects such as ants and some Coleoptera may also shift if the abundance of dead or dying insects and other suitable animals is less during the dry season in some exposed habitats. Spatial escapes of these kinds for many insects in different trophic levels of tropical ecosystems may allow sustained breeding activity in suitable (even if suboptimal) moist forest refugia during the dry season. Reproductive diapause is another form of escape in which an insect population escapes in time and waits out the dry season and until resources are once again abundant. The incidence of reproductive diapause is predicted to be generally high in specialized plant-feeding insects with chewing mouth parts and in ones which plant resources are greatly curtailed as the dry season advances in a strongly seasonal tropical region. The physiological mechanisms and environmental cues bringing into phase such diapause responses to a time of environmental stress will be different in the tropics than in the temperate zones. Tropical dry seasons are not accompanies by noticeable changes in photoperiod length on a daily basis,[45] but perhaps subtle changes in relative humidity and temperature and large changes in moisture somehow trigger diapause and the behavioral syndromes that accompany it in tropical insects. The enzyme systems involved in physiological responses to the severe tropical dry season must be qualitatively different from those regulating diapause responses in Temperate Zone insects in which other environmental cues are involved.

When a species has a broad geographical distribution encompassing both tropical and extratropical regions, it is often necessary to determine the mechanisms underlying differential adaptation to the different kinds of environments present. A major distinction is the necessity for an overwintering diapause mechanism in temperate populations of such an insect species, while individuals in populations occupying highly seasonal tropical regions require a different kind of mechanism to pass the dry season. From recent studies of Central American and Temperate Zone locust populations,[486] when such divergent phenotypes are required in the temperate vs. tropical populations, strong selection is predicted to be against hybrids along geographical borders between the two regions, since such forms presumably would not be adapted to either kind of environment. It is not a question of clinal gradation from one kind of environment into another, but rather an abrupt change. Selection, under such conditions, favors the expression of effective isolating mechanisms between the populations as a means of reducing the frequency of possible hybrid forms.

Janzen[116] has stressed the synchronization of flowering with the dry season in tree species of the tropical dry forest zones of Central America (see also Section 7.3). The timing of peak flowering in many tree species with the dry season requires that pollinators be active at this time of the year. In fact, one could argue effectively the other way around as well: it is optimal for pollinating animals to be active at this time of the year, and tree species therefore synchronize peak flowering with it. Regardless of cause-and-effect direction, selection has resulted in some insects such as large-bodied bees, butterflies, and moths being active as adults during the dry season and in pollination occurring as a result of this activity. The energy allocation tradeoff between vegetative growth and differentiation of flowers in some tree species during the long dry season is a response pattern imposing a sort of grain on the environment for various groups of insects: phytophagous insects may drop out of the community temporarily or make compensatory shifts to other communities (as discussed above), while pollinating insects of dry-season-flowering tree species respond to a grain pattern (flower resources) replacing other grain resource patterns (meristems, young leaves, etc.). Yet for the average insect species in a seasonal tropical environment, some form of selection operative is expected at least to some extent upon active stages of the life cycle throughout most or all of the year, although the direction and intensity of such selection are changing with time, especially over different successive seasons. This is a very different effect than for most Temperate Zone insect species, in which activity virtually ceases during winter.

Thus the lowland tropics as a whole provides a relatively greater opportunity through time for species of insects to respond to fluctuations in the environment, both biotic and abiotic. The response to selection is faster than in Temperate Zone communities under such conditions. Most tropical insects probably have mean generation times of less than 30 days, as evidenced for many species of tropical butterflies, with notable exceptions such as cicadas, which probably have generation times of 3–5 years in the tropics. Thus most insects can pass through about 6 to 11 generations per year compared with about 1 to 4 generations per year at northern latitudes. Of course, reduction in breeding during the dry season represents the rate-of-response-reducing effects at the low end of the generation time range for tropical insects. Nonetheless, phenotypic changes resulting from selection in seasonal tropical environments endow such species with a means of coping with at least one form of environmental perturbation. Such insect species may be better adapted to responding successfully to broad climatic changes in a region and perhaps for invading high-altitude harsh environments in the tropical regions of the world and to tropical deserts as they appear and expand.

This introductory discussion of seasonality is an extension of Chapter 6, in that seasonality in the tropics represents one form of environmental fluctuation affecting the distribution and abundance of insect species in both space and time. Below we examine some illustrative patterns.

7.2. SOME REGIONAL PATTERNS OF SEASONALITY IN THE TROPICS

The regulation of the size of insect populations is sometimes mediated by fluctuations in abiotic factors in the environment: periods of favorable weather have promoted increases in insect populations, and periods of unfavorable conditions have resulted in decline. Density-independent regulation of insect populations through severe and intense fluctuations in abiotic factors such as rainfall and temperature enforce fluctuations in the census histories of insect populations. In general, Temperate Zone insect populations exhibit considerable fluctuations in size from year to year over several successive years, with outbreaks occurring in years of favorable growing conditions (Fig. 7.2). In Temperate Zone agroecosystems, weather has been shown to locally influence the ability of some endoparasitic insects to regulate the populations of targeted pest insect species,[487] and such effects vary considerably among different localities in a region. In general, biotic control agents are predicted to be less successful in regulating pest populations at higher latitudes, principally because of the variable effects of weather and other physical factors in the environment.[104] In these cases, climatic factors exerted positive or negative influences on population growth over several generations of the insect during a single year, resulting in years of outbreak and years of low abundance. In the tropics, however, lowland regions tend to have fairly stable temperature and rainfall conditions from year to year[45] even though rainfall, for example, may fluctuate considerably from day to day (Fig. 7.3). In the lowland tropics insect populations in general may be subject to less pronounced patterns of temporal variation in temperature than in the temperate zones. There may be considerable diurnal fluctuations in air temperature in the tropics (Fig. 7.4), but month-to-month evenness in the lowlands is usually high.[45] Seasonal shifts in temperature are less likely, with the major shift being in the rainfall regimes.[116] Therefore, insect populations in lowland tropical regions are less likely to be caught in sudden, unpredictable shifts in temperature of the kind typical for the temperate zones. In some insect populations in the temperate zones, shifts in temperature interact with fluctuations in

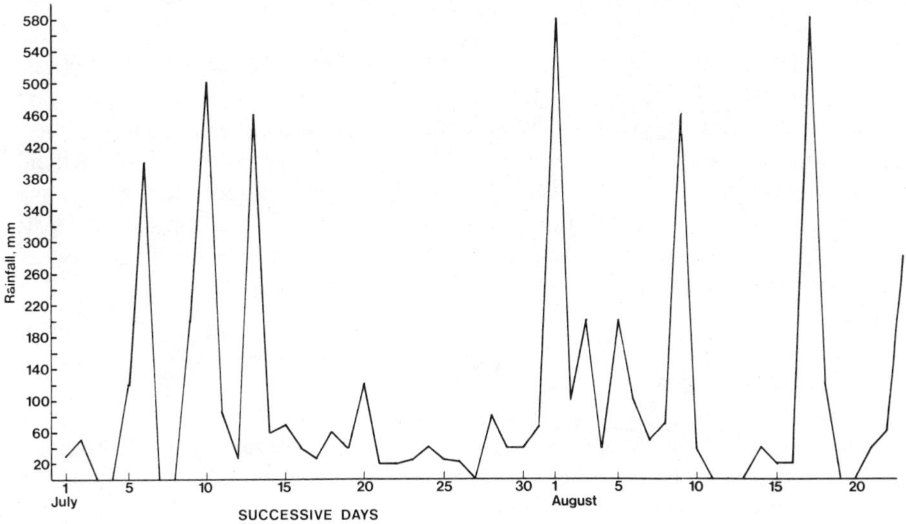

FIGURE 7.3. An example of daily variation in rainfall during the rainy season within the premontane tropical rain forest zone of northeastern Costa Rica. Day-to-day variations in rainfall may influence the activity of insects such as butterflies and cicadas, which often depend upon sunny conditions for reproduction.

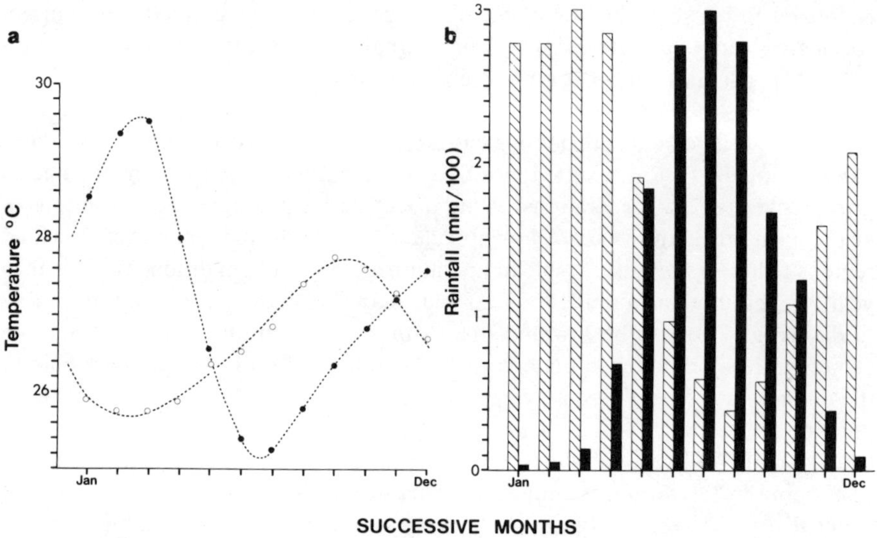

FIGURE 7.4. (a) Monthly temperature at Puerto Ayacucho (●) mean calculated from 24-hourly values for 1958–64), and at Manaus, Brazil (○) (b) Monthly rainfall at San Fernando (solid) (total: 1491 mm) and at Manaus, Brazil (hatched) (total: 2095 mm) for 1931–60. On the ordinate for the rainfall graph, mm/100 means a factor of 100 to adjust the scale shown (0–3). [Modified and redrawn from W. Schwerdtfeger (ed.), *Climates of Central and South America,* Elsevier, Amsterdam (1976).]

population density to determine patterns of reproduction.[487] In some Temperate Zone butterfly species the first killing frost date is unpredictable and results in considerable density-independent mortality of immature stages. Seasonal impacts of temperature changes of these kinds are less likely to occur in the lowland tropics but may affect high-altitude insects. Some tropical lowland regions have very stable monthly temperature regimes (Fig. 7.5) but fluctuating rainfall patterns reflecting seasonality (Fig. 7.3). The result is a seasonal pattern of flowering periodicity in some Brazilian forests (Fig. 7.6), where many tree species flower during the drier months. Different regions within the wet tropics may exhibit very different peak periods of rainfall on an annual basis, reflecting the localized effects of wind patterns and topography.[16,45]

Over relatively short or long distances different climatic and vegetation regions in the tropics may have similar total annual rainfall, but the monthly distribution is different: lowland tropical wet forest regions (rain forests) tend to have uniform rainfall each month but with one or more short dry seasons or *veranillos,* while lowland tropical dry forest regions have an extended period of little or no rainfall, the dry season or *verano,* and high elevational regions have dry seasons of varying intensities but generally far cooler temperatures. In Costa Rica, the length and intensity of the dry season differ considerably between

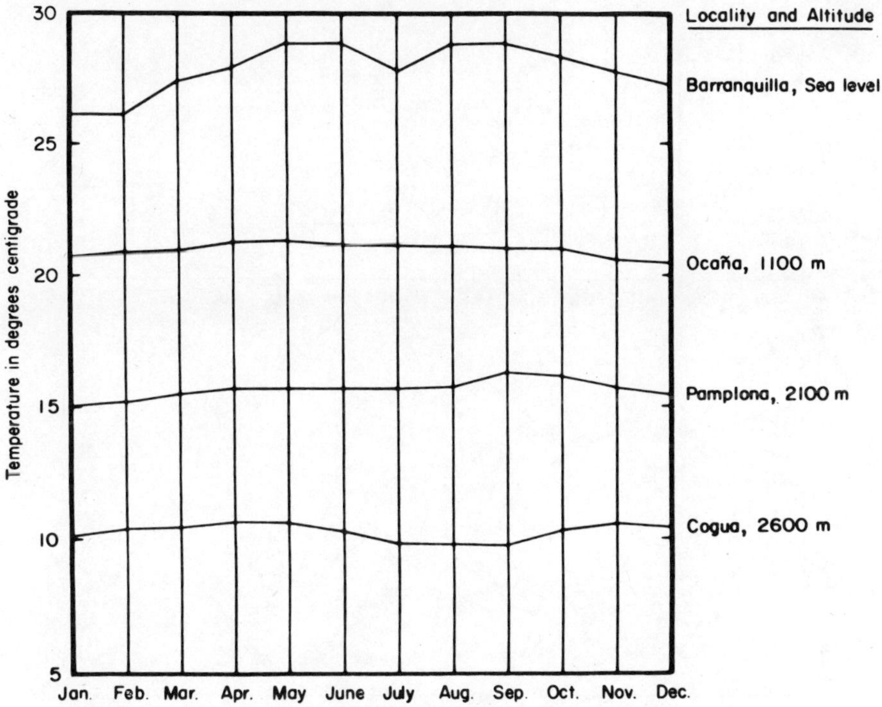

FIGURE 7.5. Along altitudinal gradients in the American topics, the greatest monthly variation in temperature occurs at or near sea level and least on mountain tops, as exemplified here for Colombia. [From G. W. Miskimen, *Biotropica* **4**:85–92 (1972).]

tropical lowlands and mid-elevation areas (Fig. 7.7). Depending upon the region and locality, the annual dry season in seasonal regions may have predictable or unpredictable beginning and termination dates, a property of considerable impact on insect populations where generation times are generally less than 30 days. In some Temperate Zone insects such as crickets, the details of how the life histories of individual species have been shaped by selection for exploiting specific microhabitats have been analyzed,[(489)] and given the overall greater taxonomic diversity within orders in the tropics, such patterns of life history synchronization with seasons should be even more apparent in strongly seasonal regions. Being relatively large-bodied, crickets and other acoustical orthopterans probably exhibit considerable behavioral and ecological specializations, giving rise to both rainy and dry season forms.

Schwerdtfeger[(45)] provides an extensive and exhaustive treatment on the origins and patterns of climatic patterns in Central and South America. The reader is urged to consult this work for detailed analyses of climate in the New

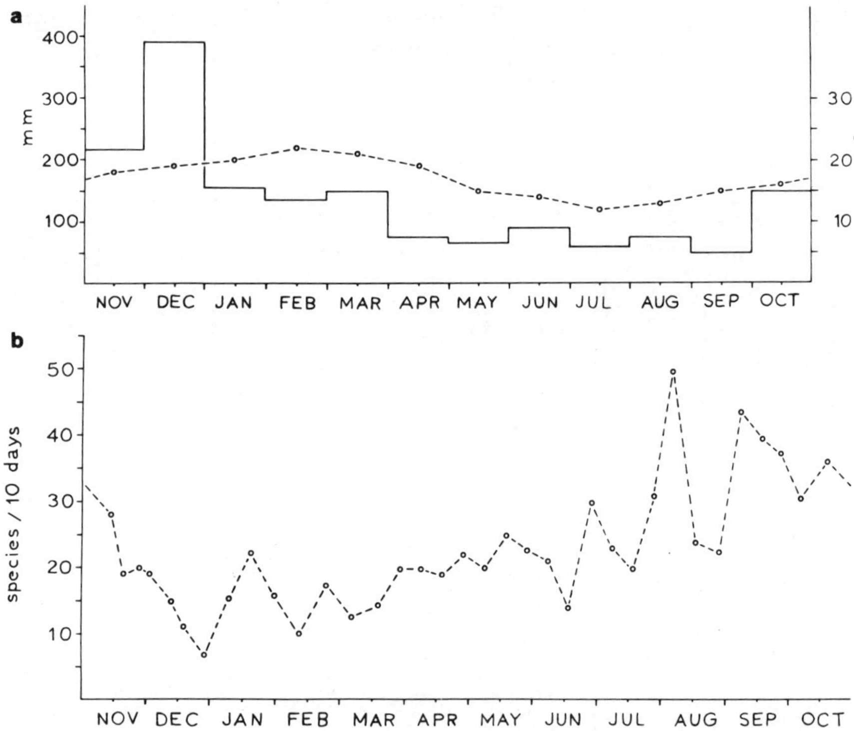

FIGURE 7.6. Annual distribution of mean monthly temperature (dashed line and circles) and rainfall at Reserva Nova Lombardia, Santa Tereza, Espirito Santo, Brazil (a) and flowering phenology for the same period (b) for a representative sample of flora. [From J. F. Jackson, *Biotropica* **10:**38–42 (1978).]

World tropics. An important general contribution from Schwerdtfeger's book, however, is an appreciation for the considerable horizontal and vertical variation in temperature and rainfall patterns in the tropics, a form of broad-scale temporal and spatial heterogeneity. This heterogeneity is often preserved from year to year in that the lowland tropics in particular tend to have fairly predictable annual rainfall and temperature regimes as do some high-elevation regions. In some portions of South America the seasonal distribution of rainfall, for example, is remarkably similar over large areas such as Surinam.[45] Larger areas such as Venezuela exhibit greater heterogeneity in monthly rainfall patterns.[45] Temperature fluctuations in tropical lowlands are on the order of a few degrees throughout the year (e.g., Fig. 7.5). Tropical regions, however, are characterized by large diurnal fluctuations in air temperature (e.g., Fig. 7.3), a feature of the physical environment that promotes separation of activity patterns of groups of organisms

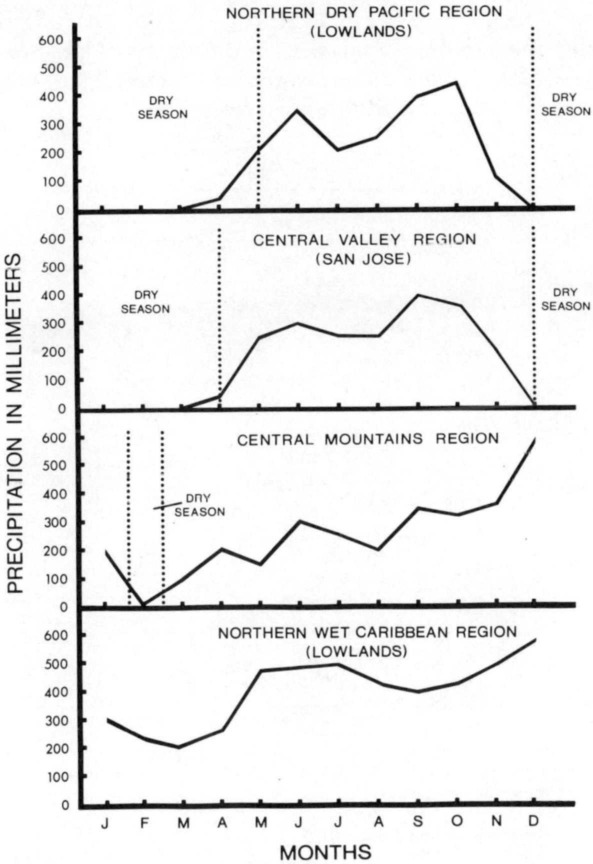

FIGURE 7.7. Variation in the length and intensity of the annual dry season at some representative elevations in Costa Rica. Ten to thirteen years of data are generalized for these regions.

over the day–night cycle. Tropical nights in the lowlands remain warm enough to permit insect activity at night throughout the year, a feature less pronounced in the northern latitudes. The diurnal patterns of insect activity in the lowland tropics may shift in different seasons for aerial forms due to differential effects of air dryness on body size,[134,152] which in turn may alter the diets of some insectivorous birds (Table 7.1).

Temperature regimes in the tropics over the 24-hr daily period thus allow for the evolution of different peak periods of activity in insects, while temporal and spatial variation in rainfall regimes in the tropics promotes the divergence of whole assemblages of plants and insects and whole periods of activity or inactivity (rainy and dry seasons). High-altitude regions in the tropics may be more

TABLE 7.1

Average Abundance of Major Prey Taxa in Stomachs of *Chactura spinicauda, C. brachyura,* and *Stelgidopteryx ruficollis* and among Flying Insects at Different Seasons[a]

	Insects flying[c]	Prey of *Chactura*[b] Individuals	Prey of *Chactura*[b] Species	Insects flying[c]	Prey of *Stelgidopteryx* Individuals	Prey of *Stelgidopteryx* Species
Dry season (Costa Rica)						
Diptera	69.9	6.5 ± 2.4[d]	5.5 ± 1.5	65.5	1.5 ± 0.6	1.2 ± 0.4
Hymenoptera	9.2	69.0 ± 13.7	36.6 ± 3.5	13.5	26.2 ± 1.1	14.7 ± 1.7
Coleoptera	10.7	39.8 ± 9.7	28.3 ± 4.3	12.6	8.7 ± 0.8	8.2 ± 0.9
Total	100	155.3 ± 22.4[e]	80.5 ± 8.2	100	38.0 ± 0.8	25.7 ± 2.6
Wet season (Costa Rica)						
Diptera	75.1	5.6 ± 2.4	4.3 ± 1.1			
Hymenoptera	12.7	98.6 ± 18.3	33.1 ± 2.8			
Coleoptera	10.2	37.6 ± 19.3[f]	15.6 ± 1.2			
Total	100	151.1 ± 22.9[f]	59.5 ± 3.6			
Wet season (Panama)						
Diptera	69.6	10.8 ± 3.3	9.3 ± 2.6			
Hymenoptera	19.6	207.0 ± 91.0	25.5 ± 3.8			
Coleoptera	6.5	27.8 ± 7.0	19.5 ± 4.3			
Total	100	253.5 ± 92.9	60.3 ± 8.7			

[a] From Hespenheide.(791)
[b] *C. spinicauda* (Costa Rica) or *C. brachyura* (Panama); numbers of prey based on six stomachs examined for each season and site for *Chactura*, four for *Stelgidopteryx*.
[c] Proportions of insects in samples at site where birds were collected; see text.
[d] Mean ± standard error.
[e] Includes termites of a single species and 27.6 ± 8.7 individuals.
[f] If nitidulid beetles of a single species found in one stomach are excluded, the number of Coleoptera individuals per stomach is 18.3 ± 1.6 and the total number of individuals per stomach is 134.8 ± 19.7.

uniformly stable in terms of monthly temperature regimes than adjacent lowlands.(45) Miskimen(490) presented climatic data showing that altitude is a critical factor in determining the amount and effectiveness of rainfall in the tropics. Rainy seasons usually begin and end at about the same dates along an elevational gradient from lowlands to mountain tops, but evapotranspiration varies.(490) In terms of physiological effects on plants, the effective rainy season is greatest and most uniform at high altitudes along such a transect, and if so, can support a lush growth of vegetation in some areas.

Several groups of insects in Neotropical environments have rather distinct bimodal daily activity periods, often coinciding with dawn and dusk. Several genera and species of cicadas exhibit highly structured chorusing bouts at dawn and dusk, possibly to maximize acoustical communication in still air,[150] although some of these forms also chorus during other hours of the day.[215] The interplay of physiological constraints related to dessication; exertion of energy associated for foraging, mating, and predator avoidance; and maximization of activity at times when predators are least active, may represent the major selective responses by such insects that result in bimodal, dawn–dusk activity periods. Although some Amazonian simulid flies exhibit clear-cut bimodal biting periods each day, others in the same region do not.[491] Different species of simulids are affected differently by changes in air temperature, light intensity, and humidity, and most species probably have a basic circadian rhythm for biting activity that is modified to varying degrees by exogenous factors. Different species, or different populations of the same species in different regions or habitats, may exhibit differences in response patterns to sets of exogenous factors.[491]

Very high altitudes in the tropics, such as those of the Andes in South America, are cool and have distinct wet and dry seasons at some elevations. Biologically, the effects on insects of seasonality at these altitudes and under cool-temperature regimes may be less than at comparable lowland areas in the tropics. General discontinuities in temperature regimes on tropical mountain tops create ecological barriers for insects[46] much more intense than in mountains of the temperate zones. This is not to say that insects do not respond to seasonality at high elevations in the tropics: they do. But the cool-temperature regimes of these regions lower overall species richness of insects through adverse effects on development and resource heterogeneity. The overall pool of insect species responding to seasonality is less at high elevations (mountain tops) in the tropics, so seasonality is less of a selective factor in shaping the partitioning of insect communities by seasons under these conditions. Long-term studies of high-altitude insects in tropical regions are needed to examine these ideas, and more data are needed on what the forests in disturbed mountain regions such as the Andes were like in the past. Tropical mountain tops are somewhat analogous in effect to the temperate zones in that both promote lower equilibrium number of species in communities and tend to have individual species with larger populations that are physically controlled.[104]

Tropical dry seasons influence insects and other organisms in the following ways.

1. Steadily increasing dryness, from a cessation of rainfall, reduces the food supplies normally available to some species residing in a particular community. Tropical communities with low or herbaceous canopies, i.e., exposed, will experience the effects of the dry season more rapidly than plant species forming the understory of forest. Although the forest canopy may be similar to the open pasture or field in this regard, the spectrum of species present is different as are perhaps the physiological tolerances of the plant species.

2. The dry season may reduce the intensity of competition among ecologically similar species in the habitat if some of the individuals of species are more vulnerable to dessication than others.

3. Likewise there may be a relaxation of biotic control agents such as predators and endoparasites in open habitats as the dry season progresses, since these organisms are generally much smaller than their hosts. The result might be that former host insect populations experience a surge of growth in the dry season if dessication can be avoided and food supplies remain abundant.

4. For many orthopterans, lepidopterans, and other groups, the incidence of egg and juvenile mortality resulting from pathogeneic fungi and other microorganisms may decrease markedly in the dry season as a result of these control agents becoming dessicated, especially in open habitats. Such an effect would contribute to increasing population growth in these insects during the dry season, everything else being equal (which it is not).

5. Small and soft-bodied insects (adults or immature stages) may decline in the dry season in exposed habitats due to water loss constraints on individuals.

6. A portion or all of the membership of a population of either a small-bodied insect species or an insect species whose food supply has dried up in exposed habitats will exhibit lateral or vertical shifts (mini-migrations) to moist forest refugia in lowland dry forest regions where forest cover is discontinuous and scattered. Such movements result in considerable density-independent mortality and increase the concentrations of insect species occupying forest refugia for the dry season. At the beginning of the rainy season such contracted populations may again expand as food supplies and other resources (nesting sites, etc.) are added to the surrounding communities. But the intervening processes may form a sort of deme selection in which drift will change the genetic makeup of population by the end of the dry season. Thus dry seasons and enforced formation of isolated to quasi-isolated demes make up an evolutionary filter determined largely by drift effects (especially if the species enters into a reproductive diapause in the dry period).

7. In some insect species, "dry-season morphs" may increase in frequency in the population while the frequency of "rainy-season morphs" declines as the dry season progresses. Dry-season morphs may be in reproductive diapause, have different thermoregulatory habits and physiology from wet season morphs, be active at different times of the day, and perhaps utilize different sets of resources from rainy season morphs. The distribution of such morphs is regulated by a form of frequency-dependent selection in a polymorphic population experiencing periodic fluctuations in the environment.

8. The insect faunas of dry forest lowlands adjacent to mountains may represent fluctuating subsets of the mountain insect faunas that expand into the lowlands at the beginning of each rainy season, and a portion of which retreats

back into the mountains in the following dry season (and perhaps several generations later). Such shifts in lowland insect faunas may be greater than for insect faunas associated with lowland rain forests adjacent to other mountain slopes in the tropics.

9. In some groups of insects, there might be evidence of considerable seasonal selection among ecologically similar species in which some species are active in the adult stage during the rainy season and other species are similarly active in the dry season in the same habitats. A good example is provided by annual peak emergence periods of some cicadas in northwestern Costa Rica.[139]

Seasonality may differ greatly between dry forest and rain forest lowlands in the tropics. First, the dry season of tropical rain forest regions tends to be less pronounced than that of dry forest regions, increasing the uncertainty of the environment for insects and other organisms to synchronize life cycles and life histories. The beginnings and ends of dry seasons in the wet tropics are less predictable events, as is the total period of dryness. How insect species interpret the dry season in terms of maximizing fitness will thus differ between the two regions. For example, the frequency of courtship and oviposition may actually increase during the dry season in wet tropical regions because many butterfly species are more active as adults in sunny weather (although others are active in cloudy and misty weather) and the incidence of successive sunny days is greater in the dry season. Many butterfly species at 200 m elevation on the Caribbean watershed of northeastern Costa Rica exhibit increased oviposition during the short dry season between January and March (A. M. Young, unpublished data). The stress of the dry season on physiological functions related to water stress in insects and their resources (i.e., plants and plant products) in tropical rain forest regions is probably nil since humidity remains in many pockets of vegetation and generally more forest cover acts as a moisture source for insects (although rain forests are disappearing rapidly in all tropical regions). Thus in rain forests many insects may actually synchronize reproductive activity with the dry season in an opportunistic fashion, while the opposite response may be true for dry forest regions.

The butterfly fauna of Brazil exhibits considerable variation in both numbers of forms active throughout the year in a given region and the estimated numbers of individuals per species.[492] Cycles of seasonality in the tropics, particularly at low and intermediate elevations, tend to reduce populations of butterflies and to a lesser extent the number of species active in the part of the wet season.[492] Clench[493] found that the greatest abundances of species of butterflies on Andros Island, Bahamas, occurred during the early phase of the rainy season (June) when rainfall was highest. Clench noted that some butterflies, particularly pierids, exhibited a seasonal change in phenotypic appearance and also that adult preferences for species of flowers changed with seasons. Subtropical and tropical changes in climate will alter the distribution and quality spectrum available for foraging insects such as butterflies, and switches in patterns of flower visitations are expected.

The large Neotropical cicada *F. mannifera* again provides some interesting insights into responses within populations in regions with different climatic conditions. This robust insect (Fig. 3.15) is distributed in forests in Central America generally below 800 m elevation and is found in both wet and dry forest regions. In northeastern Costa Rica adults exhibit a distinct diurnal singing pattern in which males chorus mostly within a 15- to 20-min period at dusk, and to a lesser degree at dawn.(150) But in northwestern Costa Rica, adults confined to small patches of forest sing intermittently throughout the day (A. M. Young, unpublished observation). Analysis of songs and external morphology of adult cicadas from both regions currently under way (A. M. Young and T. E. Moore) suggest in a preliminary fashion that the species is experiencing very different environments. The data suggest that chorusing behavior in this cicada, a major component of fitness in cicadas in general, varies considerably between seasonal and nonseasonal regions in the tropics, a phenomenon perhaps largely regulated by the differential effects of temperature and other regimes of the physical environment. The observed differences in daily chorusing patterns may be due to evolutionary divergence within the species or else represent differing response patterns in a flexible phenotype. The system warrants further study. Adult *F. mannifera* from the dry forest region have light-green prothoracic markings and are smaller (Fig. 3.15), while those from wet forest have brown prothoracic markings and are larger, suggesting a possible pleiotropic response of gene pools to different forms of selection in the regions studied. There is some merit to long-term field studies of single species in the tropics. The studies of Baldry and Molyneux(494) of the tsetse fly *Glossina medicorum* in the Northern Guinea Savannah region of Africa during the dry season indicate that populations of this insect exhibit definite dawn–dusk cycles of flight activity in certain kinds of forest habitats. Such cycles, in small flying insects, may be adaptive in permitting maximal activity at those times of the day when water stress conditions are minimal.

Depending upon the relative amount of exposure to the sun, different habitats in the tropics will contain organisms that respond to different degrees to seasonality. Plants and insects of the primary forest understory may exhibit less seasonality in peak adult activity periods and synchronization of populations with one season or another than plants and insects in adjacent young secondary habitats and primary forest canopy. Data are lacking on precisely how information about seasonality is translated into a cue for a change in population dynamics in insects in the tropics. The species composition and population densities of various species change in foliage-inhabitating insect communities between rainy and dry seasons.(72,134) Large-bodied insects such as cicadas exhibit fairly distinct allochronic adult emergence periods in tropical regions with a dry season.[70,71,135,136,138,139]

The flowering periods of many tropical trees are also allochronic to varying degrees,[117] and large-bodied pollinating insects are active in the dry season.[116] It is of interest to speculate that changes in the physiology of individual plants within a species population in response to a shift in season may function as a cue to many insects exploiting these plants. The nymphal stages of cicadas in the tropics probably last 3–5 years, as suggested by collections of nymphs: three to five size classes of nymphs occur together in the ground for some species studied (A. M. Young, unpublished data). Although subterraneous nymphs presumably feed on relatively nutrient-poor fluids of the root systems of various plant species in a habitat,[366] some cicada species may exhibit a preference for the root crowns of certain trees.[71] Dense accumulations of nymphal skins from the last molt occur around the base of adult-size trees in many tropical forests.[137] If such patterns reflect actual preferences for the root systems of those trees, is there some physiological change within the xylem fluids,[495] and do cicada nymphs, preparing for the final molt, use these changes as a time cue for emergence? The correlations of emergences with seasons in the tropics are there,[70,71,135,136,138,139] although the mechanisms are not known. Other candidate cues such as changes in soil moisture content and hydrostatic conditions of soil come to mind as cues, but there is probably less of a gradient in soil moisture content established vertically in primary tropical rain forests than in other, more exposed habitats. Yet cicadas exhibit very seasonal emergence patterns in primary forests.[70,71]

In the lowland and mid-elevation regions of the tropics in particular, the observation that many species of insects occur in such regions suggests a possible relationship with the distribution of sizes for individuals within populations of individual species. In other words, when species richness is high, the level of richness is in part a function of the size distributions of these species, a distribution undoubtedly correlated with the degree of within and between heterogeneity. As the result of about 1000 sweep samples in Costa Rica and Massachusetts, Schoener and Janzen[416] have studied the size distributions of insects in the tropics and Northern Temperate Zone. They examined how the number of species and size distributions of insects varied between these two regions. Although the range of size distribution for insects is greater in the tropics, there is a skew toward larger insects as well in this region.[416] Yet it may be difficult to generalize that insects are bigger in the tropics, because the pattern may be patchy in terms of habitats and regions sampled within the tropics. We would expect a greater range of habitats in the tropics, and within habitats a greater range in resources for insects, but such ranges in resources may vary considerably in different habitats and regions. Within- and between-habitat heterogeneity in the tropics is great, and such a pattern of resources promotes a greater number of species and greater range in body sizes among species.

Tropical dry seasons influence the rates of leaf drop and of drop of dying or dead plant products, fruit and seeds, and dead insects into forest streams and rivers.[460,496,497] The result is that nutrient patterns available to aquatic insects and other invertebrates may vary at different times of the year and in different kinds of waters varying in nutrients.[496,497] Shifts in seasons in central-Amazonian inundation forests influence the distribution of forest litter invertebrates, causing migrations or seasonal dispersal.[497] Periods of flooding result in migrations by some invertebrates to other sites. The rainy season not only increases flooding in tropical forest flood plain regions but may also lead to an irregular but continuous pattern of litter production due to falling debris from the forest, as studied by Puig[498] in French Guyana. The result of course is a pattern of pulsing changes in the distribution and abundance of organic materials for ground cover, cryptozoic, or subterraneous insects and other arthropods in such habitats.

Both seasonal and long-term changes in forest litter composition will influence the insect communities exploiting this resource. During rainy periods litter production may increase,[498] and this may be a time of optimal population growth for litter-inhabiting insects, including ant species. A greater incidence of fruit drop and seed drop taking place[101] at certain times of the year than at others may signal an increase in the abundance of ground-dwelling and seed-eating Orthoptera in tropical forests. Thus an understanding of tropical seasonality can be approached from the standpoint of influences on the ecology of individual species in a habitat or from a consideration of how communities respond to potential shifts in resources. Several workers [43,101,116,118,499] have measured the phenological patterns of flowering and fruit production in some tropical forests with a consideration of synchronization with seasons and with activity of pollinators. The annual pulsing of flowering and fruit production represents a predictable fluctuation in the resources available to insects and other animals[500] in tropical forests. Forests with a strong dry season exhibit distinct phenological differences among tree species for flowering,[117] whereas the patterns, while still present, are less abundant in tree species in comparatively nonseasonal forest.

Brown and Benson[204] reported distinct annual fluctuations in the abundance of various species of ithomiine butterflies in Brazil and provided evidence that the maintenace of a multiple mimetic polymorphism in a heliconiine butterfly, *Heliconius numata,* was due to environmental heterogeneity. Potentially mimetic species of ithomiines were far more abundant, and populations fluctuated greatly each year, conditions making *H. numata* a sort of "pseudo-Batesian" mimic of these butterflies.[204] Ithomiine butterflies in tropical rain forest in Costa Rica display a conspicuous increase in abundance during the first half of the rainy season (Fig. 7.8), a phenomenon perhaps related to increases in new host plant biomass for oviposition and larval development. Even though this

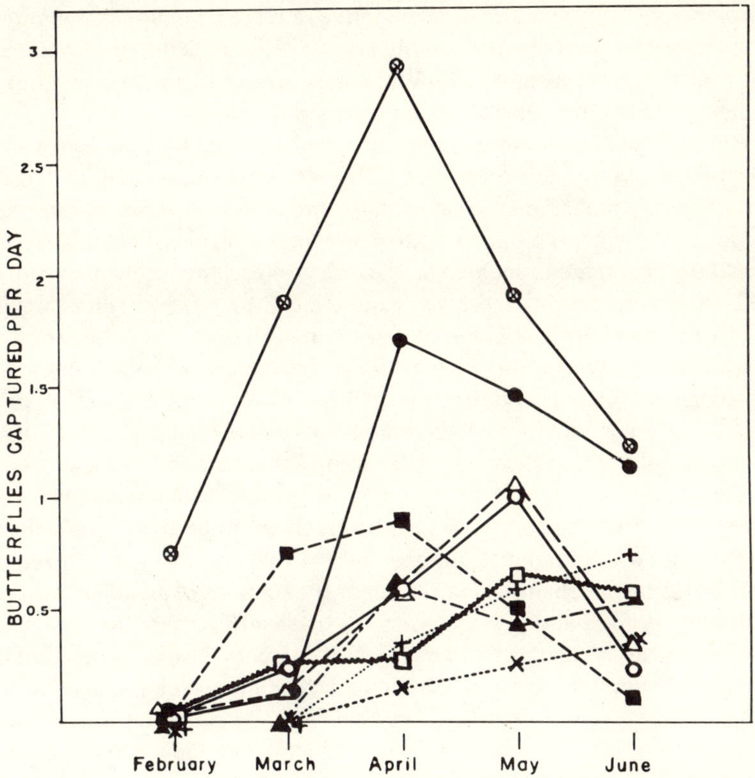

FIGURE 7.8. Abundance of butterflies near Rincon in southern Costa Rica, measured or number of captures per day. The various lines and symbols depict changes in abundance of various species. [From K. S. Brown, Jr., and W. W. Benson, *Biotropica* **6**:205–228 (1974).]

region (Osa Peninsula) is relatively nonseasonal, there is a short dry season. Other groups of butterflies may exhibit increased reproductive activity in the dry season within tropical rain forest zones, depending upon the condition of adult and larval resources: the availability of meristem tissues suitable for oviposition and larval development on host plants during the dry season, and an absence of high adult mortality resulting from desiccation, increased rates of predation, etc., function together to maintain stable or slightly depressed breeding populations and in some cases actually elevate abundance.

In northeastern Costa Rica the number adults of many butterflies markedly declines along forest edges near the end of the rainy season (October–December) each year, as emergences of most cicadas decline.[70] This observation presents a curious situation in terms of some tropical insects escaping from too much wetness rather than too much dryness in the dry season. If being active (courting, mating) in very rainy weather is not adaptive, adult emergence can be suppressed during the wettest months of the rainy season. The reduction in emergence may result from a synchronization of life histories around this period or from annual irregular responses to wetness during this period.

Alternatively, it would be interesting to determine whether in some insects in the tropics a rainy season diapause-like mechanism exists that is locked into a cue of excessive moisture and that actually arrests development in early stages of life cycles. As with reproductive diapause mechanisms in insects of strongly seasonal tropical regions such as lowland dry forest, the mechanisms of control would be different from those of Temperate Zone insects: moisture or lack thereof, in critical amounts, and perhaps some physiological cues from plants, may govern such diapause-like systems, if they exist, in tropical insects. Some field studies of tropical butterflies (e.g., that of *Parides* by Cook *et al.*[444] in Trinidad) indicate that periods of very heavy rains may result in increased mortality of adult butterflies. The heavy rains associated with the wet season in Ethiopia are a major cause of mortality in gelada baboon populations, with annual variations in birth rate varying inversely with the severity of rainfall.[501] Localized shifts of individuals among several subpopulations in the region provided a buffering mechanism against major fluctuations in overall annual population density in the baboon population. In tropical lowlands, the heaviest rains often come near the end of the wet season. This may also be a period when the vegetation consists primarily of mature leaves and in which population densities of folivores and other herbivores may be high. Such conditions may favor a reduction in reproductive output in some insects since food supplies are reduced and saturated with herbivores. Conditions may also be unsuitable for courtship and oviposition. Such effects may be most prevalent in lowland tropical rain forests.

One of the major difficulties concerning our current knowledge of how tropical seasonality cycles affect insect populations is a lack of data over many successive years. Explaining the extent to which year-to-year changes in the intensity of dry seasons in particular regions result in changes in abundance and distribution of insect species in certain kinds of communities and habitats requires such long-term studies. How such correlative patterns change with altitude in the tropics should also be studied, since harsh environments may suppress the effects of seasonality on fluctuations in population size.

The monarch butterfly, *Danaus plexippus* (Danaidae), illustrates an example of a tropical insect that each year attempts a major recolonization of northern latitudes, a migration prompted in part by the expansive geographical distribution of more than 100 species of the larval host plants, milkweeds (Asclepiadaceae), into the Northern Temperate Zone.[98] Tuskes and Brower[98] suggest that clustering behavior in overwintering congregations of this butterfly results in random mating during the spring and before the northern migration begins. Such mating ensures a generalized monarch genome that permits ecological flexibility in recolonizing northern latitudes, a condition made essential by the physiological divergence among milkweed species in terms of the chemical properties of these host plants underlying defense against herbivores. Such flexibility in a species optimizes the chance for successful colonization of a relatively less predictable environment than that from whence the founders came.[(104)] The ecological factors in the evolutionary history of tropical populations of the monarch that resulted in the migratory syndrome being built into the genome of this species remain obscure. On the other hand, the geographical divergence and spread of the larval host plants into northern latitudes would result in a form of selection on milkweed herbivores to exploit the expanded resource base under individual selection of an initially directional form. An intriguing possibility, although exceedingly difficult to confirm, is that monarchs originated in those regions of tropical America that were climatically unstable at the time the species evolved. Alerstam and Enckell[(502)] suggest that migration in many bird species resulted from species originally occupying tropical regions of high environmental instability, such as dry forests and savannas found today in North and East Africa. Such regions, at least for birds, are unpredictable in terms of providing sufficient food and other necessary resources during the breeding season. Analogizing tentatively to the monarch, the question is asked: did the monarch originate in the tropical dry forest lowlands of the Neotropics and evolve the migratory habit in response to a decline in larval host plants during the long and severe dry seasons of these regions? Is what we see today in some dry forest butterflies such as Opler's *Eurema daira,* which migrate vertically into nearby higher elevations during the dry season to reach wetter habitats and survive the dry season, an intermediate step in the evolutionary filter the monarch passed through early in its evolutionary history?

From studies on annual emergence curves for various species of saturniid moths in northern latitudes, Waldbauer[(503)] suggested that the observed polymodal emergence curves of these species are caused by unpredictable environmental conditions: a pattern of peak emergence variation within a single generation in these univoltine species permits survival if early spring conditions are favorable one year, late spring conditions favorable the following year, etc. The adaptation is expected to turn up at the marginal areas occupied by insect species at northern latitudes.[(503)] Waldbauer analogizes the situation to the model of patch structure developed by Levins and MacArthur[(481)] in the sense that the environment each year is coarse grained in terms of optimal periods for adult emergence: there is an early emergence patch and a late emergence patch, and fitness is maximized in the population by the species exploiting both patches.

Could a similar mechanism exist in tropical insects? Overall tropical environments are probably less temporally unpredictable than comparable Temperate Zone environments, even for regions with strong seasonality such as lowland dry forests. Within a single generation there is probably an absence of selection for polymodal emergences in most tropical insects, since the environment is essentially fine grained with respect to optimal emergence periods. Such predictions apply to both seasonal and nonseasonal species since the shift between seasons in the tropics is probably very predictable each year for most localities. The existence of many overlapping generations throughout the year in many insect species in the tropics functions as a form of ecological buffer against any unpredictable changes in the beginning or termination of a season. Overall, polymodal emergences of the kind discussed by Waldbauer[(503)] are probably less numerous in tropical insect faunas. A further buffer in the tropics against environmental uncertainty through time is the edible condition that most kinds of resources remain in to some degree throughout the year, particularly with regard to living and dead plant tissues. Pollinator populations may be more susceptible to the fluctuations in the boundaries of season from year to year or over a run of several years.

For insect species occupying relatively restricted habitats, population densities may undergo annual cycles of abundance in direct relation to the degree of seasonal influence on habitats. Vepsalainen[(504)] found a wing-length polymorphism (dimorphism) in *Gerris* species occupying small ponds subject to periods of increased population densities of these insects. Because the populations are food limited, selection for long-winged morphs occurs when densities are high as a means of colonizing new ponds. Populations cannot exceed the carrying capacity of the habitats, and therefore density is relaxed by colonizing episodes. Short-winged morphs are adapted to low-population-density conditions and allow the species to build up populations. Density-dependent selection favors the maintenance of morphs in *Gerris* populations in regions which experience environmental catastrophes resulting insects vacating many ponds. Such events are followed by colonizations, and when the ponds are filled, selection occurs for short-winged forms until density becomes critical relative to food supplies. At this time long-winged morphs become frequent. In this system disruptive selection favoring both morphs in a population periodically shifts to directional selection for short-winged morphs if the population is isolated.[(504)] If some degree of genetic variation is preserved following directional selection, the long-winged forms can again increase in frequency if favored by selection. Vepsalainen calls this phenomenon a morphism cycle and analogizes it to taxon cycles.[(91)] Such considerations may have a more general application for the analysis of insect populations in the tropics.

In strongly seasonal regions, a form of density-dependent selection may occur for different morphs adapted to different seasons. Some butterflies of lowland dry forest in Central America, for example, may have dry- and rainy-season morphs. P. A. Opler (personal communication) has suggested such an adaptation in the pierid *E. daira* in lowland Guanacaste Province, Costa Rica: a rainy-season morph is relatively mobile and breeds in various kinds of habitats throughout the rainy season. During the early part of the long dry season, populations of the butterfly consist of both morphs, with a gradual increase in frequency of the dry-season morph. This morph is capable of withstanding increasingly dry conditions and enters into a reproductive diapause for the dry season. The rainy-season morph migrates both horizontally and vertically into moist forest habitats, where it continues to be active. The increase in frequency of the dry-season morph is a response to decreasing availability of resources needed by the rainy-season morph. The reverse form of selection is predicted for the rainy season. The adaptation represents a facultative response by a butterfly species to a very seasonal tropical environment. The information available on this system is preliminary (P. A. Opler, personal communication) and warrants further study. Similar density-dependent selection may occur for other insects in very seasonal tropical regions. It is expected to turn up in species whose populations contain sufficient genetic variation and phenotypic flexibility to permit the existence of different morphs, with different sets of ecological traits, and for which a drastic change (usually a decrease) in the abundance of resources occurs in the dry season.

Insect species occupying tropical habitats where resources are very patchy in space and where environmental catastrophes may result in considerable density-independent mortality of individuals in breeding populations may provide another general case of such density-dependent selection in the tropics. Species of insects forming small colonies associated with highly dispersed patches of resources (host plants, fruits, certain kinds of litter, etc.) in regions where periodic floodings occur are candidates for such ecological adaptations. If the patch size is large enough to support a colony or small breeding population and if biotic control is low or density independent, populations or colonies may exceed the carrying capacity of the environment unless colonizing morphs increase in frequency under such conditions. The implication is that insect communities associated with certain kinds of resources and habitats may experience periodic cycles of morphisms if selection pressures can reverse and residual genetic variation is present within populations of each component species for adaptive response. Insect populations may undergo periodic and cyclic genetic tracking of the environment in response to fluctuations in abiotic and biotic factors. Biotic control agents such as parasitoids may experience cyclic changes in abundance, producing fluctuations in abundance in host populations. If population density is high and genetic variation is present, selection may favor morphisms adapted for colonization and integration into the community, respectively. Mixed life history strategies of this kind permit ecological flexibility and the timing of generations with changes in the environment.[(505)]

The studies of Masaki[(506)] on Temperate Zone and subtropical cricket species reveal some interesting information on which life stages are apt to enter a diapause in the tropics. Diapausing nymphs are found in some subtropical crickets, whereas Temperate Zone species tend to have diapausing eggs.[506] Subtropical forms do not have an egg diapaüse. The evolution of an egg diapaüse may represent a greater evolutionary commitment associated with a complete disruption of the annual growing season, whereas nymphal diapause is a less intense response to the less intense seasonality of subtropical regions. Photoperiodic regulation grows less intense as one moves toward the equator, and this gradient may provide the regulatory cue for the growing lack of egg diapause as one moves toward the tropics. In regions of the tropics where a pronounced annual dry season occurs, reproductive diapause of adult insects may be the most common form of responding to the interruption of the growing season. Reproductive diapause would represent the next step from nymphal diapause in the scheme developed by Masaki[506] and extended here for the tropics. Reproductively diapausing adult insects in the tropics may have a greater resiliency for resuming breeding activity once the rainy season begins. In Temperate Zone regions, the severity of winter temperatures precludes activity in adults or larval forms in most interrestrial insects and therefore limits the expression of such an adaptation in temperate regions.

Mooney *et al.*[507] studied the pattern of defoliation of leaves of the Temperate Zone desert shrub *Diplacus aurantiacus* by the larvae of the checkerspot butterfly, *Euphydryas chalcedona,* in California. The region has a severe summer drought in which these plants lose all but their terminal leaves. These remaining leaves have high concentrations of a leaf surface resin which impedes the development of the larvae, and the larvae enter into a diapause to "wait out" the summer until the plants flush out again with new leaves suitable for consumption. The timing of diapause in the butterfly's population dynamics coincides with the period of least suitable food availability. In seasonal tropical regions, similar hebivore–plant interactions may occur, e.g., many plants lose their leaves and cease new growth, leaving only older leaves. The evolution of a diapause mechanism to escape not only a severely reduced food supply during the tropical dry season, but also perhaps in many cases unsuitable remaining plant tissues, would allow the species to wait out the dry season without extinction. Alternatively, the species could migrate to a more suitable environment where suitable food plants are found during this period. Such a strategy would be expected for species which exploit a food plant that undergoes severe reduction of edible parts during the dry season and cannot undergo a reproductive diapause. A host of other factors, such as flight behavior, adult resource availability, and distance to more favorable habitats will be some of the phenotypic constraints determining whether the diapause or migration strategy evolves.

In some groups of insects such as carabid beetles, extensive habitat surveys in the temperate zones indicate that ecologically similar, taxonomically closely related genera and species often occur together in the same habitats,[508] suggesting that interspecific competition may not be as prevalent for some groups as generally believed. Similar surveys in the tropics may also indicate that the high incidence of closely related and ecologically similar species of insects in some habitats is accounted for in terms of other kinds of ecological interactions than competition. In Temperate Zone regions, fluctuating patterns of physical features of the environment may impose a density-independent mortality on insects. Coupled with a limited amount of biotic mortality, such density-independent mortality regulates the abundance of individual insect species. In the lowland tropics, however, the causation may be different: intense predation and parasitism act in a density-dependent fashion to keep insect populations below the carrying capacity of the environment, precluding the expression of interspecific competition arising from the limitation of resources. This mechanism, of course, may be only one of several operative in tropical insect communities as well as to some degree in Temperate Zone communities as well. There may also be resource limitation of some species, especially in highly seasonal regions where temporal fluctuations in food supplies may be high throughout the year. Various behavioral mechanisms may ensure that a population of a particular insect species does not become too high on a particular resource. As described above, the ability to form colonies of breeding individuals within a population results in a fractionation of the population over many resource patches. In the case of relatively few resource patches or an insect with considerable dispersal ability among patches, selective oviposition on unoccupied patches where larvae or nymphs are absent, and the avoidance of oviposition on occupied patches, distributes the population's membership equitably over resource patches. Such behavior has been reported for *Heliconius* butterflies[(283)] and chrysomelid beetles.[(509)]

7.3. SOME SEASONAL DISTRIBUTIONS OF TROPICAL INSECTS

Wolda[(510–512)] summarized several years of light-trap data on the seasonal abundance of some insect taxa (Homoptera, Orthoptera) in different tropical forest habitats in Panama. His surveys of insects were conducted in areas with distinct wet and dry seasons. Wolda found that many insect species exhibit marked changes in abundance correlated with seasonal patterns of rainfall. Many species of insects increased in abundance in the wet season, presumably as the result of new foliage being present.[(510)] Wolda's data point to the effect of increasing the harvestable productivity in terms of new foliage on the abundance of folivores in tropical forests. Less seasonal tropical rain forests may have a steady, even distribution of some insect species throughout most of the year.[(70,71,513)] The taxonomic and ecological complexities of insect communities in the tropics[(47)] suggest that different groups may respond differently to the effects of seasonality or lack thereof. For example, some folivores such as lepidopteran larvae and chrysomelid beetle larvae may exhibit rather marked fluctuations in abundance between seasons because the availability of new leaves also changes. Taxa that are specialized feeders on young or new leaf tissues may exhibit greater fluctuations than those feeding on mature leaves. Phloem and xylem feeders may exhibit fewer responses although be affected in other ways.[72]

Because of such inherent ecological complexities in tropical insect communities, most studies of seasonal changes in abundance of insects in the tropics have dealt with relatively few taxa.[514–516] Seasonal synchronization in the phenological patterns of flowers and fruit[101,118] may dramatically affect the seasonal abundance patterns of pollen-feeding and nectarivorous insects and alter patterns of niche breadth with respect to feeding habits. Clearly we need to examine the population dynamics of insect species in seasonal tropical environments, since some recent studies with vertebrates under such conditions[517] indicate that individual folivore species change their foraging habits in different seasons. Seasonal changes in the susceptibility of vegetative parts of various plant species to attack by insects[474] may result in fluctuations in population size for various insect species in seasonal tropical habitats. Young[315] discovered marked declines in the early stages of the butterflies *Anaea* (*Zaretis*) *itys* and *Anaea* (*Memphis*) *morvus* (Nymphalidae) during the dry season in northwestern Costa Rica and attributed the decline in populations to a reduction in larval host plants (Lauraceae, Flacourtiaceae) (Fig. 7.9). A reduction in both mature and young leaves during the dry season[116] reduces the food supply for the folivorous larvae (Fig. 7.10) of these butterflies, and a small pool of adults presumably passes the dry season in reproductive diapause, primarily in riparian forest habitats.[315]

The breeding populations of other butterfly species may also decline steadily as the dry season advances in such a region,[190,299,315] resulting in relatively small and fractionated adult populations left to colonize once the wet season commences. An abundance of flowers in the tropical dry season provides a food supply for the adults of many butterfly species.[116] Seasonal changes in the species composition and abundance of *Drosophila* as seen in Brazil reflects cycles of expansion and contraction of resources needed by these insects.[74] Not all groups of insects exhibit such population reductions in the dry season. Janzen[72] reported that some insects increased in abundance during the dry season in secondary vegetation. Not only are large-bodied insects such as many Orthoptera capable physiologically of remaining active during the dry season, but their populations may increase since some kinds of preferred foods (seeds, pollen, etc.) may be more abundant this time of year. Population increases in bees and wasps may also occur during the dry season in response to the abundance of flowers. It is the interaction of body size, age distribution in the population,

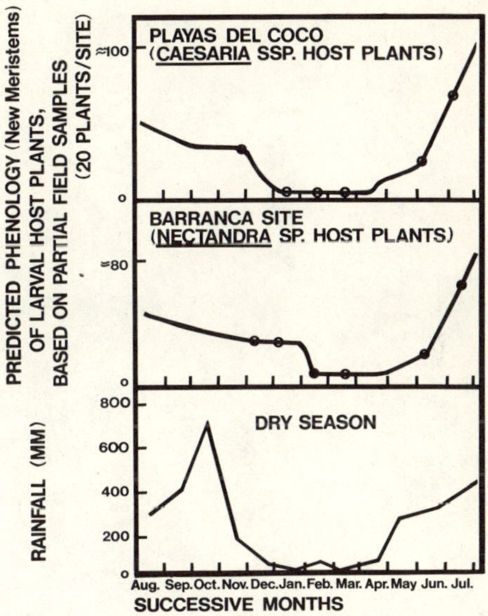

FIGURE 7.9. Seasonal phenology of new meristems for host plants of *Anaea* butterflies within the tropical dry forest zone of northwestern Costa Rica.

kinds of food supplies, changes in food supplies in response to seasonality, and dispersal abilities of individuals in an insect population that determines the response strategy of an insect species to the pronounced dry season in some lowland regions of the tropics. Information is needed on the natural history of individual species in particular habitats to determine such response patterns. Tropical insect communities have traditionally been thought of as diverse in tropical rain forest regions,[518] but such diversity may also be high in tropical dry forest, where the alternation of wet and dry seasons, with intervening short periods of dryness within the wet season,[116] promotes a dynamic and shifting pattern of community structure in response to seasonality. In tropical rain forest, insect and other invertebrate species may have relatively smaller populations that exhibit fewer fluctuations in size or density than similar species in seasonal regions.[519]

In one of the few studies of its kind in the tropics, Brasell *et al.*[520] analyzed the heterogeneity in mineral content in litter between rain forests and adjacent agricultural habitats in tropical Australia. Brasell *et al.* found significant differences in the concentrations of various minerals and nitrogen between the two habitats even though annual litterfall was similar in both. In an *Araucaria* plantation, they found a seasonal shift in the litter concentration of calcium during the fruiting period of the tree. Such data suggest that the composition of litter, rather than its abundance, may vary considerably between forest and agricultural sites in the tropics and that plant phenologies of fruiting may influence the chemical composition of litter. Such effects in turn may influence the cryptozoic and

FIGURE 7.10. Caterpillars of the nymphalid butterfly *Anaea* (*Zaretis*) *itys* in lowland Guanacaste Province, Costa Rica on the food plant (*Caesaria* spp., Flacourtiaceae). Older instars (A and B) rest on leaves while younger ones (C and D) perch on the bared midrib of an eaten leaf, tying leaf fragments and fecal material to the midrib to enhance crypsis.

terrestrial insect faunas associated with such habitats. Depending upon nutritive needs for individual species and the temporal patterns of breeding, populations of these organisms may shift in response to fluctuations in food supplies. Massive annual floodings of rain forest flood plains may also result in significant changes in the nutrient profiles of litter and soil and in the distribution of arthropods found there.[496,497] Janzen[332] pointed out that nutrient-poor soils in the tropics have fewer plant species.

Although the forest floor litter in the temperate zones is generally characterized by various stages of decay, the decomposition of leaf litter in tropical forests, particularly in humid or wet regions or in the wet season in seasonal regions, is relatively fast, with at least one full decay cycle within a year.[521] Less than 10% of the total fallen litter is carried over into a second year,[521] and in some stages of secondary succession, rates of decomposition may even be higher (J. Ewel, personal communication). The leaf litter of a mature tropical forest will be very heterogeneous owing to high local species composition of the flora, and turnover rates of litter from different species are therefore expected to vary considerably owing to different chemical and physical properties of the dead plant tissues. Such a mosaic of litter may promote the maintenance of complex communities of litter organisms, including insects such as collembolans and ants. Because turnover rates are generally high, such communities depend upon a continuous fall of new litter in order to survive. Thus the degree of residentiality of a litter-inhabiting ant colony in the tropics will be determined largely by the turnover rate in the litter as related to rates of replenishment and growth rate of the colony. In turn, the activity of the colony may hasten the rate of turnover of the litter making up the essential microenvironment for the ant species. The communities of dipterans that utilize most litter as breeding substrates, such as many Neotropical Ceratopogonidae,[522] will also be influenced by the same properties of leaf litter. Yet our current knowledge of the structure and dynamics of tropical forest litter communities, in relation to (1) the degree of physical and chemical heterogeneity of the local litter patch, (2) rates of turnover for various species of trees contributing to the litter, and (3) feedback effects of the litter organisms on the availability of litter and rates of nutrient release from it, is meager. The diameters of fallen branches on the floors of tropical forests also determine the rate of wood decomposition by some termites, and fallen branches form another major but rapidly turned-over component of tropical litter.

R. E. Buskirk and W. H. Buskirk[523] analyzed fluctuations in abundance and species composition of foliage-inhabiting arthropods in the understory of a lower-montane rain forest in Costa Rica for 19 successive months and found that both peak abundance and greatest species diversity occurred in the late dry season and early wet season. They also found that arthropods with the largest body sizes were present in greatest numbers at about the same time. They explain the observed patterns by suggesting that the within- and between-season succes-

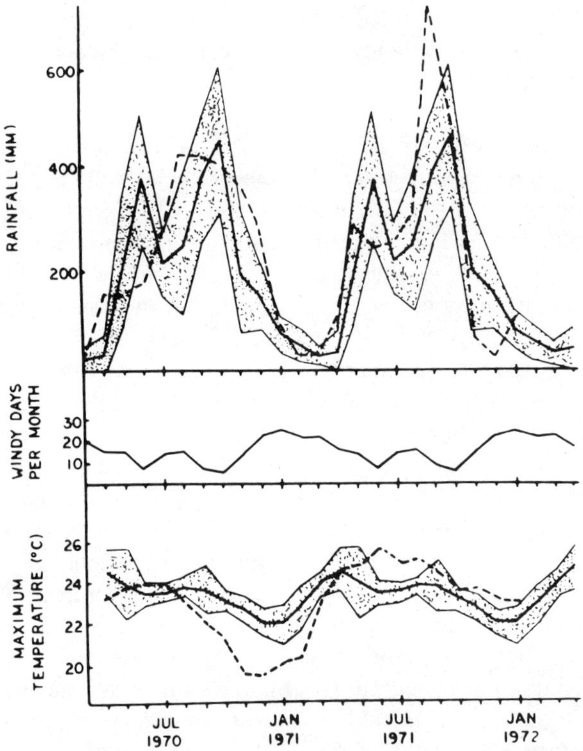

FIGURE 7.11. Monthly rainfall, temperature, and wind patterns at the Monteverde Cloud Forest Preserve in Costa Rica. Shown are monthly mean values (solid lines) with one S.D. (shaded area) for 1956–69, and monthly means for 1970–72 (dashed line). [From R. E. Buskirk and W. H. Buskirk, *Am. Midl. Nat.* **95:**288–298 (1976). Copyright 1976 by the American Midland Naturalist and University of Notre Dame Press.]

sion of arthropod species is due to a succession of whole subfaunae, each associated with a specific set of environmental conditions relating to the plant community and climatic factors. The region experiences a distinct dry season accompanied by increase in winds but no consistent change in temperatures (Fig. 7.11). The observed increase in abundance of arthropod abundance and body size at the end of the dry season and early into the wet season (Fig. 7.12) from the suitability of results large-bodied insects for the dry season and the increase in plant productivity with the first rains. The data of the Buskirks illustrate the synchronization of whole arthropod communities with seasonality in the tropics. The lower-montane rain forest environment experiences considerable seasonality in rainfall, which in turn results in changes in levels of harvestable productivity

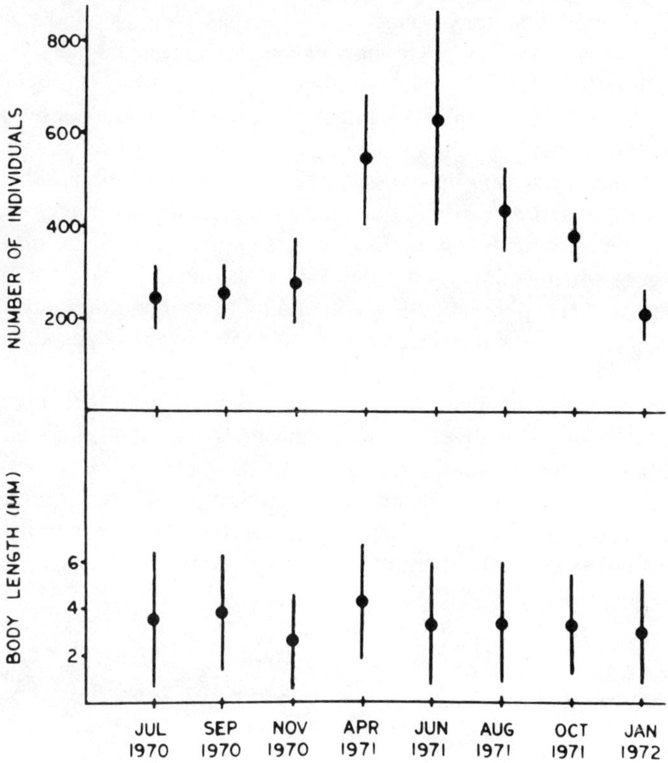

FIGURE 7.12. Changes in the abundance (above) and body size (below) of arthropods as seen in 200-sweep samples (means one S.D.) in the Monteverde Cloud Forest Preserve, Costa Rica. [From R. E. Buskirk and W. H. Buskirk, *Am. Midl. Nat.* **95:**288–298 (1976). Copyright 1976 by the American Midland Naturalist and the University of Notre Dame Press.]

in the plant communities. This in turn changes the community of herbivorous insects associated with the vegetation. Predatory arthropods were found to lag behind the phytophagous groups in cycling through the seasons, a pattern expected for most communities.

Malaise trap sampling of the insects at the interface of grassland and forest habitats in the Nairobi National Park over a 5-year period revealed that peak abundance periods of various taxa of phytophagous insects and their associated parasitic Hymenoptera occurred during periods of heaviest rains in this seasonal region of Africa, although many other groups of nonphytophagous insects exhibited fewer distinct seasonal fluctuations in population sizes and diversity.[524,525] The seasonal periods of heaviest rainfall were characterized by an increase in the abundance of individual phytophagous species as well as in the number of species.[525] Many of the nonphytophagous species sampled appeared to be univoltine, with only one peak period of abundance each year.

The time required for an insect population to exhibit a response to a change in seasons will depend in part upon the duration of the life cycle, generation time, and whether the dry season was passed in a diapause condition. Thus an insect experiencing a reduction in population size during the dry season may not show a recovery response immediately during the early wet season: there may be a lag, the duration of which is directly proportional to the mean generation time and the degree to which the population's membership is synchronized reproductively with the dry season. Diapausing forms may take even longer to exhibit the recovery response.

Bates[526] analyzed patterns of seasonal abundance of mosquitoes at Villavicencio, Colombia, over a 2-year period. Bates found that adult populations of most species peaked in abundance in the early wet season (Fig. 7.13). These species were active nocturnally and therefore must susceptible to desiccation during the dry season. But a few diurnally active species inhabiting the forest habitat had rather uniform distribution throughout the year. These species are

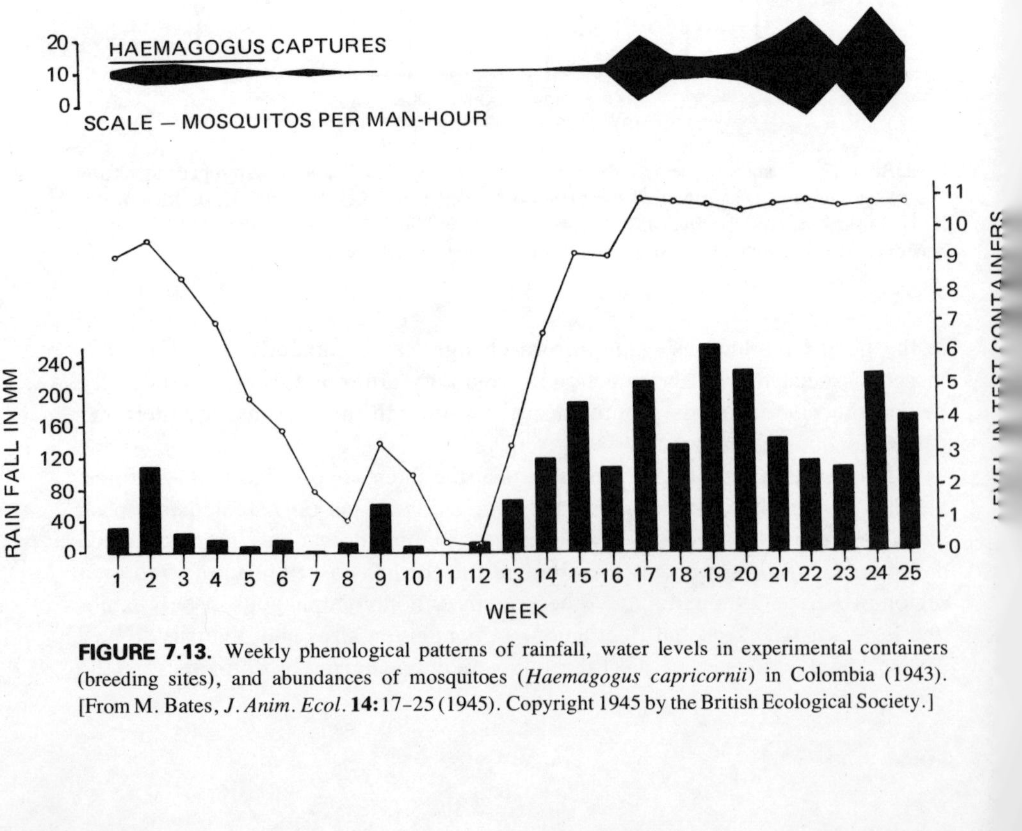

FIGURE 7.13. Weekly phenological patterns of rainfall, water levels in experimental containers (breeding sites), and abundances of mosquitoes (*Haemagogus capricornii*) in Colombia (1943). [From M. Bates, *J. Anim. Ecol.* **14:** 17–25 (1945). Copyright 1945 by the British Ecological Society.]

apparently more adaptive to somewhat dry conditions associated with daytime activity in the forest. Bates suggested that it was the uniform and predictable annual dry season that regulated the abundance cycles of most species of mosquitoes at this site. But because different groups of mosquitoes have quite different ecological requirements in the tropics,[526] not all groups respond to seasonality in the same manner. In addition to desiccation effects on adults, even in a region with a mild dry season such as Villavicencio, species which breed in small standing pools of water may experience considerable reduction in breeding sites during the dry season. On the other hand, species that breed in running waters may exhibit less of a depression in abundance if streams do not dry up completely.

Other insects associated with small pools of standing water may exhibit population declines if suitable breeding sites are decreased in abundance during the dry season. Thus even though the annual dry season at a premontane tropical rain forest site in Costa Rica is short and erratic in intensity, there is a marked reduction in the abundance of adults of the giant damselfly *Mecistogaster coerulatus* in the understory of primary forest during the dry season (A. M. Young, unpublished data). These insects breed in tank bromeliads and small pools of rainwater collecting in depressions on the trunks of fallen trees.[527] Such resource patches may be less abundant in the dry season. The distribution of tank bromeliads in tropical forests is regulated by many factors, of which availability of light is an important one,[528] and many bromeliads are therefore found in or near the forest canopy, a microenvironment subject to increased exposure during the dry season.

Insect species that act as vectors for various diseases of man may exhibit considerable variation in abundance and life history as a result of seasonal effects on the local environment. Dalmat[529,530] studied the distributional patterns of *Simulium* black flies in Guatemala and found marked seasonal fluctuations in the abundance of adults and larvae. Low densities of larvae occur at two different periods: in the late wet season when heavy rains dislodge larvae from their substrates in small streams as the result of flooding, and during the dry season when small streams dry up. Fly species in larger, stable streams tend to exhibit less fluctuations, since these habitats do not change markedly with seasons. Dalmat suggests that populations of black flies, the vector of onchocerciasis in Central America, are regulated by seasonality for species and populations associated with small streams.

In some tropical regions, broad-scale changes in habitats may result in considerable change in the composition and trophic organization of communities. Sometimes these changes are caused by the impact of seasonal rainfall regimes on tropical ecosystems. For example, Irmler[460,496,497] has examined the changes in the macroinvertebrate bottom fauna in inundation forests in the central Amazon region. The inundation forest represents a unique environment in terms of both physical and biotic features.[531] The canopy of the forest is a source of nutrient materials for the bottom fauna of aquatic organisms, including macroinvertebrates, through fall of leaves, fruits, and twigs. Three different water types occur in the region, and each has inundation forests associated with it: white-water type, mixed-water type, and black-water type. Each individual forest type has a unique fauna.[496,497] On an annual basis, the inundation forest passes through various phases linked to the prevailing seasonal rainfall pattern: (1) the inundation phase, which is the transition from dry to wet conditions in the forest; (2) the immersion phase (flooded condition); and (3) the dry phase, in which the water level recedes due to lower levels of water in the rivers from the dry season.

Irmler[496,497] found that various macroinvertebrates exhibit different patterns of passing through the annual cycle of water abundance fluctuation. Some insects, such as cicindelid beetles, exhibit highly synchronized seasonal life history patterns related to the alternation of wet and dry phases of inundation forests in central Amazonia. Irmler found that the cicindelid *Pentacomia egregia* completes larval development in the dry season, and the adults occupy the forest canopy during the wet season when the water level is high in the inundation forest. Irmler[496,497] found that inundation forest associated with black water have a greater diversity of terrestrial soil invertebrates (mostly decomposers) than white-water inundation forest. Black-water rivers are believed to be produced from nutrient-poor white sand soils as the result of the vegetation growing on these soils being exceptionally rich in secondary substances.[332] Such rivers are rich in humic acids such as tannins and other phenols while poor in nutrients.[332] The result is that inundation forests on black water tend to have low productivity and biomass because of poor nutrient content.[332] White sand soils tend to have low diversity stands of trees, while nutrient-rich soils have higher diversity of plant species. Janzen suggests that plant species growing in low-diversity areas require more effective chemical defenses against herbivores. Irmler's discovery of a lower biomass of macroinvertebrates in white-water inundation forest agrees with Janzen's predictions in terms of the litter and soil nutrient components being less than in other kinds of inundation forests. Lower litter diversity and biomass cannot support as large a community of animals as nutrient-rich litters in tropical forests. Furthermore, secondary compounds passing through white-water soils may repel many organisms.

Irmler[496,497] noted that inundation forests in central Amazonia support a high diversity of carabid and staphylinid beetles (about 319 species studied) and that different ecological classes of these insects are based on how they respond to the seasonal cycle of the region. During the inundation phase some beetles migrate to higher areas as the rivers flood, only to be eliminated as inundation continues. Many others do not migrate and synchronize their activity with the dry phase. Within the dry phase distinct phenological groups appear at different times as the water level fluctuates. Some species active in the dry phase switch habits from trees to soil and the reverse in the wet season.[496,497] Some species live in the soil throughout this period, while others remain in the canopy. The dry phase represents a period during which a formerly inundated region can be colonized by beetles from elsewhere and also during which ecological release of some arboreal species can exploit resources on the ground or in the soil. Some species synchronize their life cycles with the dry phase if they are permanent residents of the inundation forest.[496,497] Carabids and staphylinids may exhibit a mixture of generalized and specialized ecological traits in terms of utilizing a highly fluctuating environment such as the ground and vegetation of river-edge flood plains periodically inundated. It is interesting to speculate if the invasion of arboreal habitats by these insects and others may have arisen in response to selection pressures from inundation cycles in central Amazonia. Irmler[496,497] suggests that insect communities associated with inundation forests exhibit considerable change in composition and abundance of component species and that they are therefore somewhat unstable. Such conditions may provide selection for occupying more stable habitats such as forest canopy (relative to the inundated forest floor), and generalist species occupying fluctuating and seasonal tropical environments such as inundation forests may eventually specialize to the forest canopy, where resources may be more predictable through time.

Not all species are expected to display such evolutionary response patterns, since alternative strategies for adapting to the seasonal cycle are available along with the genetic and phenotypic flexibility to exploit new habitats. Nutrient-rich inundated tropical forests are expected to support a greater number of insect species overall than white-water, low-diversity stands of vegetation, and presumably more selection favors transition from ground-dwelling to arboreal existence in the former type of forest: the greater diversity of the forest results in increased selection for niche specialization if resources are periodically depleted during cycles of inundation–emersion–dryness. Irmler and Furch[426] found that some litter-feeding organisms such as cockroaches in inundation forests are faced with food supplies that vary considerably in accessibility and nutritional content at different times in the annual inundation cycle. This is yet another variable to consider as a form of selection pressure: the quality of resources fluctuates with the cycle and may periodically reduce fitness to the point where it becomes optimal to switch to arboreality or other habits.

The same or closely related species occupying different areas within a tropical region may exhibit different adaptations to prevailing conditions of seasonality. In some instances individuals will migrate to areas far from the breeding sites, the pattern determined largely by prevailing winds. In tropical Africa some species of tabanid flies enter into a dry-season diapause, while others migrate to other areas.[532] Adult tabanids generally exhibit very narrow temperature and air moisture preferenda, ecological traits that determine the level of adult activity at any one time. Bowden[532] examined the phenologies of several species of tabanids in Africa and found that some species occur in areas where adult emergences are regulated by successive bouts of wet and dry weather. These species, *Tabanus guineensis* and *T. taeniola,* pass the dry season in reproductive diapause and wait for the wet season to resume breeding.[532] The flies construct mud cylinders to pass the dry season.[533] Adult emergence occurs after a heavy rain, and the timing of emergences in these species is regulated by the moisture status of the soil where the flies pupate. Bowden found that adult emergence is triggered by heavy rainfall about 4 or 5 weeks before the emergence occurs. The time span is needed to bring soil moisture to the appropriate level to trigger emergences.

Tabanids in tropical Africa may require periods of alternate wetting and drying for successful pupation and emergence.[532] Reproductive diapause in tabanids is broken by the heavy forest rains and eggs mature followed by oviposition in moist soils. Shifting bouts of wetness and dryness in the dry season represent an unpredictable environmental fluctuation with an oscillatory period less than the developmental time of these insects. Thus the reproductive diapause ensures that breeding takes place when the wet season is underway and soil moisture gradually increases. Mating and fertilization take place during bouts of wetness when adults emerge. Greathead found that periods of wetting trigger the emergence in only a portion of a given brood of bombyliid flies that parasitize acridid eggs in Africa, so that the emergence is staggered over a long period. The emergence is synchronized with the oviposition of Acrididae. In tabanids, emergence is delayed long enough to permit soil moisture to reach the point at which it would be suitable for development of the eggs and larval survival requirements.[532] The diapausing flies exhibit a delayed response to the early rains, which is very adaptive in that the delayed response by vegetation to early rains also delays the appearance of large wild herbivores bitten by the adult flies. Bowden points out that as the wet season continues soil moisture increases, resulting in an increase in feeding by the larvae and an eventual large hatch of adult flies beyond the first rains. The response delay in adult emergence diminishes as the wet season progresses because soil moisture stabilizes and the cue is always there for larvae and pupae to complete development. The overall result is that adult flies continually emerge throughout the wet season, and the number of overlapping generations increases at the same time. Synchronization in this system is therefore largely restricted to the dry season as a means of timing egg and larval development with the first rains of the new wet season and timing emergences of adult flies with the return of large grazing animals that tabanids bite for blood meals.

Other tabanids exhibit migratory behavior in which migrations follow prevailing winds that eventually give rise to rainfall at some locality.[532] The system is similar to that of migratory locusts in Africa responding to the dry season and its effect on food supplies.[532] Many Acrididae require periods of wetness for oviposition, with dry conditions in the tropics promoting dispersal rather than oviposition.[534] Yet prolonged wet conditions tend to increase mortality of both eggs and nymphs,[534] so optimal conditions tend to be periods of alternating wetness and dryness. Locust populations occupying such areas are capable of building up to very high densities before migrations begin. Density-dependent emigration is a response to increasing population densities and dwindling food supplies, and individuals under such conditions prepare for emigration through physiological changes promoting dispersal and nullifying reproduction. The density-dependent regulation of migratory locust populations in the tropics fits the familiar model developed to show the interaction of increasing densities and decreasing food supplies that regulates population size. Predators, parasites, and pathogens on locusts do not operate in a density-dependent fashion, so population regulation is intrapopulational and related to changes in food supply as well. Migrations in locusts result when the dry season reduces the effective vegetation cover that can be harvested by juveniles, and as density increases, migratory behavior ensues.[534] African tabanids migrate but not necessarily as a density-dependent response to the dry-season effects on food supplies. The available data suggest that adult tabanids follow moisture gradients in migration, as they are sensitive to moisture levels.[532]

One of the interesting facets concerning fluctuations in the populations of insect vectors of diseases in tropical regions is the degree to which populations are regulated by other organisms. For example, levels of pathogenic attack by various microorganisms in larval black fly populations in Guatemala are generally low and highly varied among different black fly species.[39] Since many of these fly species are vectors of onchocerciasis, there is interest in whether these organisms can regulate black fly populations. It appears, however, that natural levels of infestation are low. The larvae of some Lampyridae are predators of snails that are schistosome vectors in Liberia,[535] although few data are available on the extent to which these insects regulate the snail populations. Lampyrid larval populations under natural conditions may have very patchy distributions, thus precluding a capacity for long-term effective control in a habitat, but data are needed to confirm such a prediction. There are efforts to establish populations of selected nematode taxa as control agents for *Anopheles* mosquitoes in El Salvador.[536] Many other environmental factors, such as host distribution and feeding behavior, affect the degree to which a population of mosquitoes infests a particular area,[457,537] and effective control programs must examine as many of such factors as possible and how they change with seasonal fluctuations in the environment. Approaches emphasizing genetic control of targeted mosquito populations in the tropics[538] should examine the effects of seasonal cycles in rainfall upon the ability of selected genotypes (i.e., those with male-linked translocation heterozygotes) to survive at different times of the year or at least during periods when breeding is maximized. In an interesting study, Whitlaw and Chaniotis[539] found that the distribution of triatomine bugs as vectors of Chagas' disease in Panama is determined by the distribution of *Scheelea* palms. Seasonality enters into the picture in the way fluctuations in rainfall influence the interactions of vector or control insect species with their resources, including host abundances and water levels in species with aquatic immature stages.

Although directed movement or migration into more favorable habitats is one known strategy of escape in the tropical dry season in some insects,[540,541] forms incapable of such movement may enter into a period of arrested development or diapause. Seasonal changes in ambient temperature are diapause induction cues in some tropical insects,[524,525,542] and rainfall changes may induce breeding activity in some forms.[543,544] Diapause is terminated in the mosquito *A. gambiae* by increase in rainfall.[545] Induction of diapause in the red locust, *Nomadacris septempfasciata,* is caused by changes in photoperiod near the equator,[546] although such a mechanism is probably less prevalent in most tropical insects in seasonal regions. Physiological changes in host plants provide the cue for larval diapause in some Lepidoptera[547] as well as for the termination of adult diapause in the desert locust, *Schistocera gregaria.*[548] Denlinger[525] discusses the synchronization of life history patterns of insects with seasons in the tropics. Seasonal response systems in tropical insects must be viewed in terms of (1) effects of different seasons on food supplies, (2) phenotypic constraints associated with long-range mobility or lack thereof, (3) availability of nearby riparian forest refugia for colonization during the dry season within a dry forest zone, and (4) phenotypic constraints associated with the feasibility of diapause in different stages of the life cycle.

Owen and Chanter[515] present evidence that seasonal changes in the abundance of Ichneumonidae in Kampala, Uganda, correspond to a similar phenology in the availability of lepidopteran larvae which are hosts. Peak abundances of both ichneumonids and lepidopteran larvae occur early into the wet season, and lags a bit behind the flushing out of the vegetation of the region. The region has two seasonal peaks in rainfall each year, but the insects exhibit a marked increase in abundance only at the peak in rainfall following the dry season, a period of reduced insect populations. Owen and Chanter suggest that insects synchronize their emergences with the average wettest months rather than having maximal emergences in the absolute wettest months. They point out that such abundance correlations with rainfall phenology at this locality are taken with caution since data is lacking on the interactions of the two groups of insects. More information is needed before the abundance of lepidopteran larvae can be concluded to determine the abundance of ichneumonids. Owen and Chanter point out an interesting distinction of their Malaise trap data from similar studies in the temperate zones: in the tropics the abundance of some individual species increases rather than the total number of species at a time of peak abundance, while in Temperate Zone communities the number of species present increases. Owen[514] also reported the same phenomenon in tropical Sphingidae.

Young[71] noted that some species of cicadas in premontane tropical rain forest stagger emergences among bouts of wetness and dryness but that a species may have a peak period of emergence in either the dry or wet season. Young suggested that a minimal and critical number of wet or dry days were needed to trigger emergence and that a season of peak emergence was really a composite of smaller emergence periods of this kind so that the emergence accumulates as the season progresses. Although samples are small, one might argue from such observations that in the tropics overall species richness remains about the same throughout the year but that diversity goes up and down as a few species dominate in the community in response to peak emergences being synchronized with a particular season. On a per-unit-of-contemporary-time basis, progressive and seasonal changes in tropical environments may not be as drastic as in Temperate Zone environments. In the latter a greater turnover in resources may be available to certain groups of insects as the seasons change, whereas in the tropics the effect, on the average, is more quantitative than qualitative. This is not to say, however, that seasonal phenologies in resources do not occur in the strongly seasonal tropics. There is ample evidence that they exist.[116,118] The issue at hand is that different groups of insects and other organisms may exhibit very different types of phenologies in response to tropical seasonality and that the crucial evidence resides in the effect of seasonality on resources and the type of physiological control mechanisms underlying the timing of emergences or lack thereof.

Young[138] found definitive evidence of seasonal adult emergences in various species of cicadas in the central highlands of Costa Rica. Along tree-lined stream edges in otherwise heavily disturbed habitats in this region, one large-bodied cicada, *Q. gigas,* experiences peak emergence each year during the dry season, and there is a good correlation in the abundance of nymphal skins in successive years (Fig. 7.14a). A similar pattern was found for a medium-sized cicada, *Z. smaragdula,* at this site (Fig. 7.14b), but peak emergence is during the long wet season. Emergences of other cicadas at other localities also vary considerably at different times of the year (Figs. 7.15 and 7.16). In lowland tropical dry forests of Costa Rica and elsewhere, cicada species vary considerably in both the size and breadth of the emergence each year and the time of year for peak adult emergence (Fig. 7.17). A major component of fitness in cicadas is the chorusing by males for mating,[150] and highly synchronized seasonal emergence patterns in these insects may reduce possible crossing of courtship signals while optimizing transmission of songs for species active at a particular time of the year.[150,215]

The various genera and species of cicadas studied in Costa Rica by Young[70,71,135–139] exhibit spatial distributional patterns in which the nymphal skins

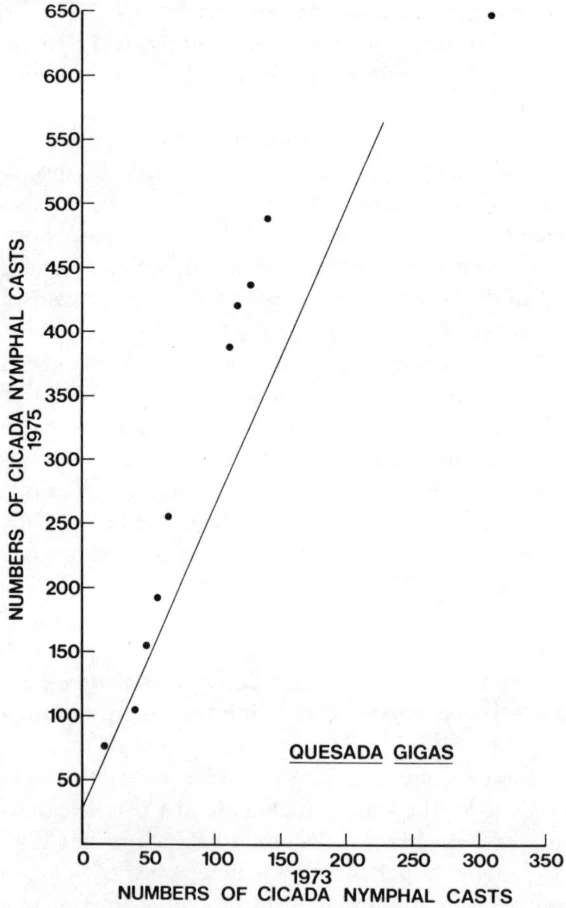

FIGURE 7.14. The relationship between abundances of cicada nymphal casts (skins) in the same study plots at one locality in the central highlands of Costa Rica in different years for two species of cicadas. [From A. M. Young, *Am. Midl. Nat.* **103:**155–166 (1980). Copyright 1980 by the American Midland Naturalist and University of Notre Dame Press.]

of two or more species accumulate beneath large forest trees, particularly Leguminosae, but different species exhibit peak adult emergence at different times of the year. As discussed earlier, the mechanisms of synchronized seasonal emergences in tropical cicadas have not been studied. The possible adaptive significance of allochronic emergences among species in the same habitats, in addition to optimizing courtship, may also minimize competition for oviposition sites. The females of various tropical cicadas tend to prefer similar or the same oviposition sites in the forest understory.[70,71] In lowland and premontane tropical rain

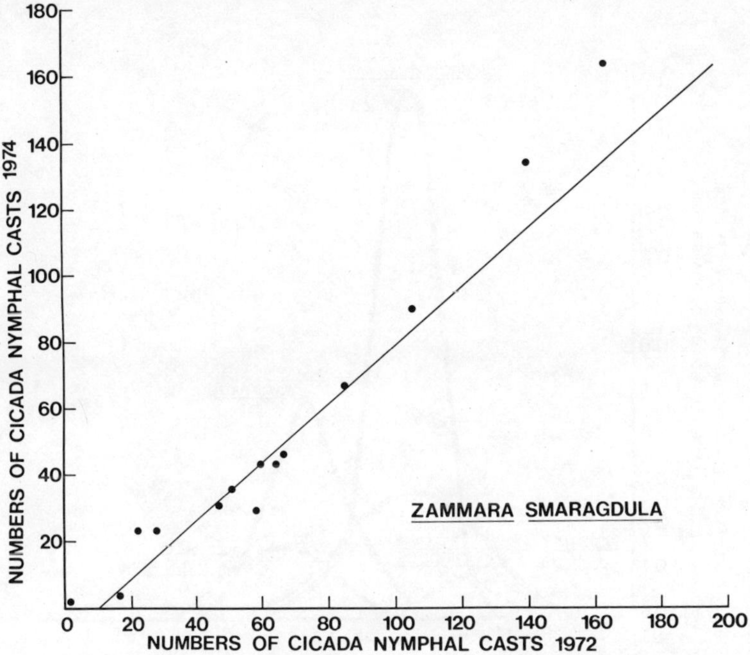

FIGURE 7.14. *(Continued)*

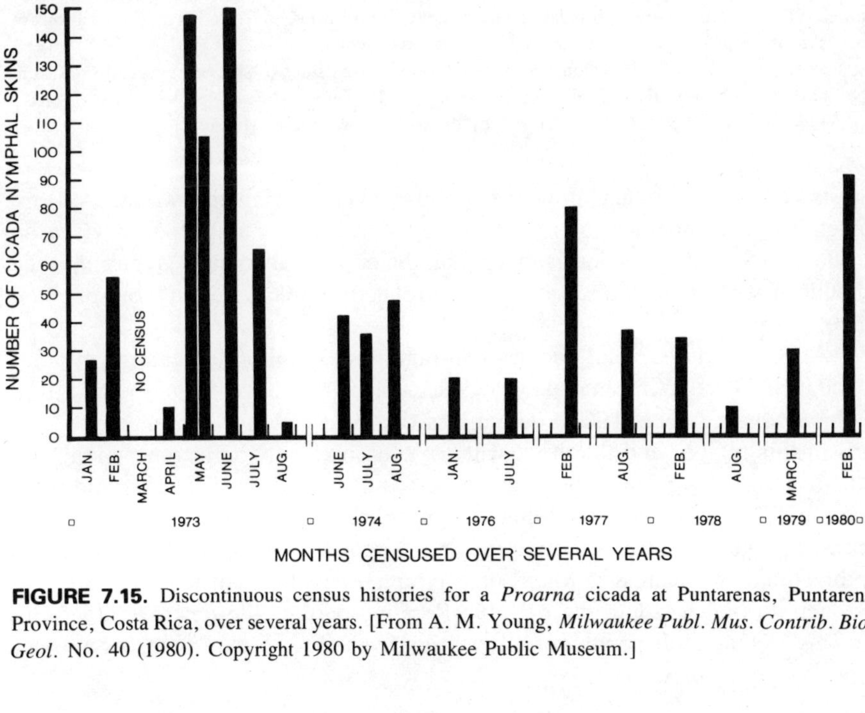

FIGURE 7.15. Discontinuous census histories for a *Proarna* cicada at Puntarenas, Puntarenas Province, Costa Rica, over several years. [From A. M. Young, *Milwaukee Publ. Mus. Contrib. Biol. Geol.* No. 40 (1980). Copyright 1980 by Milwaukee Public Museum.]

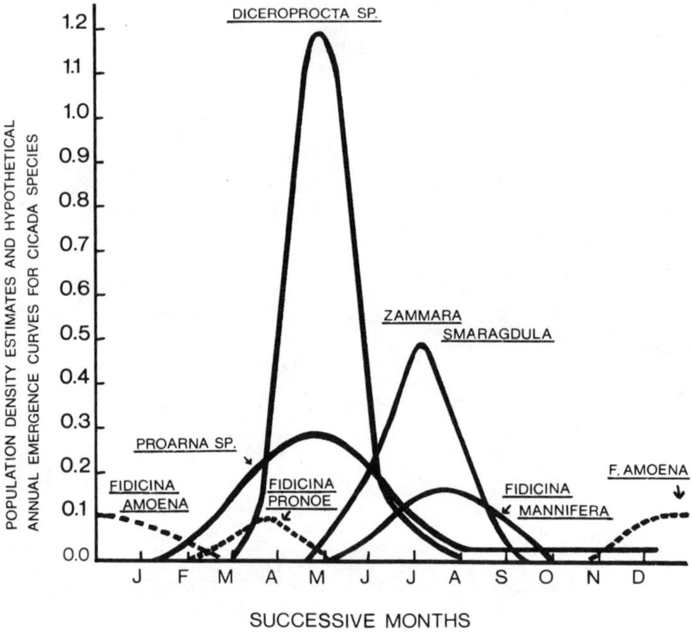

FIGURE 7.16. Predicted annual emergence curves for adults of several species of cicadas (Homoptera: Cicadidae) in the lowland tropical dry forest zone of northwestern Costa Rica. The positions of the peaks of these curves are based on actual observed densities, but the end points in several instances are hypothetical projections. The dashed-line curves for two species represent data from other localities and projected for this region. [From A. M. Young, *Milwaukee Publ. Mus. Contrib. Biol. Geol.* No. 40 (1980). Copyright 1980 by the Milwaukee Public Museum.]

forests, alternating bouts of wetness and dryness during the wet season are sometimes accompanied by emergences of a small number of dry-season cicadas.(71) Such observations suggest that the cue for emergence is related to the amount of wetness in the soil or to some related indirect effect of soil moisture status.

At Khartoum, several species of dragonflies display fluctuations in adult abundance related to the annual cycle of seasons.(549) The activity of the intertropical convergence zone (ITCZ) determines the local patterns of rainfall and dryness in this region, and the dragonflies synchronize adult emergences with hot but still moist conditions following the period of heavy rains.(549) Happold studied some 13 species and found that most appeared after the rains but also disappeared as conditions became cool and dry. Apparently periods of heavy rain fill up breeding sites near Khartoum, and nymphal development is accelerated. In the vicinity of Khartoum the periodic flooding of the Nile destroys dragonfly nymphs so that adult populations appearing in the area are migrants from other

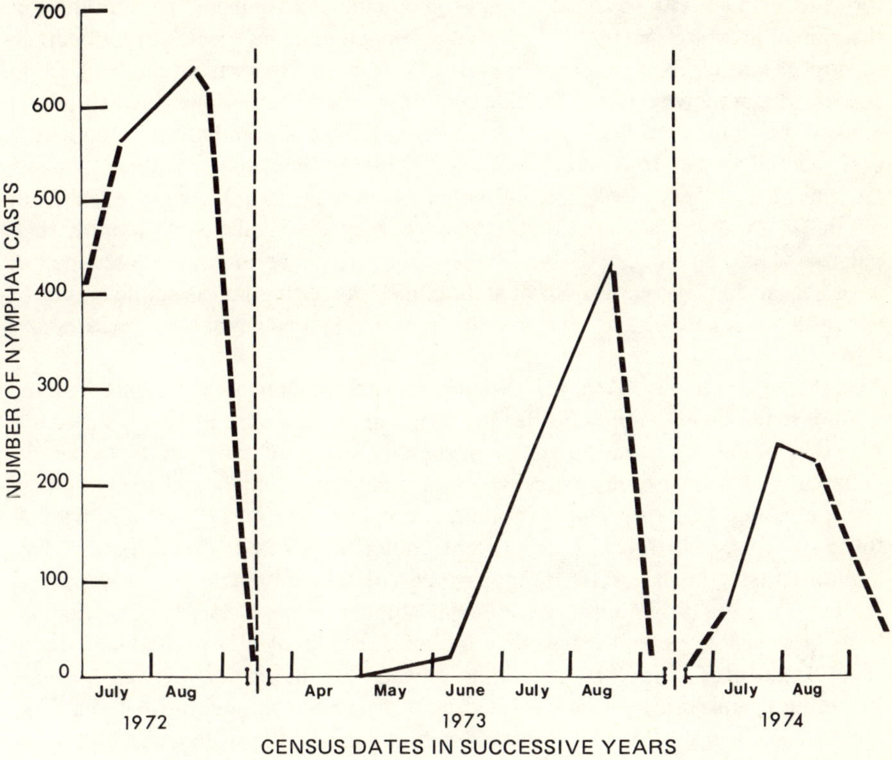

FIGURE 7.17. The wet-season adult emergence pattern for the cicada *Z. smaragdula* (Homoptera: Cicadidae) in the central highlands of Costa Rica, based on discontinuous samples of nymphal casts (skins) in successive years. The solid line connects actual data points for each year, and the dashed lines show projected or hypothetical beginning and termination points for each emergence. [From A. M. Young, *Am. Midl. Nat.* **103:**155–166 (1980). Copyright by the American Midland Naturalist and the University of Notre Dame Press.]

areas. Conditions are generally either too dry for breeding in small bodies of water or too flooded so that the Khartoum odonate fauna comes from more favorable areas upstream.[550] The annual northward movement of the ITCZ brings in some species from more southerly locations. Happold's data illustrate the considerable instability in the insect fauna of a desert tropical locality in which insects require an aquatic environment for development. While the Khartoum area itself is relatively hostile for colonization by dragonflies as a result of harsh fluctuating conditions over the year, further instability in the adult populations turning up there is a result of shifting seasonal patterns of rainfall elsewhere. Temperate Zone dragonfly species exhibit highly synchronized life histories associated with seasonal changes in aquatic environments and duration of the annual growing season.[551–555] Many Temperate Zone species of Odonata at the higher latitudes are either summer or spring species in terms of the adult flight season. But at more southern latitudes within the Temperate Zone, the distinction tends to be obliterated because overwintering nymphal populations contain several age classes.[554] In the nonseasonal lowland tropics similar patterns of overlapping generations, even more pronounced, may stagger emergences of adult dragonflies throughout most of the year. The species believed to exhibit such patterns would be those breeding in relatively permanent bodies of water such as swamps and forest streams. Tropical dragonflies breeding in less stable bodies of water in seasonal regions may exhibit rather disjunct annual emergence patterns.[556]

In the Temperate Zone of southern Australia, annual emergences of the pasture scarab beetle, *Aphodius tasmaniae,* a univoltine species, vary considerably, depending upon the extent of dryness at different times of the year.[557] Outbreaks of adult beetles occur when unusually dry winters and wet summers occur at about 4-year cycles. Wet summer conditions promote increased hatchability of eggs, and dry winters prevent young larvae from drowning.[557] Dry autumns result in drought conditions that kill off first-instar larvae, and wet winters promote drowning and pathogenic fungal attack of larvae. The fluctuations from year to year in seasonality and incidence of attack by pathogenic fungi regulate the sizes of adult hatches. In tropical areas, emergences of many insects, including scarab beetles, are expected to be staggered throughout the year, except for species with long life cycles and very specialized ecological requirements. The tropical fauna is expected to contain a high number of multivoltine species so that populations are somewhat buffered against unexpected changes in climate. Halffter and Matthews[558] state that population cycles in tropical dung beetles are synchronized with rainfall regimes: new adults generally appear at the beginning of the wet season and do not breed until the end of the rainy season but before the dry season. The dry season, in strongly seasonal regions, is passed in the egg stage, and the adults die off.[558] But in regions without a distinct and long dry season, population structure is usually very different and adults are equally abundant throughout the year.[558] As tropical dry seasons tend to be more predictable in terms of duration and intensity, dung scarabs and other insects can adjust their life histories to seasons of stress without appreciable losses due to mortality. But in Temperate Zone systems such as Maelzer studied for pasture scarabs, annual fluctuations in rainfall result in either considerable mortality to some stage in the life cycle or increased survivorship in others. Even in univoltine tropical forms such mortality may be relatively lower than in Temperate Zone forms, since the annual cycle of seasonality tends to be more precise, particularly in the lowlands.[116] Thus the tropical dry season may regulate the annual cycle of abundance and population structure for many insect species in highly seasonal regions, but it does so in a relatively more uniform or even manner from year to year.

Seasonal tropical regions do contain some insect species capable of displaying outbreak conditions.[511] The advent of the dry season may alter community structure in such a way that one or a few herbivorous insect species become abundant. If the dry season alters the abundance and foraging habits of insects and other organisms that feed on these herbivores, and if the herbivores exploit plant species that remain evergreen to some degree in the dry season, outbreaks may occur since biotic control on populations is diminished while food supply is still present (even though perhaps somewhat reduced in total biomass). Kenyi[360] noted outbreaks of the lepidopteran *Belenois aurota* on *Capparis tomentosa* thickets during the first rains following the dry season in the Ruwenzori National Park in Uganda. The first rains represent a period of increased new growth of plants, increasing the food base for herbivorous insects, and the adult butterfly population passed the dry season in substantial numbers.[360] Kenyi attributes thc observed severe defoliation by the butterfly larvae to the prior partial or complete defoliation of thickets by elephants during the dry season, an event which by increasing desiccation lowered the abundance of various predatory arthropods dwelling beneath or in leafed thickets. Thus the series of biotic control factors normally present in thickets by the time of the first rains were diminished and herbivore populations increased, causing defoliation. Other changes in the assemblage of animals found in thickets in the dry season were also noted, such as lower frequency of nesting by various bird species. In this system, the annual variation in the grazing habits of elephants and other factors regulate the degree to which particular thickets are defoliated during the dry season. Heavily defoliated thickets are ideal candidates for outbreaks of selected herbivorous insects at the beginning of the wet season since (1) biotic control agents are no longer present in defoliated bushes and (2) these bushes exhibit pronounced flushing out of new foliage at the time of the first rains.

Some relatively thin-cuticle cryptozoic arthropods such as millipedes may exhibit facultative changes in cuticular physiology in response to the alternation of seasons in tropical regions. Such mechanisms permit these species to continue to be active in both seasons without appreciable losses in fitness. Some genera of millipedes such as *Cingalobolus* and *Aulacobolus* in India exhibit changes in the composition of the cuticle in different seasons.[559] The epicuticle consists of lipoproteins that function during the dry season to conserve water in millipedes. Molts during the early wet season result in the deposition of cuticle without the highly impervious epicuticle, since transpiration rates are lower in the wet season and water loss is not a problem. Such synchronized cycles of cuticular organization may also be present in many other arthropods, including insects in highly seasonal tropical regions. The mechanism would be most adaptive in relatively cryptozoic forms that feed on detritus and litter that is generally present throughout the year even though in varying degrees of dryness. In other words, a relationship may exist between general feeding habits and ability to remain active in a particular habitat in the dry season for arthropods, including some insects. In species that can remain active, physiological changes such as periodic switches in cuticle permeability may permit activity to continue under very different conditions of dryness, so long as food is present.

Hoffman[560] has demonstrated that the butterfly *Colias eurytheme* in the Temperate Zone exhibits sex-related changes in wing pigmentation in response to photoperiod and temperature. In this species the males are orange and the females are white. The orange color results from the deposition of nitrogen-containing pigments in the wing scales. Male butterflies exhibit sensitive changes in color in response to changes in photoperiod but not to changes in ambient temperatures.[560] Low temperatures, on the other hand, promote some deposition of color pteridines in the white females. Hoffman suggests that the effects promote the appearance of light-colored butterflies when the weather is hot and dry (summer), an adaptation to thermoregulatory tolerances. Hoffman[561] concludes that in very predictable seasonal environments, physiological adjustments to changes in the environment function to eliminate losses in fitness that could arise from genetic tracking. It is a restatement of the phenotypic flexibility response pattern that permits organisms to make short-term adjustments to different environmental conditions.[99]

In highly seasonal tropical lowlands, the majority of adult butterflies seen in open habitats during the dry season tend to be light-colored, while the forms (various satyrids and nymphalids) that aggregate in shaded strips of forest along streams tend to be darker (A. M. Young, personal observations). Dark-colored species such as *Anartia fatima* that remain active in open habitats often exhibit daytime aggregating behavior under leaves in shady places.[562] During the dry season in El Salvador the adults of *Metamorpha stelenes* tend to be lighter in color than during the wet season.[317] The condition is caused by changes in the breadth of the light-green portions of the wings and to the darkness of the green color. In less seasonal lowland tropical areas, the color mix of butterflies in open secondary habitats tends to be greater during the short dry season (A. M. Young, personal observations). The general implication from such observations in the tropics is that butterflies may experience a high incidence of thermoregulatory stress during the dry season, a condition somewhat analogous to Hoffman's summertime *Colias* populations in Kansas. Of considerable interest, however, is the control mechanism. With relatively few generations per year and unpredictable boundaries to seasons in the temperate zones overall, the optimal strategy, as predicted by Hoffman[561] may be the physiological response for changes in pigmentation. In the tropics, however, where there may be more generations per year and seasons tend to have predictable boundaries, the genetic tracking strategy may be optimal. Comparisons of temperate and tropical populations of

Colias and other butterflies of seasonal zones may elucidate different or the same mechanisms underlying the regulation of thermal stress in the dry season or summer. Emmel and Leck[(563)] reported that butterfly species occupying clearings in Panama showed a marked decrease in abundance in these habitats, whereas the forest fauna increased during the same period (Fig. 7.18). They interpret these results in terms of physiological stress from increased dryness in the clearing during the dry season and the forest becoming an increasingly important moist refigium as the dry season advances. The data illustrate the dynamic state of tropical butterfly faunas in seasonal regions.

Owen *et al.*[(516)] describe the seasonal abundance of some butterflies in Africa in garden habitats. Further preliminary evidence from the nymphalid genus *Anaea* from the northwestern tropical dry forest zone of Costa Rica indicates that annual fluctuations in adult abundance are a function of phenological patterns in the growth of larval host plants.[(315)] In a very deciduous forest site two species of understory trees, *Caesaria* spp. (Flacourtiaceae), cease vegetative growth of new mcristems during the long dry season, during which time eggs and larvae of *A.* (*Zaretis*) *itys* are absent. At a semideciduous forest site (Barranca

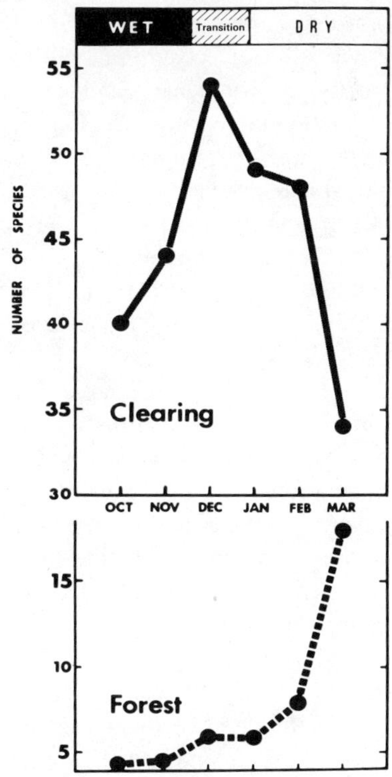

FIGURE 7.18. Relationship between tropical seasonality and the abundance of butterfly species in a forest and forest clearing in Panama. [From T. C. Emmel and C. F. Leck, *J. Res. Lepid.* **8:**133–152 (1970). Copyright 1970 by the Lepidoptera Research Foundation.]

site) the understory tree *Nectandra* sp. (Lauraceae), the larval host plant of *A.* (*Memphis*) *morvus*, also ceases growth of new meristems, lagging somewhat behind the *Caesaria* host plants at the more deciduous site (Fig. 7.9); in both instances adult butterfly populations dwindle as the dry season progresses and presumably adults enter into a reproductive diapause.[(315)] Populations of *M. peleides* in such forest habitats also dwindle as the dry season advances[(190)] (A. M. Young, unpublished data), presumably since the growth of new meristems slows down and ceases. Many tropical butterflies selectively oviposit on new meristems, and the young larvae feed on them.

7.4. MOLDING OF ECOLOGICAL TRAITS IN TROPICAL INSECTS BY THE DRY SEASON

Tropical terrestrial regions are conceptual patchworks of heterogeneous climatic regimes and vegetation cover. The continuous interplay of general weather patterns and localized unusual or unique weather effects generates this patchwork in space and time. Thus although one readily recognizes the lowland tropical dry forest zones of Central America and elsewhere, upon closer examination one begins to distinguish considerable heterogeneity over short distances: topographic effects interact with wind patterns to produce rain shadows and other patchy environments. Such effects may result in marked discontinuities in the vegetation cover of the region, and superimposed upon this is the degree of both natural and man-induced disturbances of habitats that add more heterogeneity. In a manner similar to that of Lewontin and White,[(54)] who conceptualized an adaptive landscape for *Moraba* grasshoppers in which peaks and valleys formed a highly textured contour surface with peaks of different fitness levels, one may cautiously envision a tropical region as being composed of different environments, each assigning a characteristic level of fitness for individuals of a particular insect species.

Thus although the large cicada *F. mannifera* occupies different habitats at different actual elevations within the tropical dry forest zone of Costa Rica, the population size and density and the average fitness of individuals within different populations may vary considerably in different habitats. And the degree to which a population is exposed to seasonality will also vary. Local effects on the general seasonal phenology of rainfall of the region may shorten or lengthen the adult emergence period in cicadas and other insects, possibly resulting in different populations being somewhat out of synchrony with one another in terms of peak courtship periods. The effect is especially pronounced if patches of habitat and insects are widely scattered.

Differences in local vegetation in different patches may result in different sets of insects and other animals found there, altering the sphere of selection pressures in different patches. Host tree associations of cicada nymphs, predators, or pathogens on nymphs and adults[136,564] and interactions with other cicadas may also vary considerably. Young[139] noted that different sites within the dry forest zone of northwestern Costa Rica vary considerably in the numbers of cicada species found. For some groups of highly mobile insects such as butterflies and moths, such a patch structure may result in considerable turnovers in insects associated with different habitats in different seasons. The migrations of various butterflies vertically to higher elevations in the dry season and a presumed return to the lowlands in the following wet season suggest that butterfly faunas associated with various habitats at different elevations may be far from static. The response mechanisms to survive the dry season shape the ecological traits of such species, as well as those that remain in a particular habitat during wet and dry seasons. In this section we will examine how insects and other organisms respond to tropical seasonality, with an emphasis on synchronization of individual life histories and changes in whole communities. In some species of African mosquitoes with very seasonal adult emergence periods at a particular locality, emerging adult populations are often highly congregated, with mating taking place in less than 4 days—the maximal life-span of individuals.[200]

The timing of flowering in tropical plants must be related to the optimal time of year for pollination. Distinct seasonal patterns of flowering in the tropics[116,118] are correlated with changes in rainfall. Temperature and humidity may also regulate flowering in some tropical plants.[565] The observation that many tropical plants are triggered into flowering by short-day cycles[566] suggests that plants are sensitive to small shifts in photoperiod, conceivably of the kind associated with day-to-day variations in cloud cover during the wet season. Janzen[116] was one of the first workers to conceptualize the adaptive significance of plants and animals synchronizing their life cycles with the tropical dry season. His argument was based on observations of the synchronized flowering of many tree species with the dry season in the lowland tropical dry forest zone of Costa Rica. As the dry season progresses, trees lose most of their leaves and vegetative growth is reduced. These trees burst into flowering, and the flowers tend to be large and conspicuous. Pollinating insects such as bees and wasps can locate the flowers more readily during the dry season when foliage density is reduced by deciduousness. It follows that pollinating insects also synchronize their life histories with the dry season in order to maximize utilization of resources (nectar and pollen). Some of these insects track the phenologies of various tree species in the sense that different tree species are visited at different seasons. Thus there is a definite structure to communities in seasonal tropical environments as well as relatively nonseasonal ones.[117] The pollination systems of tree species in both wet and dry tropical forests are very diverse, involving a variety of animals such as insects, birds, and bats.[117] Many overstory trees of tropical forests produce large quantities of relatively small flowers, each with a small food reward.[117]

These trees are visited and pollinated by bees, moths, butterflies, wasps, and beetles[117] and exhibit distinct seasonal flowering patterns. Understory trees and shrubs tend not to have the mass-flowering strategy; bloom times are considerably longer than in most overstory trees[117] and do not necessarily adhere to a seasonal pattern.

In lowland tropical dry forests, many tree species flower in the middle of the long dry season and are pollinated by medium- to large-sized bees.[117] Frankie[117] also points out that other pollination systems predominate in this zone: moths and small bees opportunistically pollinating various tree species in both wet and dry seasons. In some tropical communities occupying habitats with distinct seasons, closely related bee species show allochronic activity periods, with one species active in the wet season and the other in the dry season.[567] But such differences in abundance of bees between seasons was not due to shifts in the availability of pollen but rather to differential effects of rainfall on activities of the bee species. One species tended to collect a greater amount of pollen from a few plant families, while the other species showed a more even distribution of pollen-collecting for many plant families (Fig. 7.19). Frankie also discusses the organization of pollination systems within seasons and in the transitional periods between seasons. The greatest diversity of pollination systems in the dry forest zone occur at the middle of the dry season.[117] Frankie points out that the life

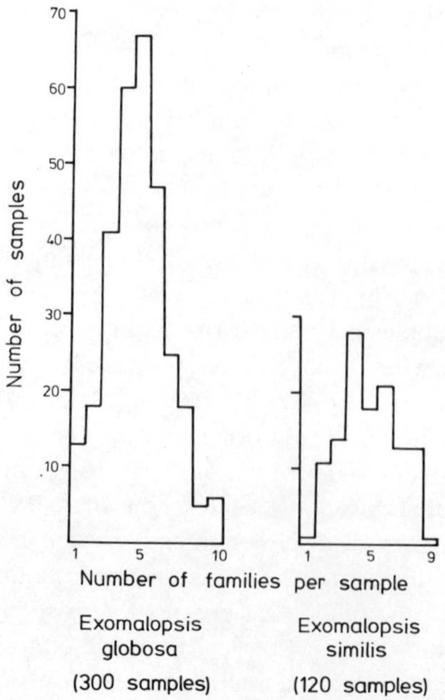

FIGURE 7.19. Pollen-resource differentiation between closely related bee species at the same locality in Jamaica. [From A. Raw, *Biotropica* **8:**270–277 (1976).]

cycles of pollinating moths such as Sphingidae may be synchronized in order for adults to be most active at the early phase of the wet season when many moth-pollinated plant species are in bloom.

Other species may be active in the middle dry season as the result of a buildup of larval populations thriving on various shrubs and trees in riparian forest communities of the region: many of these plants put out flushes of new growth in the dry season in these moist habitats, providing a continuous food base for some insects. Many moths involved in tropical pollination systems tend to be large and can therefore withstand some amount of dryness. Larval populations may peak twice a year in tropical dry forest, at the beginning of the wet season and at the beginning of the dry season (citation of D. M. Windsor in Frankie[117]). These peaks are maintained by the different vegetative growth cycles of various trees and shrubs in selected habitats of the region; the early wet-season peak is presumably the result of the plant communities overall responding by massive leafing out at this time, while the early dry-season peak is presumably the result of many insects synchronizing the last wet-season generation with the end of that season (see Fig. 7.20). For cultivars such as *T. cacao* and other endemics of tropical rain forest, flowering is maximized during the wet season, and a build up of insect pollinator populations follows (Fig. 7.20). Such plant species, when cultivated in seasonal environments as shown in Fig. 7.20, do not maintain high levels of flowering during the dry season, and pollinator populations also decline at this time. This effect is expected to be most pronounced for pollinating insects of small body size, where the rate for a species to drop out of the pollinating guild is directly proportional to the length and severity of the dry season (expressed by the number of days or weeks of absolute dryness), and the magnitude of the deleterious effects of dryness on body size and physiology. Large-bodied insects may be less drastically affected and, in some regions with long dry seasons, such pollinators may increase in abundance in the dry season. In nonseasonal tropical regions or habitats, both flowering and pollinator abundance may be more or less evenly distributed over the year.

The dry-season synchronization of activity is a response to the impending contraction of food supplies for larval populations that occurs in the dry season. One might suggest that the early dry-season flush is smaller than the early wet-season flush of insects, since more new foliage is available in the latter and predators, parasites, and some pathogens may lag in response to the appearance of phytophagous forms that are hosts. But there is also evidence that some species of vespid wasps begin rearing brood in the dry season,[568] and these wasps are predators on a variety of soft-bodied insects used to feed larvae. Such information is indirect evidence that prey insect populations are available during the dry season, and the greatest densities are found in moist habitats.[134]

Frankie *et al.*[118] found that some solitary bees forage in great abundance at mass-flowering overstory trees. Some bees visit many different individuals of a

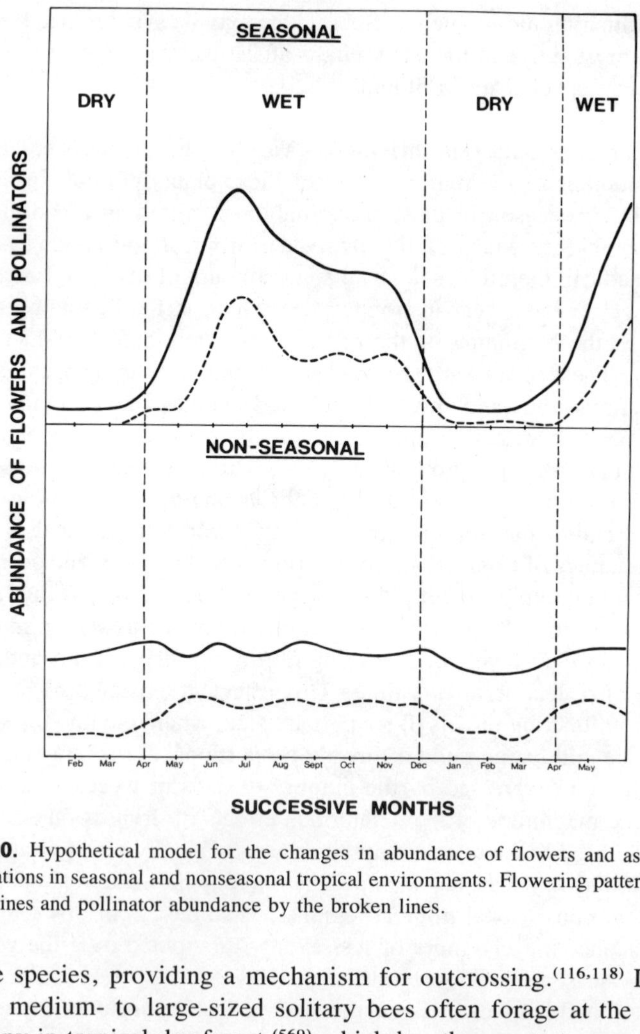

FIGURE 7.20. Hypothetical model for the changes in abundance of flowers and associated pollinator populations in seasonal and nonseasonal tropical environments. Flowering patterns are shown by the solid lines and pollinator abundance by the broken lines.

certain tree species, providing a mechanism for outcrossing.[116,118] During the dry season medium- to large-sized solitary bees often forage at the top of the forest canopy in tropical dry forest,[569] which has the greatest concentration of flowers and lower densities of other bees such as *Trigona* species.[117] Tropical bees exhibit diverse means of collecting pollen and exhibit specialized to generalized foraging patterns.[570] Schlising[570] analyzed the insect visitors to flower of *Ipomoea* vines during the dry season in the tropical dry forest zone of Costa Rica. These vines exhibit mass flowering during the dry season at this site. He found that although the composition of visitors varied at three different habitats, there was evidence of considerable structure in the timing of foraging for insects visiting the flowers throughout the day. Bees belonging to four different families exhibited distinct nonoverlapping foraging periods throughout the day. Species visiting flowers early in the day generally had longer visits, and as air temperature increased, visits diminished in length, presumably as a result of increased thermoregulatory problems. These vines and another studied, *Aniseia* (also Convolvulaceae), are pollinated by these bees. Thus pollinating assemblages of bees associated with certain plant species during the dry season exhibit considerable temporal organization on a daily basis. Such behavior is undoubtedly linked to physiological constraints associated with desiccation in bees, differences in nutrient requirements (qualitatively and quantitatively) among bee species (nectar composition may change over a daily cycle,[572,573] and the reduction of possible competitive conflicts over a patchy and seasonally abundant resource.

Data on the succession of flowering seasons among 10 plant species in the Monteverde cloud forest in Costa Rica that are pollinated by 14 species of hummingbirds suggest that these plants compete for pollinators and that the allochronic flowering periods lower competition for pollinators.[574] The simultaneous flowering of several plant species in the same area may distract hummingbirds from making an optimal number of visits to each species and thus lower fitness in these plant populations. The succession of Sphingidae in the relatively nonseasonal tropical rain forest near Puerto Viejo in Costa Rica[575] suggests that flowering periods in some sphingid-pollinated plants may also be staggered throughout the year for similar reasons, although more data are required to confirm such a prediction. Seasonal succession of sphingids also occurs in Amazonian rain forest.[576] The staggering of flowering times in many plants species is likely to turn up in less seasonal tropical forests where both plant and pollinator populations may have even distributions in abundance throughout most of the year. Yet the marked suppression in cicada emergences near the end of the wet season in lowland tropical rain forest[71] suggests that this period (1–2 months) may suppress activities in other insects as well. More seasonal areas are more apt to exhibit the bimodal peak abundances in insects (Frankie[117]). Again, a dearth of data from long-term studies prevents us from conforming or rejecting such ideas.

As with pollination, different regions in the tropics contain seed predation systems that may or may not be synchronized with seasons. D. H. Janzen has explored the ecological traits of seed predation systems involving insects and vertebrates in both wet and dry forests of Central America.[29,362,363] In seasonal tropical regions as well as some nonseasonal regions, different tree species will exhibit distinct fruiting periods each year. For example, in the lowland tropical dry forest zone, peak fruiting may occur in the early to middle phase of the wet season for tree species with mass flowering in the dry season. Smythe[101] has summarized the major coevolved features of seeds and seed predators. Smythe[101] found on Barro Colorado Island in Panama that tree species with large-seeded fruits tended to have very seasonal fruiting periods while those with small-seeded fruits tended to have rather nonseasonal periods. Competition for dispersal is suggested to be greatest among tree species with distinct seasonal patterns of fruiting.[101] In lowland tropical forests where different tree species exhibit distinct fruiting periods each year,[101,517] organisms which use these fruits as resources may adjust their life histories to them, particularly if they are specialists or very limited generalists (feeding on a few kinds of fruit). Morton[577] discussed the synchronization of avian breeding cycles in frugivorous species in tropical forests with fruiting periods. A wide variety of insect groups exploit fallen, rotting fruits as food supplies and breeding sites (butterflies, beetles, flies, etc.). As it is known that various volatile products of fermentation attract many different groups of Lepidoptera,[578] it is not surprising that a variety of sweet-to-foul-smelling rotting fruits and seeds on the floors of tropical forests attract a variety of insects, and in some cases, the resources emitting these odors are the major food sources for insects. Many groups of Neotropical butterflies regularly exploit such resources.[186] Depending upon the degree to which the insect is a specialist on a certain kind of fruit and the phenotypic constraints arising from life cycle and tolerances to subtle fluctuations in temperature, air humidity, etc., the population structure of such an insect species may be very viscous or fluid. Young[175,186] noted considerable localized movement patterns in adult butterflies feeding on rotting fruits replenished experimentally in a lowland tropical rain forest. Patchy spatial distributions in small insects such as drosophilid flies may arise as a function of locations of rotting fleshy fruits and changes in such distributions throughout the year. In other words, the distribution of fruit-feeding insects on the forest floor is very dynamic and changing as fruit distributions change. Changes in fallen fruit distributions will be a function of the seasonality of fruiting in tree and shrub species in the forest. In lowland tropical rain forests, a predicted organization of such insect communities is one characterized by relatively low densities of insects associated with an accumulation of fruit beneath an individual adult tree (in the fruit shadow) and the utilization of many different resource patches (different types of fruit) per unit area. If an insect species is a specialist under such conditions, the distribution is very patchy and each patch may become a distinct subpopulation with its own breeding structure.

Sterculia apetala trees are found in lowland tropical dry forest in Central America and develop large crops of pods that mature in the dry season.[365] The seeds fall to the ground and are fed upon by *Dysdercus* cotton stainer bugs, which may kill virtually all seeds in forest habitats.[365] Individual trees in pastures produce seed crops that stand a greater chance of surviving from bugs but die from desiccation. Various small mammals such as squirrels and monkeys are the major dispersal agents of the seeds, as they carry off pods from trees and drop some of the seeds elsewhere, a behavior resulting from the exposed use of these animals to hairs from the inner walls of the seed pods.[365] The cotton stainer bugs opportunistically feed on the abundant seed crop of *Sterculia* in the dry season but presumably switch to other foods at other times. In southern Ghana, for example, *Dysdercus* bugs exploit several different tree species throughout the year, moving to different species as they come into fruiting and returning to *Hibiscus,* which flowers and fruits throughout the year when the other species cease fruiting.(470) *Hibiscus* in the Savanna zone acts as a reservoir for cotton stainer populations to build up and migrate from when suitable host trees are fruiting. In the forest zone, *Sida* trees make up the predictable reservoir in terms of flowers and fruits for these bugs.(470)

A major component of the adaptive strategy of coping with the tropical dry season in the apparently migratory bug *Dysdercus bimaculatus* is the exploitation of those host seeds with the highest amount of energy.(579) Nutrient availability places a constraint on the evolution of optimal body size in relation to selection favoring either migratory behavior or diapause to pass the dry season.(579) Migratory behavior, even if relatively localized, permits an insect to exploit all available energy-rich food resources and becomes the primary strategy, since diapause involves a more drastic reorganization of the phenotype.

These examples of pollination systems and seed predation systems illustrate how insects synchronize population cycles and feeding habits with a seasonal succession of resources in the tropics. Many insect species exploiting plant parts or products as food resources are adapted to exhibit facultative switches in response to a changing spectrum of available resources if the average fitness of individuals within breeding populations is to be preserved through periods of environmental fluctuation, even if predictable.

Rockwood(225,226) analyzed the flexible foraging behavior of two species of *Atta* leaf-cutting ants in the lowland tropical dry forest region of Costa Rica. During the wet season, foraging was primarily diurnal, while during the dry season it was nocturnal. Rockwood suggests that these changes in foraging behavior might have been related to effects of seasonal changes in rainfall on vegetation cover and other features of the habitats. Various factors also influence the selection of leaves for cutting by *Atta,* young leaves generally being preferred,(580) and shifts in foraging patterns may be related to availability of preferred leaves in a habitat. Rockwood found that the greatest quantities of leaves and flowers were harvested at the beginning of the dry and wet seasons, periods during which new growth and flowers were abundant in the habitat. Mature leaves are only minimally harvested in both seasons.(225) Rockwood's data provide another example of how seasonality affects the availability of optimal food resources for specialist insect species: the supply of such resources (new plant tissues and flowers in this case) varies in different seasons, and the insects display facultative changes in foraging habits to compensate for the changes. The growth of *Atta* colonies depends upon the selectivity of food resources brought to the nest for fungus gardens by the ants.(581) These data point to the existence of a phenomenon now generally accepted by biologists: the tropics are not uniform in space and time in terms of the availability of resources required by individual species fragmented into breeding populations, and whole assemblages of organisms are organized in specific ways that may change both through cycles of contemporary time such as seasonality and through evolutionary time.

Temperate Zone assemblages of insects such as *Drosophila* species occupying the same region or habitat may contain species that exhibit different peak activity periods corresponding to different seasons.[(582)] Some species may peak in annual abundance during the spring and summer, while others peak in late summer and autumn, and some species may exhibit different preferences for habitats.[(582,583)] In these species there is considerable dietary overlap with niche separation manifested in spatial and temporal separation of populations. Pipkin *et al.*[(584)] found mixtures of specialized and generalized species of flower-feeding *Drosophila* species at several tropical sites. The flower-feeding habit is synchronized with the seasonal flowering periods of the preferred plant species, and some flies enter into a diapause in the dry season when these flowers are not available.[(584)] Although such species exhibit specific morphological and behavioral adaptations for flower feeding, on a per species basis there appears to be considerable flexibility in feeding so that some species can be polyphagous while others, more restricted, are monophagous. Monophagous species are ideal candidates for enering into reproductive diapause during seasons when their preferred host plants are not blooming in sufficient numbers to maintain breeding populations.

A similar system may be present for phytophagous insects in seasonal tropical regions: if a portion of the host plant spectrum utilized by a particular polyphagous insect species in a seasonal region remains lush to some degree during the dry season, the population may continue to be active, barring effects of other ecological constraints, but perhaps reduced in size proportional to the loss of resources. Such a situation may be expected for some species that utilize host plants with new growth in riparian forest habitats. If the host plant distribution of a monophagous species is such that a major portion of the population goes deciduous in the dry season, the optimal response strategy would be to evolve facultative diapause or migrate to a higher elevation where conditions are moist and the host plant population is still evergreen. Thus we might expect a greater incidence of reproductive diapause (or pupal diapause) or migratory ability in monophagous insect species in very seasonal lowland tropical regions than in polyphagous species, assuming that plant distributions include habitats where some degree of evergreenness is preserved throughout the dry season.

Monophagous-feeding *Drosophila* in the tropics exploit predictable and abundant flower species[(584)] sufficient to maintain breeding populations of these species. Polyphagy seems to appear when any one potential resource type is neither predictable nor abundant. For some insect groups, one might expect that polyphagous species are more prevalent in strongly seasonal regions where some resources of these particular groups are only seasonally abundant and predictable in distribution. Ecological specialization in some fungus-feeding *Drosophila* species is generally limited to species exploiting mushrooms available in different seasons and habitats throughout the year, even though competitive interactions may reduce niche size in such species. As preferred resources fluctuate in quality and abundance temporally, there will be a greater tendency for insect species exploiting such resources to become ecologically flexible and possess a broad niche with respect to those resources. Any single tropical forest ecosystem will contain species of plants with varying intensities of physiological response to seasonality, even in regions where the dry season is relatively short. Thus tropical forest ecosystems represent highly fluid mosaics of resources through contemporary time, requiring or promoting both specialist and generalist forms within different communities, depending upon the degree of fluctuation in properties of the resources affecting fitness in insect populations. Such mosaics may be more oscillatory in very seasonal tropical regions.

The sweep-sample studies of Janzen and Schoener in the tropical dry forest zone of Costa Rica[(134)] show that insect species are more abundant in moist habitats than in dry habitats during the dry season and that insect communities in close proximity may have disjunct distributions of species depending upon the position of the community in a moisture gradient. As the dry season becomes more intense, such discontinuities are expected to increase.[(134)] Yet in all communities examined, Janzen and Schoener found a greater number of insect species during the dry season than in comparable Temperate Zone communities sampled in the growing season. Thus even a very dry tropical habitat during the dry season tends to support more insect species than Temperate Zone communities. Information on the natural history of component species in such tropical communities would reveal why species richness is maintained during a stressful season.

Size of suitable habitat, and abundance and distribution of essential resources within the habitat, are two major determinants of the success or failure of an insect species invading the habitat in response to selection or changes in environment elsewhere. If the habitat is too small or the resource base within it too small or of low quality during periods of peak reproduction (in a seasonal environment), the invading species will go extinct within a short period. Duffey[(585)] found that a habitat in Britain was too small to maintain a breeding population of the large copper butterfly, *Lycaena dispar*. If the resource base or habitat is too small, the population cannot become large enough to (a) buffer against mortality of various sources and (b) allow for sufficient breeding activity to produce the next generation. In the lowland and mid-elevation tropics, one can conceptualize tropical forest habitats as being dynamic mosaics of natural and artificial (human-induced) light gaps interspersed with forest, and an ever-changing size distribution will be associated with such habitats. The likelihood for colonizing species to survive in them will be determined in part by the size and quality (in terms of resources) of each habitat at any one point in time. Turnover of insect faunas associated with such heterogeneous environments is expected to be high.

Insect species associated with plants in wetter tropical regions may remain active in the short, often erratic dry season characteristic of many tropical rain forests. Perhaps some of the best evidence for such sustained activity is that from the analysis of population size fluctuations in the butterfly *H. ethilla* in Trinidad.[451] There were remarkably stable population sizes throughout most of the year, with only a slight depression in the few dry-season data available (Fig. 7.12). Ehrlich and Gilbert[451] ascribe the observed high stability of the butterfly population to a combination of factors, including high adult longevity, high reproductive output, predictability of adult and larval resources, and an absence of severe fluctuations in the environment, i.e., rainfall. The relatively mild dry season of this region has runs of wet days, during which many insects can remain active. The observed continued activity of many butterflies in forest understories throughout the year in tropical rain forests[186,563] results from (1) the generally weak effect of the short dry season on the flora and (2) the preservation of wet conditions in the understory, an environment generally shielded from the direct effects of the dry season. Forest-edge populations of *Parides* butterflies in tropical rain forest zones are generally active throughout the year (A. M. Young, unpublished observation; see also reference 586). Cook *et al.*[444] found sustained adult activity of *H. charitonius* in a mid-elevation tropical rain forest zone throughout the year, even though the study period included the short dry season. The extent to which the annual dry season is intense determines in large part the degree to which arthropod communities associated with rotting carcasses are cycled through the ecosystem and the organisms expediting the recycling.[191]

The tropical cash crop *T. cacao* is an understory tree species with small, cauliflorous flowers generally pollinated primarily by tiny midges of the dipteran family Ceratopogonidae.[406,587] Because of the general lack of dense populations of these midges in many cacao plantations in both the New and Old World wet tropics, pollinator abundance is believed by some workers to be a limiting factor in fruit set.[522,588] An interesting feature of cacao phenotypes is that they occur in both seasonal and nonseasonal tropical climate regimes, although a distinct wet season is needed for normal levels of flower production. Population sizes of midges in some cacao plantations may be determined in part by the abundance of suitable breeding and development sites on the ground cover beneath mature trees.[588–590] In nonseasonal cacao-growing regions throughout the year, midge populations may exhibit small oscillations and fruit set, although low, tends to be the same (A. M. Young, unpublished observations). There is some evidence, however, that in seasonal regions of Africa where cacao is grown, massive flowering occurs early into the wet season, and fruit set at this time is very low. According to Myers[(591)] the native habitat of cacao is the understory of Guiana rain forest, and because of this origin, the species does best in cultivation where daily moisture fluctuations are minimal, a property requiring the use of shade cover in plantations.[(592,593)] In the Gold Coast, the dry season is a limiting factor for adequate growth of young cacao trees.[(594)] Because the early stages of these particular groups of ceratopogonid midges require very moist conditions for successful development and because adult midges are subject to desiccation very easily, pollinator populations may lag behind flowering periods following a dry season. Fruit set in subsequent flushes of flowers later in the wet season may be higher since midge populations have the opportunity to recover from the dry season. In seasonal environments, cacao production provides an example of how pollinating insect populations may be severely curtailed in abundance by the end of the dry season and how the ensuing asynchrony between flower abundance and pollinator abundance may result in very low fruit set. The variance in fruit set from tree to tree for a given strain of cacao in a single plantation is expected to be greater at the beginning of the wet season in a very seasonal region than in subsequent periods of fruit set in the wet season. Pollination systems involving larger-body insects may not exhibit such behavior, since these insects may survive the dry season without appreciable losses in population size or density.

Keep in mind, however, that for cacao and other trees of tropical forests, production of mature fruit is influenced by factors other than pollinator abundance. It is interesting to note that cacao originally evolved as an understory tree in Amazonian rain forests,[(595)] where there was probably a well-established equilibrium between pollinator abundance and abundance of flowers in small clumps of the tree. The environment was essentially nonseasonal, and pollinator populations were probably stable throughout each year. The "invasion" of cacao as a cultivar in more seasonal tropical regions has been only partly successful from a commercial viewpoint, since seasonality of rainfall imposes annual periods of environmental stress on pollinator populations under these conditions.

CHAPTER 8

DYNAMICS OF ORGANIZATION OF INSECT COMMUNITIES IN TROPICAL ECOSYSTEMS

8.1. NOTES ON TROPICAL ECOSYSTEMS

The relatively few studies of tropical ecosystems available have emphasized the pathways of nutrient cycling in some tropical forests[(596)] or descriptive and experimental studies of biotic organization of communities.[(29,65,435)] Okwakol[(597)] demonstrated in an African termite *Cubitermes* a relatively high rate of removal of organic materials from soil, suggesting that in habitats where termites are abundant, cyclic depletions of the nutrient content of soil are frequent, thus imposing some constraints upon other litter organisms. To what extent, for example, termites compete with collembolans for certain nutrients under such conditions may indicate food as a limiting factor in collembolan populations. It is clear, regardless of the interactions, that the litter mosaic of tropical forests is very dynamic in terms of turnover of accessible nutrients and breeding substrates. Historically, much of tropical biology has focused upon expedition-type collecting of specimens for museums and universities, development of cultivars and agronomic systems, and medically related entomology. The entomological literature, for example is full of detailed descriptive and experimental studies on insects as vectors of various tropical diseases and the effects of changing weather patterns on predicting outbreaks. Much of the earlier work in the tropics has deployed a "targeted" approach: in a particular study, an emphasis was placed on few taxa or single species. Surprisingly, a need remains for detailed natural-history studies of tropical organisms to feed the conceptual and theoretical modeling of how communities and whole ecosystems are put together.

In this book, I define an ecosystem as the sum of abiotic and biotic components of a distinct environment or habitat such as a forest. Within that environment, there are one or more communities of organisms. Each community is associated with a particular trophic level in the ecosystem, and each occupies a distinct microenvironment or microhabitat. The community of decomposer organisms lives in the forest floor litter microhabitat. A community is composed of a distinctive but dynamic set of species each maintaining a breeding population. Sometimes communities contain transient species that do not maintain breeding populations in them. It is necessary to be able to recognize transient and resident species in communities. Depending upon the kinds of resources exploited by the species in a particular community and the temporal and spatial distributions of these resources (i.e., their patchiness), some resource patches may accumulate over evolutionary time a variety of species exploiting different parts or products of the resource. A group of closely related and ecologically similar species, a guild, exploits a specific portion of a resource, although different sets of species may occur together on the same resource. Thus seedlings of *P. macroloba* in lowland and premontane rain forests on the Caribbean watershed of southern Central America may have several distinctive sets of species feeding on them. Both the number of sets and their species composition may change over ecotonal and geographic gradients within the distribution of the tree species. Thus the seedlings support one or more guilds of phytophagous insects, and the community is the collection of such guilds in a particular area of habitat. Usually the species within a community are taxonomically and ecologically similar, and they usually exploit the same structures or substances from the resource. Different subfamilies are represented in the rain forest understory butterflies exploiting juices liberated from rotting fruits molding on the forest floor.[186] Although within this community there is a guild of satyrid butterflies, many other satyrids do not exploit the same resource. Guilds are almost always composed of closely related species.

In the tropics, ecosystems are generally more complex because the vegetation cover is generally more diverse in lowland areas, there is greater structural complexity to habitats, and consumer and decomposer trophic levels are more complex in terms of numbers of species. The vertical stratification of a lowland tropical rain forest,[394] for example, imposes a vertical gradient in light which differentially attracts different kinds of insects and other animals. The rain forest is the ecosystem, and different insect communities are found at different levels because plant and debris resources change vertically. The convergence of different groups of butterflies, for example, at different strata may determine the kinds of mimicry complexes found at different levels of tropical rain forest.[174] Horizontally, tropical regions display a great deal of heterogeneity in the organization and biotic complexity of plant communities in different successional stages. Each successional stage may be considered as a separate ecosystem, with a unique set of communities of plants and animals, with a different array of structural complexity.

Connell[65,435] has provided evidence that tropical rain forest ecosystems may not be as stable as generally believed: a lot of turnover in species results from a variety of environmental perturbations. Tropical forest ecosystems, primary or secondary successional stages, are dynamic in that there are continual localized changes within the species composition and structure of communities and guilds within them. Because such complex ecosystems contain interactions among species that are specialized or generalized, an understanding of tropical communities, especially in terms of plant–insect interactions, is achieved by examining some examples of how various kinds of insects interact with resources in tropical habitats. In one sense the stability of an ecosystem is determined by the degree of stability and nutrient profile of soils. If soil is nutrient poor but supports a stand of trees with well-developed chemical defenses that leach out into the soil from plant parts, the ecosystem may be stable though low in diversity.[332] Nutrient-rich soils may promote a greater diversity of plant species, each with its own characteristic "defensive compound shadow," and it is this sort of soil profile heterogeneity that endows some degree of instability in tropical forest ecosystems. Instability also arises from natural perturbations such as tree falls and floodings. Yet at the same time that various factors promote instability in tropical forests, many others promote stability and the resulting accumulation of species-rich ecosystems. Some of these factors are explored in the sections to follow.

8.2. PATTERNS OF ORGANIZATION IN TROPICAL FORESTS: SOME COMMENTS

Pires *et al.*[325] measured the abundance of tree species in an Amazonian rain forest and emphasized the high number of tree species found per unit area and the very different tree species distributions that may exist within different regions within Amazonia. Although previous workers maintained that much of the great diversity of tree species in tropical forests was the result of the interplay of stable and benign climate with ecological interactions such as competition,[63] Janzen[29] provided evidence suggesting that specialized interactions between seed and seedling predators (herbivores) and tree species generated that patchiness characteristic of many tree species in tropical forests. Janzen proposed that each tree species has a host-specific set of herbivorous insects that kill large amounts of seeds and seedlings beneath the parent trees, while the few survivors are those carried off by various vertebrates. The net result is a lowering of population density for each tree species for both adult and preadult age classes, and an increase in the distance between parents and offspring. Such a system results in a greater number of tree species found over relatively small areas of tropical forest. Janzen proposes that the system is most prevalent in lowland tropical rain forests, and becomes less likely as one moves into dry forest and into the Temperate Zones. Severity of physical factors such as temperature and rainfall fluctuations tend to lower the opportunities for ecological specialization by herbivorous insects and seeds and seedlings through evolutionary time. Janzen's model also predicts that specialized coevolved relationships between herbivorous insects and seeds and seedlings will keep population densities of individual tree species below abundance levels which otherwise result in competition among species. An important implication of the model is the historical component: such specialized interactions have evolved over long periods, yet strong stabilizing selection should not be operative to preserve these traits.

The integrity of the tropical rain forest flora, so to speak, depends upon the maintenance of sufficient genetic and phenotypic flexibility to cope with changes in the coevolved association. The relationships are not static, since coevolved associations involve the expression of chemical defense barriers by plants to herbivores, and herbivores must change physiologically or biochemically to overcome such changes if the species is to continue exploiting that resource. Such coevolved associations are expected to be most frequent for species of insects that lack alternative strategies of exploiting other resources without going extinct locally or suffering appreciable losses in fitness. The greater the frequency of coevolved associations, which are more likely to evolve under continually favorable physical conditions of the environment[(104)] as long as they are nonoverlapping, the greater the evolutionary accumulation of species in a habitat or region. The data are impressive for many groups of tropical butterflies.[(6)]

Going back to the formulation of MacArthur for adding additional species to tropical communities (Chapter 2), herbivore specialization of a coevolved form provides space in the ecosystem for other plant species that are poor competitors.[(29)] In turn, these new species in the ecosystem can provide new resources for other insect species and develop their own coevolved associations. Janzen points out that tropical herbivores are apt to be effective in overcoming the chemical defenses of plants and that the more effective such interactions become, the less likely it is that plant species will survive in a habitat. Thus extinction rates of plant species are predicted to be higher in the tropics than in Temperate Zone regions, and tropical rain forest communities are therefore very dynamic over time. Connell[(65)] shows that the great species richness of tropical rain forests is a nonequilibrium condition of the sort discussed by Janzen[(29)]: species abundances are well below an equilibrium that would result if each species were to realize the carrying capacity of the habitat. Perturbations will result in a lower diversity equilibrium state if coevolved associations between plants and their herbivores are altered, thus making it possible for some species of trees to become dominants. In the evolution of lowland tropical rain forests, the overstory tree species make up a dominant feature of the ecosystem, and these trees were pioneers relative to the rest of the forest (understories). Federov[(598)] remarks that the spatial distributions of overstory trees are great, while those of understory trees tend to be lower. Understory trees are specialists physiologically for shaded conditions and are more recent members of the community.

At present, we have too few data on the extent to which many tropical forest trees have host-specific seed and seedling predators, and comparing these associations between overstory and understory tree species in a habitat would be especially revealing. Early stages of succession in the wet tropics are characterized by large, monospecific clumps of trees, a result of the absence of host-specific seed predators in these habitats. Either such species never evolved such associations with insects, or else the associations were lost as the result of changes in the vegetation cover of a region. Monospecific clumps of trees may be stable over long periods of time and thereby provide predictable resources for some phytophagous insects. But one would expect the frequency of outbreaks of a few insect species associated with monospecific clumps of trees along the borders of primary forest to be relatively rare. The combination of benign physical environmental conditions and the high predictability of the food base (leaves, etc.) for these phytophagous species results in predictable resources for some predators and parasites of these insects, and effective biotic control may result. The control system may be less stable in highly seasonal tropical regions, and some outbreaks are expected under such conditions. Depending upon the constraints on life history traits shaping *r*, an insect species feeding on a large food base may not exploit the food base and thereby provide space for other species. In this manner monospecific stands of plants in lowland tropical rain forests may accumulate phytophagous species, either as a community associated with that plant species or as separate guilds on the plant.

Studies show that insect outbreaks are related to the aftermath of typhoons. A typhoon may drastically alter the resource base for an insect species, as indicated in an analysis of outbreaks in the spruce bark beetle *I. typographus* in Hokkaido.(599) One of the limiting factors operating on populations of this insect is the availability of fallen trees of the host species as the insect tunnels beneath the bark of these trees. Equilibrium maximal densities of the beetle population are estimated by counting entrance holes of burrows, and the average is 60–70 burrows per square meter of bark on fallen trees.(599) Yet after intense storms and typhoons, the density of fallen trees increases tremendously and the density of entrance holes of the beetle exceeds 350 per square meter.(599) Thus outbreaks result from density-independent factors affecting the resource base of the insect, that is, by increasing the resource base and number of insects responding in a density-dependent fashion to an increased food supply. The data illustrate how large-scale abiotic factors can disrupt the equilibrium population densities of insects in complex ecosystems. A recent analysis of light-gap dynamics in tropical forests(337) suggests that tree falls occur frequently and alter the vegetation of such ecosystems. Depending upon the size of the area affected and the tree species in it, and the life history properties of insect species in it, and the life history properties of insect species in the area, outbreaks of tropical insects may occur as the result of such environmental perturbations.

The ebb and flow of coevolved associations between insects and plants in tropical forests is mitigated by the genetic structure of both insect and plant populations. Outcrossing is expected to be frequent in tropical forest trees as a means of providing a mechanism for maintaining genetic variation to cope with changing selection pressures.[40,499] The evolution of nectar reward properties is shaped to a large degree by the presumed adaptive significance of outcrossing in tropical forests.[216,217] Pollinating insects such as bees exhibit a diversity of methods for collecting pollen,[600] and their foraging strategies are molded by energy requirements and the spatial distribution of nectar and pollen sources.[40] In some groups of bees there is considerable ecological specialization for exploiting specific groups of plants, a classic example being orchid bees.[601] The degree to which a bee species is specialized for visiting certain kinds of flowers and other resources, the spatial distribution and abundance of these resources, effects of seasonality on resources, and the nesting requirements (including how nest sites are determined by proximity of resources[21]), will determine the population structure of such species. Different species within a single genus or subfamily are often very specialized in the tropics, suggesting that evolutionary divergence in foraging habits evolved as a means of lowering competition. If so, then bees tend to be at equilibrium population sizes in the tropics.

Along with hummingbirds and some flies, some tropical bees are opportunistic feeders on other products such as the excreta of coccoids.[602] Such products may be supplementary sources of nitrogen needed for dietary optimization. Some tropical seed-eating forest floor grasshoppers are opportunistic predators on insects at lights (A. M. Young, unpublished observations). Different genera and species of sand wasps in South America exhibit different prey preferences and nesting habits.[603] Such differences may reflect ecological diversification in response to an abundant and diversified supply of prey insects and microhabitat differences. Limiting distributions of resources, for example, optimal nesting sites in the case of Neotropical sphecid wasps such as *Trypoxylon,* may necessitate the evolution of male guarding behavior if strong competition exists for nest sites or mates.[604] Strong intermale competition in these wasps at nesting sites suggests that females develop a mating strategy maximizing mate choice. Optimal mate choice systems may be more prevalent in tropical insects than originally suspected, since the intensity of selection pressure from multiple sources, including predation and parasitism, is expected to be high. This condition favors the maximizing of opportunity to promote the maintenance of the best suited alleles within breeding populations. Limiting resources such as food supplies, nesting sites, and mates are expected to play major roles in shaping the mating strategies of tropical insects. Although clearly many different kinds of insects in the tropics are specialists, there may also be considerable phenotypic flexibility in such species, expressed as opportunistic feeding on other resources. Thus although biological organization in tropical ecosystems is very evident, many species in associated communities will exhibit varying amounts of behavioral and physiological flexibility.[99] Such capacities may be significant in permitting species to switch feeding habits in response to changes in community or guild structure. There can be facultative switches associated with predictable changes in resource abundances (seasonality, floodings), or obligatory switches associated with addition of new species to the community and possible competition. We need to gather more data on whether insect species in different communities are at equilibrium population sizes or densities.

8.3. NOTES ON THE DYNAMICS OF TROPICAL INSECT COMMUNITIES

The majority of insects found in tropical habitats exploit plants and plant products as food resources. An observation of a particular insect species feeding on a particular plant species is to be interpreted in terms of (1) the degree to which the relationship is the original coevolved association or one developed from the original one in the past,[436] (2) the degree to which the insect is a specialist on this plant species, and (3) the spectrum in fitness within the insect population on each plant species if the insect is polyphagous or exhibits broad monophagy.[275] It might also be advantageous to know something about the parts or substances of the plant used as the resource and the extent to which the resource is shared with other species. Such measurements should form a basis for continued research on the natural history of tropical insects.

At the conceptual level, phytophagous insects regulate the biomass and diversity of plants in an area.[605] Severe defoliation of plants by insects, especially in areas where species numbers are high and there are seasonal periods of peak abundance, results in losses in fitness in plant populations,[474] which in turn should result in counterselection to reduce defoliation in future generations. Some amount of defoliation contributes to the maintenance of low-density populations, along with host-specific seed predation, and when plant species are regulated in this way, diversity within communities increases over evolutionary time.[606] Patch structure for a given plant species will be generated by the effects of dispersal agents and biotic mortality agents affecting the survivorship of adults and juveniles, and also by various density-independent factors such as flooding or fires. In lowland tropical dry forests, fire is a major factor contributing to patch structure of vegetation. Coevolved interactions between plants and herbivorous insects can reduce the likelihood of competition among plant species as a source of patch structure: host-specific predation by insects on seeds and seedlings reduces the likelihood for some tree species to become dominants in tropical forests,[29,65] and insects grazing on leaves and other vegetative structures may promote diversity by reducing population sizes of adults. In specialized habitats with several ecologically similar invertebrate species, reduction in niche size of some resident species can result from competition with other species.[607,608] The evidence for such niche reduction would be nonmicrosympatric distributions in which each species occupies a different microenvironment.[608]

In some Temperate Zone communities, plant diversity and structure correlate positively with diversity of Homoptera[414]; similar patterns are expected in tropical communities. But for some insect taxa, host nonspecificity in the tropics is increased[418] in response to individual resource types being too patchy in spatial distribution.

Depending upon the parameters of population growth and dispersal abilities within populations of individual resident species in tropical communities, there may be specialization or lack of it. Long-distance pollination of some trees in tropical forests by euglossine bees[40] makes it possible for these insects to specialize on these resources, whereas insects with more restricted movement patterns may become generalists on several alternative resources distributed as small patches in a relatively small area of habitat (where *small* is defined relative to the dispersal prowess of the individuals). Specialization can evolve if patch size is very large relative to dispersal ability and population growth rates. Karr[609] demonstrated with comparisons of temperate and tropical avian faunas of forest and field habitats that the higher diversity of birds in the tropics is due to several different factors such as reduced energy requirements for breeding, greater vertical stratification of faunas, and increased abundance of insectivores. Further heterogeneity in avian distributions in the tropics results from convergence of many species to exploit fruits that are distributed in a patchy fashion within forests.[609] Bird species specializing in insects as food are less abundant in the tropics.[43] Insectivorous bird species specializing in larger insects are more abundant in the lowland tropics than in the temperate zones or at higher elevations in tropical zones.[43] But in seasonal tropical lowlands many bird species feed on both fruit and insects[610] because seasonal fluctuations in insect abundance reduce insects as a food source at certain times of the year. Thus selection pressure from avian predators may be greatest on some insect populations at the beginning of the wet season in strongly seasonal regions of the tropics, but the degree to which population increases at this time satiate such predators is not known. In less seasonal regions, predation pressure from birds may be more evenly distributed throughout the year.

Sweep samples of foliage insects in various tropical habitats reveal that groups such as Homoptera tend to be very diverse on herbaceous vegetation.[72] Working in the Temperate Zone, McClure[281] concluded that leaf hoppers exhibit considerable niche diversification by utilizing different portions of the same plant. Microhabitat differentiation is well documented in Homoptera in salt marsh grasses in the Northern Temperate Zone.[73] If the species within a community are in equilibrium, resource partitioning results in coexistence.[281] Frequency-dependent competitive ability within a community may result in coexistence of species if high densities attained by a species reduce its fitness more than that of a competing species.[611] Although in the nonseasonal tropics potentially competing species have little or no opportunity to shift abundance levels allochronically, so that an unstable equilibrium of coexistence is maintained,[612] such effects may occur in seasonal areas. Markedly allochronic emergence patterns of cicadas in the tropics[71,138,139] may or may not be indicative of such interactions compensating for losses in fitness from competition. It is necessary to determine what, if any, resources utilized jointly are in sufficiently limited abundance to invoke competition among microsympatric species.[71] Although the adult populations of several species of cicadas with markedly seasonal emergence cycles can be very dense,[139] there is little evidence to date of the operation of density-responsive predators or pathogens, even though a pathogenic fungus may kill off large numbers of *P. biolleyi*,[136] and some species of birds in lowland and premontane wet forest successfully stalk singing males of the large and loud cicada *Z. smaragdina*[71] (And A. M. Young, unpublished observations). Such mortality appears to be patchy. Similarly, army ant swarm raids may take heavy tolls on the immature stages of paper wasps,[193] but such effects must be very patchy and thereby not regulating prey populations in a density-dependent fashion. Other factors, such as the degree to which a habitat has been modified by human disturbance, can influence the likelihood for outbreaks of herbivorous insects associated with a particular plant species surviving in the remnant habitat.[613]

Regions in the tropics that have repeated and localized catastrophes such as floods or fires may provide sufficient density-independent mortality of individuals in populations of two or more competing species to result in another form of unstable equilibrium of coexistence that fluctuates considerably each year. Such a form of nonequilibrium coexistence was originally conceptualized by Hutchinson.[612] If it exists in the tropics, it is probably very patchy in effect and perhaps not repeated over successive years. It seems probable that examples of coexistence among several insect species on the same host plant or other resource in the tropics are explained by microhabitat diversification on the resource and the maintenance of population densities at low levels due to biotic control agents. However, few data confirm such a prediction. Rolled-leaf hispine beetles exhibit both monophagous and polyphagous feeding habits on various Zingiberales.[614] Polyphagous species are those utilizing several different hosts in a region where no one host is abundant. The number of rolled-leaf hispines correlates positively with the number of common host plant species (Fig. 8.1). Those species of

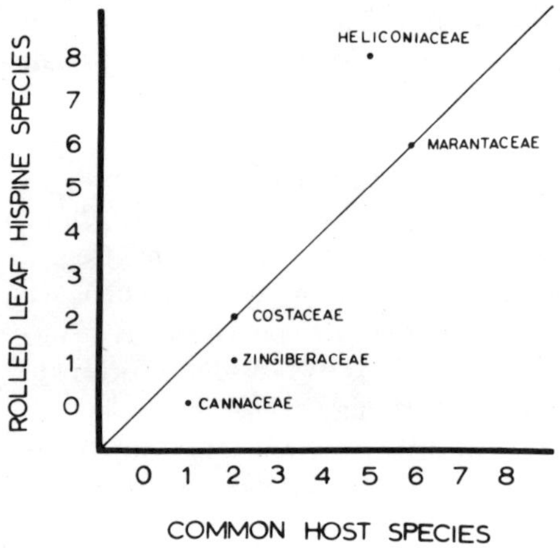

FIGURE 8.1. The abundance of rolled-leaf hispine beetle species as a function of the frequency of common species in their host plant families, at Finca la Selva, in northeastern Costa Rica. [From D. R. Strong, Jr., *Biotropica* **9:**156–169 (1977).]

Zingiberales that are widely distributed grographically have a greater number of hispine species feeding on them because of considerable geographic replacement of species.(614) But the number of hispines coexisting in a habitat is a function of the abundance of the host plant, sizes of host plants, and diversity of host plant families.(614) When many common host plants are in an area, there will be many species of these beetles, presumably since the rolled-leaf microhabitat is a limited resource for these specialist insects. On the other hand, Slater(615) found no such correlation between the commonness of a host plant and its utilization by various species of blissine bugs in the tropics. The host plant associations of blissines appear to reflect the evolutionary history of the host plant groups in that the insects diversified for feeding on three major groups of monocotyledons early in their evolution.(423,424,615)

A significant result of studies such as that of Strong(614) is that mechanisms of species coexistence in tropical habitats are determined largely by within-habitat ecological associations and are not related to the geographical distribution of species. Host plant range may provide the evolutionary opportunity for the geographical replacement of species, but these species will in many instances have similar ecological roles as their counterparts in other portions of the range. The localized coevolution between an insect and its host or resource will represent a biochemical and physiological commitment on the part of both partners in the interaction, although some degree of genetic or phenotypic flexibility remains in the gene pools. Janzen(363) and Wilson and Janzen(616) found considerable evidence that seed-eating bruchid beetles are host specialists in the dry forest lowlands of Costa Rica, each species usually having only one host (range of one to six host tree species). In some cases, two or three species of bruchids were on a host seed. Because of this almost 1 : 1 correspondence, it appears that the number of host seed tree species determines the number of bruchid beetle species.

Gilbert and Smiley[329] summarize data on the host plant associations of ithomiid butterflies to provide additional evidence on the frequency of coevolved associations between insects and plants in the tropics. The results of Young[88, 270, 290, 298, 301, 302, 310] provide additional evidence of such a pattern. Drummond,[617] working in Ecuador, reported 53 ithomiines co-occurring with 44 potential host plants. Of 27 species whose life cycles were determined, 81% were limited to a single host plant, while others used two to five larval host plants.[617] Drummond did not find a correlation between the size of larval host plant populations and the number of ithomiine species found on them. Gilbert and Smiley[329] believe that such patterns of resource partitioning do not arise from competitive interactions among the larvae of different ithomiine species on the same host plants but rather as the result of the coevolved associations of each ithomiine species with a particular host plant species within the Solanaceae. The degree to which the larval host plants studied were in fact different in terms of secondary defensive compounds underlying host plant specialization was not determined. The Solanaceae as a group are well known for high concentrations of alkaloids as defensive substances,[271] a general trait underlying the coevolved association of the Ithomiinae with the group.[6] Some primitive ithomiine genera are coevolved with Apocyanaceae[270, 273] even though they co-occur with several solanaceous-feeding ithomiines.

Gilbert and Smiley[329] discuss the generally low abundance of immature stages of ithomiines on larval host plants and show that there is a positive correlation between the abundances of individual ithomiine species and the abundances of their host plant species. The butterflies do not appear to be food-limited in the larval stage.[617] Gilbert and Smiley explain the pattern by assessing the effective density of larvae per unit of leaf area as the primary variable influencing the foraging behavior of predatory ants and wasps: ithomiines using an abundant host plant species are expected to attain greater levels of abundance before a functional response is elicited in the predators.[329] Gilbert and Smiley argue that rare ithomiines may have similar abundance levels of larvae when assessed in terms of leaf areas occupied, but because the host plants are rare, these species are limited before host plants are defoliated or inaccessible to ovipositing females. The result is that such ithomiines never build up on such a patchy host plant and therefore populations remain at low densities unless the abundance of the host plant increases in the habitat. Communities of ithomiine butterflies should also be examined with regard to the kind of habitat they occupy and the degree to which certain potential host plant species increase or decrease in abundance as a result of perturbations in the habitat. In general, however, the "indirect host limitation" idea of Gilbert and Smiley[329] shows that a highly specialized ithomiine exploiting a rare host plant species will have population densities considerably lower than those of another specialized species exploiting an abundant host plant in the same habitat. Gilbert and Smiley[338] also report that ithomiine species richness increases in mid-elevation forest habitats since these areas have a greater abundance of naturally disturbed sites invaded by various Solanaceae.

As pointed out by Gilbert and Smiley, such findings agree with those of Janzen,[29] who found greater densities of beetles and moths at mid-elevations. Janzen concluded that plants at mid-elevation sites in the tropics can increase net productivity of harvestable biomass because cool nights reduce respiratory losses. Each host plant species therefore presents a larger food base that makes space for additional specialist species to coexist. This idea is an extension of MacArthur's theory as to how new specialist species can be added to tropical communities through increases in productivity.[16] But for ithomiines, Gilbert and Smiley[329] provide evidence supporting the view that increased diversity of host plants, rather than increased biomass, is the primary factor in increasing species richness of ithomiines. The natural history data on various ithomiines from various elevational sites in Costa Rica collected by Young[88,270,290,298,301,302,310] support the view that both geographically and locally ithomiines appear to be larval host plant specialists. The observation by Young[88,89] of severe defoliation of *S. rugosum* shrubs by the ithomiine *H. euclea* in premontane tropical rain forest provides further support for the prediction that specialist ithomiines exploiting an abundant host plant will have larger adult populations. In this particular case no functional response was observed by predatory insects such as wasps or ants, and larval densities reached outbreak conditions and many individual host plants were heavily defoliated.[88] During periods of wet weather, clusters of eggs were killed by an unidentified pathogen but no predation on eggs or larvae was observed. Young suggested that recent changes in the forest edge habitat at this site resulted in rapid expansion of the host plant population, providing a greatly increased food base for *Hypothyris*.

The observed absence of other ithomiines on this abundant host plant was attributed to the plant having unusually tough leaves, a trait overcome by the communal feeding habits of the gregarious larvae. A similar system occurs in the specialized association of the ithomiine *M. isthmia* with certain solanums found in pastures and along roadsides: these plants have thick, spine-covered stems and leaves that form a formidable physical barrier to many insects with the exception of a few aposematically colored Orthoptera and some chrysomelid beetles. Cluster oviposition and gregariousness in the larvae are two conspicuous behavioral traits of *Mechanitis* and *Hypothyris* generally not found in other ithomiines, and it was suggested[88,221,222] that they are adaptations to breaking down tough leaf tissues by group feeding. Larvae may grow faster in large groups (Fig. 8.2). These host plants establish ecological barriers[6] to colonization by other ithomiines with single

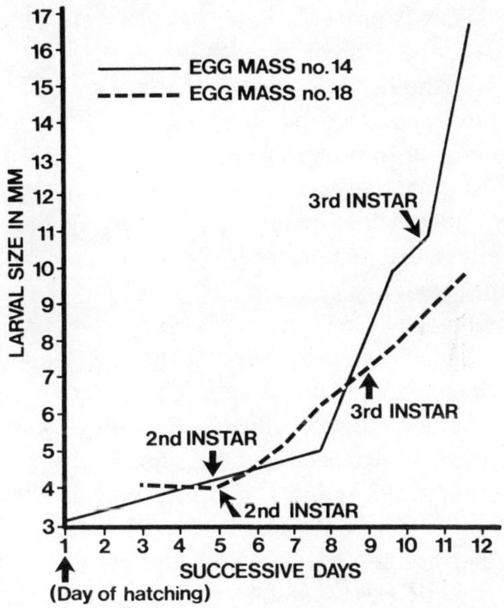

FIGURE 8.2. Rate of development for the Neotropical butterfly *M. isthmia* (Ithomiidae) as seen for two egg masses (clusters) in the wild. Although this sample is very small, of interest is the faster rate of larval development for the clustered group of larvae from No. 14, the larger egg mass. [From A. M. Young and M. W. Moffett, *Dtsch. Entomol. Z.* **26:**21–38 (1979).]

oviposition and solitary larvae as well as other insects. In the *Mechanitis* system, however, considerable mortality of eggs and young larvae results from predation by orthopterans and wasps.[220,221] The complete absence of such predation in *Hypothyris* suggests that this plant species possesses some ecological traits that deter searching behavior by such predators. Aggregative behavior in the early instars of Lepidoptera larvae is believed to be related to the physiological vulnerability of young larvae to food supplies and weather conditions, while such behavior in older instars is related primarily to evasion of biotic mortality agents such as predators.[618] In some aggregating insects, groups break up in response to the presence of a predator.[619] Not all tropical plants and their associated herbivores are subject to the same kinds of effects as described by Gilbert and Smiley[329] in terms of functional responses by predators as a means of effective biotic control of herbivore populations. Nothing is known about the chemical defense system of *S. rugosum* and whether leaf toughness is the only defensive trait of this plant. The plant may produce other signals that deter predators and parasites, so that wind-blown pathogenic microorganisms are the only biotic mortality agents operative on *Hypothyris* populations at this locality. Further studies are needed to compare the intensity of predation and parasitism in selected herbivorous insects associated with a variety of plant species in the tropics. If predators and parasites are generalists, the availability of more accessible food supplies at any one time may shift biotic control from other herbivore populations, possibly resulting in outbreaks. The effect of pathogenic microorganisms is probably largely density-independent and determined largely by day-to-day patterns of rainfall.

Another example summarized by Gilbert and Smiley[329] illustrates the difficulty in making broad generalizations about the strategies and patterns of resource utilization in tropical insects. Gilbert's analysis of local patterns of larval host plant exploitation by 14 species of the satyrine genus *Euptychia* in Trinidad revealed a high percentage of polyphagous species present in an area does not reflect the number of *Euptychia* species.[329] In fact one site studied had 10 species of this genus and only one larval host plant species, and another four sites had twice as many butterfly species as host plant species.[329] But the distribution of specialist species of the genus (monophagous species) tracks that of the selected host plant species. The pattern breaks down for polyphagous species as explained above. Gilbert and Smiley, citing some unpublished studies of their colleague Michael Singer, suggest that unusual features of host plant clumps (grasses and sedges), coupled with unusual oviposition behavior of *Euptychia* species, determine the patterns of several butterfly species on relatively few host plant species. Gilbert and Smiley also hypothesize that a lack of "chemical distinctiveness" among various grass species determined the patterns of host plant utilization by the butterflies. They analogize the *Euptychia* patterns to the organization of stem-boring guilds studied in the Temperate Zone by Rathcke.[620] At several sites in lowland and premontane rain forest in Costa Rica the butterflies *M. peleides* and *M. granadensis* exploit at least one larval host plant together, namely, the forest edge woody vine *M. seemanii,* even though *peleides* exploits many other leguminous species in the same region and habitats (A. M. Young, personal observation). In Costa Rica, several species of *Morpho* butterflies partition the tropical rain forest environment in terms of habitat type, larval host plants, and, to some degree, the height adults fly above the ground (Table 8.1). All of these species have closely related allies in South America, and in most instances one might expect similar patterns of environmental partitioning. Yet the reasons for such patterns remain obscure, especially since adult population densities of these butterflies are generally quite low.[175,176,190]

At least three species of troidine swallowtail butterflies overlap considerably in using the larval host plant *Aristolochia constricta* in tropical rain forests in Costa Rica (A. M. Young, personal observation). Given the interesting findings of Gilbert with other butterfly–plant systems in the tropics, it would be interesting to explore the organization of *Morpho* and troidine swallowtail butterfly communities, with an emphasis on the relationship between the number of potential host plant species in an area and the number of allied butterfly species. Many species of *Morpho* are in Central and South America, and host plant records suggest considerable ecological diversification within the genus.[176] Critical variables, as emerging from the studies of Gilbert,[338] include patch structure of host plants, habitat species densities of the butterflies, habitat species densities of the

TABLE 8.1
The Species and General Habits of *Morpho* Butterflies Found in Premontane-to-Lowland Tropical Wet Forest Region[a] of Northeastern Costa Rica[b]

Species	Major adult habitat(s)	Flight height range	Relative abundance	Larval food-plant taxa
M. peleides	Various secondary; large forest light gaps	1–3 m	High	Several Leguminosae
M. granadensis	Advanced secondary; large forest light gaps	1–3 m	Low	Probably few Leguminosae
M. amathonte	Primary forest	1–6 m	Intermediate	Probably few Leguminosae
M. theseus	Primary forest	6–30 m	Loa	Probably few Menispermaceae
M. cypris	Primary forest	10–30 m	Low	Probably few Leguminosae

[a] Defined here as 0–200 m above sea level.
[b] From Young.[(331a)].

host plants, the degree of distinctness among host plant patches of a particular plant species in terms of features attractive or repulsive for oviposition in one or more butterfly species, and the degree of chemical and physical distinctiveness among patches of different host plant species co-occurring in the same habitats.

Benson[(282)] has shown that there is considerable resource partitioning in heliconian butterflies for patterns of exploiting their passifloraceous larval host plants. In these butterflies, populations are generally limited by the abundance and distribution of the larval host plants, and there is considerable competition among heliconian species for oviposition sites.[(282)] Because of such effects, ecological shifts have resulted from selection pressures of interspecific competition as a means of reducing losses in fitness where species overlap in distribution. Benson shows that such resource partitioning among heliconians at several localities has taken place with respect to (1) plant species utilized, (2) habitat of preferred host plant, and (3) specific part of plant used for oviposition and development of early larval instars. The organization of heliconian communities or guilds is similar in regions where faunistic complexity is high or low.[(282)] In tropical rain forest regions considerable larval host plant specialization is evident among sympatric heliconians, although at one subtropical site in Brazil where the number of heliconian species is low, the species tend to be generalists exhibiting considerable overlap in larval host plants. Yet each species exhibits different habitat preferences, making it a patch generalist, thus minimizing competition for individual host plants.[(283)] This subtropical site is strongly seasonal, which results in periodic reductions in the harvestable productivity of each larval host plant, and each heliconian species tracks the fluctuations in host plant abundances through life history adjustments. The seasonal fluctuations in host plant cycles impose the generalist strategy upon these heliconian species as a means of maintaining breeding populations through periods when one or more host plant species may be scarce.[283] But competition is prevented under such conditions by distinct patterns of habitat separation among the species, reducing resource partitioning in terms of host plant species.

A fine equilibrium appears to be established at each locality between the number of host plant species, host plant population sizes, habitat separation among host plant species, and the number of heliconian species found there, to the point that Benson *et al.*[283] conclude that heliconian communities are at saturation levels. Competition is believed to be the mechanism excluding other species from being added to the communities.[283] Gilbert[338] suggests that the monophagous habit of many specialized herbivores as the result of coevolution, implies that populations of the specialist butterfly will increase as the host plant population increases. Another case cited by Gilbert, an alternative pattern, is the herbivore species generally adapted to feed on many members of a single plant family: when a host plant becomes abundant in a habitat, it is likely to be colonized by several herbivore species. Gilbert points out that most heliconiine butterflies and *Passiflora*-feeding flea beetles fall between these two extreme conditions in terms of the relationship between host plant abundance and insect abundance. Thus an insect may be a coevolved specialist for feeding on several different host plant species within a certain family yet feed on only one or two as the result of ecological factors.[329,338] To some extent it is true that increases in the abundance of a larval host plant increase the number of insect species feeding on it if these species are generally preadapted to exploit several members of the host plant family. Gilbert and Smiley[329] show a significant correlation between numbers of herbivore species on individual species of *Passiflora* as a function of the local abundance of each plant species. Higher harvestable productivity of *Passiflora* results in a greater number of herbivorous insect species on it, even if many members of such a guild are monophagous.[329] Gilbert and Smiley[329] have examined other ecological factors affecting the numerical relationship between host plant diversity, host plant abundance, and herbivore species numbers, and the reader is referred to this work for a detailed analysis.

In developing his arguments with regard to heliconiine butterflies exploiting the Passifloraceae, Gilbert stresses the role of biogeographic events in determining the broad and restricted geographical distributions of various heliconiine species. Turner[621] reported that heliconiines experienced considerable adaptive radiation while isolated in the glacial forest refugia areas of South America. During such times the group underwent considerable habitat separation as evidenced in their present-day distributions in some cases. Cyclic refuge isolation resulted in some species spreading out over much of the range of potential habitats, while others were more restricted.[621] Brown *et al.*[622] suggest that localities in South America adjacent or close to Quaternary refugia have a high number of species to colonize new habitats. Some sites extremely rich in terms of numbers of heliconiine species are within the overlap zones for the expanded faunas of at least four surrounding forest refugia.[468,478,623–626] Forest plants experienced considerable diversification within Quaternary refugia, and many of these host plant species (e.g., in the family Passifloraceae) could have resulted in the rich diversity of heliconiiine larval host plants found in the overlap zones today.[622] These biogeographic heterogeneous effects make it necessary to select sites where refugia influences are similar in order to examine the effects of local ecological factors regulating herbivore communities associated with particular groups of plants. Gilbert[338] examines the role of such ecological factors, including resource variety, habitat structure, climatic predictability, and local topography in determining the local species abundances of heliconiine butterflies in relation to how such factors have influenced the local *Passiflora* communities and adult resources of the butterflies. Thus although strong coevolution between groups of insects and their resources is a common phenomenon in the tropics, it is also necessary to examine the role of a variety of ecological factors in shaping the organization of local herbivore communities or guilds and, whenever possible, to recognize the influence of biogeographical events.

In Chapter 5 the concept of habitat selection in tropical insects was discussed, yet clearly some of the considerations here on the maintenance of insect communities with a set number of species also relate to strategies of habitat selection pointed out by Benson's studies of heliconian butterflies and their larval host plant associations.[282] The degree to which a resource is in limited supply will clearly mold the strategy of habitat selection for an individual herbivorous insect species that is a specialist on that plant. The intensity of habitat selection will also be a function of the quality of individual resource patches, and any one habitat is expected to exhibit a spectrum of quality determined by the condition of the resource on a patch to patch basis. Any one herbivore species exploiting the resource should maximize the choice of high quality resource patches while minimizing the colonization of poor-quality resource patches. If several ecologically similar species are converging upon the same resource, there will be either (1) intense competition within a habitat between these species for the high-quality patches, perhaps resulting through evolutionary time in resource partitioning along different resource parts or by divergence onto different host plant species, or (2) segregation of the herbivore species into different habitats where the same host plant species is found, thereby reducing effects of competition. Furthermore, Janzen[93] has suggested that even herbivores exploiting different parts of a single plant, or different individuals of the same host plant species in a particular population of that plant, are, in effect, in competition.

In an interesting study of the food-limited and specialist Temperate Zone aphid *Pemphigus,* Whitham[627] found considerable evidence of variation in resource patch (size range of *Populus* leaves) resulting in exceedingly high levels of colonization for optimal patches and very low levels of colonization for poor-quality patches. Whitham[627,628] found that fertilizer-treated individuals of one host plant species increased the breeding success of the aphids, indicating that nutrients were a limiting factor affecting these herbivores. Overall resource quality can be improved by fertilizer treatment for this host plant species. Yet for a second species of *Populus,* the breeding success of this aphid species could not be significantly increased by adding fertilizer.[627] This second species of *Populus* was the preferred host of the aphid, and there was evidence of considerable intraspecific competition for optimal resource patches. In another paper, Whitham[628] conceptualizes the *Populus* host plant population as a heterogeneous environment or habitat in terms of the great variation in quality of individual leaves for colonization by *Pemphigus*. As habitat quality is increased, so does the density of competitors, and there is a clear pattern of adjustment in densities in different habitats of varying quality so that average fitness is about the same in all habitats.[628] In tropical species in similar systems of habitat quality variation, it would be interesting to determine how population densities of a single herbivorous insect species are regulated in terms of the effects of interspecific competition on optimal resource patches. The greater density of closely related and ecologically similar species in many tropical habitats, and the higher incidence of ecological specialization, suggests that the maintenance of an adequate level of average fitness within a population is a function of competition between those herbivore species that have populations at or approaching the carrying capacity of the habitat. Such effects may be less significant in species where biotic control agents such as predators and parasitoids keep populations well below such levels.

A growing body of evidence shows that mate recognition in some butterflies is the result of ultraviolet reflection cues rather than color cues in the visual spectrum.[629] If generally true, visual coloration in butterflies, especially in tropical forms, may be more of an adaptation to thermoregulation and sheltering behavior and for active or passive escape from visually hunting vertebrate predators. Neotropical *Morpho* may exhibit pursuit invitation behavior toward such predators, depending upon the degree of blue reflection from the wings and flight behavior.[207] In El Salvador and presumably elsewhere, anoles successfully knock down ithomiid butterflies with their tails and devour them (A. Muyshondt, personal communication) and wing color patterns in these butterflies may be more effective against other kinds of predators. Some butterflies undoubtedly use visual colors in courtship as well. Turner[630] has given a detailed discussion of the evolution of butterfly mimicry and other forms of defense against predators, with an emphasis on tropical faunas. Many rain forest understory butterflies in the great families Brassolidae, Satyridae, and Morphidae, have conspicuous eyespot markings on the ventral sides of the wings. These butterflies typically fold their wings when feeding or resting, and the colors are generally cryptic to the human observer, making the insects blend into the forest floor or shadows of understory vegetation. Many of these butterflies exhibit a peak of feeding near dusk in lowland and premontane tropical rain forest,[(188)] and several individuals of a species may accumulate daily at rotting, fermenting fruits on the forest floor. The early stages are cryptic and those of some species are associated with monocotyledons (Fig. 8.3). Home range movements may be involved in such feeding behavior, with individuals returning over successive days to feed at the same resource.[(175,186,190)] Young[(188)] suggested that eyespot markings are enhanced in their appearance to humans, and perhaps to predators other than butterfly collectors, when streaks of sunlight near dusk hit the wings of feeding butterflies. Stradling[(631)] suggested that such eyespot markings in some butterflies together with other wing markings resemble the heads of amphibians and reptiles, two possible groups of predators on these insects. If interpreted as sucy by predators, these animals may avoid feeding on resting butterflies. Young[(188)] noted a decrease in the size and incidence in eyespot marking for butterflies active in the higher strata of tropical rain forests and suggested that selection would not maintain such eyespot phenotypes if ‘‘eyespot-sensitive’’ predators were less abundant in such microhabitats.

In the context of Pimentel,[(632)] the degree to which an herbivorous insect species affects the fitness of a plant species, as a result of specialized predation, depends upon the degree to which both plant and insect enter into cyclic patterns of frequency-dependent selection for traits underlying the coevolved relationship. Tropical forest ecosystems may consist of many such coevolved relationships, each one within the many communities and guilds present constantly changing in response to reciprocal selection from each partner. Thus a given plant species may possess a certain spectrum of monoterpene compounds that function as chemical defense compounds, and there is a range in variation for effectiveness against specific herbivores. The habitat consists of many plant species, each endowed with a spectrum of defensive compounds genotypes and phenotypes that may qualitatively differ among species.[(256)] As selection pressures from herbivorous insects shift with respect to a particular tree species (and such shifts may be frequent given the high species densities of herbivorous insects in most tropical terrestrial habitats below 2000 m), directional selection may segregate a relatively rare toxic genotype of the plant in response to a change in the coevolved relationship. Subsequently, strong stabilizing selection (Chapter 2) maintains and perfects the new toxic phenotype, which then increases in abundance in the plant population. But the more abundant the formerly rare toxic phenotype becomes in the population, the rate of increase being a function of the life history pattern of the plant species in that habitat and the intensity of preda-

FIGURE 8.3. Life stages of the brassolid butterfly *Opsiphanes tamarindi* associated with *Heliconia* and *Musa* in Costa Rica. (A) Egg, (B) third- and (C) fifth- instar caterpillars, (D) pupa, (E) newly-eclosed adult on pupal shell. [From A. M. Young and A. Muyshondt, *Stud. Neotrop. Fauna* **10:**19–56 (1975).]

tion on it by one or more herbivore species, the greater the contact with herbivorous species and the more rapid the counter (evolutionary) response by the insects. In this manner, there is a frequency-dependent overcoming of the new defensive barrier by the insects, and if biotic control agents are inoperative in this system (i.e., predators and parasitoids of the herbivores) or if interspecific competition is nil, the insects will become pest species and outbreaks may occur. The new toxic phenotype is now ineffective in deterring attack by the herbivores. Thus an optimal strategy for a tree or other growth form of plants in tropical forests in particular, where plant predation rates can be at least potentially very intense, is to segregate a range of toxic phenotypes at each generation, with some proportion of these toxic phenotypes remaining rare within the population.

The strong coevolution in Central America between some Cucurbitaceae and the luperine chrysomelid beetles that feed on them is mediated by the use of bitter substances, cucurbitacins (oxygenated tetracyclic triterpenes), in the plant tissues as feeding stimulants by these herbivores, even though these substances probably originally evolved to deter such predation or plant parasitism.[633] Such studies also point out that although some coevolved feeding associations between plants and insects have evolved in the tropics, the chemical or physiological mechanisms underlying the interaction have been retained through the evolutionary radiation of both the host plant group and the herbivore group into other kinds of environments, including those promoting the domestication of the plants as food sources.

The observation in nonseasonal lowland tropical regions that outbreaks of herbivorous insects on plants in natural ecosystems are relatively rare suggests that most plant species in such environments maintain breeding populations with a high degree of genetic variability with regard to defensive compound chemistry and other traits relating to defense against herbivores. Such properties, in conjunction with disjunct spatial distributions, fluctuations in physiological cycles (season or nonseasonal) regulating vegetative and reproductive growth, and the probable limitation of herbivore populations by predators and parasitoids, lower the likelihood of outbreaks in natural ecosystems. Communities and guilds of plant-feeding insects, as emphasized in the studies of Gilbert,[338] Strong,[634] and Janzen,[47] are well organized in space and time, and such properties reflect both the highly coevolved interactions between plants and herbivores and the interactions of both plants and their herbivorous insects with other components of tropical forest ecosystems (i.e., interactions between plant species, interactions between insect species including competition and predation, other features of food chain associations).

Some surveys of avian species in lowland tropical forests indicate a high proportion of insectivorous species relative to that found in Temperate Zone forests.[609] Some studies indicate that predation on herbivorous insects by birds in forest ecosystems reduces the population densities of these insects markedly.[635] In tropical forests such predation by birds[610] may take on even greater significance in terms of regulating insect populations: many insects have relatively low population densities to begin with, because plant species tend to have patchy distributions and many insects are specialists on a few plant species,[47] and under such conditions of low population densities, avian predation may act as an even greater selective agent and source of biotic regulation of populations.[635] Thus it is necessary to examine, whenever possible, the food chain associations for herbivorous insects in determining the degree to which individual species are resource limited or regulated by systems of predation.

Chapter 9 considers how systems of agriculture in tropical regions have altered such coevolved associations and the resulting high structuring of insect and plant communities in the tropics. Detailed studies on the ecological traits associated with coevolved interactions between plants and their specialized seed-eating insects in both natural and disturbed tropical habitats are given in Janzen.[29,362,363] The reader is encouraged to examine these papers for a comprehensive understanding of the dynamics of such coevolved systems in the tropics and their implications for the organization of tropical forest ecosystems.

Some portion of community structure in tropical forests is determined by selection pressures promoting the association of certain kinds of insects with plants. Chapter 4 included a brief discussion of the highly coevolved mutualisms between plants and ants[33,373,376] that emphasized that ants replace chemical substances as the major line of defense against herbivores in plants. The studies of Rehr *et al.*[262] showed that chemical defense in acacias turns up when the mutualist ants, *Pseudomyrmex* species, are absent. The analysis of the functional role of extrafloral nectaries of *Bixa* for attracting ants indicated that herbivory is less on plants with actively patrolling ants. The greater incidence of such mutualisms in the tropics indicates that (1) the benign physical environment has been conducive to the evolution of such coevolved relationships, and (2) in plant lineages where chemical defenses have not evolved, possibly as the result of evolutionary history or other forms of contemporary selection tending to erode genetic variation that would permit such a system to evolve, intense herbivory has resulted in the evolution of mutualisms with ants. In some cases, the relationship has evolved to the point where the activity of the resident ant species results in an increased nutritional supply for the plant.[376]

In an interesting study of ant–plant mutualisms in the Temperate Zone, Inouye and Taylor[636] found that the effectiveness of ants as herbivore deterrants may not always be as effective as chemical defense. They studied the relationship, along an elevational gradient in Colorado, of various ant species with the aspen sunflower, *Helianthella quinquenervis,* a plant with conspicuous extrafloral nectaries that produce large amounts of sugar and amino-acid-rich nectar. At high elevations the ants are effective in lowering the amount of herbivory by various insects on seeds prior to dispersal and are effective in preventing oviposition by various flies. But at lower elevations, where such seed predators and ants are more dense, ant protection was less effective in reducing predispersal seed predation. Other plant species in the same habitat appeared to be better protected by chemical defenses than by ants.[636]. Compared with some tropical systems such as *Acacia* × *Pseudomyrmex* or *Cecropia* × *Azteca*,[33,374,375] in which the plants provide a nesting site for the mutualist ants, Inouye and Taylor's Colorado system lacked ants actually living in the plants.[636] Perhaps under such conditions, the relationship is not as well evolved to the point that selection has resulted in ants actually living in the plants. But such a relationship may not evolve if reductions in fitness are not appreciable (unlikely, given Inouye and Taylor's data) and if other genetic and phenotypic constraints within the plant populations preclude the expression of the necessary physiological and morphological changes in phenotype needed to accommodate the ants. If the average plant species within certain habitats in the lowland and mid-elevations of the New and Old World tropics experiences greater selection pressures from herbivory (insect and vertebrate) than plant species in comparable Temperate Zone environments, either there will be a greater incidence of highly coevolved associations of insects with plants with many other insects screened out by the chemical defense systems of each plant species,[6] or ant mutualisms will evolve in lineages where chemical and physical defenses are not effective against herbivores, reducing fitness, and there will also be a greater incidence of the evolution of extrafloral nectaries.

The evolution of extrafloral nectaries, as seen in many tropical plants, may represent intermediate stages in the evolution of plant–ant mutualisms if such plant species are genetically flexible enough to evolve the traits necessary for such mutualisms and if the selection pressures favoring such an interaction are sustained through enough generations. If, on the other hand, the plant species is incapable of evolving such a mutualism (Janzen[33] has stressed the extreme phenotypic commitment by both partners that are involved in the evolution of such mutualisms.) due to genetic and phenotypic constraints affecting the overall average phenotype within the population, the expression of extrafloral nectaries may be the end point in herbivore defense strategy for that species (although the system is not static and subtle changes in the phenotype related to effectiveness of extrafloral nectaries will continue). The likelihood of an ant–plant mutualism evolving also depends upon the available pool of ant species, the degree to which each species is already ecologically specialized, and energy requirements related to maintenance of colonies. The incidence of plant–ant mutualisms in the tropics may be higher than what is known today.

The integration of the tropical forest ecosystem is highly dependent upon the breeding systems of the various plant species found there. Primarily through the efforts in recent years of workers such as D. H. Janzen, H. G. Baker, K. S. Bawa, and G. W. Frankie, there is a growing body of information on the breeding systems of tropical trees and shrubs in different kinds of tropical forest environments.[118,218] The subject, although fascinating and extremely relevant to matters of community structure in the tropics, is too vast to summarize here, as the breeding systems and associated pollinators are varied and diverse, so that only a minimal treatment is given.

Van Der Pijl[334] has emphasized the overall diversity of pollination systems seen in the tropics and has attempted to examine general evolutionary patterns within the angiosperms in the context of pollinators and dispersal agents. The striking diversity in floral morphology and physiology seen in tropical floras indicates that some sets of pollinating insects may be specialized ecologically to exploit unusual substances secreted by flowers. Some groups of tropical bees, for example, may exploit noncarbohydrate or nonprotein substances such as fats or lipids secreted by certain plant taxa. Although such insects may visit a variety of other plant species for pollen and nectar (e.g., Neotropical solitary bees), the feeding niche or nest provision niche component may include these kinds of flowers. The degree to which the exploitation of such unusual resources is carried out by a particular bee species will be a function of many factors, including the ecological role of the substance in the biology of the insect, accessibility of the resource, and competition among species for the resource. Different groups of tropical pollinating insects may have spatial population structures permitting the exploitation of such resources opportunistically or as a regular component of the diet. In the tropics, there is a greater degree of specificity by pollinators and animals acting as dispersal agents of seeds for certain plant groups, although in other cases the relationships are more generalized. Floral structures are generally designed to attract certain kinds of pollinators while excluding others in many specialized plants, and exhibit considerable nonspecificity in other groups. Floral structure, seed dispersal, life table, and resulting patchiness in distribution of the adult population of a plant species together constitute the primary determinants of the optimal breeding system for that species. A large percentage of plant species in the tropics are pollinated by a variety of insects. Insects and other animals are attracted to flowers by nutritional rewards associated with retrieval of nectar and pollen, and such rewards vary considerably in quality and quantity among different plants and within the same plant species at different times of the day, etc.[216,217] Some insects, e.g., certain tropical Lepidoptera, also obtain substances from vegetative parts and seeds of various plants[177–179] and do not enter into the pollination association. The insects may use these substances as precursors in the synthesis of courtship-related substances in one sex or the other,[177] and such systems represent a different form of coevolution between insects and plants.

The relationship of pollinators to a particular plant species must in part be viewed in terms of the role of flowers as attractants, sources of rewards (energy), and ecological screens for permitting only certain kinds of insects and other animals to visit them. The role of a particular kind of flower as a food source for some insects will be a function of its phenotypic traits, including the morphological and physiological. In turn pollinating insects evolving into specialists for a particular flower will possess specific morphological, behavioral, and ecological traits that optimize retrieval of required nutrients from the flowers. Such specializations are diverse in the tropics, where both insects and floras are diverse, locally and over geographical regions.

A specialist insect species on a particular flower species or cluster of flower species will exhibit regular foraging patterns for exploiting that resource(s), since individuals will receive virtually a full complement of a nectar or pollen diet from that resource(s). Yet there will also be a spectrum in the amount of specialization closely related insect species will possess for exploiting a particular set of flowers, and some species will be generalists in some geographical zones, drop out completely in others, or become specialists in still others. Questions of pollinator interaction with tropical plants must be viewed at different levels of taxonomic organization: one can ask about the regional proportions of plant species pollinated by bees, butterflies, sphingid moths, or dipterans (and others), perhaps over an altitudinal gradient in the tropics and expect to find distinct patterns for different zones. Within a given zone, one will expect to find a wide range of specialization by insects to the flora of that zone: a set of bee species in the humid lowlands may be associated with a different set of plants from the same or similar set of bees in the mid- or high-elevation zones, and the proportions of plants pollinated by these bees may also change over such a gradient. The number of sphingid-moth-pollinated plant species may be greater at mid-elevation sites than in adjacent lowland sites, with constraints being set by the composition of the floras at such sites, both in terms of floral structures represented and in terms of the availability of larval host plants. Within a given zone, each pollinating insect species may have primary and secondary food sources on daily foraging routes, and the composition and ratio of such resources may also change among different zones. In the lowland tropical rain forest zone, the average *Starchypheta* bush may have a greater number of butterfly species associated with its blooms[637,638] than a similar bush of the same species at a mid-elevation site. Such differences will have great impact on the population structure and foraging behavior of the individual insect species in both zones.

Studies in tropical pollination biology therefore stress the community of pollinating organisms associated with a particular plant species, the population structure of that species and flowering phenology, and the degree to which that particular plant species fulfills the food requirements of each pollinating insect species. We also need to examine the impact of joint vertebrate and insect pollinating systems associated with individual plant species. Such considerations illustrate that one explanation for the great diversity of insects associated with flowers in the tropics is the regional diversification of floras and associated pollinating insects. The other explanation is more ecological than regional: diversity of such insects is also determined by the composition of communities within a particular region and the degree to which component species are specialists or generalists in terms of exploiting food sources supplied by flowers.

The majority of pollination systems in the tropics involve birds, bats, butterflies, moths, flies, beetles, bees, and wasps.[216,217] Various aspects of pollination systems were already discussed briefly on page 340, and also in terms of effects of tropical seasonality (Chapter 7). Baker *et al.*[218] believe that plant species in the early stages of tropical succession are self-compatible while those of primary forest are predominantly self-incompatible. Mass flowering and self-incompatibility seemed to be linked as indicated Frankie *et al.* in their study of bee pollination of the papilionaceous legume tree *Andira inermis* in lowland dry forest of Costa Rica.[499] The authors found that two daily peak periods of nectar flow occur in this tree and that pollen is released during the first period. About 70 species of bees were collected from the flowers, and most of the solitary bee species exhibit limited movement among conspecific trees. Low intertree movement accounted for the observed patterns of fruit set.[499] Mass flowering provides a large resource base in some tree species of tropical forests for many bee species, and competition for these resources (nectar and pollen) is minimized due to satiation effects relative to patch size of the resource. Apparently, however, there is sufficient movement by foraging bees among conspecific individual trees to maintain a certain level of fruit set. Self-incompatibility appears to require mass flowering in order to attract a sufficient number of bees to ensure some low frequency of exchange with conspecifics in the same habitat. Self-incompatibility functions in tropical forest trees as a means of outcrossing[638a,639] presumed necessary for providing adequate genetic variation to allow the tree to respond to shifting selection pressures associated with herbivory and other ecological factors.[40] Mass flowering may also function to minimize potential competition for pollinators by providing a sufficiently large resource base for many species. The high local diversity of various Leguminosae and other groups of forest trees in tropical forests, and the similarities in floral structure and peak flowering periods among these trees,[117] can result in considerable interspecies competition among trees for a set of pollinators. Pollinator constancy is one means of reducing competition and allowing each tree species to have a complement of pollinators[640,641] when the potentially competing tree species co-occur in the same habitat. Mass flowering in several closely related tree species at the same time in the same habitat presents many large resource patches of regular size to pollinators, a condition that reduces competition for pollinators. Both conditions may be within colony segregation of resource patches by a bee species or different colonies of the same species may forage at different resource patches, resulting in increased resource availability for other bee species foraging in the same habitat. Although it is necessary to examine nest locations of individual bee species, foraging distances, and how workers within each nest divide up resource patches, if they do at all, the high diversity of bee species observed by Frankie *et al*[499] suggests such mechanisms to be operative.

Closely related plant species in tropical forests may exhibit considerable divergence in floral characters as a means of partitioning the local pool of potential pollinating insects. Schemske[642] found that two sympatric species of the shrub *Lantana* (Verbenaceae) in tropical rain forest possess markedly different floral characters, including inflorescence size and corolla length. The result is that different sets of lepidopteran pollinators visit the flowers of the two species and competition for pollinators is minimized between these species.[642] A recent survey of the amino acid composition of nectars from floral and extrafloral nectaries of many temperate and tropical plants revealed that (1) amino acid complements vary considerably between species for both kinds of nectars, (2) extrafloral nectars were always different from floral nectars of the same plant species, and (3) certain kinds of amino acids are invariable and consistently found in extrafloral nectaries, a condition suggested to be related to nutritional requirements of protective guard insects such as ants and wasps.[218] Such patterns of specificity at the biochemical level function to partition pollinator communities in environments where species densities of plants are high.

Other forms of behavioral and physiological changes in flowers, such as the color change syndrome studied by Barrows[643] in *Pachyptera hymenaea* (Bigononiaceae) in Mexico, in which flowers changed color even after nectar and pollen ceased being produced, build in further specificity in pollination systems in such habitats. Barrows suggested that patches with mixed flowers attract greater numbers of pollinating insects than patches of dark flowers (ones with nectar and pollen) alone. The level of synchronous blooming in patches results in a more effective signal for pollinators, thereby increasing the likelihood of pollination in young inflorescences with pollen and nectar still present.

In some tropical forest ecosystems, the primary pollinators of many trees, shrubs, and vines are the stingless bees (*Trigona* spp.). A. Wille (personal communication) of the University of Costa Rica has been studying the abundance and specificities of these bees for various plant species in tropical rain forests and suggests that these insects are the major pollinators of many plants in tropical forests. Stingless bees exhibit flexible foraging habits and preferences for inflorescences, and make seasonal shifts, habitat shifts, in foraging easily. Such responses enable these insects to exploit a wide variety of plant species both in primary and in secondary successional stages in tropical forests.

In terms of the overall organization of pollinator communities in the lowland and mid-elevation regions of the tropics, the ranked order of abundance and diversity is believed to be as follows: stingless bees (principal pollinators), various large-bodied bees, lepidopterans, other insects such as flies and beetles, and various vertebrate groups such as bats, mice, and hummingbirds. This ranked order probably follows a reverse gradient in specificity for certain groups of plants, although far more data are needed on the specificities of pollinating insects such as small flies, beetles, and wasps, groups often assumed but not confirmed in most cases to be generalist pollinators. The composition of local pollinator communities, especially in the case of insects with their complex life cycles, will be limited by the availability of other resources such as nesting sites in the case of bees, or larval host plants in the case of the Lepidoptera. An analysis of pollination systems within a particular region or forest ecosystem must consider these other resources as affecting pollinator abundances and population cycles. Abiotic factors such as winds, air moisture content, and temperatures will also affect the composition of pollinator communities, especially along altitudinal gradients or within localized rain shadows on mountain slopes. Such effects will influence the overall organization of pollinator communities.

H. G. Baker and I. Baker(217) have discussed the role of nectar and pollen composition in molding the specificities of pollinating animals to flowering plants. Hummingbird-pollinated flowers tend to have nectars rich in sugars, whereas wasp-pollinated flowers are generally low in sugars. Large bees tend to be associated with sucrose-rich nectars and small bees with hexose-rich nectars; hummingbirds and butterflies also exhibit preferences for sucrose-rich nectars. Amino acid concentrations tend to be high in lepidopteran-pollinated flowers in tropical forests. Pollen composition also influences pollinator choice. H. G. Baker and I. Baker(216,217) recognize two major groups of pollen: those with starch and low lipid content, and those with no starch and high lipid content. Bees tend to prefer starchless pollen, while hummingbirds prefer starchy pollen. Small pollen grain size and low starch content are characteristic of bee or fly pollens, and starchy pollens tend to be larger. Starchy pollens can be harvested by some species of *Heliconius* butterflies as a source of amino acids for egg maturation(31) and perhaps also to increase energy reserves, from the breakdown of starch components of collected pollen, for sustained flight associated with oviposition behavior.

Ceratopogonid midges in the Canadian region exhibit considerable specificity for feeding on the pollen of certain plant species.(245) The Ceratopogonidae (Diptera) are known for their diverse feeding habits, which include pollen feeding in some lineages. Pollen feeding may be a link in why these midges are the principal pollinators of tropical trees such as *T. cacao* and *Hevea brasiliensis*.(644) Many dipterans require environmental sources of amino acids for proper egg development, and this may also be the case for *Euprojoannisia* and other genera associated with cacao flowers in the New and Old World tropics.(522,587–589) Successful pollination of cacao requires the deposition of composite aggregates of pollen grains (pollen balls) of the correct variety on the style of an open flower (Fig. 8.4), so pollination in this system depends upon the frequency of deposition of aggregates, the size distribution of pollen aggregates, and the relative effec-

FIGURE 8.4. (A) Buds and open flower of *T. cacao* (Sterculiaceae); (B) buds, flowers, young and mature fruit of *T. cacao* showing cauliflorous habit.

tiveness of insect pollinators in depositing pollen aggregates.[645] Some studies in Africa indicate that a variety of insects, in addition to ceratopogonid midges, are involved in the pollination of cacao.[646] A review of the systematics of *Forcipomyia* (*Euprojoannasia*) midges involved in the pollination of cacao is given by Bystrak and Wirth.[406] Life cycle studies of these midges (Fig. 8.5) have been conducted in both the field and the laboratory,[587,647–652] and analysis of natural breeding sites has also been undertaken.[590,649,651,653,654] Considerable interest focuses upon the environmental factors affecting the abundance of pollinating midges in cacao plantations.[522,588–590,647,648,651,655] The success of cacao as a major cash crop of the tropics depends upon environmental factors such as pollinator availability and disease and upon intrinsic genetic and physiological mechanisms determining fruit set.[565,644,656–658] The structure of the cacao plantation habitat, and the kinds of insects and other arthropods found in it, influences the healthiness of cacao trees.[659–661] Mutualisms involving ants, including those associated with cacao, are often successful, although a major focus concerns ant–plant associations.[662,663] From systematic studies of the genus *Theobroma* (Cuatrecasas),[595] it is evident that the group has been successful in undergoing evolutionary differentiation in tropical forests of Central and South America.

Bawa[638a] has examined the breeding or sexual systems of tree species in lowland tropical forests and has noted the high incidence of dioecious species associated with these ecosystems. Trees bearing small flowers pollinated by a variety of small bees are usually dioecious. Dioecism is an adaptation in plants that permits outcrossing. The separation of sexes on different plants ensures some level of gene flow among plants. Hemaphroditism in conjunction with self-incompatibility also promotes outcrossing. Self-incompatible flowers possess structural features to minimize pollen interference, i.e., to prevent pollen from the anthers of a flower ending up on the stigma of that same flower. If this were not so, incorrect pollen grains would compete with space on the stigma for correct pollen grains arriving on insect pollinators from other flowers. Accumulation of the incorrect pollen grains on the flower lowers the likelihood of successful zygote formation. Self-pollination generally includes the situation in which incorrect pollen results in partial growth of the pollen tube and thus represents a different level of incompatibility from pollen interference. Sexual or breeding systems in tropical plants must be viewed in terms of constraints imposed by the particular pollination systems involved for that species, and the optimal partitioning of resources (energy) to female and male functions (regardless of monoecism, dioecism, hermaphroditism, etc.), so that female and male components of fitness are optimized each generation. As Janzen[42] has indicated, a tree full of hermaphroditic flowers may set few fruit but have a high fitness in terms of performing the male function, i.e., be a pollen donor, and selection would favor a breeding system, given the type of pollination system present, that optimizes both male and female components of fitness within the breeding popu-

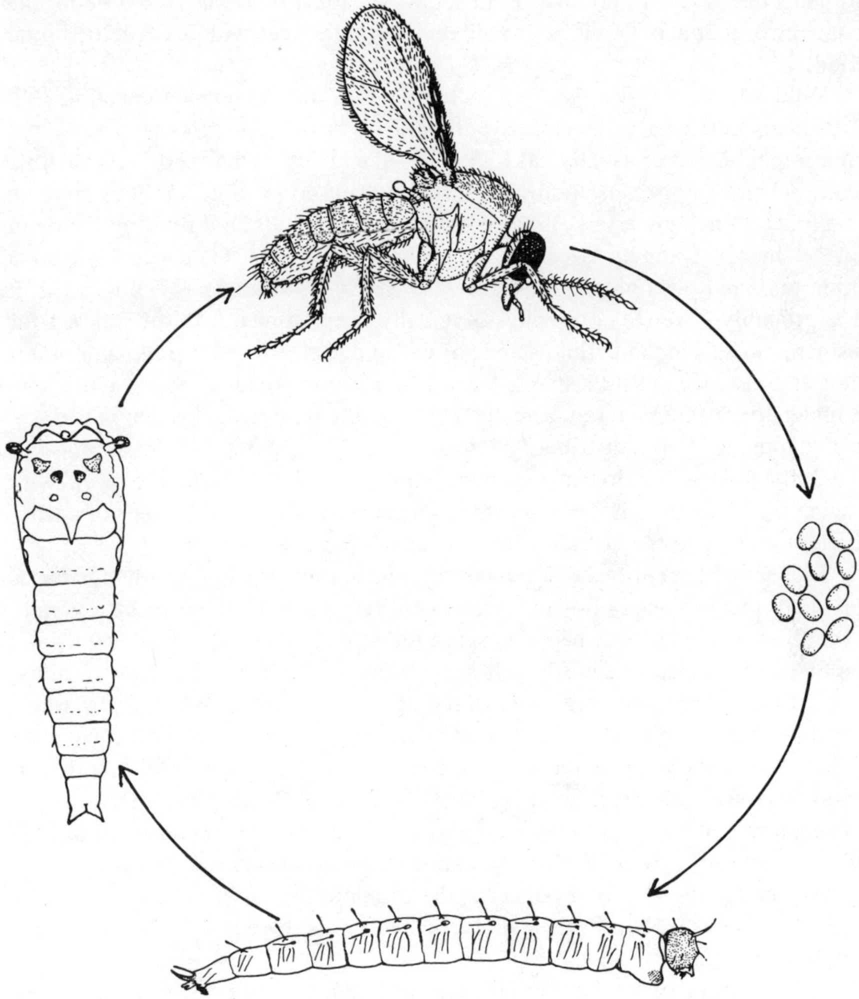

FIGURE 8.5. Schematized life cycle of *Forcipomyia* midges (Diptera: Ceratopogonidae), the pollinators of *T. cacao* in both the New and Old World wet tropics. Midges are about 2 mm long and undergo life cycle in moist, rotting organic debris. (Illustration courtesy of Dr. Willis W. Wirth, USDA and Smithsonian Institution.)

lation of the tree species under consideration. Level of fruit set, therefore, is not the only measurable component of fitness for the population. In *T. cacao,* for example, self-incompatibility exists because there is often partial growth of the pollen tube when incorrect pollen lands on the stigma of the same flower.[656]

Estimates of successful pollination in such trees must thus take into account the frequency of partial pollen tube growth on individual trees within a specified time period.

Wild species of *Theobroma,* including *cacao,* are self-incompatible,[595] while many artificially developed cultivars of *cacao* are self-compatible.[656] Self-compatible cacao usually has higher levels of fruit set than self-incompatible cacao, presumably because pollinators visiting self-compatible varieties are able to pollinate more flowers. Pollinator abundance may not be a limiting factor in fruit set in self-compatible cacao. In its original wild state, cacao evolved a certain maximal physiological level for flowering and fruit set. Ecological fruit set is probably lower than the physiologically programmed level of fruit set for most trees in self-incompatible stands of cacao due to effects of pollinator abundance and diversity. Wild cacao is known to occur in small aggregate clumps in the understory of tropical rain forest,[595,644] and the pollination system must have been adapted to such a distribution. Purseglove[644] suggests that the question of apparently low pollination of cacao in plantations should be approached with viewing the plantation as a composite aggregate of many small clumps of cacao, a pattern similar to the spatial arrangement of cacao in the wild.

Richards[394] mentions that many trees and shrubs of wet forest understories bear small inconspicuous flowers, many cauliflorous and pollinated by a variety of small insects other than bees, wasps, moths, and butterflies. The origins and adaptive significance of cauliflory is still obscure.[394] Such pollination systems, as in cacao, are probably highly buffered against shifts in seasons by the forest overstory. Yet when forests are cleared away or destroyed by other means, such systems are vulnerable in the direction of lowering pollination and fruit set, thereby increasing the likelihood for local extinctions. Cacao probably represents an exception that has been developed into a cultivar for human consumption, but still the matter of low pollination in exposed environments remains.

An interesting question concerns the relationship between physiologically programmed levels of fruit set and levels determined by ecological factors such as pollination and disease.

K. S. Bawa (personal communication) has opened the possibility that cacao and other tropical trees may exhibit a form of "functional andromonoecism" in which some trees produce a greater abundance of physiologically and morphologically bisexual flowers but which function as a source of pollen only. The strategy of the tree under such conditions is to maximize the output of male gametes. Such trees are characterized by many flowers but relatively few fruits. The tree is physiologically set for relatively low fruit set but produces an abundance of flowers, which function as a source of genetic material for pollination of other trees in self-incompatible systems. Functionally andromonoecious plants may exhibit either large, showy flowers or small, inconspicuous flowers (K. S. Bawa, personal communication). Those with large, showy flowers are most likely pollinated by bees; the others, by relatively small insects such as midges. The absence of morphological andromonoecism in tropical plants such as cacao may be due to selection favoring the maintenance of functional pistils: in functional andromonoecism any flower has the capacity to be pollinated, while in morphological andromonoecism only a small fraction have this capacity, possibly greatly disrupting pollinator activity; investing in extra pistils, so to speak, increases the option for selection of zygotes in future generations. Such a system is adaptable to changing selection pressures affecting other portions of the phenotype, such as interactions with herbivores. In hermaphroditic flowers such as cacao, there is physiologically an equal distribution of reproductive effort between male and female portions of the flower before fertilization, but if there is functional andromonoecism, there is greater portion of flowers with the male function because the tree only ''needs'' a number of fruit determined by the number of functionally hermaphroditic flowers (i.e., those producing fruit). Under such considerations, pollinator abundance would be less a limiting factor in fruit set than if functional andromonoecism did not exist.

Alternatively, an absence of functional andromonoecism in self-incompatible cacao suggests pollinator limitation of fruit set since levels of fruit set are very low. Again, optimal fruit set will be determined by the physiological state of the tree, but ecological factors such as pollinator deficiencies in the habitat may lower fruit set even further than what normally occurs when pollinator populations are in some form of equilibrium with abundance of flowers in a patch of trees. In such a system, increases in pollinator populations may result in increased fruit set if pollinator populations are synchronized through time with flowering periods. The observed lack of fruit set in the majority of flowers per tree is the result of insufficient pollination rather than of functional andromonoecism. Such distinctions are especially meaningful when examining fruit set patterns in cultivated tropical trees such as cacao or at in trees in disturbed habitats. In natural habitats one might expect some form of quasi-equilibrium between pollinators and flowers, particularly in species that are self-incompatible and require certain kinds of pollinators.

The maintenance of fruit set in self-incompatible plants such as cacao may depends in part upon the movement patterns of pollinating midges relative to the abundance of flowers and degree of spacing among flowering trees. For example, in a small clump of wild cacao trees in flower, the average pollinating midge may encounter a greater number of flowers on different trees within the clump, which results in relatively greater fruit set than for a similar midge in a cacao plantation. In the plantation habitat, midges may be distracted by a far greater number of flowers by surrounding trees, which may result in a lower fruit set than in wild clumps. Thus although satiation points may be the same in both habitats, midges in the first situation visit a greater number of flowers per tree than in the plantation and total fruit set is higher. The size, density of individual plants, and synchrony of flowering within resource clumps of different sizes may have profound effects on the movement patterns of pollinating insects and the percentages of flowers pollinated per tree. Intertree distances may be on the average shorter in clumps of wild cacao than in planted cacao in plantations. If so, pollinating midge efficiency may be greater in wild clumps than in plantations since less distance is needed to reach different trees. Many tropical pollinating insects may exhibit density-dependent foraging if their preferred resources are of relatively few kinds per unit area of habitat, thereby precluding diet switches in response to rarity of any one preferred type. But it is also predicted that such pollinators would switch to other resources as the preferred one becomes too patchy.(15,640) As patch size decreases, it is less desirable to be a patch specialist.(664) Pollinators should exhibit considerable patch selectivity when there is a broad spectrum in patch quality and size.(166) Charnov *et al.*(665) show that a pollinator or patch specialist will abandon that patch when its quality falls below the average quality for all patches of that resource in the habitat or region.

Givnish(666) has reviewed the various hypotheses regarding the evolution and adaptive significance of dioecism and monoecism in plants. He presents information relating to when dioecism is adaptive to plants. One of his predictions is that dioecism should appear in a tropical tree species when flowers or floral rewards are large relative to the costs of pollen and seed production and in which flowers are adapted to specialized pollen vectors such as some bees and moths. The preponderance of trees in tropical forests indicates that dioecism should be common, but not from the standpoint of promoting outcrossing.(666) The reader is encouraged to read the review of concepts in Givnish.(666) Givnish points out that the study of Hubbell(28) shows that dioecious species in tropical forests are no more clumped in their spatial distribution than hermaphroditic species, suggesting that the clumping of individual trees is not necessarily correlated with breeding systems. Givnish(666) has also emphasized that the selection pressures resulting from the seed dispersal and seed predation systems influence breeding systems. The diversity in growth form of tropical vegetation indicates that tropical ecosystems consist of plant species with diverse breeding systems and that one must examine the breeding system of a particular group in terms of many ecological factors. A cornerstone of such an effort is an examination of the various factors affecting fitness within a breeding population: herbivory, seed dispersal, pollination, and ecotypic effects. The studies of Bawa(638a) indicate a diversity of breeding systems in tropical trees alone.

The "fruit investment strategy" is expected to vary considerably among different plant species. It is that portion of available resources allocated to the development of fruit as opposed to the development of flowers. When this physiological potential is high in a given plant species, pollinator abundance is more likely to be a limiting factor in production since the plant is capable physiologically of producing a relatively high number of fruit. Frequency-dependent selection is associated with the allocation of resources to fruit production. Fruit set is limited by the amount of resources required to support the development of fruit, pollinators, and mortality of flowers and fruit. In the case of simultaneous hemaphrodites such as cacao, excess production of flowers increases the number of pollinator visits and the amount of pollen dissemination. In such a system, pollinating insects limit the distribution of pollen in the breeding population. The flowering strategy, including the abundance of flowers relative to fruit set, is believed to be a compromise between fruit production and pollen donation. Although it is traditionally maintained that dipterans and other nonhymenopterous and nonlepidopterous insects are inefficient pollinators, we need to examine the pollination systems of tropical trees and shrubs pollinated by these insects in light of current theory and interest in tropical pollination systems in general.[638,639] Mesler *et al.*,[667] for example, working in coastal redwood forests of Northern California, found high fruit set in two plant species and attributed it to high abundance of pollinating fungus gnats (Sciaridae and Mycetophilidae).

Janzen[42] notes that flower-to-mature-fruit ratios of from 100 : 1 to 500 : 1 are common in large tropical trees in natural habitats and indicates that such trees are physiologically programmed to observed levels of fruit set. Pollination appears to be a challenging process for tropical trees in that a variety of factors influence fruit set.[42] For example, Janzen suggests that a large pool of stigmas is necessary in order that the correct pollen, particularly in self-incompatible systems, ends up in the proper amount on at least a minimum number of stigmas through activity of pollinators. It is also suggested that a large flower crop is necessary to satiate pollinators that may be of limited numbers owing to competition with other plants.[42] Janzen suggests further that each tree species may possess a strategy of functioning as pollen donors rather than capturing pollen and developing zygotes, including hermaphroditic species. Thus individuals of a certain tree species may have in fact reproduced more because they provided pollen for other individuals of that species, even thought fruit set is low or absent. Viewed in this light, low pollination may not be due to limited pollinator supplies in natural habitats, but this may not necessarily be the case in crop species such as cacao. The cultivation process of tropical tree species such as cacao might have altered the average ratio of flowers to mature fruit, with an even greater skew toward flowers because the agricultural habitats of such species may not be optimal for pollinator populations. The arguments developed by Janzen would be most applicable, therefore, to cacao in its wild state, and with some offset in agricultural habitats due to insufficient abundance or spectrum of resources required by pollinating midge populations. Individuals that are pollen donors and little else may occupy habitats suboptimal for seed bearing.[42] Janzen also suggests that for species in which there is seed predation, greatest fitness is achieved when large seed crops are produced so that such predators are satiated and a portion of seeds survive. If there are periods when large seed crops cannot be produced, the optimal strategy of such a tree may be to be a pollen donor.[42]

Pollinating insects in tropical ecosystems may be specialized for exploitation of one or few resource types if the quality of these resources is high and if their distribution and abundance are predictable.[411,481] Pollinators actively seek out and select high-quality resources and not necessarily on the basis of their abundance. Specialized pollinators are believed to be abundant in the tropics,[334,668] and seasonal tropical regions may support a great wealth of pollinating species if plants partition the environment into different peak flowering periods,[116] thereby permitting a larger pool of species to co-occur with allochronic activity periods. Seasonal tropical regions may necessitate timed and synchronized switches to different flowering species by a pollinator guild as the resource spectrum changes and nectar or pollen rewards fluctuate in abundance accordingly. Any single plant species may offer a spectrum of resource patches to a set of pollinators, with considerable variation in patch quality. Depending upon the amount of competition, if any, for high-quality patches among member species in a pollinator guild, the exploitation of low-quality patches may result from such interactions and if persistent may result in evolutionary or facultative switches to other resources. Competition among pollinating species associated with a particular plant species or group of plant species may result in one or more of such species experiencing a temporary contraction in home range movements as related to exploiting a certain number of resource patches in limited supply, although overall diet preferences may not be altered and may expand in response to competition.[341] But if competition persists, and if the species in competition possess sufficient genetic or phenotypic flexibility, selection may favor the evolutionary switch to an alternative and abundant resource to alleviate losses in fitness due to competition on the original resource base.[669]

Because many tropical plants have distinct flowering periods and are a limited resource, closely related or ecologically similar pollinating species are expected to partition available resources.[433,642,670] If there is considerable variation in patch size for nectar or pollen rewards, pollinators will exploit the most abundant patches since nutritional intake is faster than for associations with rare patches. Some tropical butterflies as well as Temperate Zone species exhibit considerable constancy in daily exploitation of flower patches, and such home range patterns are known in the tropics for some large bees.[40] Trapline home range movements are known for tropical bees and butterflies,[31,40,338] and such pollinators may exploit several different species along the daily trapline route. In some nonseasonal tropical forests, there may be a gradual temporal pattern of flowering in some plant species, making the spatial distribution of such resources constant throughout most of the year. Successional patterns of preferred resources may impose shifts in foraging behavior among pollinating insects, thus changing through time the organization of pollinator guilds and communities in response to temporal patterns of flowering of resources.

Tropical forests undoubtedly consist of diverse mixtures of patch structures among the many plant species found there, each one having a specific kind of breeding structure. Some species may have disjunct breeding populations consisting of widely scattered individuals, i.e., patchy distribution, while others may be characterized by large patches of several individuals. Patch structure can also be assessed in terms of flowering strategies: some tropical trees exhibit mass flowering in which there is synchronized blooming of a large number of flowers per tree, while others stagger flowering throughout the year. Mass flowering makes available large patches of resources for pollinators, while staggered, low-level flowering provides for small patches at any one point in time. Mass flowering invokes a density-dependent response in foraging by many tropical bees, while small patches promote the trapline-type foraging.[40,499] Levin[640] predicts that in the absence of strong selection distinct patterns of local differentiation in tropical plants with patchy distributions will be low. Outcrossing promoted by long-distance pollinators in such species prevents local differentiation. It is stressed that many tropical tree species of low adult densities have very skewed sex ratios,[638a,639] which reduces effective densities of breeding individuals even more.[640] Density-dependent foraging in tropical pollinating insects is expected to occur in plant species with large patch sizes. Levin[640] discusses the evolutionary implications of patchiness in plant populations as related to foraging behavior of pollinators.

A growing body of evidence indicates that various kinds of floral parasitism involving insects are prevalent in tropical forest ecosystems.[671,672] Flowers, with their nectar and pollen rewards being predictable resources, are targets for a variety of insects and other animals, many of which are pollinators.[216] Ants form a conspicuous component of the rain forest fauna,[673] and some of these species routinely visit flowers. Nectar or pollen robbers take away rewards for pollinators and thus reduce the likelihood of pollination in some or all individuals within the patch or breeding population. Thus intense and routine nectar or pollen robbing reduces the average fitness and alters the breeding dynamics of the target plant species. Barrows[671] distinguishes three kinds of floral robbers of *Lantana camara* (Verbenaceae) in Costa Rica: nectar foraging, nectar-foraging perforating, and pollen foraging. A guild of floral robber insect species may therefore accumulate on various plant species, depending upon factors that include (1) selection pressure arising from limited resources for potential floral robbers, (2) abundance of preferred hosts for plundering, (3) the pollinator guild or community associated with the plant species, and (4) the floral structure and reward physiology, including diurnal–nocturnal rhythms of production. Additional factors include the predictability and abundance of the resource in space and time. Nectar robbery decreases cross-pollination and seed set in *Lantana*.[671] In this system the stingless bee *Trigona fulviventris* is a nectar-foraging perforator of yellow flowers of this species.[671] Perforating behavior by nectar robbers damages the flowers and removes the nectar reward for pollinators. Nectar robbing of four-o' clocks, *Mirabilis jalapa,* by worker paper wasps (Fig. 8.6) can result in considerable performation damage to corolla tubes,[674] a factor that may reduce breeding success in the plant population.

Nectar and pollen robbing in tropical plant groups may be much more prevalent than indicated by the amount of information available to date. McDade and Kinsman[675] found that pollen robbing by *Trigona* bees on hummingbird-pollinated *Aphelandra* (Acanthaceae) in Costa Rica did not reduce pollination. Flowers pierced by various insects and other bird species produced less nectar after perforation in addition to removing nectar in *Aphelandra*.[675] When much damage is frequent within a population, pollinators avoid visits to damaged flowers or increase foraging time on damaged flowers since the nectar reward per flower is less.[675] Such floral parasitism may result in reduced fruit and seed set in affected populations. McDade and Kinsman make the interesting suggestion that floral parasitism may result from human disturbance affecting plant populations and that early successional communities in the tropics may have a greater incidence of such relationships.

If the natural distributions of resources of potentially floral parasites are altered as the result of such disturbance, there may be convergence to exploit available high-energy and nutrient-rich resources of plant species characteristic of such communities. Early successional communities in the tropics are characterized by large patches of principally herbaceous species, and such distributions provide readily available resources in large supply for floral parasites. *Trigona* bees are expected to turn up frequently as floral parasites, given their opportunistic and flexible foraging habits.[21] Introduced plant species in tropical regions may be ideal targets for floral parasites[671,672] either affecting or not affecting pollination.

When optimal resources are patchy in space, the population structure of a species exploiting these resources will adjust to the patchiness[676,677] or new, more locally abundant resources may be used. In Costa Rica, the introduced ornamental vine *Thunbergia grandiflora* (Acanthaceae), originally from India, does not set seed but is visited by *Trigona* bees and hummingbirds, including nectar and pollen thieves, as discussed by Young.[677a] At some sites currently being studied, at least two species of hummingbirds routinely visit the large, showy lavender flowers, and several species of *Trigona* exhibit two kinds of thieving: (1) bees enter the corolla and bite through the tubular anthers and collect the released pollen grains, and (2) bees land at the base of the calyx, perforate the corolla tube, and collect nectar. In studies conducted to date, more than 50% of the open flowers within a patch of the vine are so damaged at any one time when flowering is high. The striking absence of fruit set in this intro-

FIGURE 8.6. Nectar-robbing behavior by a social paper wasp, *Epipona querini* (Vespidae: Polistinae: Polybiini), on the inflorescences of four-o'clocks, *Mirabilis jalapa* (Nyctaginaceae), in Costa Rica. (A) Wasp lands and seeks base of corolla tube; (B) and (C) wasp bites through calyx and corolla, leaving large hole after taking nectar.

duced vine in Central America is possibly due to a lack of effective pollinators. In the New World tropics, hummingbirds and *Trigona* bees have converged to exploit the nectar and pollen rewards, but the vine reproduces only by vegetative growth. Within a patch, flowers are present throughout the year, making a predictable resource for hummingbirds and insects, as discussed by Young.[677a] These preliminary studies of the lack of effective pollination of the introduced ornamental *T. grandiflora* in Costa Rica, and the associated rampant floral parasitism, suggest that introduced plant species in tropical regions may be subject to rapid exploitation by endemic insect species. If resources for some of these insect species are limited, there might be short-term facultative switches to novel resources or long-term evolutionary switches if the novel resource remains and becomes abundant. Likewise, cases may exist where introduced pollinating insects displace native pollinating species if the former have a competitive advantage over resident species. For example, there is some evidence that the Africanized honey bee is displacing resident pollinating stingless bees in Neotropical regions.[678] These studies point to the high instability of interactions between some pollinators and their resources in the tropics.

Tropical forest ecosystems are highly structured assemblages of organisms, with characteristic patterns of vegetation in distinct successional stages. An understanding of the guilds and communities of insects associated with different plant species, and the trophic organization of insect communities, is obtained by examining the diversity and patch structure of plant species in different successional stages. In this section, we have touched on a few illustrative aspects of how ecological properties of plant species determine the structure of insect communities and guilds. Detailed studies of the natural history of individual insect species are needed to determine the role of abiotic factors (temperature, rainfall, light intensity, etc.) and biotic factors (competition, predation, mutualism, resource limitation, etc.) in order to explain the population structure and dynamics within different communities.

The considerations discussed by Janzen[42] of fruit and seed crop patterns of variations on individual large trees in tropical forests indicate that tropical organisms may possess inter-individual patterns of variation within species that are adaptive to specific forms of selection pressures. Spatial variation in the distribution of fruits or seeds in the canopy of a tree or on the forest floor may influence the manner in which such resources are exploited by insects and other animals. Lowland Neotropical rain forests exhibit continuous production of fruits and seeds throughout most of the year,[101] even though there may be a succession of fruiting species of large trees over the year. Yet within the fruit or seed crop of an individual tree of a particular species, a spectrum of variation may exist in weight, position, and other traits affecting attractiveness as a resource for insect species. In forests of the Old World tropics, some tree species are monocultures due to growing on poor-quality sandy soils,[332] and such large trees exhibit mast fruiting. These trees are believed to possess well-developed systems of chemical defenses against herbivores, at the same time show pulsing of fruiting and generally low productivity due to poor nutrient profiles of the soil. Such monoculture forests have a low biomass of insects.[332]

A high species richness of insects, both herbivores and other trophic groups, is found in mixed-species tropical rain forests with high harvestable productivity.[47,412] Pollinating insects are more diverse in Neotropical mixed-species rain forests, thus providing the raw materials for the recycling of plant resources in the form of seeds and fruits and their breakdown products.[42] Specialized coevolved interactions between plants and insects provide the basic ecological fabric of Neotropical rain forests, with plant diversity made potentially possible by the high nutrient profiles of volcanic soils. Janzen[57,679] points out that when mixed-species tropical rain forests are altered, the coevolved interactions may be dissolved although some of the participants may persist outside of the context of such interactions. The annual succession of vegetative growth and sexual reproduction of plant species in tropical rain forests provides a diverse resource spectrum for insects and other animals. The species richness of herbivorous and pollinating insects, as well as the predators and parasitoids exploiting them, is determined by the plant species richness of the habitat. The review of various complex ecological interactions between plants and insects by Janzen[41] illustrates the interdependence of these organisms in determining the overall structure and composition of tropical forest ecosystems.

8.4. SOME ECOLOGICAL FACTORS GENERATING PATCHY DISTRIBUTIONS OF INSECTS IN TROPICAL HABITATS

The membership of an insect population will have a spatial distribution in a habitat determined to a large extent by the distribution of resources. Population growth is a function of the size of the existing population and the effects of intrinsic and external factors.[680,681] If the resources are widely scattered as small patches, and if the breeding population of the insect species being investigated possesses life history traits[92] so that reproductive effort is distributed over more than one patch at any moment in time, the insect population is fractionated over some or all patches of the resources. Thus we have defined one parameter affecting patchiness, namely, the spatial availability of those resources influencing fitness.

From previous considerations in this book, we also know that a variety of other ecological factors affect patchiness of an insect population.

1. When there is a quality distribution associated with a spectrum of neces-

sary resources. Selection in most situations will favor the exploitation of optimal resource patches even though other patches of the same resource exist in the habitat.

2. When there is habitat selection, in which one or more contingent habitats differ considerably in resource quality and abundance. Selection should in most instances favor the exploitation of high-quality habitats by an insect species, with or without gene flow between habitats.

3. Optimal quality or quantity of resource patch or optimal habitat selection reflects the physiological capacities of the species, but other ecological factors may alter the choice of potentially optimal resource or habitat patches. Intraspecific and interspecific competition may result in a density-related dispersal of individuals over a spectrum of resource or habitat patches, some of which may be suboptimal in terms of providing resources that maximize average fitness.

4. Coactions (Chapter 2) such as predation and parasitism may vary considerably among different resource or habitat patches, but an optimal patch from the standpoint of food or resource quality may not necessarily be an optimal patch in terms of low predator or parasite density. Selection in most instances should favor choice of patches where predation or parasitism is low on prereproductive or young reproductive age classes, assuming that the loss in fitness from such mortality is less than loss of fitness from selecting resources or habitat of suboptimal quality in another dimension of the niche.

5. Some habitats or resource patches may be better for the insect in terms of microclimatic or other abiotic factors affecting fitness. Selection should favor patterns of site specificity and degree of residentiality in part due to variations in microclimate and abiotic conditions in general.

For all of these classes of ecological factors influencing the degree of patchiness in an insect population, the extent to which these factors fluctuate through time must also be considered, especially in terms of tropical seasonal regions and degree of regional and habitat stability (see the discussions of light-gap dynamics in tropical forests in Hartshorn[(337)] and Denslow[(682)] for some recent thinking and evidence indicating considerable instability in lowland tropical rain forests). The degree to which individual host plant species, for example, will be patchy resources for phytophagous insects and targets for the predators and parasitoids of these insects depends in large part upon the successional stage of the community.[(683,684)]

Miller[(685,686)] states that competition between species results in either (1) an increase in niche size and concomitant elimination of some ecologically similar species in the same habitat or (2) a reduction in niche size with increased ecological diversification, so that overall diversity increases in the community and the incidence of specialization is greater. Populations at or near equilibrium levels with the available resources in the habitat may enter into competitive interactions with other species. But if there are considerable temporal fluctuations in the availability of resources, the intensity of such competitive interactions will vary considerably within such a guild or community.[463,464] Vertical distributions of the populations of closely related species of scorpion flies in a Temperate Zone habitat exhibit patterns in which more aggressive species occupy the higher quality strata in terms of feeding sites.[184] In the Temperate Zone, scorpion flies maximize foraging above the ground level, a behavior exhibited by many tropical ants that are scavengers (Thornhill; [184] Thornhill's citation of C. R. Carroll, personal communication), and Thornhill[184] suggests that the essentially extra-tropical distribution of scorpion flies may be due to competition with arboreally foraging ants in the tropics. If so, the implication is one of complete displacement of different insect groups as a result of competition for food supplies in arboreal microhabitats in the tropics.

Baldry and Molyneux[494] have shown that the dipteran *Glossina medicorum* in West Africa exhibits considerable preference in terms of habitats where trypanosome infections are high. Although the species is widely distributed, different populations vary considerably in the extent of infection at any one time of the year and the differential effects are correlated with the type of forest habitat.[494] In some species of *Simulium* flies that function as vectors of onchocerciasis, the distribution and abundance of the aquatic larval stages are determined largely by the availability of host arthropods such as crabs and mayfly nymphs, to which the larvae attach themselves.[687] An analysis of the habitat requirements of the host arthropods is often needed to determine the potential for outbreaks of the flies in a particular region of Tanzania.[687] Such studies and others concerning correlations between resource quality, patch size, and equilibrium population sizes for a species[677] show that population growth and size are directly related to the quality of resource patches. Dethier[279] has considered the problem of host plant populations limiting population size in the Lepidoptera in a general way: reduced food supplies result in mass dispersal of larvae when food supply is very low, which in turn results in considerable mortality of larvae associated with dispersal. Yet other studies of Lepidoptera in particular suggest that larvae disperse from individual host plants as a result of diminished quality of the resources provided,[688] even though these species may have a strong coevolved association with a certain host plant species.[689] Some tropical butterflies exhibit specific forms of home range behavior determined in part by the distribution of larval and adult food supplies.[479] The carrying capacity of the habitat may be very different for closely related species, and density-related effects may influence population size in different ways.[81,82] In some parts of Uganda, the dipteran *Simulium neavi* and two other species, vectors of onchoceriasis, have larval populations requiring completion of the life cycle in crabs, but considerable mortality is associated with crab-locating behavior.[687]

The results indicate that shifting distributional patterns of the crabs, even in confined conditions, make for an unpredictable "resource" for the fly larvae, contributing considerably to regulation of larval populations.

One kind of patchy environment characteristic of lowland and mid-elevation tropical rain forests is the tank bromeliad, so named because its structure results in the accumulation of rainwater. The water is held long enough to allow the buildup of invertebrate communities. There are many species of such bromeliads, and each species may have a characteristic vertical distribution in the forest habitat. Because tank bromeliads harbor communities of microorganisms and potential disease-vector dipterans, there is a medical interest in the species breeding in these patchy microhabitats. The tank bromeliad, with its mini-aquatic environment, has been analogized in the literature as a microecosystem composed of invertebrate herbivores that feed on the plant's tissues, algae, and other plants growing from it; decomposers exploiting accumulated organic debris such as twigs and fallen leaves; and carnivores that feed on the organisms in these other trophic levels. The system is very open, primarily as the result of vertebrates (birds, lizards, frogs, snakes) feeding on the invertebrates in tank bromeliads. The studies of Laessle[(407)] on the physical and chemical properties of the water in tank bromeliads indicates considerable heterogeneity among bromeliads in a region or habitat, suggesting heterogeneous selection pressures or nongenetic physiological response characteristics for organisms inhabiting these microhabitats. Laessle mentions that a variety of insect larvae, such as Diptera and Odonata, thrive in such environments along with many other arthropods and other invertebrates. Calvert[(690)] studied the association of nymphs of giant Neotropical damselflies (*Megaloprepus* and *Mecistogaster*) with tank bromeliads in Costa Rica. In some instances, what appear to be rather simple interactions between arthropods in the tropics, turn out to be more complex when studied further. Thus although the large, familiar orb-spinning spider *Nephila clavipes* constructs elaborate, sticky webs to attract a variety of prey, the spider itself also attracts a variety of parasitic and predatory or scavenging flies that form a small assemblage of arthropods interacting in specific ways.[(408)] Further complexity is added by the exploitation of prey lodged in these webs by the giant damsel fly *Megaloprepus coerulatus*[(527)]

Vandermeer *et al.*[(691)] studied the dynamics of protozoan communities of tank bromeliads in Costa Rica, showing evidence of competitive exclusion of *Paramecium*. Because of the large biomass provided by these epiphytes in tropical rain forests, and because of the association of many different organisms with them, many of which are probably highly specialized to such microhabitats, the distribution and abundance of tank bromeliads in tropical rain forests may account for an appreciable portion of the animal diversity found in the tropics. For example, an analysis of 11 large tank bromeliads sampled once in the Costa Rican highlands (Bajo La Hondura) revealed a total of 65 species of inverte-

TABLE 8.2

The Distribution and Abundance (Number of Individuals) of Aquatic and Terrestrial Arthropods[a] among 11 Tank Bromeliads at Bajo la Hondura Pasture, Costa Rica, February 1975[b]

Bromeliad	Height above ground (m)	Diameter[c] (cm)	Aquatic arthropods		Terrestrial arthropods		Totals		Diversity (H')
			Number of species	Number of individuals	Number of species	Number of individuals	Species	Individuals	
1	2.0	40	0	0	5	35	5	34	0.428
2	2.8	99	3	31	2	2	5	33	0.607
3	4.0	78	7	23	0	0	7	23	0.576
4	2.0	18	3	3	4	4	7	7	0.845
5	4.0	25	3	3	5	5	8	8	0.905
6	1.6	40	0	0	4	118	4	118	0.149
7	1.6	40	4	23	6	15	10	38	0.813
8	1.8	72	1	10	6	33	7	43	0.467
9	1.5	51	2	7	2	12	4	19	0.939
10	2.0	30	0	0	2	223	2	223	0.144
11	2.4	80	1	8	5	94	6	102	0.377

[a]The following taxa were present and censused (A, aquatic; T, terrestrial): *insects:* Diptera (A), Neuroptera (A), Coleoptera (T), Orthoptera (T), Homoptera (T); *other arthropods:* Isopoda (T), Arachnida (T), Annellida (T), Diplopoda (T), Chilopoda (T), Collembola (T).

[b]Unpublished data (A. M. Young, instructor) collected by students in the Lawrence University (Appleton, Wisconsin) winter term ''Tropical Ecology'' program.

[c]Diameter as measured from tips of leaf axils.

TABLE 8.3
Changes in the Biomass of Insects Attracted to Experimental Baits[a] in the

Time interval	Diptera Total S^c	Diptera Total[d] N^e	Hymenoptera (excl. ants) Total S	Hymenoptera (excl. ants) Total N
8:30–9:30 A.M.	20	116	3	15
9:30–10:30 A.M.	19	129	3	7
10:30–11:30 A.M.	17	91	10	42
11:30 A.M.–12:30 P.M	18	117	8	64
12:30–1:30 P.M.	12	69	7	80
1:30–2:30 P.M.	17	51	7	105
2:30–3:30 P.M.	19	84	7	185
3:30–4:30 P.M.	18	66	6	156
4:30–5:30 P.M.	21	77	4	158
5:30–9:00 P.M.	1	3	0	0
	$\bar{X} \pm$ S.D.	$\bar{X} \pm$ S.D.	$\bar{X} \pm$ S.D.	$\bar{X} \pm$ S.D.
8:30–9:30 A.M.	1.81 ± 1.40	10.54 ± 11.23	0.27 ± 0.46	1.36 ± 2.83
9:30–10:30 A.M.	1.72 ± 1.55	11.72 ± 12.67	0.36 ± 0.50	0.63 ± 0.92
10:30–11:30 A.M.	1.54 ± 1.63	8.27 ± 10.37	0.90 ± 1.04	3.81 ± 7.44
11:30 A.M.–12:30 P.M.	1.63 ± 1.36	10.63 ± 12.55	0.72 ± 0.78	5.81 ± 9.42
12:30–1:30 P.M.	1.09 ± 1.30	6.27 ± 8.58	0.63 ± 0.80	7.27 ± 14.96
1:30–2:30 P.M.	1.54 ± 1.12	4.63 ± 4.67	0.63 ± 0.67	9.54 ± 19.27
2:30–3:30 P.M.	1.72 ± 1.27	7.63 ± 10.23	0.63 ± 0.67	16.81 ± 45.58
3:30–4:30 P.M.	1.63 ± 1.74	6.00 ± 6.51	0.54 ± 0.68	14.81 ± 24.75
4:30–5:30 P.M.	1.90 ± 1.92	7.00 ± 8.02	0.36 ± 0.67	14.36 ± 32.59
5:30–9:00 P.M.	0.09 ± 0.30	0.27 ± 0.90	0.00 ± 0.00	0.00 ± 0.00

[a] Eleven types of bait, including 8 odoriferous ones (meat, banana, pineapple, orange, papaya, tomato, jelly, eugenol) and controls (salt water, mud) were used to examine species richness and abundance of insects in the forest understory and their distribution throughout the day.
[b] The locality is Finca la Tirimbina near La Virgen (220 m elevation), Heredia Province, and the habitat is understory (bait height range 1–2 m) in mixed primary and advanced secondary tropical rain forest.

brates and 649 individuals, of which more than 50% of the latter were isopods (Table 8.2). The communities were found to consist of recognizably terrestrial and aquatic insects and other invertebrates (Table 8.2). Although the sample size is small, the data indicate considerable variation from bromeliad to bromeliad in terms of species diversity in the communities, when using the diversity measure:

$$H' = (C/N)\ N \log_{10} N - n_i \log n_i$$

There are no apparent correlations between diversity and size of bromeliad or height above the ground. All 11 bromeliads were sampled in a single pasture dotted with various trees. Those bromeliads with very low diversity measures

Understory of Premontane Tropical Rain Forest in Northeastern Costa Rica[b]

Ants		Coleoptera		Overall total	
Total S	Total N	Total S	Total N	S	N
3	70	9	44	35	245
11	331	8	24	42	491
13	1,090	9	25	49	1,248
16	1,816	14	30	56	2,027
13	1,270	11	32	43	1,451
8	441	17	41	49	638
5	771	20	55	51	1,095
12	1,059	17	33	53	1,314
8	411	16	51	49	697
3	504	16	74	20	581
$\bar{X} \pm$ S.D.	$\bar{X} \pm$ S.D.	$\bar{X} \pm$ S.D.	$\bar{X} \pm$ S.D.	$\bar{X} \pm$ S.D.	$\bar{X} \pm$ S.D.
0.27 ± 0.46	6.36 ± 19.47	0.81 ± 0.98	4.00 ± 4.57	3.18 ± 2.27	22.27 ± 25.79
1.00 ± 1.00	30.09 ± 87.70	0.72 ± 0.78	2.18 ± 2.67	3.81 ± 2.89	44.63 ± 88.52
1.18 ± 1.32	99.09 ± 305.66	0.81 ± 0.75	2.27 ± 2.61	4.45 ± 3.58	113.45 ± 305.51
1.45 ± 1.43	165.09 ± 509.55	1.27 ± 1.42	2.72 ± 3.60	5.09 ± 3.30	184.27 ± 513.62
1.18 ± 1.07	115.45 ± 359.84	1.00 ± 1.09	2.90 ± 3.36	3.90 ± 3.36	131.90 ± 369.87
0.72 ± 1.19	40.09 ± 29.33	1.54 ± 2.20	3.72 ± 5.58	4.45 ± 3.35	146.30 ± 58.00
0.45 ± 0.68	70.09 ± 225.53	1.81 ± 2.99	5.00 ± 7.61	4.63 ± 4.03	99.54 ± 268.36
1.09 ± 1.04	96.27 ± 303.12	1.54 ± 2.16	3.00 ± 4.42	4.81 ± 4.19	119.45 ± 323.48
0.72 ± 0.64	37.36 ± 113.68	1.45 ± 1.80	3.72 ± 4.79	4.45 ± 4.10	62.45 ± 142.99
0.27 ± 0.46	45.81 ± 150.63	1.45 ± 1.69	6.72 ± 8.13	1.81 ± 1.83	52.81 ± 148.55

[c] Number of species.
[d] Totals, *N* values, are best considered as "bits" rather than as individuals because some or all of the actual individuals present during a particular time interval may occur in subsequent time intervals, thereby biasing the actual number of individuals present, as no marking was done.
[e] Number of individuals.

(such as Nos. 6, 10, and 11) were characterized by a very high abundance of either isopods (for Nos. 6 and 10) or dipteran larvae (No. 11) (Table 8.2). Bromeliads such as No. 5 with a high diversity were characterized by a low number of individuals, high number of species, and in some cases, approximately equal representation of both aquatic and terrestrial forms. The composition of individual communities was determined, of course, to a large degree by the degree of wetness or dryness and the amount of litter in each bromeliad.

In an ongoing study with about 14 months of data obtained to date, "artificial" plastic-cup bromeliads established in two cacao plantations in Costa Rica reveal some interesting distributional patterns of insects and other invertebrates.

The cups were anchored to branches of cacao trees, one per tree and to 20 trees in each plantation, perforated in the bottom for drainage but initially filled with a mat of washed leaf litter. The two plantations are very different in terms of degree of shade cover over cacao trees and the heterogeneity of shade cover trees: one plantation (La Tigra) has a mixed canopy of various wild tree species, and the other (El Uno) has a rather uniform cover of *Hevea* rubber trees. Eleven of the 20 artificial bromeliads censused in La Tigra after 14 months had thriving ant colonies, a total of four species present, but only one species in a cup. The colonies had larvae and pupae. The remaining cups had low densities of roaches, isopods, and a few other invertebrates. In the El Uno plantation, 10 of the cups had ant colonies, but species different from those in La Tigra, and the colonies were generally smaller than those of La Tigra. A few isopods were found in the other cups along with one or two species of spiders. One cup, clogged and filled with rainwater, had a belastomid bug and many mosquito larvae. Such data, while preliminary, indicate that experimentally introduced artificial arboreal habitats are rapidly colonized by ants and a few other invertebrates. Privat[649] found that ceratopogonid and cecidomyiid midge larvae breed in natural bromeliads in Costa Rica, but none of these were found in the artificial bromeliads in this study. Colonization of natural or artificial bromeliads in the tropics is a function of the suitability of individual bromeliads for certain organisms and the degree to which a community builds up in each one.

Short-term temporal variation in abundances of insects at resources in the understory of tropical rain forest can be studied by placing a series of "experimental" and "control" food resources (baits) in the habitat, replicating these, and observing the numbers of species and individuals per species on the baits over a diurnal cycle. Sometimes the data generated by such a study provide an estimate of temporal evenness in the species composition and abundances for the community of insects and other arthropods exploiting, or attracted to, these food types. Because the system is experimental and combines both odoriferous, nutritive baits and nonodoriferous, nonnutritive ones, the degree to which the observed patterns of insect distribution approximate natural distributions over the naturally distributed spectrum of resources remains unknown. Yet the data can reveal some interesting comparative differences among various taxa[186] and thereby tell us something about the relative differences or similarities between species associated with the resource spectrum studied.

When a series of experimental and control food resources was placed in the understory of a mixed primary and secondary forest within the premontane tropical wet forest region of northeastern Costa Rica during the short and erratic dry season, the predominant groups of insects found on the baits were Diptera, Hymenoptera, and Coleoptera. Each of these orders was represented by several subfamilies, genera, and species, and the number of species and individuals (abundances) remained markedly stable from 8:30 A.M. to 5:30 P.M. for all orders, lumping the data from all baits (Table 8.3). A marked decline was found for most groups, save for some ants, after dark (Table 8.3). Virtually all groups peaked in species abundance and numbers of individuals between 10:30 A.M. and 1:30 P.M., and ants made up the greatest biomass of insects present. The nine experimental baits yielded more than 80% of all insects censused, with the remainder on controls (Table 8.4). While the distribution of flies was remarkably even over the experimental baits, the Hymenoptera were bimodally skewed, with two groups, predatory wasps and bees, being associated with the meat and eugenol baits, respectively (Table 8.4). Eugenol, while not a food source, tests for the effect of a strong odor, mimicking the odor components of some flowers, while the other experimental baits contain both odoriferous and nutritive substances.

Ants were heavily associated with papaya in terms of species richness, but the greater abundances were found on meat (Table 8.4). Coleoptera, mostly Staphylinidae, were greatest in species richness on fruit baits such as pineapple, orange, and tomato, with greatest abundances being on pineapple and tomato

TABLE 8.4
The Distribution and Abundance of Various Groups of Insects on Baits for the 11:30 A.M. to 12:30 P.M. Time Interval for the Premontane Tropical Rain Forest Data from Costa Rica

	Diptera		Hymenoptera (excluding ants)		Ants		Coleoptera		All insects	
Baits	% of all *S*[a] (18)[c]	% of all *N*[b] (117)[d]	% of all *S* (8)	% of all *N* (64)	% of all *S* (16)	% of all *N* (1816)	% of all *S* (14)	% of all *N* (30)	% of all *S* (56)	% of all *N* (2027)
Meat	16.6	5.9	25.0	34.3	18.8	93.6	7.1	3.3	16.1	85.3
Banana	11.1	7.6	0.0	0.0	6.3	0.4	7.1	10.0	7.1	0.9
Pineapple	16.6	29.6	12.5	40.6	0.0	0.0	21.4	26.7	12.5	3.4
Orange	16.6	4.2	0.0	0.0	6.3	0.1	28.6	3.3	14.3	0.8
Papaya	16.0	17.8	12.5	10.9	31.3	4.2	7.1	10.0	17.9	5.2
Tomato	11.1	25.4	12.5	1.6	0.0	0.0	21.4	20.0	10.7	1.8
Jelly	11.1	8.4	12.5	1.6	12.5	0.5	0.0	0.0	5.4	0.5
Eugenol	0.0	0.0	25.0	10.9	6.3	0.1	7.1	3.3	10.7	0.9
Salt water	0.0	0.0	0.0	0.0	6.3	0.7	0.0	0.0	1.8	0.6
Water	0.0	0.0	0.0	0.0	6.3	0.1	0.0	0.0	1.8	0.0
Mud	0.0	0.0	0.0	0.0	6.3	0.3	0.0	0.0	1.8	0.3

[a] S, Number of species.
[b] N, Number of individuals.
[c] The numbers in parentheses refer to the number of species found on baits for each insect order.
[d] The numbers in parentheses refer to the number of individuals for all species found on the baits, for each insect order.

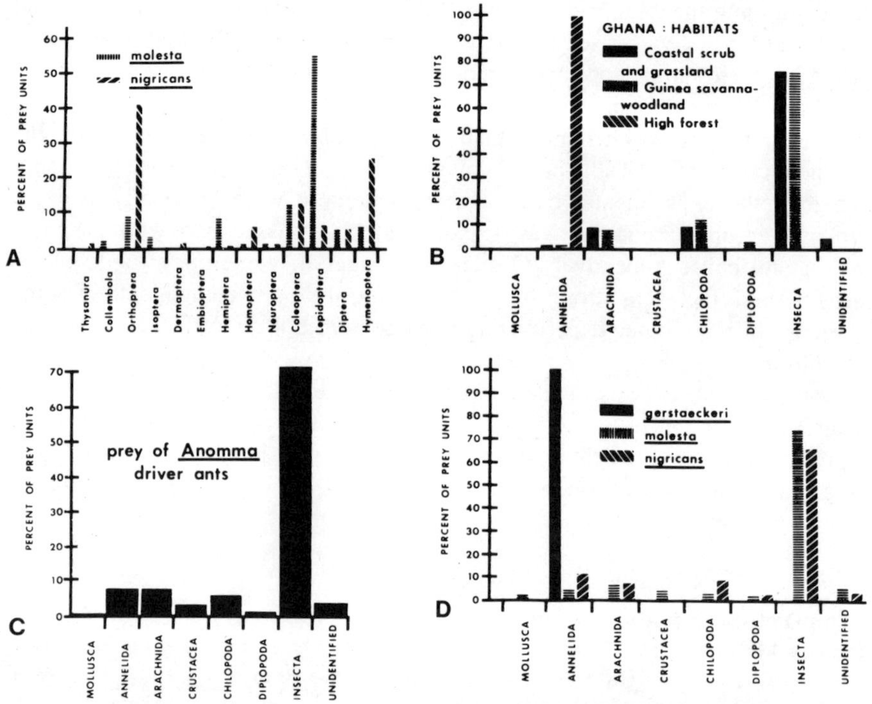

FIGURE 8.7. Prey selection patterns by African driver ants, *Anomma*. (A) Insects make up the major kind of arthropods taken as prey in all habitats and by different species of these ants. (B) Resource partitioning for insect prey by two species of African driver ants, (C) and by three species. (D) Different species of these ants forage in different habitats, resulting in the observed pattern of prey resource partitioning. [From W. H. Gotwald, Jr., *Ann. Entomol. Soc. Am.* **67:**877–886 (1974). Copyright 1974 by the Entomological Society of America.]

(Table 8.4). As expected, control baits were virtually empty in many instances save for a low occurrence of ants (Table 8.4).

The data indicate that some insects, e.g., ants, are patchy in distribution in that both species and numbers of individuals peak on certain food resources and not on others. And although there is a diurnal replacement of some species in all of the orders studied, the data show that overall species richness and biomass of this "synthetic community" associated with the baits are stable over the period of 1 day. In some instances, such as the interaction of Coleoptera with tomato bait, species richness is inversely related to abundance, but in several other associations of insects with baits, there appears to be a positive relationship between these two components of community structure (Table 8.4): as species accumulate at baits, so do their numbers or population sizes. On a per hour basis, the range in number of species, for all groups of insects studied, was from 20 (9:00 A.M.) to 56 (12:30 P.M.), while for numbers of individuals for all species and orders was from 245 (9:00 A.M.) to 2027 (12:30 P.M.). Of the maximum number of individuals seen at one time (2027), about 90% were ants, and after 9:00 A.M., ants made up more than 50% of all insects on baits. Since 60–90% of all ants were found on the meat bait, it appears that these forms were predatory. These preliminary data point to the need for further long-term and replicated studies of this approach.

In some predatory insects such as African driver ants, while there is often a preference for a general category of prey among all species, different species in different habitats will exhibit distinct prey preferences within the generalized category, as exemplified by the study of Gotwald.[430] The explanation for such differences is that the abundance of different kinds of prey will vary in different habitats (Fig. 8.7). It is the interaction of habitat type with prey types and predators that mold the distinctive features of divergence in resource exploitation in these insects.

CHAPTER 9

INSECT SPECIES IN AGRICULTURAL HABITATS IN THE TROPICS

From the summaries in the previous chapters concerning the well-established patterns of species richness of insects in lowland and mid-elevation tropical forests, it is clear that various kinds of crop lands adjacent to mixed tree species forests in the tropics will be invaded to varying degrees by insect taxa in space and time. It is evident that mixed species tropical forests of the New World represent complex and dynamic assemblages of organisms and that a major evolutionary and ecological interface is the sets of coevolved interactions between insects and their host plants. These interactions impose an unstable equilibrium in the distribution and abundance of species in tropical forest ecosystems. For some insect species, there may well be the case in which breeding populations in natural ecosystems are kept well below the carrying capacity of the habitat by carnivorous predators and parasitoids. Janzen[72] has amply shown that such organisms tend to be most diverse in the very same habitats where both herbivorous insects and plants are also diverse, namely, lowland and mid-elevation tropical rain forests where the harvestable productivity from plants is presumed, but not demonstrated, to be high. Yet for other species such controls may not be as effective and insect populations under such conditions may approach an equilibrium states with available resources. It is not hard to imagine that insect species in productive tropical ecosystems exhibit both forms of population structure in space and time.

The floral and faunistic components of tropical forest ecosystems, especially the insects for the faunistic part, are very old and have had considerable evolutionary opportunity to develop coevolved interactions. Janzen[41,47,57] has stressed the dynamic state of tropical insect–plant interactions and the adaptive role of variations in resources from the viewpoints of both insect and plant for a specific kind of interaction. We know that the occurrence of natural successional stages in plant communities in the tropics is a gradual affair, with specific sets of ecological traits associated with each stage.

9.1. PLANT–INSECT INTERACTIONS AND TROPICAL AGRICULTURAL HABITATS: SOME COMMENTS

The occurrence of successional stages implies that natural tropical forest ecosystems undergo change, mostly the result of natural perturbations such as floods, mud slides, hurricanes, and large- and small-scale tree falls. Undoubtedly such events alter the structure and composition of insect communities associated with the vegetation affected. Compared with adjacent forests, tropical fields left alone tend to build up distributions of plant species with large patch structure and relatively low diversity, and such habitats may have a high harvestable productivity. The result is that herbivorous insects under such conditions may build up large populations on dense host plant patches, especially if carnivorous predators and parasitoids cannot regulate herbivore populations under these conditions.[72] Such herbivore species are well adapted for intraspecific competition, and the patch structure of plant species might be such that interspecific competition is minimized by abundant food supplies. The result is an evolutionary accumulation of many herbivore species in secondary succession habitats. Because insects can evolve rapidly to meet changes in the environment, and because lowland and mid-elevation tropical insect faunas can be very rich,[72,412] these organisms become integral components of agricultural habitats in tropical regions.

Studies of tropical forest ecosystems show that the bulk of nutrients are kept in the flora and fauna, not in the soils. Thus when tropical forests are cleared away by humans for agricultural endeavors, sustained crop yields for food-type crops generally last only a few years. Furthermore, such crop species are vulnerable to intense attack by herbivorous insects. Monocultures of food crops in the lowland and mid-elevation tropical regions have a different spatial arrangement of the vegetation from natural forests: individuals of the same plant species are packed closely together in monocultures. Janzen[57] explains that a plant species has three lines of defense against attack by herbivorous insects: (1) periods of unfavorable weather reducing the insect populations or resulting in a period of inactivity each year, (2) a complex of carnivorous predators and parasitoids that regulate the abundance of the herbivorous forms, and (3) the scattering of individual host plants of a given species in the habitat, making it difficult for herbivorous insects to locate them. The combination of host plant spatial heterogeneity and carnivorous predator–parasitoid complexes is probably sufficient to regulate herbivorous insects on many plant species in the Temperate Zone since the intervening winters decimate insect populations and the annual growing season is short.[57] Furthermore, unpredictable patterns of fluctuations in weather at the boundaries of Temperate Zone growing seasons provide for density-independent regulation of insects. Yet in the lowlands and mid-elevations of the tropics weather fluctuations are generally predictable and many regions are nonseasonal, conditions favoring a steady buildup of insect populations over a long annual growing season.

Regions with a strong dry season are somewhat analogous to Temperate Zone regions in that some insect populations decline during this period, and Janzen[57,692] points out that farming successes are highest in the lowland seasonal tropics. The absence of annual density-independent control of insect populations by weather in the nonseasonal or mildly seasonal tropics necessitates host plant populations having a patchy spatial structure in which individuals or small clumps of a species are scattered throughout the habitat in order to counter any attack by herbivorous insects.[41,57] With the notable exceptions of tropical cash crops such as coffee, cacao, spices, and rubber, which to a large degree have well developed chemical defenses against herbivores, dense plantings of food crops in the nonseasonal or mildly seasonal tropics provide the opportunity for herbivorous insects to build up large populations. Food crops are characterized by palatable foliage or fruit, and palatability in the eyes of humans may also be palatability in the eyes of herbivorous insects.[42] In such environments herbivores are believed to be essentially free from population regulation by carnivorous predators and parasitoids, since these insects are not adapted to feeding on high densities of any one kind of insect species.[57]

The question of whether species richness of carnivorous predators and parasitoids of herbivorous insects is high in most tropical terrestrial habitats is an open one at this time. Janzen *et al.*[412] have shown that parasitic Hymenoptera are more abundant (numbers of individuals) and more diverse (numbers of species) at a mid-elevation, mildly seasonal site in Costa Rica than at comparable lowlands sites, and ascribes this pattern to a greater biomass and richness of herbivorous insects being associated with an increased harvestable primary productivity of this site. Lowland habitats have relatively lower harvestable productivity and fewer herbivore species, and such conditions may select for increased generalist behavior in carnivorous predators and parasitoids. Thus when tropical forests are cleared and replaced by fields or monoculture food crop plantings, a greater incidence of population outbreaks of herbivorous species is predicted, since (1) the food supply is increased per unit area of habitat, (2) the host plant in question may have few or no effective chemical defenses, and (3) the carnivorous predator–parasitoid complex may be less effective in controlling such populations.

The behavior, feeding physiology, and life history characteristics of carnivorous predators and parasitoids might be adapted to the exploitation of many kinds of prey or low densities of any one kind of prey, so that considerable time might be needed to reestablish an equilibrium between abundance of herbivores and such biotic control agents in disturbed or agricultural habitats dominated by monocultures or simple polycultures of palatable crop species. Although some tropical crop species have vesicular–arbuscular mycorrhizal associations with their root systems, which augments growth rates,[693] such systems and other factors such as fertilizer application may do relatively little to offset the losses in fitness due to intense herbivore attack.[57] Janzen[57] discusses various aspects of the inability of monoculture plantings of palatable crops in the nonseasonal or mildly seasonal tropics to protect themselves against herbivorous insects. Selective breeding within crop species reduces the spectrum of variation in chemical defenses that normally occurs in wild breeding populations of the species, reducing its defenses against herbivores.[57] Pesticide applications in the tropics may kill off what biotic control agents there are to exploit targeted ''pest'' species, thus only intensifying the abundance of the herbivores.

Janzen[692] has delineated some of the major features of why it is difficult to grow food crops in the tropics, and the reader is referred to that discussion. Carroll[694] has pointed out that tropical agroecosystems may be very complex in terms of the kinds of interactions among plants, insects, and microorganisms, and predicting levels of harvest yields requires a knowledge of these components and how they fit together. Such considerations indicate a need for long-term, in-depth studies on the details of agroecosystems. Carroll,[694] in his examination of the corn–bean–squash agroecosystem, points out that harvest yield will be a function of subtle interactions of predator–prey and host–pathogen competition, and mutualisms among the organisms in such systems. Although considerable controversy exists as to the points of origin for several food crop species in the New World tropics,[695] apparently cultivation of such plants is very ancient, suggesting considerable opportunity for an evolutionary buildup of complex communities of organisms, including insects, centered around these plants as resources.[694]

The conversion of tropical forests to pastures and open savannas involves the erosion of the soil,[696] setting the stage for colonization by relatively few species such as some grasses. Tropical grasslands in turn may have a very high harvestable productivity promoting large biomasses of various taxa of herbivorous insects.[697] Such communities can be very stable even though less diverse than other communities in tropical habitats. The clearing of lowland tropical dry forests in Central America to establish cattle pastures has provided a new environment for the diversification of swollen-thorn *Acacia* species and establishment of continuous populations of these plants.[372] Such changes in the geographical distribution of these plants result in changes in the community or guild of insects feeding on them, not to mention the distribution of *Pseudomyrmex* ants. In some Temperate Zone agricultural habitats dominated by a single crop species, increased densities of the plant result in a lowering of the number of herbivorous insect species associated with the crop.[698] Dense plantings are characterized by a greater number of herbivore insect species and few carnivorous predator species, and similar patterns are expected for food crop monocultures in the tropics.[57]

The establishment of successional habitats in the lowland tropics in Africa has provided the opportunity for ecological diversification in wood- and debris-feeding insects such as termites. In some regions, termites form a major component of the insect biomass in agricultural and secondary successional habitats and in which they increase plant productivity by enriching the soils with nutrients.[699,700] Mielke[700] points out that indigenous peoples of termite zones in Africa have coexisted well with these insects, although approaches to modern agricultural technology include programs aimed at eradication of termites. Again, as with Janzen's discussion,[57] there appears to be a conflict with attempting to convert tropical habitats into massive food-generating machines for the world. In Africa, the extensive clearing of tropical forests and the resulting establishment of mixed forest and savanna ecosystems has resulted in new assemblages of insects such as butterflies exploiting these new environments.[701] In fact, it is suggested that some butterflies appear to be better suited for the exploitation of resources in such ecosystems than they were in their original environments,[701] but such conclusions are difficult to confirm without an analysis of the whole communities involved. Weedy gardens in the tropics are believed to be suitable habitats for many butterfly species, although such open systems of "farming" butterflies do not necessarily take into account the interplay of genetic structure of populations in relation to optimal distributions of resources and effects of mortality factors. Clearly we need more data on the long-range effects of such open systems on the maintenance of butterfly diversity in selected habitats.

The absence of severe annual changes in weather conditions in many regions, the ability for plant species to accumulate in certain kinds of tropical soils, and favorable, long growing seasons together provide a huge obstacle for monoculture and simple polyculture for tropical food crops, when viewed in terms of herbivorous insects exploiting wild and introduced plant species. On a different level, material from earlier chapters has made us realize that tropical pollination systems in natural habitats have evolved in response to a variety of selection pressures influencing the spatial distribution of a plant population, optimal reproductive periods, and availability of pollinators. Much of that discussion focused on the often specialized coevolved associations between plants and pollinators, even though some pollinating insects such as stingless bees may be fairly opportunistic in foraging tactics. As a result of selective inbreeding and agronomic requirements centered around harvesting, the pollination systems of tropical crops are often specialized and the challenge becomes the need to find the optimal pollinating insect.[702] Janzen[42] believes that animals pollinating monocultures of plants in the tropics are drawn from the communities of pollinators associated with adjacent natural forests, and since the crop cannot move away from the site, it depends upon some portion of these communities for reproduction. The matter becomes critical if the object of harvesting is fruit or seed.

Many current and past field studies of insect pollination in the tropics have emphasized the often great specificity between pollinators and plant species in natural forest ecosystems.[334] On the other hand, some flowers attract a variety of opportunistic insect visitors such as stingless bees. A good example of our present lack of information on the pollination systems of some rain forest under-

TABLE 9.1
Amount of Pollen on the Bodies of Insects Captured on Pigeon Pea Flowers in India[a]

Species	With (+) and without (−) pollen loads	Number of insects examined	Mean number of pollen grains	Mean number of pigeon pea pollen grains	Mean percentage of pigeon pea pollen
Hymenoptera					
Apis dorsata Fab.	+	11	3,045	3,045	100
	−	2	1,250	1,250	100
Apis florea Fab.	+	2	250	250	100
Megachile bicolor (Fab.)	+	9	82,334	81,056	98
	−	4	2,875	1,500	69
Megachile lanata (Fab.)	+	9	107,333	47,389	75
	−	3	1,334	1,167	67
Megachile conjuncta Smith	+	6	51,500	41,417	86
	−	1	0	0	—
Megachile spp.	+	12	39,666	39,000	94
	−	8	7,875	7,000	83
Anthophora sp.		1	16,500	16,500	100
Xylocopa spp.	+	2	50,500	49,500	98
	−	3	2,000	1,667	91
Polistes sp.		11	1,090	1,045	99
Delta sp.		2	2,000	50	27
Rhynchium brunnem Fab.		3	1,500	1,333	92
Diptera					
Musca lusoria Wied.		6	1,500	1,250	88
Carcelia sp.		1	0	0	—
Lepidoptera					
Catochrysops cnejus Fab.		15	534	467	87
Lampides boeticus L.		15	167	100	93
Syntomis cyssea Stoll.		1	500	500	100
Terias hecabe		3	833	833	100
Homoptera					
Clavigrella horrens D.		2	2,000	1,750	90
Dysdercus sp.		1	0	0	—

[a] From Williams.[704]

TABLE 9.2
Foraging Behavior of Different Species of Bees on "Normal" Pigeon Pea Flowers in India[a]

	Species			
	A. dorsata	*A. florea*	*Megachile* spp.	*Xylocopa* spp.
Number of bees observed	26	61	81	20
Percentage of bees with pollen loads	92	67	78	50
Mean number of flowers visited per plant	1.9	—	1.8	2.7
Mean number of plants visited per row	1.4	—	1.4	1.3
Mean number of plants moved per plant change	1.2	—	1.8	1.1
Mean number of rows moved per row change	1.4	—	1.4	3.5
Time spent per flower visit(s)	14.7	25.6	3.8	3.6

[a] From Williams.[704]

story plants is illustrated by a recent studies of Soderstrom and Calderon[703] that reported discovering two species of phorid and two species of cecidomyiid flies that were the possible pollinators of certain grasses in Brazil, Colombia, Panama, and Venezuela. Their data require a reexamination of traditional views of pollination systems in tropical grasses and other plant groups as well. When a plant species is grown in monoculture, it may attract a wide range of insect visitors, simply because it provides a concentrated and large resource adjacent to a mixed-forest environment. Depending upon the floral design and physiology of that particular plant species, some degree of specificity displayed will be in terms of which insect visitors actually pollinate the flowers. For example, Williams[704] censused the great variety of insects visiting the flowers of the crop species pigeon pea, *C. cajan,* in India and discovered that although the flowers attract many different insect taxa, a smaller portion bear pollen grains of this plant species, and an even smaller percentage of these insects, mostly bees such as *Megachile,* actually cross-pollinate the flowers (Table 9.1). Williams's data illustrate the high specificity that can exist for cross-pollinating insects in a tropical food crop species, even though the flowers attract a variety of insects. Interestingly, Williams noted an abundance of thrips on the flowers of this species and cites other reports in which thrips were found to have exceedingly high loads of pigeon pea pollen on their bodies. Yet the possible role of thrips in the pollination of pigeon pea remains obscure. Williams also demonstrated that some species of *Megachile* bees discriminate between cultivars and wild forms of pigeon pea, which differ in flower structure (Table 9.2).

9.2. INSECT POLLINATION IN AN AGRICULTURAL HABITAT IN THE TROPICS: THE CASE OF CACAO

In previous discussion of tropical seasonality and pollination systems (Chapters 7 and 8), there was a discussion of cacao pollination. Cacao as a cultivar in both the New and Old World tropics (although the origin is New World) falls into the class of very low fruit set compared with the proclivity of the mature trees to produce flowers. General reviews of the biology of cultivated *T. cacao* are given in Cuatrecasas[595] and Purseglove.[644] The emphasis in previous chapters concerning cacao has been the possible role of pollination as a limiting factor in fruit set (Fig. 8.4) and the effects of strongly seasonal patterns of rainfall on asynchrony between flowering and peak activity of pollinating midges. It was stressed that cacao in its original wild state as an understory tree of lowland tropical rain forests in South America had a patchy spatial distribution along streams with several trees to a patch and that an equilibrium was established between abundances of flowers and pollinating insects through time. Such an equilibrium is not unexpected in natural tropical forest ecosystems, as illustrated by recent studies on the pollinator specificity and abundance in relation to fruit set in large tropical trees.[117]

From a consideration of recent thought and data on the evolution of breeding systems in tropical trees,[638] it would appear that cacao, in its wild state, opted through evolutionary time for low fruit set relative to flower production, with individual trees being physiologically limited for low fruit set due to other phenotypic constraints affecting overall fitness in the rain forest environment. The breeding system of the tree was a self-incompatible system in which tiny certatopogonid midges and possibly other insects exchanged pollen between individuals within a clump, perhaps between nearby clumps in the forest understory. According to one view,[394] the cauliflorous condition of cacao and other tree species of tropical rain forest understories is believed to be due to physiological constraints. Although cacao as a cultivar experiences considerable damage from herbivorous insects and pathogens in some regions,[705] the chemical defense systems of the foliage of this tree species are generally known to be well developed and probably very functional in the original habitat.

Although such a species, therefore, stands a good chance for agronomic success in monoculture or simple polyculture (with crop shade trees such as *Hevea* rubber), Old and New World cacao plantations are plagued today with diseases and defoliation generated by insects, the result of colonization by insects from surrounding forest habitats.[705] Overall, cacao has been reasonably successful as a major cash crop grown in simple polyculture. Efforts to increase fruit set in cacao plantations should include the detailed study of the breeding systems in both plantation and natural habitats for this tree species. This effort is necessary to determine whether some proportion of the trees in an area are functionally andromonoecious, being pollen donors for other trees. Relatively few data exist on the intertree variation in fruit set in cacao in either wild or cultivated cacao stands, so that we have little information about the breeding system of the tree species. Once the breeding system is determined, including the disclosure of whether or not there is functional andromonoecy (see Chapter 8 for a discussion of this topic for cacao), meaningful statements about the effects of cultivation on the pollination system of cacao will be possible.

Pollinator populations may be limited in cacao plantations by the low abundance of suitable breeding sites for the midges (see the review of Winder[(588)]). This argument is based on the observation that experimentally-placed breeding substrates of these insects results in a patchy but rapid increase in the larval populations of the midges.[(590)] We have few data on the life tables of pollinating midges, although the high densities of predatory arthropods and *Anolis* lizards in some cacao plantations[(475)] and the observation that salticid spiders successfully capture adult midges on cacao trees (A. M. Young, unpublished data), indicate that various biotic mortality agents may regulate midge populations. The courtship and mating behavior of these midges (several genera and species; see Table 9.3) is probably not characterized by large mating swarms. Thus adult population densities are suspected of being low, and adults are not abundant at cacao flowers.[(652)] Young[(590)] noted in a preliminary way that patches of breeding substrates with ant colonies generally lack ceratopogonid midge larvae (at least later instars), while patches with high densities of midge larvae lack ants.

Leston[(476)] has described the "ant mosaic" associated with cacao plantations in Ghana. Ants and other predatory arthropods readily invade cacao plantations from adjacent mixed-species forests, as these polyculture agricultural habitats contain both food supplies and nesting sites. The rate of invasion of cacao plantations by ants, for example, will be a function of (1) resident ant species in the surrounding forest and the degree to which these species are in competition for resources with one another; (2) the spectrum of intercolony variation within a species in the surrounding habitat, so that some colonies may be more prone to move into agricultural habitats than others[(706)]; and (3) the suitability of the agricultural habitat in terms of prey, other resources, and nesting sites. In this regard, a significant factor in the agricultural habitat is the amount of shade cover and the degree of shade tolerance in potential invading ant species. Similar arguments can be generated for the communities of other arthropods, and their associated diseases in the case of vectors, that may invade from the surrounding habitat. Leston[(705)] showed for cacao trees interspersed with wild shade trees in Ghana that the crop canopy insect fauna is a subset of the fauna associated with the forest canopy, having the same dominant ant species and some fugitive species of ants from surrounding savanna habitat. Leston suggested that changes in the light regime are the primary factor resulting in

TABLE 9.3

The Distribution and Abundance of Various Genera and Species of Ceratopogonid Midges in Two Cacao Plantations (La Tigra and El Uno) in Northeastern Costa Rica[a]

Plantation	Genera	Species	Number of Individuals						
			Flowers	Foliage	Natural ground litter[c]	Box litter	Simulated bromeliads	Banana stems	Total
La Tigra	*Forcipomyia*	n. sp. nr. *argenteola* Macfie	3	—	3	33	—	146	213
	Forcipomyia	*sexvittata* Wirth	—	—	7	10	—	123	140
	Forcipomyia	*cinctipes* group	3	—	5	7	—	20	35
	Forcipomyia	species-1	—	—	1	3	—	4	8
	Forcipomyia	species-2	—	—	1	—	—	2	3
	Forcipomyia	species-3	1	—	1	—	—	5	7
	Forcipomyia	species-4	—	—	—	4	—	10	14
	Forcipomyia	new species	—	—	—	—	—	20	20
	Forcipomyia	species-5	—	—	2	—	—	4	6
	Dasyhelea	*borgmeieri* Wirth & Waugh	1	1	3	—	—	—	5
	Dasyhelea	*williamsi* Wirth & Waugh	—	—	—	—	—	6	6
	Dasyhelea	species-1	—	—	1	—	—	2	3
	Atrichopogon	species-1	1	2	2	1	—	—	6
	Atrichopogon	species-2	—	—	1	1	—	—	2

	[illegible]	species-1 (*bulla* group)	—	—	—	1	—	—	1
	Stilobezzia	species-2	—	1	1	—	—	—	2
	Fittkauhelea	*amazonica* Wirth & Blanton	1	—	—	2	—	4	6
	Culicoides	*hylas* Macfie	—	—	3	2	—	7	12
	Culicoides	*glabellus* Wirth & Blanton	1	2	—	—	—	—	3
	Culicoides	*diabolicus* Hoffman	1	—	—	—	—	—	1
		Total individuals:	14	10	60	67	0	368	519
		% of total sample:	2.7%	1.9%	11.5%	12.9%	0%	70.8%	
El Uno	*Forcipomyia*	*genualis* (Lw.)	1	—	6	18	—	18	43
	Forcipomyia	*cinctipes* group	—	—	8	1	—	2	11
	Forcipomyia	species-1	—	—	—	2	—	—	2
	Forcipomyia	species-2	1	—	2	4	—	—	7
	Dasyhelea	*borgmeieri* Wirth & Waugh	—	—	1	—	—	—	1
	Culicoides	*hylas* Macfie	2	6	—	—	—	—	8
	Culicoides	*foxi* Ortiz	—	1	2	—	—	—	3
		Total individuals:	3	7	19	32	0	20	81
		% of total sample:	3.7%	8.6%	23.4%	39.5%	—	24.6%	

[a] From A. M. Young.[590a]

[b] Specimens were obtained by collecting from cocoa flowers and foliage, and sorting through rotten organic substrates (August 1978–October 1979). Approximately 15% of total field time was devoted to examining flowers, another 10% to sweeping the foliage of cacao trees, and 60% to examining samples of rotten organic substrates (ground litter, litter in simulated bromeliads, and banana stems).

[c] Controls.

shifts in the insect fauna in such agricultural habitats. Forest degradation is followed by an almost immediate reduction in herbivorous Coleoptera, which can no longer maintain breeding populations on a scattered food base, and complete removal of forest results in the establishment of savanna insect faunas associated with cacao. Such degraded forests are rapidly colonized by herbivorous forms such as Lepidoptera, then followed by carnivorous predators. The overall species richness of degraded forests is well below that of natural forests.[(705)]

Janzen[(42)] believes that monoculture forest habitats, presumably naturally occurring ones, draw their pollinators from surrounding mixed-species forests, and this may also be true for agricultural habitats. The ceratopogonid midges in general require moist conditions for breeding, and adults (Fig. 8.5), owing to their small body size, also position themselves within the moist regions of vertical and horizontal gradients in air humidity in different kinds of habitats. Little is known about the complete feeding habits and nutritive requirements of adults for genera and species assumed or proven to be pollinators of cacao.[(588,589)] The studies of Downes on Nearctic ceratopogonids clearly point to specialized forms of pollen-feeding in adult females of some genera and species.[(245)] Yet the niche breadth with respect to pollen types exploited within a habitat is not known for any species anywhere. Richards[(394)] alludes to the pollination of some tropical rain forest understory trees and shrubs by nonspecialized dipterans among other nonhymenopterous insects, yet in-depth studies of the tenacity for certain forms to pollinate only certain tree species, and the mechanisms underlying the interaction, are not known for most understory trees, although there is considerable interest in this area.[(42)] The effectiveness of any one species of ceratopogonid midge for pollinating cacao will be, in part, a function of the physiological niche breadth for food resource types and the realized or ecological niche breadth at any one time and within a particular natural or agricultural habitat. A common shade tree for cacao is *Hevea* rubber, whose flowers are also pollinated by these midges. The relationships between the pollinating midge communities associated with the overstory of rubber and the understory of cacao in such a simple polyculture have not been studied.

In the case of the well-known cultivar *Vanilla fragrans* (Orchidaceae), individual vines have a significantly higher growth rate in plantations on Puerto Rico where the overstory consists of wild trees left there when the land was underbrushed than in similar plantations where all of the vegetation was cleared and shade trees were replanted.[(707)] Puerto Rican vanilleries utilize legume trees such as *Erythrina* and *Inga* to provide the shade necessary for suitable growth of the vanilla vines, but because some species of planted shade trees are very susceptible to varieties of root rot, highest yields of beans are obtained with natural shade cover left for shading vanilla vines.[(708)] In a manner similar to that discussed by Leston[(705)] for optimal cacao production in Ghana, retention of a natural forest overstory when planting vanilla or cacao functions to keep down herbivorous insect populations and other organisms that may otherwise reduce an introduced shade cover to a leafless condition very rapidly, perhaps attacking the crop plants themselves. Natural shade cover may also promote the activity of pollinating insects such as bees by preserving a portion of the habitat required for nesting, optimal thermoregulatory conditions, and retention of landmarks used in marking home ranges. The amount and kind of shade cover over cacao in Ghana has been shown to be a major factor in determining the intensity of attack by capsids on cacao trees.(705) When cash crop species are relatively inedible to many herbivores, and when the shade is provided by a natural overstory, the least chance appears to exist for serious outbreaks of pest insects on the crops. Planted shade trees may not respond in the same manner to seasonal cycles in rainfall as do species of the natural forest and therefore may go deciduous and injure the crop population through increased desiccation.(707) This effect may increase the susceptibility of individual trees or vines to attack by certain groups of insects.

Even if cacao trees turn out to utilize a breeding system characterized by functional andromonoecy, one must ask questions concerning the adaptive role of such a system within the context of the natural habitat of this rain forest tree species. The answers to such questions will not be possible without in-depth knowledge of the population dynamics of the pollinating insects and their natural history. If the average cacao plantation does not provide adequate abundance of suitable resources and physical conditions required for optimal population growth of these pollinators throughout most of the year, and if replenishing pollinator populations annually from adjacent forest habitats is not possible, fruit set in cacao may be lowered even further for individual trees functioning as pollen acceptors under the assumption of functional andromonoecy or other breeding systems. If conditions within plantations can be demonstrated to limit midge populations and if fruit set is lowered because of such effects, questions can then be asked about the optimal design of cacao plantations as a means of increasing harvestable productivity (beans) without necessarily increasing the area of terrain occupied by plantations. In most existing plantations, flowers losses are very high, as indicated in a recent census of flower drop in a Costa Rica plantation (Table 9.4). Flower drop in *T. cacao* refers to the loss of buds or open flowers on healthy trees, generally the result of pollination failure or other factors (e.g., removal by *Atta* and *Ectatomma* ants). Two different plantations within the same region but differing in terms of complexity of shade cover exhibit different phenologies of flowering throughout the year (Fig. 9.1). The El Uno plantation has a monoculture canopy of *Hevea* rubber that goes partially deciduous during the short dry season of this region, causing an increase in the amount of sunlight hitting the cacao trees below, thereby resulting in a burst of flowering at the beginning of the rainy season (Fig. 9.1). The more complex

TABLE 9.4

Some Estimates of Flower Drop for Cacao Trees in the La Tigra Plantation in Northeastern Costa Rica[a,b]

Tree no.	Initial Number of buds marked[b]	Number of marked positions (pins) present later[c]	Number of new pods present[d]	Number of flowers/ buds present	Number of empty positions	Flower drop (loss) (%)
1	320	310	3	15	282	About 95%
2	560	534	1	10	523	About 99%
3	285	280	5	23	252	About 97%
4	483	476	9	17	450	About 95%

[a] From Young.[590a]

[b] The position of individual buds on each of four healthy trees was marked by gently inserting a colored pin next to each bud and recording of pins used per tree; pins were inserted firmly to minimize their loss by falling out, but none were inserted at the base of a bud so as to minimize injury to the bud cushion.

[c] The initial marking was done on 23 March 1979, and the census of pins, and presence or absence of new pods or new flowers, was taken on 27 June 1979. Slight losses in pins, through falling out, etc., were expected and found.

[d] New pods present included healthy pods and aborted pods present at locations of pins on each of the four trees censused. Trees censused were similar in size and healthiness, and all were located within 10 m of the others.

natural canopy of the La Tigra plantation, consisting of at least eight species of forest trees, sustains considerable flowering throughout the year because the trees are patchy in distribution and exhibit differential effects to dry periods (Fig. 9.1). Pronounced differences in flowering patterns between different cacao plantations in the same general climatic region of the wet tropics differentially affect insect pollinator populations and fruit set.

Within small samples of mature and flowering trees in each plantation, peak fruit set occurs in the first month of the rainy season, slightly sooner in the La Tigra plantation, presumably because flower production is already high at this time while lagging in El Uno (Fig. 9.2). Increased fruit set at this time may be due to increased abundance of pollinating midges in both plantations resulting from wetter conditions favoring their breeding and survival. The effects of seasonality discussed here are minimal in the sense that these plantations occur in a region of premontane tropical rain forest where the dry season each year is relatively short and erratic in length and severity. Severe tropical dry seasons (e.g., in Ghana; see Chapter 7) may reduce midge populations so drastically that there is considerable asynchrony between pollinators and flowers at the beginning of the rainy season.

The two plantations are strikingly different in terms of the laundry list of pollinating midge genera and species, and numbers of individuals, collected in each (Table 9.3). The greater species richness and abundance of midges in the La Tigra plantation is suggested[590] to be the result of more suitable resources and increased habitat complexity (litter substrates). The existing mosaic of different kinds of rotting organic debris present in a cacao plantation will have a profound effect on the species of ceratopogonid midges present since different groups exhibit different preferences for breeding sites.[653] Young[590a] conducted a preliminary field study in which sets of known breeding substrates of these midges (see Winder[589] for a review) were distributed in both plantations in equal numbers and with similar patterns of distribution. After remaining undisturbed and being allowed to decay for several months, the substrates were censused for larvae (only larger instars censused) and pupae of midges. The details are summarized in Young,[590a] but it is interesting to note here that although the results are tentative, midge abundance (for a few species only) is increased significantly in rotting slices of banana trunk in the La Tigra plantation but not for any treatment in the El Uno plantation (Table 9.5). Furthermore, mature cacao trees adjacent to patches of experimentally placed banana trunk slices have increased fruit set (Table 9.6), although once again these findings are at best tentative owing to small sample sizes. Young[590a] suggests that ecologically more complex cacao plantations may support a greater species richness of pollinating midges and the abundance of some species may be limited by available breeding sites, perhaps as a result of competition among species. To what extent midge populations are kept down due to predation and parasitism is not known at this time.

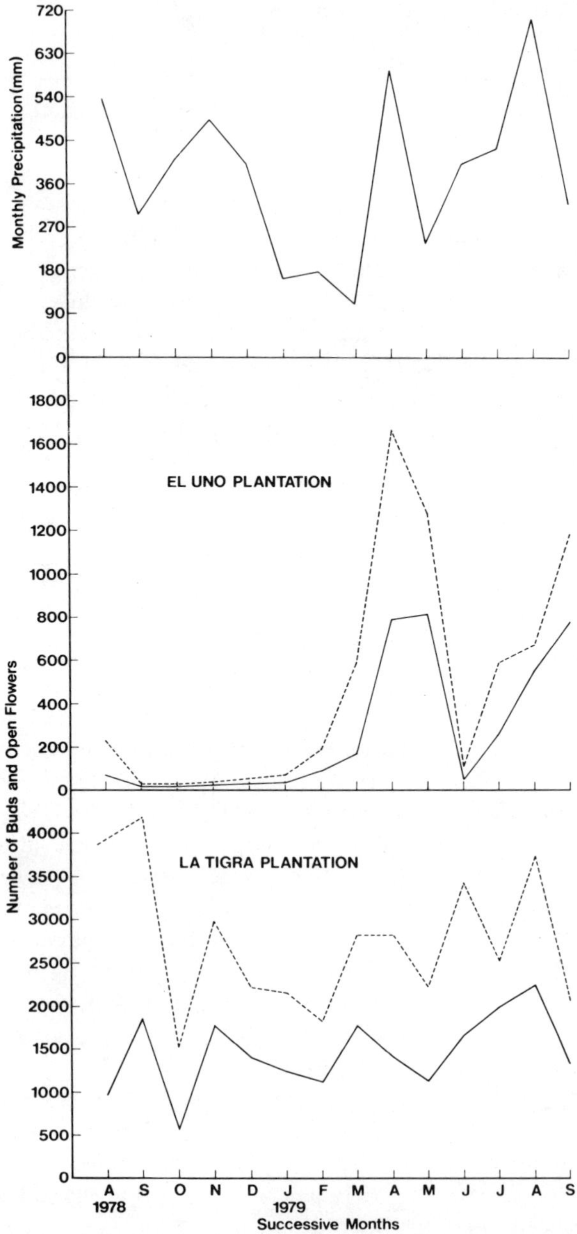

FIGURE 9.1. Monthly phenological patterns of rainfall and production of flowers (buds and open flowers) in an open cacao plantation (La Tigra) and shady plantation (El Uno) in Costa Rica. Broken lines represent buds, and solid lines represent open flowers. The same number (20) of marked trees were censused in each plantation to generate these data. [From A. M. Young, *J. Appl. Ecol.* **19**:46–67 (1982). Copyright 1982 by the British Ecological Society.]

It has also been observed that predominantly one species of stingless bee, *Trigona jaty,* collects large quantities of pollen from cacao flowers but does not contribute to pollination and fruit set.[405] In fact, the bee might be robbing pollen that would otherwise be used both as a reward (food) and source of pollination by ceratopogonid midges. The extent to which such bees take pollen from cacao trees will be in part a function of the proximity of other suitable pollen supplies in

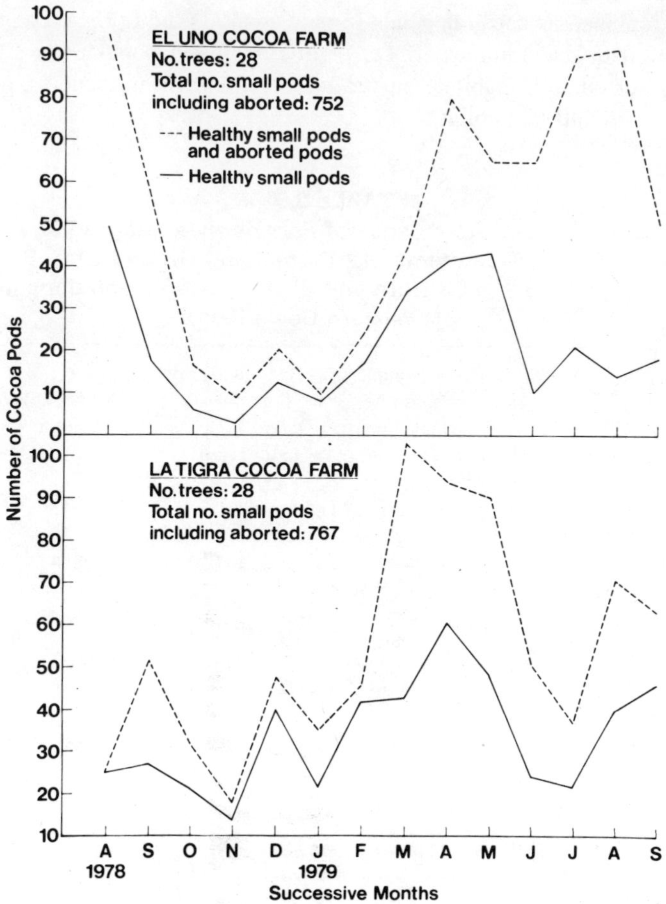

FIGURE 9.2. Phenological patterns of fruit production in two cacao plantations within the same general climatic zone of Costa Rica. Broken lines represent healthy small pods and aborted pods, and solid lines represent healthy small pods. Overall production of small and aborted fruit (pods) is strikingly similar for these samples between the two plantations, although a seasonal pattern is somewhat more evident in the exposed plantation (La Tigra). [From A. M. Young, *J. Appl. Ecol.* **19:**46–47 (1982). Copyright 1982 by the British Ecological Society.]

the surrounding habitat. For example, in another cacao plantation in Costa Rica, strips of blooming *Schistocarpha oppositifolia* (Compositae) attract large numbers of *T. jaty* but the bees do not visit, to any appreciable extent, cacao flowers adjacent to these patches of weedy species. A critical bit of information needed to determine whether pollen-robbing by such bees in cacao groves contributes to losses in fruit set is the determination of ecological fruit set versus physiological fruit set capacities of individual trees. While the studies described above are preliminary, they do suggest that the degree of successful pollination in a cash tropical crop such as cacao depends in part upon (1) the floristic complexity of the shade cover and indirect effects of this on insect populations and (2) the effects of surrounding habitats on the influx of insects influencing pollination within the agricultrual habitat.

TABLE 9.5

The Distribution and Abundance of *Forcipomyia* Midges (Larvae and Pupae) in Various Treatments of Arboreal and Ground Litter Breeding Substrates in the La Tigra and El Uno Cacao Plantations in Northeastern Costa Rica[a]

Treatments	Total volume of substrate sampled on each census[b]	Total number of *Forcipomyia* larvae + pupae in all censuses	Number of *Forcipomyia* per unit volume, all censuses
	La Tigra		
Natural cacao leaf litter (control)[c]	1.20 m³	51	About 11 midges/m³
Thickened ground cacao leaf litter[d]	2.00 m³	57	About 7 midges/m³
Arboreal cups of litter[e]	1.52 m³	0	0
Rotting banana stem slices[f]	0.14 m³	334	About 119 midges/m³
	El Uno		
Natural cacao leaf litter (control)[c]	1.20 m³	16	About 3 midges/m³
Thickened ground cacao leaf litter[f]	2.00 m³	25	About 3 midges/m³
Arboreal cups of litter[e]	1.52 m³	0	0
Rotting banana stem slices	0.14 m³	20	About 7 midges/m³

[a] From Young.[(590a)]
[b] There were four censuses of natural litter, thickened litter, and arboreal cup litter, and two of banana litter.
[c] Samples took 10 collections of leaf litter, each 2 m × 2 m × 3 cm.
[d] Samples were obtained from square wooden frames, 2 m × 2 m × 20 cm deep, replicated 10 times, and placed beneath cacao trees; all replicates were then censused at each census and the litter replaced. The litter was replaced after each census.
[e] Twenty plastic cups were placed in 18 cacao trees and filled with rotting cacao leaves, rotting banana stem fragments, or rotting cacao pods. The litter was replaced after each census.
[f] One hundred ninety-six slices of banana stems were distributed among seven plots for each experiment.

TABLE 9.6
The Effects of Different Rotten Organic Substrates of Pollinating Midges on Fruit Set in Cacao Trees in Two Plantations (La Tigra and El Uno)[a]

Treatment	Number of trees examined	Total number of new fruits present	Mean (± S.D.) number of new fruits
		La Tigra	
Rotten banana stems	14	61	4.4 ± 3.9
Control	14	18	1.3 ± 1.6
Box litter	20	24	1.1 ± 0.8
Simulated bromeliads	26	33	1.7 ± 1.4
Control plots of natural litter	24	27	1.4 ± 1.2
		El Uno	
Rotten banana stems	14	18	1.8 ± 1.4
Control	14	22	1.6 ± 1.3
Box litter	20	19	1.5 ± 1.4
Simulated bromeliads	24	20	1.3 ± 1.4
Control plots of natural litter	24	24	1.5 ± 1.2

[a] From A. M. Young.(590a)
[b] Controls were deliberately increased to maximize taking into account any effects of inter-tree variation in fruit set. For a two-tailed test with 95% confidence limits, the only significant difference was between the mean fruit set of trees in plots of rotten banana stems and the two groups of control trees in La Tigra (t = 2.69, p = 0.001; t = 3.86, p = 0.001). All trees were censused in September 1979.

9.3. GRAZING AND RELATED ACTIVITIES BY INSECTS IN AGRICULTURAL HABITATS IN THE TROPICS

In Temperate Zone monoculture fields, periodic grazing is essential for the maintenance of stable populations of herbivorous Coleoptera such as chrysomelids, because the maturation of host plants is accompanied by a reduction in foliage quality as food for these insects as the growing season progresses.(709) Grazing stimulates the growth of new meristem tissues, a high-quality resource for these herbivores. In strongly seasonal tropical areas, heavily grazed pastures or mowed habitats may also enhance the maintenance of insect populations associated with the plants and, to some degree, the carnivorous predators and parasitoids exploiting them. In the mildly seasonal or nonseasonal wet tropics grazing, effects may be more dispersed throughout the year. In some Temperate Zone guilds of closely related species of leaf hoppers associated with Kentucky bluegrass, effects of mowing on population structure within a species depend upon whether the species is a fugitive; different species exhibit different responses.(710)

The studies of Scriber and Feeny[477] with selected Temperate Zone lepiddopterans illustrate the selective force of larval feeding efficiency in the regulation of host plant selection in herbivorous insects in general. Host plant choice will be a function of the ease with which the insect can extract the nutrients from the target tissues once the plant is located in the habitat. Ease of extraction of harvestable nutrients will be a function of, among other things, physical toughness of the tissues as a challenge to the feeding process, the kinds and concentrations of defensive compounds in the tissues, the water content of the tissues, and the kinds and concentrations of nutrients in the tissues. As an individual plant grows, all of these factors are expected to change to some degree, and the initial condition of the plant and its temporal changes through the growing season determine the quality of the plant as a food base to insects; quality is expected to be different for different insects feeding on the plant. From the viewpoint of the insect species, suitable or optimal quality is interpreted as the ease with which the harvested plant material can be converted into new insects, either directly (mobilization of nutrients for production of eggs, courtship traits, etc.) or indirectly (use of nutrients to provide energy for avoiding carnivorous predators and parasitoids, for locating suitable shelters, etc.). In Neotropical butterflies such as *Morpho* and *Parides* it is not uncommon to observe gravid females placing viable eggs on exceedingly small seedlings of host plants in natural habitats (A. M. Young, unpublished observation),[311] host plant individuals too small to allow complete development of the larvae. Apparently such plants are very nutritive or nutrients are easily obtained from them, and the oviposition behavior is correlated to allow exploitation of such plants. A critical piece of data lacking is the proportion of such plants used by the average gravid female butterfly in a particular habitat (compared with oviposition on mature or large plants).

Scriber and Feeny[477] observed that *Battus philenor* larvae feeding on small seedlings of *Aristolochia* in Texas pine forests actively move among individual plants as they are depleted. Dispersal of this kind may also occur in other butterflies and is not necessarily related to density of insects on host plants. Rather, dispersal of larvae, usually low densities per plant in the Neotropical species studied, is a response to decline in quality of individual host plants as they grow. The short annual growing season in the temperate zones imposes a progressive change on host plants in terms of the factors influencing suitability as food for herbivorous insects. An analogous condition is predicted for the strongly seasonal lowland tropics, where the annual growing is also shortened by the intervening dry season (analogous in effect, for a large part of the plant communities, to the northern winter). In the tropical rain forest regions, the growing season is extended throughout the year for most plant species, and the depletion of suitable tissues (i.e., new meristems) is expected to be slower than in seasonal regions. Not all plant species will respond to monthly changes in rainfall patterns, and each species will also have a breeding system necessitating some energetic tradeoffs with vegetative growth tactics. The more-or-less continuous rainfall pattern of many tropical rain forest localities ensures that new vegetative growth will occur, even at minimal levels, throughout most of the year. In strongly seasonal regions, moist forest refugia may continue to exhibit vegetative growth among individual plant species even though surrounding vegetation has gone deciduous. The former situation, mentioned in Chapter 7, provides a nucleus of food supplies for some herbivorous insects during the long and severe dry season of such regions.

In subtropical or mild temperate regions such as southeast Australia, annual environmental conditions may regulate potential pest populations by a multiplicity of density-dependent and density-independent factors. Detailed studies of the pasture scarab, *Aphodius tasmaniae,* (Maelzer[577,711–713]) provide an example of how an insect becomes a pest species in monoculture or simply polyculture agricultural habitats, with some implications for similar phenomena in the tropics. This medium-sized beetle feeds in the larval stage on the root systems of grasses and clover, and pest conditions have resulted from improved pastures being established following European settlements in this region: improved pastures mean the expansion of existing pasturelands and the addition of clover as a food source for livestock. What has really happened is that not only has the food base for *A. tasmaniae* been progressively increased in total size over the postsettlement years but the food quality for the insect has been improved: both this species and an allied pest species, *A. howitti,* exhibit a preference for feeding on clover, and growth rates of immature stages are faster on clover than on grasses.[712,714] Clover was introduced as a food plant for livestock and was rapidly exploited by the beetles as well. High-density patches of the larvae of both species impede the carrying capacity of pastures for livestock, and the addition of fertilizers to the soil has been of great benefit to the beetles in improving the quality of the food supply. Because the larvae are effective herbivores of the root systems of pasture clover and grasses, the harvestable productivity of pastures for livestock dwindles in areas where infestations are high. Patches of high infestation tend to occur in the same areas in both beetle species in successive years, perhaps as the result of unfavorable weather conditions impeding oviposition-related flights of mated female beetles to colonize other areas. The populations of both species are greatly influenced by prevailing weather conditions, especially the overall climatic pattern of monthly amounts of rainfall.

High rainfall during the summer months promotes the growth of the insect populations through increased growth of pasture host plants and sufficient moisture to trigger flight activity in the beetles. *A. howitti* appears to have a polymorphism underlying oviposition behavior; a large batch of eggs is deposited by the average female before feeding on dung, and a second batch is deposited after feeding on dung. Dung feeding is followed by considerable flight activity before the second and smaller batch of eggs is oviposited into cracks in the soil around healthy clover plants. The presence of dung, however, is not an absolute for survival as *A. howitti* can occur at high densities in ungrazed turf. It appears that the introduction of grazing into the original habitat of the beetle, and the development of dung heaps as a patchy but often abundant resource, has altered the feeding and oviposition behavior of *A. howitti* and to some extent of *A. tasmaniae* as well: fresh dung heaps may be nutritively rich resources for the development of early larval instars and ovaries, if the adult females obtain nutrients such as nitrogen, which is then used in egg production in a manner somewhat analogous to the exploitation of pollen nitrogen in some Neotropical butterflies.[31]

The predilection of larvae for clover roots may also represent a more efficient feeding strategy[477] in that nitrogen may be concentrated as a resource for rapid mobilization to build larval and future ovariole proteins. Both beetles are univoltine, and years of high abundance are years of heavy and steady summer rains but light or no winter rains. Winter rains promote the development of pathogenic fungal populations that kill off the larval stages. During a "good" year, populations are distributed in a somewhat patchy manner, with the greatest densities of larvae occurring in pastures about 2–3 years old and with dense vegetation including clover. Areas of high and low densities are usually interspersed, and although there is some evidence that intraspecific competition among *A. tasmaniae* larvae occurs in high-density patches, there is little evidence that such behavior generates a density-dependent mortality shadow for the populations. But Carne[714] shows that considerable density-dependent mortality exists among larvae of *A. howitti* and that a variety of other biotic control agents (i.e., fungal pathogens, predation by carabid beetle larvae, predation by birds and thyrnnid wasps) and abiotic factors (desiccation and drowning) generate density-independent mortality in the larval populations. For both species the distribution of suitable oviposition sites is considered a limiting factor since gravid females exhibit considerable site selection when ovipositing.

Pastures in the wet tropical lowlands may have high harvestable productivity above ground and undoubtedly provide a food base for a variety of pasture scarabs. Yet variations in annual abundance of such insects are predicted to be low, and overall population densities of immature stages will also be low: the imminently wet conditions of such regions probably kill off large numbers of immature stages either directly through drowning or through promoting various entomophagous pathogens in the soil. Spatial distributions of populations or subpopulations for a species are expected to be very patchy due to differential mortality and not necessarily due to quality of resources such as host plants. Even though the introduction of livestock into wet tropical lowlands in recent years in Central America may provide an expanded but limited food or oviposition base for such insects, some studies indicate that cow pats are ephemeral and patchy resources.[715] Because this resource is very temporary the movement patterns of individual beetles among a collection of cow pats is best described as stochastic,[715] since the quality distribution attached to the resource is changing rapidly. Some studies have also indicated that different kinds of habitats in the tropics influence the colonization of dung and carrion by coprophagous scarab beetles near Leticia, Amazonas, Colombia, in two different years. A significant discovery was that the coprophagous scarab fauna of terra firma was considerably richer, in terms of number of species and individuals, than that of flooded areas, although deforested areas in the terra firma had fewer species. The species associated with dung in deforested areas tends to be different from those of other habitats where diversity is high. Because the scarab fauna associated with dung in forest clearings is reduced, dung persists longer and provides a more stable microhabitat for various other insects such as flies, enhancing the opportunities for disease organisms such as helminths to build up and infect humans in such areas.

In the strongly seasonal lowland tropics of areas such as Central America, historically speaking, livestock agriculture has existed for much longer periods than in wet areas and one might expect that dung-exploiting insects such as scarab beetles might have more diverse communities, since such communities have had more time to develop, and the density and overall abundance of livestock are greater. In the strongly seasonal tropics, where pasturelands prevail, the ecological conditions underlying the dynamics of pasture-inhabiting herbivorous insects such as pasture scarab beetles approaches those of the system described in the studies of Maelzer and Carne[711–713] for southeastern Australia: the annual severe and long dry season imposes a density-independent regulation on the populations of these insects if their host plants experience a reduction in quality as a food base for these insects and the life stages of these insects experience massive mortality from desiccation and other abiotic effects. Likewise, the often heavy rains associated with the early phase of the wet season may kill off immature stages for some forms. The relationship between seeds in cow pats and the kinds of insects associated with them remains to be studied. The overall reduction in the annual growing season in such areas, viewed from the standpoint of grasses and other forage providing a food base for immature stages of scarabs in pastures, will result in seasonal fluctuations in the population sizes and spatial properties of populations for these insects. Shifting livestock distributions between seasons will also change the abundance and density of cow pats as a resource for a variety of insects: cow pats may exhibit considerably different compositions of nutritive substances, including partially digested or undigested seeds, and thereby provide an index of seasonal changes in availability of seeds and other resources derived from plant species in grazed habitats.

Although some detailed studies of Neotropical scarabs associated with dung have been made,[558] as well as natural-history studies of individual species, there have been few published attempts to determine the role of these insects in degrading cow pats in tropical ecosystems and to analyze cow pats as a source of food for other insects. Tropical cow pats may turn out to be interesting monitors of interactions between some herbivorous insects and seeds if such forms exploit the seeds after they have passed through the digestive tracts of wild and domesticated vertebrates. The tropical dung scarab *Onthophagus gazella* of Africa and Asia has recently been introduced into the American southwest to control horn and face flies breeding in dung.[(717)] These beetles are effective in burying large quantities of fresh dung, thereby lowering the survival of the flies which breed in it on the ground surface; these parasitic forms cannot then reach the surface and infest livestock. At the same time, the beetles increase soil fertility through their processing of the buried dung. Tropical coprophagous scarabs are effective agents for reducing dung-breeding parasite of humans and livestock. The degree to which any one species is effective in doing so in its indigenous habitats will be a function of the number of species occupying pastures.

The degree to which a host plant species for a phytophagous insect is restricted in distribution will be a determining factor in the extent to which the plant population becomes saturated with the herbivore population, assuming an absence of effective biotic control agents on the insect population. If the insect is good at dispersing, and if the plant population is relatively restricted in distribution, a high probability exists that the herbivore will attain outbreak conditions. Such an effect will be pronounced if the herbivore is very host specific, or if alternative host plant species are relatively scarce in the region. Island populations of host plant species are very susceptible to such effects since their distributions are very restricted. At the same time, deliberate introductions of a biotic control agent to limit the populations of the herbivore for a plant species targeted for economic use will be effective, especially if the generation time of the control agent is considerably less than that of the herbivore. *Oryctes rhinoceros* is a scarab beetle pest of the coconut palm in Papua, New Guinea, and the larvae in natural populations in Western Samoa are killed off in large numbers by a baculovirus disease. Deliberate introduction of larvae infected with the disease into Papua, New Guinea, where the beetle still has a relatively limited distribution, quickly led to effective control of the insect in some areas, presumably from the spread of the virus into the resident breeding population from the introduced propagules.[(718)] Newly eclosed adult beetles have the virus, and these can then be released into other areas of the island where the beetles are abundant.

Secondary successional habitats and agricultural habitats such as pastures and monoculture plantations of cacao, bananas, and coffee may be stable environments in the tropics in the sense that there is little turnover in the composition of the vegetation over short periods of time. Harvestable productivity may also be high in some secondary habitats and pastures, providing enough plant biomass to maintain a high species richness of herbivorous insects. The patch structure of simple polyculture agricultural habitats such as cacao shaded by rubber or an assortment of wild trees or coffee shaded with legume trees such as *Inga* represents a predictable environment in terms of providing food and shelter for insect species. Some forms of husbandry ensure that such resources are maintained in these agricultural habitats.

Several genera and species of Neotropical cicadas undergo adult emergences around the trunks of *Inga* trees in Costa Rican coffee plantations: study plots with individual shade trees have far greater numbers of cicada nymphal skins than similar plots containing only coffee bushes (A. M. Young,[781a]; see Young[137] for a list of these cicada species). A typical pattern is that of the medium-sized cicada *Fidicina semilata* in the coffee farms at Naranjo in the Cordillera Central northwest of San Jose in Costa Rica: nymphal skins are found beneath large individuals of various species of *Inga* planted to shade coffee bushes, and few or none are found around coffee bushes away from these trees (Table 9.7). The trees are planted in even rows. The data also suggest that only a small portion of the trees have nymphal skins associated with them, so that these cicada populations are not saturating the agricultural habitat. This region of Costa Rica possesses a landscape mosaic in which there are expansive coffee plantations and small strips of natural forest along streams. The cicadas are probably moving into the coffee habitats through a courtship–oviposition–nymphal feeding association with the *Inga* trees, although the evidence is still incomplete. There is a striking pattern of many Neotropical cicadas emerging at large legume trees in both natural forest and agricultural habitats in different climatic zones of the country,[137] suggesting a possible "primary" association of the nymphs with the root crowns of these trees. This is not to say that nymphs do not feed solely on the roots of these trees, since they undoubtedly encounter palatable root systems associated with many different plants in natural forest habitats in particular. Use of the term *primary* in this context implies that (1) Neotropical cicadas were originally associated with legume trees in the evolution of tropical forests and what we are looking at is an archaic coevolved association, or (2) contemporary selection favors such associations with legume trees because they provide an efficient food source for developing cicadas. As cicadas have incredibly long life cycles, presumably as the result of exploiting generally nutrient-poor plant saps, selection may favor the exploitation of those root crowns that increase feeding efficiency, even if just a small amount. Doing so reduces the age of first reproduction and maximizes fitness under individual selection.

White and Strehl[366] have emphasized that the long life cycles of cicadas are mostly the result of nymphs' feeding on nutrient-poor xylem fluids; the rate of growth is very low, and several years may be required for complete nymphal development. The developmental time of any species of tropical cicada is not known. That the striking pattern of nymphal skins of many Central American species is associated with large legume trees in a variety of habitats, including

TABLE 9.7

Collections of Nymphal Skins of the Cicada *F. semilata* Associated with *Inga* Shade Cover Trees in a Large Coffee Plantation at Naranjo, Alajuela Province, Costa Rica[a]

Collection date	Tree species	Number of trees censused	DBH range (cm)	Number of tree trunks	Distribution of nymphal skins[b]				Total nymphal skins		
					Range (trees)	Coffee bushes within 16 m² of a tree	Coffee bushes 16 m²	Ground	♀♀	♂♂	Total
29 December 1972	*Inga* aff. *fissicalyx*	17	20–30	12	0–2	74	5	4	55	40	95
	Inga goldmanii	3	25–35	4	0–1	25	0	1	19	11	30
	Inga leptoloba	6	20–25	10	0–2	3	0	2	10	5	15
11 February 1973	*I.* aff. *fissicalyx*	17	20–30	3	0–1	6	0	5	5	9	14
	I. goldmanii	3	25–35	0	—	0	0	0	—	—	0
	I. leptoloba	6	20–25	0	—	0	0	0	—	—	0
17 April 1973	*I.* aff. *fissicalyx*	17	20–30	6	0–1	0	0	0	4	2	6
	I. goldmanii	3	25–35	0	—	0	0	0	—	—	0
	I. leptoloba	6	20–25	0	—	0	0	0	—	—	0
28 April 1973	*I.* aff. *fissicalyx*	17	20–30	0	—	0	0	0	—	—	0
	I. goldmanii	3	25–35	0	—	0	0	0	—	—	0
	I. leptoloba	6	20–25	0	—	0	0	0	—	—	0

[a]Total area is 416 m² with a 4 m × 4 m plot around each tree. Area estimate excludes areas checked beyond 16 m².

[b]Density of nymphal skins: 0.40/m².

planted shade trees in some agricultural habitats, suggests that the root systems of such trees contain better nutrients for nymphs than other tree species. The nitrogen-fixing properties of legume roots results in higher concentrations of amino acids and other nitrogen-containing substances being found in legume root systems, and root systems of other tropical trees with nitrogen-fixing nodules may also have higher concentrations of amino acids and other nitrogen compounds. Cicada nymphs feed primarily on nitrogen-containing substances instead of carbohydrates.[366,495] Selection would favor a cicada species in tree-species-rich habitats to feed on those trees with the greatest concentrations of required nutrients in the root systems, and many Neotropical Leguminosae are good candidates. The interesting question develops as to whether legume trees were the original host trees of cicadas in tropical forests or if strong contemporary selection favors such coevolved interactions. Tropical cicadas may have "primary" host tree association, since the likelihood is very high, especially in tropical forests, that nymphs encounter the root systems of many different plants and that many of these may be suboptimal in terms of nutrients available to developing cicadas. In cases where high densities of nymphal skins are associated with nonleguminous tree species, for example, *Z. smaragdina* in groves of *Goethalsia meiantha* (Tiliaceae) in secondary succession in Central American rain forests,[71] it would be interesting to determine if the root systems of these tree species have nitrogen-fixing nodules associations.

We also stress that other phenotypic constraints operate in cicada populations to determine host tree associations: the environmental cues used by gravid female cicadas in finding suitable oviposition sites. The studies of Lloyd and White[719] on North American periodical cicadas have shown that females of these species require specific kinds of oviposition sites and that eggs die in incorrect ones. Several Neotropical cicadas oviposit in dead tissues on understory trees and palms in forest habitats.[70,71] Toughness of the wood, bark, and pith are some determinants of oviposition site and hatching success.[719] If a habitat is altered in such a way that suitable oviposition sites are no longer available to cicadas, existing generations in the ground represented by feeding nymphs may be the last generations to emerge in that habitat. The root systems for developing nymphs may be the same, or nymphs may switch to other species, but oviposition site deprivation, through a change in the environment (e.g., flooding of a lowland stream clears away understory vegetation) can lead to a local extinction of the cicada population in that habitat or that portion of the habitat.

Alternatively, legume root crowns may "work" the soil in such a way as to increase its suitability for developing cicadas, and more so than around other trees, increasing the survivorship of nymphs. Furthermore, the increased logging activities in lowland and premontane tropical wet forest zones in Central and South America act as a potent form of artificial selection in leaving behind softwood trees such as legumes. Thus cicadas may be forced to exploit legumes

FIGURE 9.3. Cicadas associated with mountain rain forest in Costa Rica (Bajo la Hondura zone). (a) *C. postica;* (b) *P. biolleyi* resting on moss-covered tree trunks in this forest habitat; (c) nymphal skins of *P. biolleyi* (left) and *C. postica* (right). The scale is in centimeters.

as a response to a shrinking supply of trees in tropical forests. Finally, it is important to recognize the role of physical features of root tissues and bark in determining the feeding preferences of nymphs.

Our factual data base for the feeding behavior of Neotropical cicada nymphs is scanty at this time, and data are lacking on the egg–adult developmental time for any species. The montane tropical rain forest cicada *P. biolleyi* may have as

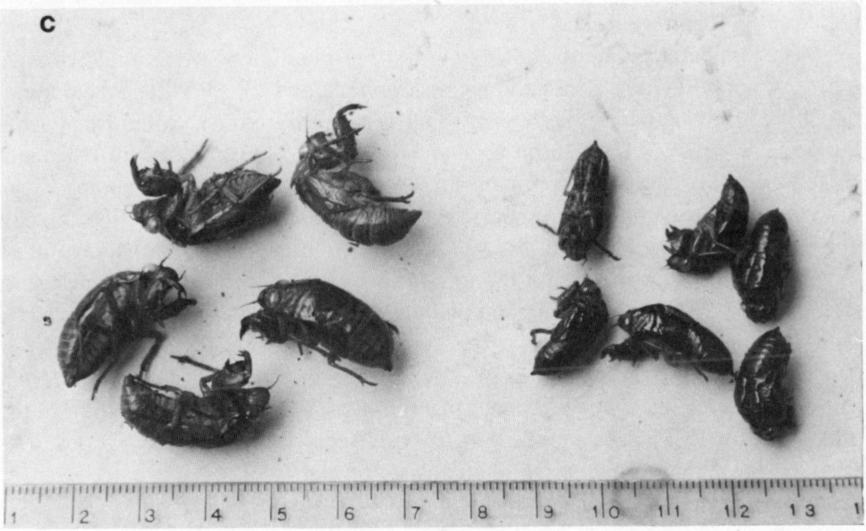

FIGURE 9.3. *(Continued)*

many as five distinct size classes of nymphs in the soil at the same time, suggesting a developmental time of at least 5 years.[136] This cicada shares the same forest habitat with another species, *Carineta postica,* and the two species are readily distinguished in the field both as adults and as final nymphal skins (Fig. 9.3). Neotropical cicadas are nonperiodical, and since emergences occur each year, there is great turnover in genetic structure from year to year and populations tend to be similar in size in successive years, especially in the wet lowlands.[70,71] Although evidence exists that some species in the premontane tropical rain forest zone of northeastern Costa Rica are invading secondary successional habitats,[71] some of these species still occupy primary forests. In some secondary habitats, emergences occur near trees and shrubs other than Leguminosae. The invasion of some secondary habitats in the tropics may involve a switch to nonleguminous root crowns as the primary host association of nymphs. Physiologically and behaviorally such transitions, in response to selection created by loss of primary forest habitat and the original host associations of nymphs with leguminous root crowns, may or may not involve a qualitative change in the nature of the assumed coevolved association: this will depend upon the degree of similarity or dissimilarity between the original host tree and the newly colonized one in terms of those physical and chemical properties related to gustation and feeding success by nymphs. Some lowland and premontane rain forest cicadas are associated with the margins of forest habitats, in some ways perhaps making them preadapted to invade earlier successional stages in the same region when the carrying capacity of the original habitat becomes constrictive in terms of maintaining breeding populations for these insects.

The deliberate planting of *Inga* in coffee plantations in Central America may have provided the ideal ecological stepping stone for some cicadas to maintain breeding populations as the original forest environment diminished from human activities. In the central highlands of Costa Rica and also in the major coffee-growing districts, a dominant stream edge tree species is the legume *Zygia longifolia,* and several cicada species emerge at these large trees.[138] The association of nymphs with the root crowns of these abundant trees, and the evolution of oviposition tactics to permit the gravid females to place eggs at or near such trees, may have been two principal adaptations that paved the way for invading the leguminous overstory of coffee plantations by cicadas. Nymphal skin densities beneath overstory *Inga* trees tend to be lower than for the large *Zygia* trees along streams adjacent to such plantations, and the adult population in the plantation for each species tends to be more dispersed than at stream edges. The stream edge habitat in such regions consists primarily of the *Zygia* trees, in an even row down each side of the stream, with an occasional individual of a different species interspersed. When the original vegetation cover of the surrounding hills was cleared to plant coffee, cicada populations might have been compressed into the stream edge environment because the surviving host trees occurred there. Under such conditions, avian predation, effects of floods on nymphs and emerging adults, might have been some of the forms of selection favoring the eventual invasion of newly established coffee plantations once the *Inga* trees reached a certain size.

The transition to exploiting the coffee plantation environment may have involved a change in population structure, namely, one that emphasized a dispersal of reproductive effort over the abundant supply of *Inga* trees: the average gravid female of a species such as *F. semilata* may distribute her egg clutches over several *Inga* trees in the coffee plantation, while in the stream edge environment there would be a greater tendency to place multiple egg clutches in the same or a few closely spaced *Zygia* trees. Also, in the coffee plantation, there would be a reduced tendency for gravid females to converge on an individual tree, because the environment is not very patchy in terms of oviposition sites. Such oviposition tactics would result in scattered reproductive effort in the coffee plantation and at the same time many empty host tree patches for each generation because the resource base is large and uniform in spatial distribution. The observed spatial patchiness of nymphal skins associated with overstory trees in coffee plantations may also result from differential survival effects on eggs and nymphs. Our present knowledge of Neotropical cicadas is mostly at the level of knowing some portion of geographical ranges, habitat associations of adults and nymphs, associations of nymphal skins with certain tree species, and data on annual peak periods of adult emergences. We greatly need to study the detailed natural history of individual species to determine some of the ecological traits underlying population structure in relation to habitat type.

In the premontane tropical rain forest zone of northeastern Costa Rica, plantations of black pepper (*Piper nigrum*) less than 15 years old have uniform borders made of plantings of *madero negro,* or *Gliridicia nigra,* a small legume tree. Within the past 5 years these trees have been used as major chorusing sites for cicadas such as *F. pronoe,* although nymphal skins have not yet been found beneath these trees (A. M. Young, unpublished observation). What might be happening here is that we are observing the early stages of colonization of these legumes as a resource by these cicadas. Alternatively, the cicadas may be using the trees merely as chorusing sites and not as oviposition sites and root crowns for nymphs. High densities of adult *F. pronoe* (as many as 30 in a 3-m-tall tree at one time) have been seen and courtship observed. The black-pepper habitat is relatively new in this region of Costa Rica and the deliberate planting of *madero negro* trees as perimeters to the plantings may represent a situation somewhat analogous to the early cultivation of coffee in Costa Rica more than 200 years ago. But apparently, if the breeding hypothesis is correct, a first generation of cicadas has not yet been produced on these trees.

Some species of nymphalid butterflies such as *Metamorpha* and *Anartia* have successfully invaded coffee plantations in Central America where larval host plants occur as brush borders along the perimeters.[(317,322)] But at least species of *Anartia* are extremely labile and are known to be colonists of a variety of weedy habitats in the American tropics.[(720)] Yet we expect that the total butterfly and moth faunas of agricultural habitats such as coffee plantations will be considerably depauperate relative to adjacent secondary successional habitats. Tropical agricultural habitats screen out through ecological time most of the herbivorous insects occupying natural habitats, an exception being the highly palatable food crops exploited by a variety of herbivorous insects. But even under these conditions, the diversity of such communities is less than in other habitats. The toxic foliage of cacao, coffee, and other tropical cash crops effectively screens out many insects, and those that occupy these environments are usually exploiting overstory or weedy species associated with perimeters and gaps. In many cases, weedy species will be very patchy spatially, limiting herbivore populations in these agricultural habitats. Many agricultural habitats are unsuitable for maintaining populations of carnivorous predators and parasitoids that exploit herbivorous insects, and in such instances, herbivore populations may build up to greater densities than normally encountered in other kinds of tropical habitats. Once again, we need to conduct comparative field studies on the natural history and population dynamics of selected herbivorous insect species in agricultural and natural habitats in the tropics.

Borrowing from the concept of local and geographical effects influencing the community of pollinating insects associated with plant species in tropical forests (Chapter 8), other kinds of insect–plant coevolved associations may follow similar patterns. One example, gradually unfolding as the result of recent field work in Costa Rica, is the interactions of cicadas with legume tree species in tropical forests. The cicada–legume interaction, postulated on the basis of the observation that nymphal skins of various species are more abundant near large legume trees in different kinds of tropical forests than around other large legume trees in the same habitats, appears to have both local and regional (geographical) traits. At the base of this discussion is a point already discussed in this chapter, namely, the belief that cicadas in tropical forests have a primary host tree, usually a legume, that is the place where eggs are placed in the environment and where the nymphs grow faster as the result of tapping into nutrient-rich xylem fluids of legume root systems. The question is then the examination of local and geographical patterns of numbers of cicada species and numbers of legume tree species exploited under the assumption of feeding association of nymphs with these trees. The following tentative patterns emerge.

1. The lowland and premontane tropical rain forests dominated by large *P. macroloba* have high densities of several genera and species of medium-to-large-sized cicadas; fewer species and lower densities are found on other legumes with a more patchy spatial distribution than *Pentaclethra*. The high densities of adult *Pentaclethra* in these forests promote high densities of cicadas associated with them.

2. On the eastern slopes of the Cordillera Central above about 800 m, the density of *Pentaclethra* drops off markedly and forests contain large trees belonging to several leguminous genera and species, each having a patchy spatial distribution. The result is low densities of nymphal skins per tree and fewer cicada species occurring at the same tree species and same individual trees in the forest. Cicada populations are more evenly distributed, at least in terms of nymphal skins through the forest. Under these conditions, the carrying capacity of the environment for the same cicada species found in the adjacent lowlands (e.g., as *F. mannifera*) is defined in terms of the population exploiting a broad range of legume tree species. The nymphs may be exhibiting an expanded form of monophagy as found in the lowlands for specialist feeding on *Pentaclethra* and a few other legumes. Then again, the legumes may be interpreted as the same resource from the viewpoint of the cicada species in different geographical regions. The mid-elevation cicada community contains fewer species and the population structure of each species is far more diffuse than for similar species in similar lowland habitats.

3. On the western side of the Cordillera Central, cicada species are specialists on the legume tree *Z. longfolia* and have secondarily, in some cases, moved to *Inga* planted in coffee plantations as a shade cover. There are fewer species in such regions than in the regions mentioned in (1) and (2) above. In areas within this region where forested valleys remain, the cicada species are

associated with forest legume tree species and densities of nymphal skins are low, approaching values seen for the mid-elevations on the eastern slopes of the Cordillera Central. The diffuse spatial distribution of legume tree species in some of these habitats promotes low densities of cicadas per unit area.

4. In the dry lowlands of northwestern Costa Rica, cicada species emerge in low densities per tree around large legume trees with patchy distribution in forest remnants. The pattern is similar to that seen in other regions where legume tree species are low in density and several genera and species are widely scattered in the forest habitat. The total number of genera and species of cicadas is low but approaches that of western slopes of the Cordillera Central.

The low densities of nymphal skins for a cicada species around individual large legume trees are believed to be the result of several factors, including gravid females dispersing reproductive effort over a greater number of individual trees in response to a small average patch size, competition among nymphs of different species at the same trees, and differential survival of nymphs at different trees. Large patches of *Pentaclethra* present gravid female cicadas with a large "target" for egg placement (in the understory beneath the host trees) and also provide nymphs of several species with a large primary food base, minimizing competition and any deleterious effects of a limited food supply. Under the assumption that many Neotropical cicadas are legume specialists as feeding nymphs (although not exclusively so), the distribution of cicada populations in tropical forests and the maintenance of local and regional patterns of cicada diversity are determined by the spatial distribution and relative abundance of large legume trees. The occurrence of secondary and agricultural habitats in many regions of the tropics has added additional levels of environmental heterogeneity for cicadas. In some cases, the populations of some species, often only a fraction of the total species of the region, can make the transition to plant species in these new habitats, but the end result is predicted to be a gradual colonization of these habitats at a rate determined for each species by many factors, including (1) the rate of removal of the original forest cover, (2) the availability of suitable new host plants in the new habitat(s), (3) if (2) is low, the rate at which a species can evolve the ability to exploit new host plants of a different physiological makeup with regard to feeding and ovipositing cues, and (4) the interactions among species if several are making such a transition at the same time or doing so at different rates. The availability of suitable oviposition sites in the new habitat will also be a potentially limiting factor in cicada colonization. The predilection of many species to chorus in sunshine, and the mobility of the adults, promote the colonization of secondary and agricultural habitats by tropical cicadas.

Xylem feeding by cicada nymphs on tree roots[366,495] probably precludes direct contact with toxic secondary substances but may result in the operation of other forms of feeding specificities by nymphs. The marked, though not exclusive, association of cicada nymphal skins with the immediate environs of canopy-size legume trees in different habitats in the tropics suggests that cicadas survive better in association with these trees than with others. Possibly cicada nymphs obtain a greater amount of nutrients from the roots of legume trees, as a result of the nitrogen-fixing association of the root systems in many groups of legumes, or perhaps the association and other factors of the roots make for better soil conditions for cicada nymphs. Perhaps the tissues of legume root systems are easier to penetrate by cicada nymphs than other root systems. The selective lumbering of hardwoods in tropical forests excludes the legume trees, making them a more abundant resource for cicadas. Natural catastrophes such as landslides and floodings provide opportunities for legume tree species to colonize new habitats in the early stages of succession. A combination of physiological properties of legumes and ecological factors promoting their high abundance in many tropical habitats (including natural and man-induced changes in the environment) may promote the exploitation of legume tree root crowns by the nymphs of most species of tropical cicadas.

9.4. TROPICAL INSECT PEST POPULATIONS

Many tropical food crops are beleaguered with herbivorous insects that from the viewpoint of humans can be considered "pests," since their feeding activity removes large amounts of harvestable tissues that would otherwise be available for human consumption. Janzen[57] has discussed some of the reasons why the regulation of insect pests of such crops is almost a fruitless task. The long evolutionary history of herbivorous insects adapting to the defensive machinery of plants in tropical habitats makes them ideal for exploiting the often inbred and low-resistance crop species: selection would favor such a niche transition as more and more forests are cleared because (1) the crop is often more palatable; (2) the insect may extract nutrients more efficiently from the crop; (3) the density of the crop is high if in monoculture or simple polyculture and therefore more accessible to the herbivore communities associated with surrounding vegetation; (4) the crop becomes a predictable resource through successive growing seasons, thereby stabilizing the populations of individual herbivore species associated with it, and a period of strong directional selection favoring the switch to the crop species (or, alternatively, nonevolutionary physiological response if the crop is similar to the wild host plant species from the insect's viewpoint) will be followed by stabilizing selection to maximize exploitation of the crop species; and (5) there may be a higher fitness for the insect species in leaving behind in its natural habitat a complex of carnivorous predators and parasitoids so that larger populations build up on the expanded and accessible food base in the agricultural habitat. It is the habitat switch and weakened defensive systems of crop species that are primary factors in causing the insect to become a pest species. Coupled with the long growing season and high density of the crop species, these factors provide ideal conditions for such insects to realize the carrying capacity of the agricultural habitat. Tropical crop species may provide sufficient resources for a variety of insects without resulting in interspecific competition, and harvestable productivity is usually high in such plants, especially if production of foliage has been maximized in the breeding program.

Levins and Wilson[721] show that damage to crops resulting from herbivory is not a linear function of the size of the herbivore population, since in some cases low densities of herbivores are sometimes beneficial to the plant species, stimulating the growth of meristem. But high densities usually result in a lowering of yield per unit time or total yield. As shown in some field studies, such as that of Hinckley[722] on the coconut rhinoceros beetle, *O. rhinoceros,* fitness of individuals within a population of an actual or potential pest species will vary considerably over different kinds of resource patches, generating considerable heterogeneity in conditions promoting outbreaks locally. Thus herbivory by a single species will vary considerably in space and time, and the intensity of such feeding on a crop will be a function of many factors. Levins and Wilson[721] stress that a pest species will damage a particular crop only to the extent allowable by constraints of community structure. They enumerate the following kinds of interactions of the pest insect with other organisms and the environment at the community level:

1. Different parasitoids or predators may utilize the same host at different stages of their life cycle.
2. Different parasitoids or predators may utilize the same host in different microclimates with or without host movement between sites.
3. Natural enemies of a pest species may also use alternate hosts, thus affecting natural enemy abundance and variability.
4. The control organisms themselves may have natural enemies (predators, parasitoids, pathogens).
5. A cluster of pest species may interact via common predators and food competition in such a way that controlling one of them may increase another and not reduce damage.
6. Agronomic interventions in the field—releasing of control agents, applying of pesticides, clearing or encouraging vegetation at the edges, etc.—are undertaken as a response to the stage of the system. The pathways and parameters linking observations of the system and intervention are themselves part of the system and affect its dynamics.
7. There is a nested hierarchy of control agents each of which becomes primary at different population densities of host.

The above sorts of interactions are expected to occur in tropical agricultural habitats perhaps to a greater degree than in many Temperate Zone habitats because the growing season is generally longer, giving more time for interactions to occur, and also because the pool of potentially interacting species is generally greater. Carroll[694] has called attention to the complex sets of interactions believed to take place in tropical agroecosystems. Under point 6 above, tropical examples are probably not controlled in any long-term, effective manner by application of pesticides.[723] But in terms of various levels of ecological interactions among organisms in tropical agroecosystems, one might anticipate that there will be many such interactions, and ones changing through time (i.e., points 1–5 and 7 in the above list). Levins and Wilson[721] discuss the role of the regional mosaic of crop lands and natural forest patches in contributing to the likelihood that some insect species become pests in some regions. In the lowland and mid-elevation tropics, there is continual and rapid turnover of habitats into agroecosystems, small or large, and such a mosaic is very dynamic.

The clearing of the lowland and mid-elevational regions of the wet tropics today imposes a dynamic state of change on insect communities associated with forest environments: colonizations and local extinction of species populations in different habitats, natural and agricultural, alter the spatial distributions of insect species. If natural habitat refugia do not remain, insects will shift to crop species. A shift to a crop as a host plant will be a function of food abundance and quality in existing refugia (fragments of the original habitat), population growth requirements of the insect species, migratory or dispersal ability of individuals in the population, and distances between refugia and crop stands. Polyphagous species utilizing several different wild host plants in refugia may move in stepping-stone manner to the crop stands and probably stand a better chance for successful colonization of crop species because of relatively greater feeding flexibility than strictly monophagous species. Other resources required by each insect species (e.g., nectar sources in the case of adult butterflies) will also contribute to the process of defining the spatial distribution of the insect population between refugia and crop stands.

The regional approach emphasizes the association of an insect species with both sorts of habitats, natural or agricultural, and the degree to which both kinds of environments are necessary in joint distribution to provide sufficient carrying capacity for the species. Interwoven with such effects will be the influences of local community structure and organization upon whether or not a particular species makes the ecological shift to an agricultural habitat, exploiting the crop species or other species (weeds or overstory trees) in it. Selection for dispersal into an agricultural habitat from a diminishing natural habitat and the likelihood for successful colonization of the new environment depend upon the kinds of interactions among species in the local community in the natural habitat and the kind of community in the agricultural habitat.

A tradeoff exists: adapt to high density of the same species or adapt to competition with other species. If the colonist species leaves behind biotic control agents that keep its population well below the carrying capacity of the new environment, both kinds of competition may prevail, depending upon the population growth characteristics of the colonist species and the amount and quality of the new food base. If the food base is very large and of high quality [i.e., can be harvested efficiently, compared with the wild host plant(s)] and the species is essentially *K*-selected, colonization will be relatively free from competition of both kinds. Yet if other insect species are already associated with the crop host plant, even if exploiting different parts or different individuals within the crop patch, there may be a very subtle form of indirect competition since resources are being depleted. If the colonist is an *r*-selected form and the food base is still large, colonization may be successful. If the food base is small (i.e., if the crop patch is small and different crop patches are widely separated and interspersed with other vegetation) intraspecific competition may occur as population densities rise on the crop plants. The negative effects of such competition may be offset in tropical regions by continual growth of the plants in the long growing season of mildly seasonal and nonseasonal regions. Interspecific competition may be intense if the crop species already has a high herbivore load and the colonist species hasn't the kind of genotypes best suited to counter such effects.

Although for most food crops in monoculture the food base will usually be large, fluctuations in quality may be frequent. Success in colonization, or promotion of rapid local extinctions within the crop stand and refugia mosaic, will also be molded by climatic factors[(721)] and a variety of agronomic practices affecting the spacing of the crop and its arrangement relative to other vegetation patches. It is known that Temperate Zone complexes of herbivorous insects associated with a particular kind of host plant are sensitive to a broad range of biotic and abiotic control factors either lessening or enhancing food abundance and quality or acting as outright mortality agents on the insects.[(724)] In the lowland and mid-elevation wet tropics, such effects may be generally constant throughout the year, with the possible exception of the last 2 months of the rainy season when rains are heaviest.

Guilds or complexes of herbivores associated with certain plants in strongly seasonal tropical lowland will show periodic reductions in some or all species within guilds or complexes in relation to detrimental effects on plants and insects promoted by seasonal cycles (e.g., Fig. 9.4). This model (Fig. 9.4) assumes that many tropical environments are fairly nonseasonal and wet. In the temperate zones, comparable species or herbivores, one exploiting a crop, the other a wild plant species in the same climatic regime, both exhibit an initial overshoot in abundance at the beginning of the growing season, following a brief population growth phase associated with the appearance of new meristems. The crop herbivore builds up and maintains a larger breeding population during the annual growing season than the herbivore exploiting a wild plant species, because the crop plant species is usually far more abundant per unit area, thereby providing a larger resource base for the insect population, and because a very abundant crop

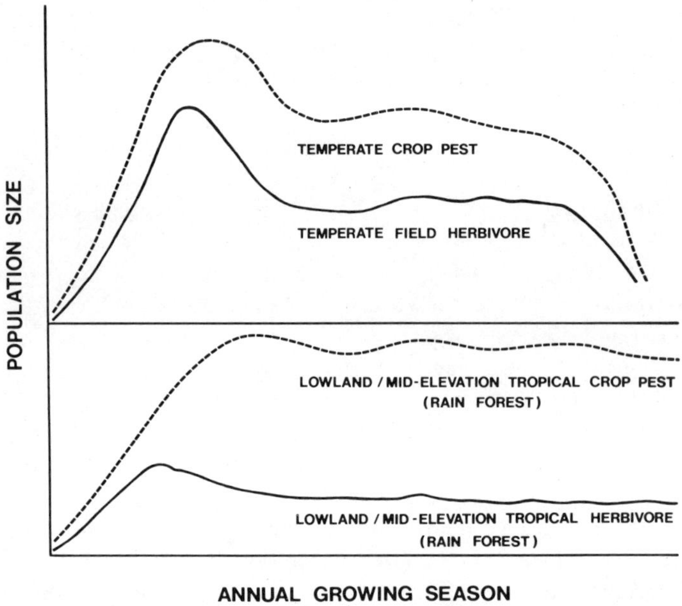

FIGURE 9.4. A hypothetical model for annual patterns of population growth and maintenance for herbivorous insects associated with forest, field, and crop habitats in the temperate zones, and lowland and mid-elevation tropics.

pest may satiate some or all of its predators. The herbivore associated with the wild plant is expected to experience more biotic interactions with other arthropods (competitors, predators) than the crop pest, and the plant will have a more patchy distribution. These ecological factors interact to keep the herbivore population, in this situation, at a lower level of numerical abundance. In both situations, population growth and maintenance are compressed into the growing season, with populations falling off rapidly with onset of winter. In the tropics, the situation is about the same in terms of relative abundance between herbivores associated with crops and those on wild host plants. But since biotic controls are predicted to be even greater in lowland and mid-altitude habitats in the tropics, there is a greater absolute difference in abundance of the herbivore between crop and natural habitats once populations stabilize within the census history. It is difficult to generalize about such effects for all insects and plants in a particular region or habitat since each species will have a different evolutionary history and different ecological requirements for maintaining breeding populations. The complex, often subtle multitude of biotic and abiotic factors affecting each individual within a single population of an insect species requires detailed natural history studies in order to determine the spectrum of variation in the ways the population as a whole responds to changes in habitat and resource availability (e.g., Fig. 9.5).

There have been, of course, attempts at developing mutant genetic lines of targeted pest species[725] as one approach to changing the impact of the insect on the crop or livestock, but such an approach is unlikely to be fruitful and promising for tropical regions, since the host organism itself will respond to such changes (i.e., perhaps become more vulnerable when the selection pressure is relaxed and thereby increase its susceptibility to another pest species). The high species densities of insects in many tropical habitats increase the chances of a

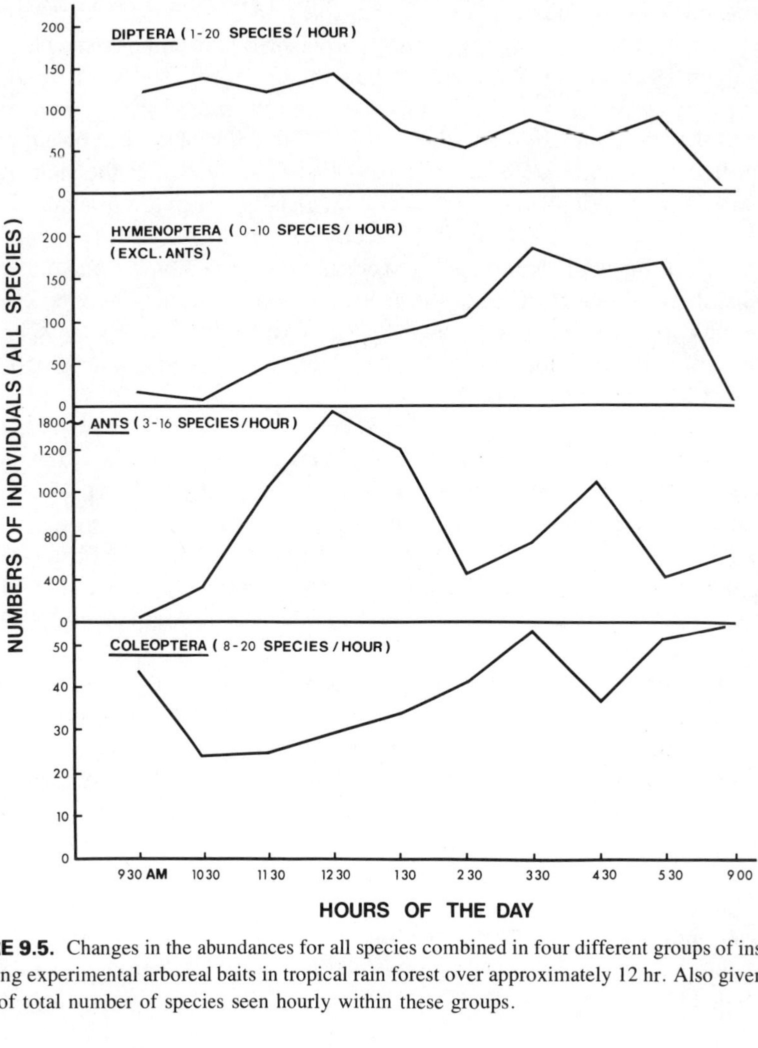

FIGURE 9.5. Changes in the abundances for all species combined in four different groups of insects exploiting experimental arboreal baits in tropical rain forest over approximately 12 hr. Also given are ranges of total number of species seen hourly within these groups.

resource being filled if another species ceases to use it or is deterred from using it. In some Temperate Zone crops, somewhat successful attempts have been made to control pest insects with insecticide applications at times other than flowering as a means of reducing the mortality of bees and other pollinators of such crops.[723] But the same bean crop as in the study of Bardner *et al.*[723] located in the lowland Central American rain forest zone would support a greater number of herbivore species than in their English field, and the populations would be high throughout most of the year as a result of the long growing season. The flowering regime would also be different. Together, these conditions buffer the herbivore guild against an insecticide, since the insects remain abundant throughout most of the year (i.e., lack the seasonal cycling of Temperate Zone crop systems) and any complex of carnivorous predators and parasites associated with them would also be knocked down by the treatment. The result would be an increase in the herbivore populations as the growing season advances. Populations of the crazy ant, *Anoplolepis longipes,* in the Seychelles can be temporarily controlled by specific kinds of toxic baits that are effective in the short run,[706] but long-term solutions to the control of such an organism are probably lacking. Such an ant species produces many colonies per unit time as an introduced species, and the manpower required to check for every colony in an area, even on an island, would be high. Short-term local control measures may be effective, but eradication of the species in the long term is not a likely prospect. The spillover effects of toxic baits on other members of the local arthropod communities have not been studied.

From detailed studies of the interactions of parasitoids with pest-labeled herbivorous insects in the temperate zones,[726–728] it is seen that density-dependent regulation of pests by parasitoids or predators works best at low densities of the host species, i.e., during periods of population census when the insect is not behaving as a pest. Parasitism tends to become obscured as an effective mechanism for regulating populations when densities soar in some systems studied.[728] Seasonal fluctuations in the abiotic components of the environment also affect the effectiveness of parasitism in regulating a host population.[727] More than one density-dependent mechanism may regulate the same population. In the humid or wet tropics in particular, densities of herbivore species in natural habitats may be kept low through the operation of several different forms of extrinsic-control agents such as parasitoids. Coupled with the energy sink associated with the physiology of detoxification and processing the defensive substances in host plants, these factors keep the herbivore populations generally low through successive generations. To handle the defensive chemistry of a host plant may require an energy allotment then made inaccessible for making more eggs. Birth rate is lowered through evolutionary time as the coevolutionary interaction of the insect and its host plant continues and changes. Such an effect, coupled with the often scattered or spatially patchy distributions of host plant populations, ensures that the density of individuals of a given species per host plant individual or clump remains low. Parasitoids and carnivorous predators may effectively keep numbers well below the carrying capacity of the plant population when densities are low. In the strongly seasonal lowland tropics, the dry season adds a dimension of density-independent mortality to both herbivores and their carnivores and parasitoids, excluding, of course, populations thriving in moist refugia. When the habitat is converted into a highly palatable monoculture of a crop species, the herbivores's phenotypic constraints associated with overcoming the defenses of a host plant are relaxed, and the thereby increasing energy available for reproduction through evolutionary time, promotes high population growth on the expanded food base. Population densities of the herbivore soar to levels that promote a complex of carnivores and parasitoids to exploit the herbivore on its new food base. However, these biotic control agents cannot regulate the populations. In the high-altitude regions of the tropics, the rigors of the physical climate would keep herbivore populations down throughout the growing season and abiotic mortality during the dry season would be high.

CHAPTER 10

BIOGEOGRAPHICAL AND REGIONAL EVOLUTIONARY-ECOLOGICAL EFFECTS ON THE MAINTENANCE OF TROPICAL INSECT FAUNAS

A Brief Perspective

A central theme in the development of a unified approach to understanding the high species richness, population dynamics, and habitat associations of insects in the tropics in the foregoing chapters has been the effects of local selection pressures on individuals in the populations of single species, and the interactions among species as the basis for coevolved sets of interactions underlying the composition of species assemblages, guilds, or communities.

10.1. HISTORICAL EVENTS AND ECOLOGICAL PROCESSES

One might be tempted to conclude that the distribution patterns of species in particular communities can be explained fully in terms of contemporaneous ecological effects. Yet there must also be a historical or evolutionary perspective to explaining the associations of insects with certain regions, habitats, and communities in the tropics. Thus one also might conduct an exhaustive field study of the ecological properties of a group of sap-feeding insects such as plant hoppers in a particular climatic zone of South America, determining that a high degree of host plant specificity is a key factor in accounting for the species composition of the group in that region. However, this is not an explanation as to why these species occur in the region, and to what degree have regional or biogeographical effects molded the initial occurrence of these forms in the region. The fact that different regions of the world have experienced different geological histories, and the absence of some major groups of insects from some Old and New World tropical regions or relatively poor representation of other groups, is an indication of the role of biogeographical events in determining the broad distributions of insect faunas, tropical and extratropical. The general observation that some regions of the terrestrial tropics contain more species than others, and more than do Temperate Zone regions, is really a question of the interplay of biogeographical events with the dynamics of local communities. The sum total of such interactions is that regions of the tropics with high species richness are characterized by faunas with higher rates of speciation or reduced rates of extinction once populations of each species come into some sort of equilibrium with resources in the habitat. Tropical regions with great species richness are also considered those with benign and predictable environments.[104]

The ecological question incorporates an attempt to account for how each species in the local community maintains a breeding population in terms of how each species exploits resources. The evolutionary question incorporates such information but goes further: it attempts to explain why these species occurred there in the first place. Geographical distributions of single species or whole groups are explained in terms of ecological and broad-scale climatic and other environmental patterns, and how such patterns change through evolutionary time, causing changes in the distributions of species and in the morphology, behavior, and ecology of species. Tropical regions consist of mosaics of different topographic and climatic regions. The large amount of heterogeneity in the climate, vegetation, topography, and soils[729] of South America points to the role of regionality in determining the distributions of organisms over the land mass. The recognition of distinctive soil regions[729] indicates that plant communities will also be very distinctive among different regions, in turn influencing the distribution of herbivorous insects associated with individual plant species or plant families. Thus we might expect that tropical mountain tops above 4000 m will have very different insect faunas from intermediate elevations of about 1000 m, and that lowland dry zones will support different insect faunas from lowland tropical rain forest zones. Superimposed upon these vertical and horizontal patterns of environmental heterogeneity over large areas are regular and irregular changes induced by shifting climatic conditions, regional catastrophes such as earthquakes and volcano eruptions, and the impact of humans in clearing large expanses of forest for agriculture and lumbering. It is also recognized that natural geological cycles in the earth result in changing distributions of tropical rain forests, expanding and contracting in a kind of pulsation cycle corresponding to cycles of climatic change through geologic time. We are referring to a variety of regional and biogeographical effects that operate in different time frames, influencing distributions of organisms in different ways.

There have been numerous in-depth studies on the systematics and biogeographical origins of specific kinds of insect groups in tropical regions[730] as well as attempts to explain phylogenetic patterns of specific groups in terms of discernible regions within the New and Old World Tropics.[731] Such surveys have emphasized the great differences in richness of floras and faunas among distinctive regions in the tropics and the correlations of such patterns with abiotic factors such as changes in topography and soil profiles.[731] These studies have provided the framework for both ecological and biogeographical studies in tropical regions.[732] Thus as developed below, it is believed that cycles of wetness and aridity during the Pleistocene promoted cycles of expansion and contraction in Amazonian rain forests. Insect species trapped in forest islands[469,623–626,733] must have experienced drastic changes in selection pressures associated with distribution and availability of resources, in turn affecting population size and degree of ecological specialization among genotypes within the local population.[734,735] Such isolation followed by periods of new contact helps to explain the present-day distribution of species within different groups throughout much of tropical South America. Brown,[622–624] for example, has discovered many distinct refugia of *Heliconius* butterflies in South America and emphasized the role of historical events such as the pulsing cycles of climatic change, in determining the overall evolution and distributions of species. We need to recognize such events and weigh this information with an examination of current ecological properties associated with the populations of individual species, their geographical distributions, their habitat associations, and the spectrum of resources utilized (e.g., larval host plants) in attempting to erect a model for the local and regional distribution of the group. Vane-Wright[736] has emphasized the importance of ecological factors in the speciation of butterflies, especially in the tropics, where more intense patterns of spatial and temporal heterogeneity of resources appear to be promoting ecological diversification in local communities. Vane-Wright stresses the added role of behavioral mechanisms such as mimicry in promoting species richness in tropical butterflies.

Cycles of climatic changes influencing the size and distribution of suitable habitats for particular species, as exemplified by Quaternary refugia model for Amazonian floras and faunas, determine the degree to which the gene pools of a species are isolated at a given locality and time. Any breeding population of an insect species will contain different kinds of genotypes, some of which are adapted to conditions in the present environment, others of which promote the colonization or migration to new environments. Shifts in directional selection will determine in part the distribution of such classes of genotypes in the population. Range extensions for a species, following the appearance of increased suitable habitat, such as during a period of rain forest expansion under the geological model, are accompanied by a reduction in cohesiveness of the breeding population and the increase in frequency of genotypes capable of local or extended migration to locate and colonize new resource patches. The environment, under such conditions, may be encountered as a series of disruptive changes in physical and biotic conditions, or as a gradual (clinal) shift in such properties. The degree to which a species in the process of range extension encounters other populations of the same species, or populations of closely related or ecologically similar species, will influence the rate at which the species undergoes range extension. Disruptive environmental gradients may present the challenge of new sets of resources and can function as ecological filters in blocking range extension if the species is not able to exploit them.

Brittnacher *et al.*[(737)] noted considerable geographical variation in the electrophoretic analyses of 16 loci in several species of the butterfly genus *Speyeria* in California. They tentatively conclude that selection may be maintaining the observed patterns of genetic divergence within each species over the geographical regions surveyed. Species widely distributed over altitudinal gradients in the tropics are expected to be experiencing very different forms of selection pressure, both within and between selection established by abiotic and biotic components of the environment. As Brittnacher *et al.*[(737)] indicate, the degree to which a local breeding population or unit of a species is residential, that is, less prone to dispersal, the greater the differentiation locally in genetic structure from population to population. Janzen[(46)] has called attention to the divergence in ecological characteristics of tropical mountains versus lowlands. Such divergences probably result within a species found in both areas, in either (1) expression of a flexible response system to environmental changes at the physiological or behavioral levels, particularly if there is considerable dispersal of individuals over the gradient, (2) local genetic (evolutionary) diversification, particularly when dispersal is reduced or absent and (3) a combination of (1) and (2). Attempts have been made to explain the geographical distribution of whole butterfly communities in terms of limitations in geographical range imposed by local, specialized habitat requirements in mountain regions.[(738)] Local specialized ecological conditions affecting a species result in ecological divergence and geographical patchiness.

Both abiotic and biotic factors will determine the spatial distribution of an insect species, both locally and regionally, although different forms of these environmental factors may be involved. For a given species, a generalized phenotype will result from regional influences, e.g., abiotic factors such as ambient temperature, humidity, and light. A variety of other abiotic factors will also influence the distribution of a species over large areas. A region can also be viewed as the collection of different habitats encompassing large land areas horizontally (e.g., the tropical rain forest flood plain region of northeastern Costa Rica) or vertically (e.g., a transect through an altitudinal gradient extending from the top of the Peruvian Andes down into the Amazonian rain forest).The distribution of a butterfly species, for example, over a region will be a function of primarily (1) the ability of the species to withstand possibly marked changes in physical features of the environment across the region surveyed, (2) the degree to which larval and adult resources are distributed across the region in question, (3) the types of interactions with other consumer species including potential competitors and predators, and (4) historical events. The distribution of the species across the region is a dynamic condition in which one or more distinct breeding populations experience a changing set of selection pressures arising from interactions with the species in the habitat or habitats occupied by the population(s). Thus one expects that the actual boundaries of the species range are changing. If the species is successful in occupying an increasing number of habitats through colonizing episodes, there will be a range extension if the original populations remain intact as well. In addition to a host of abiotic factors, the primary biotic factor in the case of insects for determining the geographical range of the species in question is the distribution of food sources. Other biotic factors, such as interactions among species within communities and trophic interactions, will be characteristic of the particular habitats occupied by the species. The broad geographical distribution of the species thus becomes a matter of determining the past and present ranges of suitable environments to be occupied by the species. This is the historical component of species richness in the tropics.

Prolonged periods of drought in some subtropical regions can result in very different responses by butterfly populations for a species widely distributed, with some populations going extinct, others remaining stable in size, and still others actually increasing in size.[735] The crucial factor appears to be the response patterns of both the insect and its host plants to drought conditions. Response patterns of this kind vary considerably with locality and habitat conditions.[735]

Janzen[46] suggests that organisms experiencing a severe tropical dry season should be better adapted to crossing a dry zone from a wet zone in the tropics than a similar organism occupying a rather nonseasonal wet zone. Changes in temperature along a gradient are probably a more effective barrier to dispersal for organisms in the tropics than absolute height of a mountain on which the gradient occurs.[67] Such a phenomenon helps to explain the frequently observed high degree of endemism in tropical mountain top insect faunas.[739] Various orders of insects often show high percentages of endemism (greater than 60%) for mountain top assemblages in the tropics.[740] Such patterns are explained in part by the distribution of suitable resources for individual genera and species and also by the effective barriers to dispersal subsequent to colonization due to temperature and temperature-related effects creating physical barriers to movement (e.g., moisture gradients). In the temperate zones, insects experience mountains as less of a temperature gradient than in the tropics because these species are already adapted to severe seasonal fluctuations in temperature and other abiotic factors, lessening the effect of spatial gradients in temperature.[46] Thus endemism is probably greater in insect faunas along altitudinal gradients in the tropics than along comparable gradients in the temperate zones. If temperature zones are equated with regions, the tropics tend to have a greater diversity of geographical zones generating a high regional diversity of insects and other organisms. The ecological factors associated with specific differences between habitats making up the geographical range of the species constitute the ecological component for explaining species richness, and much of the preceding chapters were devoted to this aspect of tropical insect diversity. The following sets of ecological factors are operative:

1. *Regional*
   Broad abiotic factors: temperatures, humidity, light, soil cover, etc.
   Biotic factors: distribution and abundance of food resources (i.e., food plants for phytophagous insects). Together, these factors select for a generalized phenotype capable of occupying a range of environments or habitats.
2. *Local (Habitat)*
   Specific sets of physical features: local soil nutrient profiles, topographic effects on moisture and light, abundance of shelters and other resources affecting fitness.
   Biotic components of the niche: food and other resources, predators and parasites, competing species. Together, these factors mold the life history pattern of the species in each habitat, determining the patterns of phenotypic allocation of resources to mating, foraging, escaping from predators and parasites, and buffering against physiologically stressful changes in temperature, moisture, etc.

Both general sets of factors determine the strategies of environmental response given by Slobodkin and Rapoport[(99)] and discussed in Chapter 1. As discussed in Chapter 5, it is also important to consider that patterns of environmental response often differ greatly between different habitats in a region. White[(741)] has considered this phenomenon as the ''area-effect'' component of speciation. Different habitats produce different selection pressures, and when a species has a continuous population over a large area, and considerable genetic and phenotypic variation exists in the population, speciation proceeds by localized area effects in which populations or subpopulations associated with different habitats diverge phenotypically. The vicariance model (refugia or core area) of speciation essentially predicts that when a species with a large, continuous breeding population eventually becomes fragmented due to the formation of new barriers over the region (e.g., geological, climatic), the resulting, isolated subpopulations may diverge into different subspecific forms and perhaps into new species. Thus the vicariance model of speciation is a logical extension of area effect speciation. Biogeographers working in the tropics are becoming increasingly interested in this approach to understanding the dynamics of speciation. Thus one envisions that biogeographical patterns of particular groups of species or individual species are really the result of evolutionary or historical events affecting regional distribution,[742] and the local effects of how various sets of ecological factors in distinct habitats determine the tenacity for a population to maintain itself through time. Surveys of insect groups along altitudinal gradients in tropical countries (e.g., Mani and Singh[740]) reveal discontinuities in faunas as a function of elevation and past geological events affecting migration and distribution of species. Yet it is necessary to look beyond such data and attempt to examine the question of abrupt or gradual changes in species composition and richness for particular groups on a regional basis.[743] Some species of alpine Lepidoptera exhibit considerable geographical variation,[744] presumably the result of differing selection pressures and overall environmental conditions along altitudinal gradients. Alternatively, the recent studies of Noonan on the carabid beetle, *Notiobia peruviana,* in the Andes of South America, support the view of vicariance origin of new subspecific forms. Noonan censused populations of this beetle over about a 4000-m-altitudinal transect from sea level to the high slopes of the Andes and found evidence to support the vicariance model of speciation in this insect.[745]

In this chapter an emphasis is placed on the ecological properties and distribution patterns of insects along altitudinal gradients in the tropics, with further emphasis on the Neotropics. There is an attempt to explain the adaptive responses of insects to altitudinal gradients as well as a brief examination of biogeographical effects as related to the evolutionary differentiation of high elevation insect faunas from the tropical lowlands.

10.2. ALTITUDINAL AND OTHER MACROGRADIENTS IN TROPICAL REGIONS

MacArthur[16] stressed that one must consider the time scale when attempting to account for changes in the flora and fauna of a region. For example, the time scale of thousands of years encompasses the Pleistocene with fluctuating periods of dryness and wetness over large areas such as Amazonian rain forest. The studies of Vanzolini and Williams[746] and Haffer[469] imply the existence of core areas or refugia in which tropical rain forest remained during periods of dryness. Many such core areas of forest islands were scattered over large regions of South America, especially below 900 m. The inference of core areas or refugia arises from present-day patterns of evolutionary radiation in different groups of plants and animals, including insects,[622–624] and from paleontological data. Core areas remained wet during periods of Pleistocene and post-Pleistocene dryness and functioned as reservoirs for many species. Different cores had different ecological conditions, presumably since communities became fragmented as forests contracted in size and patches. Physical conditions may have been uniform in each core area, but conditions varied considerably between core areas, promoting evolutionary radiation as the result of different forms of selection pressure to physical and biotic components of core areas (e.g., the vicariance model).

The existence of specific core areas cannot be proven but can be strongly infected from existing patterns of distributions of species or groups of species. The studies of Darlington[747,748] stress the relationship between size of a region and the number of species found there: on the time scale of millions of years, continental areas with mild climates accumulated more species than other areas, and such tropical regions presumably had more successful colonists to invade other regions. Competitive superiority within communities is held to evolve most frequently in larger regions where communities are larger by virtue of occupying a greater area of habitat(s), and such species may include good colonists for island invasion. Yet selection for genotypes to be optimal colonists does not necessarily include competitive superiority,[87] since newly arriving colonizing species on islands often do poorly if the island already contains ecologically similar species that are good competitors.[91] Whether we refer to actual archipelagos[742] or ecological islands on mainlands,[93,749] the initial invasion of a colonizing species into the island communities does not necessarily imply superiority in becoming integrated into the existing communities. Success will be largely a function of the existing species composition of the community[750] and the degree to which competitive genotypes are being selected for. Thus in the inferred cycles of expansion and contraction of Amazonian rain forests the original pattern of high species diversity accumulated in the lowlands of tropical South America became subdivided into refugia periodically, and each contraction phase generated a size distribution of such refugia, with species richness and community organization and complexity varying as a function of sampling error and size of the core area.

Larger core areas probably had a greater number of species and presumably experienced a greater amount of evolutionary radiation for particular taxa.[751] I refer to such patterns as an ''unstable ecological mosaic'' in the sense that each core area was different in size, species composition, community structures, and physical conditions from others, and that each must have passed through a period of nonequilibrium communities following its formation in the cycles of climatic change. Possibly such communities, because they represented sampled portions of other communities that evolved in the intact forest environments, and because of insufficient time, never attained equilibrium conditions in the core area state. Yet the studies of Brown[622–624] indicate that the existence of core areas is implied in distributional patterns of groups of closely related species such as *Heliconius* or ithomiine butterflies.

MacArthur[16] asserts that the high species richness of insects and other invertebrates in the tropics results from (1) their ability to maintain denser populations than vertebrates, (2) lower extinction rates, (3) small body size allowing speciation in relatively small portions of habitats, and (4) physiological limitations of invertebrates to invade Temperate Zone areas, thereby lowering immigration rates. MacArthur states that invertebrates in the tropics exhibit a high proneness toward speciating and that many forms are highly endemic to certain regions. Distribution patterns of endemic species of insects commonly seen along altitudinal gradients in the tropics are best interpreted as the result of steep temperature and moisture gradients blocking immigration among zones.

Ability for an insect species to successfully invade another zone on an altitudinal or horizontal gradient will also depend upon the community being invaded. If a community has saturated the resource base within a particular zone on the gradient, invasion by a colonist is improbable.[15] Such effects may be seen at the level of whole communities or faunas: where the assemblages meet along the gradient, a narrow zone of chaos[15] in the replacement area of the environment acts as a form of diffuse competition and the structure and organization of each community is maintained. The monarch butterfly may represent a species in some form of diffuse competition with the queen at montane overwintering sites in Mexico: the monarch enters the habitat temporarily but does not reproduce there, while the queen may do so. At the same time, the monarch is only a transient member of Temperate Zone communities, since it is not adapted to withstand the physical rigors of the northern winter. The monarch shuttles between two very different kinds of communities, one tropical and one temperate, but reproduces only in one, and generally does best where the queen is absent. Truly residential lowland tropical populations of the monarch are presumably of sufficiently low density (immature stages) so that competition with the queen is probably low. The original selection pressures favoring the evolution of the migratory behavior in the monarch presumably consisted of a mixture of competition from the queen under certain conditions, effects of seasonality on larval and adult resources in dry forest regions of Central America, and the great evolutionary radiation of the larval host plants into the Temperate Zone of North America.

Yet the monarch is the exception in the sense that one can recognize whole insect faunas restricted to different regions in which these biogeographical realms coincide with present or past distinct physical boundaries of continents and islands, and areas within continents. Within such realms, it is necessary to discern subregions containing distinctive assemblages of species. Going from the lowland (0—900 m) tropical rain forest on the eastern slopes of the Andes in South America to the snow line at mountain tops (about 4500 m) entails passing through several distinctive vegetational and abiotic zones,[739] each with a distinct set of insect species and with some degree of overlap in these faunas, especially at the lower elevations. Mani[739] provides considerable detailed examples of the insect groups found in the Andes at various elevations and, in some cases, attempts to explain biogeographical origins.

Current explanations for the origins of tropical biotas stress the apparent gene flow that occurred between populations on different tropical continents into the mid-Tertiary as the result of plate tectonics.[733] The advent of a modern theory of island biogeography[341] together with inferences about the effects of Pleistocene climates on Amazonian rain forest vegetation[469,623,733] has portrayed the effects of mainland (continental) island habitats (core areas or refugia) on the present-day distribution of flora and fauna in the Neotropical region. Various types of coevolution between plants and insects must have occurred in these forest environments, most notably (1) the evolution of herbivore defense systems in plants, (2) partitioning of nectar and pollen sources through specialized pollination systems, and (3) the evolution of mutualisms between ants and plants, the latter including the elaboration of extrafloral nectaries. Furthermore, the structure and organization of insect communities in tropical forest habitats, particularly in lowland (*tierra caliente*) and mid-elevation (*tierra templada*) zones, and the accumulation and cyclic fragmentation of whole insect faunas (biogeographical effects) during the Pleistocene, have also been a major factors in determining the geographical distribution of insect species within tropical regions.

Prance[733] has also emphasized the geological differentiation of Amazonia into distinct forest types, based largely upon the ecological constraints arising from the structure and composition of soils in different regions and along topographic gradients. These forest types vary in terms of degree of flooding and composition of the vegetation, establishing distinct areas for the accumulation of different insect communities. Soil type, slope, and elevation, coupled with the effects of flooding, have created the mosaic of forest types in Amazonia, of which the terra firma forest represents the major portion and one of the richest in terms of floral composition.[733] Some forest types, because of their relative isolation within others, promote considerable endemism, also the result of these types of habitats being scarce and patchy in distribution. Brown and Benson[752] discuss the impact of the *campina* forest habitat on endemism in *Heliconius* butterflies in Amazonia.

Prance[733] accounts for four distinct dry periods during the Pleistocene in which the Amazonian forest region was reduced, each time to a different degree. Such pulsations of forest reduction created a complex series of responses in insect species found in the region, conceptualized as periods of evolutionary radiation followed by reduction in species richness and increased endemism. Thus many species today have seemingly puzzling disjunct geographical or regional distributions that can be explained by the refugia or core area model. Rather than observing that a species has a rather transitional pattern of differentiation along horizontal or vertical gradients over large areas, the more characteristic pattern is that of disjunct populations, a condition explained by the refugia model. Some or all of these populations for a given species reflect the location of core areas during the recent geological past. Thus Amazonia and other regions within the Neotropical region, including Mexico and Central America, are viewed as highly heterogeneous regions in the sense that they contained numerous core areas that regulated the subsequent pattern of range expansion throughout the region. The problem for the present-day biogeographer is to reconstruct the spatial (geographical) routes followed by individual species or groups of species (genus, tribes, etc.) before, during, and following the existence of the refugia, and the amount of time involved to reach a certain kind of distributional pattern.

The biogeographer approaches such questions through the lenses of pulsations in climate and resulting effects on the size and composition of habitats occupied by the species being studied. Historical events are therefore held to be of prime significance in the determination of geographical distributions of species,[753] but it is also necessary to consider contemporary or ecological phenomena in interpreting such patterns. The refugia model generates a complex fabric of species distributions in which allopatric populations arise for each species and different spheres of selection pressures promote eventual speciation.[754] Brown and Benson[752] describe the subspecific differentiation of *Heliconius hermathena* into four distinct and separate savannalike habitats surrounded by forests. Prance[733] emphasizes the role of historical factors in the shaping of forest types and their distribution in Amazonia. Yet it is necessary also to view distributions of species in terms of ecological processes such as seed predation, pollination requirements, and niche availability. These topics were examined in previous chapters.

The message here is that historical and ecological effects, at the time the species are being studied, must be taken into account together to explain existing patterns of species. The historical component concerns the impact of past major climatic changes on floras and faunas, while the ecological component relates to interactions between species in communities and across trophic associations in tropical ecosystems. But the ecological component is not exclusive of the historical: specific kinds of ecological interactions during the past also contribute to the present-day ecological traits of individual species.

Fittkau[22–24] notes the great abundance of insects and other arthropods found in South America and discusses the evolutionary development of these faunas as a function of geographical isolation from other regions, promoting endemism. South America is divided into two major regions: the tropical Guianan–Brazilian region and the more temperate Andean–Patagonian region.[755] This tropical region, encompassing lowlands and mid-elevation zones, possesses a fauna overall that appears to be derived from that of Africa,[755] while the temperate region bears closer similarities to that of Australia and New Zealand. Ecologically, the tropical faunas are adapted to forest environments while those of the temperate zones are adapted to open savanna or prairielike environments. It is the tropical region that contains the greatest number of species, including insects, while the

Temperate Zone has fewer species. Both regions contain high percentages of endemics within different groups, although the Temperate Zone of South America is considerably lower in species numbers than other comparable Temperate Zone regions.[24] As discussed by Fittkau[24] and Prance,[733] it is the tropical region that exhibits subregional diversification into distinctive faunistic areas, largely the result of geological history and resulting topographic and soil profiles. Prance[733] outlines in detail these subregions in terms of forest types in Amazonia.

Geological and paleontological data indicate that South America, part of Central America, and the Antilles were isolated from North America by a Pacific–Atlantic marine channel during the Tertiary.[24] The land bridge of Central America seen today arose in the Pliocene, about 2–3 million years ago and served as a vehicle of faunal exchanges between the two continents. Evidence from arthropods for such exchange is seen in the ectoparasites of some mammals.[51] Contemporary mountain tops in Central America disclose ecologically isolated and diversified Nearctic forms as islands in the otherwise tropical lowlands. During the Tertiary some parts of present-day Central America were ''tropical sections'' of the Nearctic region.[24]

Illies[756] discusses the biogeography and ecology of some freshwater insects in tropical mountains and lowlands, emphasizing that aquatic communities are more buffered against fluctuating ecological conditions over short periods of time and thereby provide a more accurate view of past biogeographical events of larger time scales, accounting for present-day distribution patterns. The Plecoptera and other groups attain their greatest diversity as cold-adapted (oligostenothermal) forms in tropical regions of South America but are poorly represented in the temperate region of that continent. Some 44 genera and 13 species occur in the Neotropical region, with the Andean mountains alone accounting for 31 genera at the northern and southern ends and these faunas being separated by the desert barrier of northern Chile and western Peru. Ten other genera are endemic to mountains of eastern Brazil and two others to mountains of Guiana and eastern Venezuela,[25,756] Illies[25,756] makes an interesting biogeographical comparison of the montane Plecoptera faunas of South America and Europe. Both regions contain a high number of monotypic genera, 15 of 32 genera in the Neotropical region and 9 of 29 in Europe, and there are similar numbers of genera with 2–3 species and very few with 4–5 species in both regions. The European fauna also contains a higher number of species than the Neotropical region. Similar geological or historical events have apparently determined the faunistic complexity of Plecoptera in both regions: alternating patterns of climate during the Ice Age promoted the interplay of isolation and subsequent breakdown of distribution barriers after the advance and retreat of glaciers.

The European fauna in particular is characterized by many endemic species restricted to certain mountain regions, but so is the Andean fauna.[756] Illies[756] suggests that high mountain areas of Europe are extremely isolated and are positioned in an east–west axis, that is, against the axis of glacial pulsations and parallel to the leading edge of glaciation. During periods of glaciation each of these mountain areas was included in the movement patterns of the aquatic fauna, but following periods of glacial retreat each of these mountain areas became a refugium in its elevated parts resulting in isolated populations and evolution of endemic species. But in the Andes, the mountains are positioned north to south, the same direction as the axis of glacial movement in the Quaternary. Thus every glacial advance thrusted the fauna toward the equator along the slopes of the Andes, and during periods of glacial retreat the fauna recolonized extensive areas to the south, and to a lesser degree the north, and recovered its original distribution patterns without long-term isolation. In a comparative sense, the Andes provided insufficient opportunities to attain the degree of endemism seen in the mountains of Europe, even though pronounced cycles in climate during the Quaternary resulted in the appearance and dissolution of forest refugia. Illics[(756)] maintains that the European montane fauna is oversaturated with species and represents a unique situation from the biogeographical standpoint. Lowland Neotropical areas contain a greater number of species of aquatic insects but fewer genera,[(756)] indicating a strong ecological component to explaining the species richness of these communities. Ecological processes such as competition and niche diversification may be promoting considerable speciation in the lowland tropics. Mani[(739)] provides a detailed discussion of the tropical mountain environment as affecting the vertical distributions of terrestrial insects, stressing the evolutionary limitations of cold, dry conditions on the upper limits of faunas and also as affecting the availability of food resources.

The bulk of high-altitude insects, defined here as those forms above the natural forest line, are scavengers or omnivores on a variety of organic debris, and food chains are expected to be relatively simple compared with those in the terrestrial and arboreal communities at lower elevations. The tropical mountain is analogous to the high latitudes in that harsh conditions tend to select for generalist species, dampen overall rates of speciation, and allow successful species to develop rather large annual populations. Diurnal, rather than seasonal, cycles in temperature, together with local topographic and climatic effects such as rain and wind shadows, provide the major physical features of the high-altitude environment of tropical regions influencing the distribution of insect species. A predominance of some forms such as carabid beetles above 2000 m reflects the suitability of such regions for species that exploit a variety of similar and abundant resources (decaying vegetation, seasonally available seeds of grasses, and limited predation on other invertebrates). Body-size gradients along altitudinal transects in the mountains of tropical regions are believed to be the result of reduced dessication stress, and perhaps physiological stress associated with reduced rates of growth and less nutritive food resources at higher elevations. Carabid beetle body size shows a gradual decrease in the mean for a given species with increased elevation.[739] High-altitude insects in general appear to be specialists in the sense that they occupy a stressful-environment zone that excludes, primarily for physiological reasons rather than ecological saturation of communities, many of the forms integrated into lowland and mid-elevation communities. Yet within the zone, the species must clearly be generalist in terms of niche effects: such organisms must be able to switch to different kinds of food resources as they become available, seasonally or otherwise. Harsh environments select for generalist species.[104] The reader is encouraged to examine the detailed treatise of Mani[739] pertaining to the biogeographical and ecological properties of insects at high altitudes in tropical regions.

10.3. NOTES ON ALTITUDINAL GRADIENTS IN THE TROPICS

The study of Silberglied *et al.*[720] of the butterfly genus *Anartia* in the American tropics provides an interesting example of geographical differentiation of an ancestral form into distinct evolutionary lineages. Silberglied *et al.*[720] suggest three distinct phylogenetic lineages to *Anartia,* one of the most widespread and common nymphalid genera over lowland rain forest, lowland dry forest, and mid-elevation moist or wet forest habitats in Central and South America and the Caribbean. *Anartia fatima* and *A. amathea* are a pair of sister species that evolved from an originally widespread ancestor when populations of this form were isolated during "the Tertiary subsidences of the Panamanian isthmus, after which time the distinctive colors and patterns of the two species evolved."[720] As with other butterfly groups, there has been secondary contact between the species in Panama in recent geological history. A second lineage comprises the sister species of *A. chrysopeles* and *A. lytrea* of the Caribbean islands, with distinctive isolation on islands producing the present-day species. *A. jatrophe* (a third lineage) is a widespread species exhibiting considerable geographical variation but without speciation taking place, because adults are mobile and gene flow is high.[720]

Anartia butterflies in general possess ecological traits making many populations representative of colonizing episodes in secondary tropical habitats where the butterflies exploit weedy species of Acanthaceae as larval host plants.[322,720] Another system deserving of similar analysis is the apparently sister pair of species of the genus *Metamorpha,* another nymphalid Neotropical group associated with weedy Acanthaceae and disturbed habitats.[294,317] In Costa Rica, there appears to be some altitudinal separation of *M. stelenes* and *M. epapha,* the latter species being common above 500 m on the eastern slopes of the Cordillera Central and the former species abundant in the lowlands (A. M. Young, unpublished data, 1969–1980). Both species are capable of exploiting each other's larval host plants (several genera and species of Acanthaceae, including some also used by *Anartia*). At about 220 m in northeastern Costa Rica there is a contact zone between the species, but with *M. stelenes* the more abundant. *Metamorpha* in Central America is also a successful colonizer of coffee plantations where the larval host plants survive beneath and along the borders of coffee bushes.[316] Detailed morphological, behavioral, and ecological studies of both species, in the spirit of Silberglied *et al.*,[720] are needed to determine how closely related the species are (wing colors and patterns differ greatly between the two species, much more so than between *Anartia* species), and to what degree populations are isolated from one another along an altitudinal gradient or along horizontal gradients. Both species share several general features of their life cycles and behavior with *Anartia*. Young and Muyshondt[316] noted considerable geographical and local variation in wing color patterns and intensity of colors in *M. stelenes,* which appears to be the more widely distributed of the two species in Central America. Tentatively it is suggested that *M. epapha* represents the geographical differentiation of the genus into mid-elevation and lower montane tropical rain forests in Central America and that *M. stelenes* is the lowland form, bearing mimetic resemblance to the heliconiine *Philaethria dido,* also abundant in lowland tropical rain forests.[300] Patterns of evolutionary differentiation of the genus *Morpho* along altitudinal gradients in Central and South America[176] also deserve detailed study to determine the probable pathways to speciation.

The survey studies of mid-elevation and montane butterfly faunas in some parts of Peru[757,758] suggest the need for more exhaustive studies on the distribution and ecology of individual genera and species. Although Mani[739] states that our knowledge of Andean Lepidoptera is good,[739] few data exist on the ecology of individual species and biogeographical distributions, save for the recent studies of Shapiro.[759–761] Lamas[757] has made a survey of coastal-region Lepidoptera in Peru west of the Andes, and such information is useful in determining the uniqueness of the endemic Andean fauna. In addition to survey-type field studies on insect faunas across altitudinal gradients in the tropics, we need detailed studies of single genera and species (e.g., Silberglied *et al.*[720] used such an approach for asking questions about biogeographical patterns of *Anartia* in general) with an attempt to determine the environmental factors that influence distribution over the gradient, the degree of isolation among populations of a species, and the changes in communities of insects and other arthropods along the gradient. The recent study of Noonan,[745] a detailed examination of historical and ecological factors that determined the present distribution of forms of *N. peruviana* Deyean, represents one of the few studies of its kind.

One major factor influencing altitudinal changes in distribution of insect species in the tropics is the changes in the flora, since a majority of insects are associated with vegetation or its decomposition products. Studies (e.g., Buckley *et al.*[762]) have clarified the role of stressful environmental factors on the physiology of tropical upper montane rain forest plants. Gentry[763] discusses the transitions in plant communities along elevation gradients in Central and South America, emphasizing the higher species richness of premontane tropical rain forest than other formations such as tropical moist forest. Janzen *et al.*[412] conducted sweep-sample studies of insects and other arthropods along an altitudinal transect in the Venezuelan Andes (200- to 3600-m gradient in secondary vegetation) and found that the greatest number of insect species occurs at intermediate elevation (about 1000 m) and that species richness declines with increasing elevation beyond this point, with some species becoming very agundant at the higher elevations. Janzen *et al.*[412] also found a pattern of reduction in mean body size of insects with increasing elevation. Interestingly, Janzen *et al.* found that Diptera and parasitic Hymenoptera are not reduced in species richness with the proportion in reduction found in other groups, a pattern perhaps related to the ecological requirements and body-size distributions of these two groups: both groups are characterized by relatively small body size and therefore perhaps would not respond as much to increasing elevation as would other groups in which body-size range was much greater. Furthermore the larvae of both groups may be generalists on a variety of resource types, making these insects ecologically flexible to thrive over an altitudinal gradient. Mani[739] suggests that montane insect faunas, especially those of the upper limits, do not exhibit adaptations to seasonal changes in climate as do lowland forms. Buskirk and Buskirk,[523] however, found considerable change in the species diversity of insects with the annual seasonal cycle in lower montane rain forest in Costa Rica. Upper montane areas are presumably less seasonal than lower montane areas in the tropics. In Papua, New Guinea, species richness of moths declines drastically from 220 to 2800 m, and any higher than this there is no evidence of seasonal changes in moth abundance or diversity.[513]

The colonization of high elevations in tropical regions by various groups of insects is a function of the amount of genetic variation in already established populations, especially if an ancestral or contemporary form is highly adaptable to lower elevations. The evolutionary history of the species and the amount of time passed since the montane habitat developed are additional factors influencing colonization. Noonan[745] discovered that *N. peruviana* occurs from sea level to above 4100 m in South America and has large amounts of morphological variation, and that the observed pattern of variation is accountable in terms of the vicariance model. Mani[739] has stressed the strong transient component to the high-altitude insect faunas resulting from wind patterns and topography. Yet within the transect of about 500 to 2000 m, distinctive changes in the resident communities will be associated with living vegetation, litter cover, and soil. The studies of Campbell and Dearn[765] with grasshoppers in the mountains of Australia reveal that species occupying high-elevation regions are successful colonists because a portion of the phenotype was capable of responding to the distinctive ecological conditions of the higher areas. Clearly, as their studies show, a great many morphological characters remain invariant over altitudinal transects, while others show clinal variation. Therefore, it is important to distinguish between the portion of the existing genotype capable of responding to new environmental conditions that may be associated with a range extension over an elevation gradient and that portion that does not change.

Parsons and Bock[766] have discussed the relative capacities of various species of *Drosophila* over temperate and tropical regions of Australia to withstand very different physical conditions of the environment. They have emphasized the local differentiation of rain forest forms to be sensitive to desiccation problems, while the more cosmopolitan forms exhibit intermediate levels of preference in such gradients as ambient temperature and humidity. Their surveys of different forest habitats within the humid tropical zone of North Queensland revealed that the more floristically diverse forest habitat associated with basaltic soils supported a greater species diversity of *Drosophila* subgenera. The basaltic forest is characteristic of lowland tropical forest in this region of Australia, while higher-elevation forests are on granitic soils.[766] The range in altitudes studied was from less than 400 m to 1600 m. Parsons and Bock[766] report the six distinctive kinds of forest found along this gradient. They point out that all of the major subgenera (five) of *Drosophila* are represented by several species along this gradient from the humid tropical zone, while the nontropical fauna of southern Australia is characterized by the predominance of one subgenus. The tropical *Drosophila* fauna is maintained by different subgenera and species exploiting different kinds of food resources.[766] Miskimen[490] studied the relationship between altitude, time of flowering, and appearance of soldier beetle populations on flowers along an altitudinal transect in Colombia. Timing of flowering is increasingly postponed with increasing altitude for the various Compositae used by soldier beetles (*Chauliognathus proteus*), and the beetles appear progressively later on the flowers with increasing altitude. In populations found in the cool, moist highlands, beetles are darker in color than in the lowland tropical rain forest zone, presumably to increase heat absorption under cooler conditions.

In moving from the tropical lowlands into upper montane rain forest, insect species richness bulges at mid-elevations and then drops off, as seen for communities in secondary vegetation.[412] Decrease in species richness is the result of the interplay of (1) increasingly stressful thermal and moisture conditions with higher altitudes and (2) changes in the distribution and abundance of resources. Thus zones of upper montane forest on tropical mountain tops are similar to some Temperate Zone areas at the higher latitudes. Within the transect from lowland tropical rain forest to upper montane forest, herbivorous insects probably remain as specialists on certain groups of plants and the trophic organization of communities is not expected to change. An overall lowering of species richness in some groups above mid-elevation habitats may allow some species to become capable of thriving in these zones. Beyond the upper-montane forest zone, insect species richness probably drops off even further, and a shift in community structure, with an emphasis on scavenger and decomposer species, takes place. The feeding habits of high-altitude insects, defined as forms living above 4000 m, have been discussed at length by Mani.[739]

Wille[767] discovered that populations of the dung beetle *Megathoposoma candezei,* a species associated with human feces, decline markedly in size below 600 m in Costa Rica. In regions of the tropics where species richness is high in the lowlands, the populations of individual insect species for some trophic positions and form some groups of insects in certain kinds of habitats may be considerably smaller than populations of the same species at higher elevations, where corresponding patterns of species richness may be reduced. In order to examine such relationships between species richness for, say, the communities of dung beetles, and the sizes of populations of individual species along altitudinal gradients in the tropics, it is necessary to determine whether an optimal habitat or microhabitat for each species changes in frequency along the gradient range occupied by the species as well as the size and complexity of the community along the gradient. If a certain kind of pasture habitat preferred by a particular species of dung beetle occurs in greater frequency at higher altitudes than in the adjacent lowlands, the beetle may become more abundant at higher altitudes, especially if other species of that community drop out along the gradient. Alternatively, if the preferred habitat does not increase in abundance but the beetle does, the explanation is related to changes in the community arising from other effects of altitude.

10.4. BIOGEOGRAPHICAL PATTERNS OF THE ORIGINS OF TROPICAL INSECT FAUNAS

Of profound ecological significance in determining the composition and structural diversity of tropical floras and faunas have been the interglacial and glacial cycles of the Quaternary.[201,469,768] It now appears that the distribution of species richness in tropical insects has been generated largely by the distribution of plant communities and the cyclic effects of expanding and contracting forest refugia and other forms of vegetation (e.g., savanna patches) within the lowland and lower montane tropical rain forest zones of Central and South America.[733,769] Communities or some other unit of whole assemblages of organisms are believed to oscillate in diversity and structure as a function of such geological cycles, but an additional perturbation within the present glacial–interglacial climatic window is man's alteration of the vegetation cover in the tropics.[770] Such a perturbation, widespread and expanding, may result in a further decline of species richness than otherwise anticipated as a natural result of the glacial cycle.[770] The impact of man on the mammalian faunas of southern Africa in the geological past[771] attests to the diversity-lowering impact of human activities on faunas.

The generally rich biota of Amazonia is believed to have been the result of the periodic pulsation of the physical environment arising from cycles of climatic change within the past 3 million years[769]: the periodic fragmentation or vicariance of forests into restricted and isolated refuges (core areas), each within a pulse and among successive pulses having a distinct sphere of selection pressures and intensity of genetic drift effects. Given sufficient time, the refugia hypothesis states that such a pulsation of habitats would result in the evolutionary radiation of new species or subspecific forms. Depending upon the size and proximity of different refugia during any one pulse, subsequent periods of forest expansion (amalgamation of refugia) would give rise to patterns of endemism and clinal variation. The pulsing of the overall physical environment in the Neotropical region probably resulted from Pleistocene and possibly past-Pleistocene climatic changes.

Livingstone[769] discusses the lack of an adequate pollen record to prove the existence of refugia cycles in the Neotropical region, but others[469,622–624] have presented convincing though inconclusive evidence from animals, including butterflies. Suspected or determined tropical centers of endemism are inconsistent with marked and recent changes in the Neotropical flora as revealed by pollen analysis, and it is unlikely that all endemics have evolved since the beginning of the present interglacial period about 10,000 years ago.[769] It is also unlikely that rates of range extension of endemics from refugia or core areas are about the same for all species. The localized distribution of many plant endemics as seen today suggests that their populations were perhaps too small to show up in pollen analysis, even though present.[769] The Neotropical region is characterized by both lowland (Amazonian) refugia and montane refugia on the slopes of the Andes, with the axis of migration of groups in the latter being vertical along temperature and moisture gradients resulting from the glacial–interglacial cycles.[469] Lowland refugia were characterized primarily by lateral or horizontal expansion or contraction. The high Andes became available for colonization by plants and animals since the end of the Tertiary and with considerable altitudinal and high-altitude lateral differences at different locations along the Andes chain.[201] High-altitude environments provided refugia conditions promoting geographical isolation and evolutionary radiation, and plant species experienced large-range extensions "during glacial periods when vegetation zones were lowered."[201] Simpson[201] discusses the differences in pathways of range extension and differentiation in different parts of the Andes, but all areas had interglacial periods characterized by the isolation and differentiation of species. As revealed by the studies of Brown,[623–626] Brown and Benson,[204] Benson,[282] and

Turner,[621,772] the effects of shifting distributions of lowland tropical rain forest on the population dynamics of larval host plants (Passifloraceae) have been a major factor in understanding the present-day geographical distribution of heliconiine butterflies, most notably *Heliconius,* in Central and South America. Similar effects are expected for other groups of plant-associated insects in both lowland and montane regions of the tropics.

The widely accepted view that favorable ecological conditions in the lowland tropics promote high species richness[15] is reconcilable with the concepts of uneven distributions of species or groups of species in accordance with the forest refugia model. On the one hand, the ecological mechanisms promoting high diversity in the tropics (predation, competition, niche diversity, etc.) provide the local opportunity for species to diversify, given a sufficiently complex habitat or collection of habitats and isolation from other such collections of habitats. Yet superimposed upon the trend for the promotion of species richness through local ecological mechanisms is the idea of discontinuities in environment, communities, and individual species ranges due to cyclic geological (historical) events such as climatic changes promoting Quaternary forest refugia in Amazonia and in the Andes.[469] As discussed in Chapters 5, 7, and 8 the explanation of how high species richness for particular taxa is maintained in the tropics involves an analysis of local ecological processes determining the structure and dynamics of guilds, species assemblages associated with specific kinds of resource patches, and whole communities and the resulting trophic interactions between different communities forming an ecosystem. Yet we also need to consider the regional patterns of species distributions and recognize that such distributions are also a function of regional environmental or geological events promoting or discouraging the associations of complex species systems. Thus the fact that the Neotropical region contains a very high number of endemic sphingid moth genera and species[751] is explained in part by the ecological refugia model, which accounts for why genera and species are diverse in some regions and not in others. But there is also the ecological explanation: sphingid species within an area partition the environment in distinct ways (seasons, larval host plants, adult hosts).

Turner[772] has outlined the major pathways of adaptive radiation of the heliconiine butterflies, particularly *Heliconius,* that occurred in South America during Pleistocene periods when forest refugia were in existence. Species pairs such as *H. melpomene* and *H. cydno* apparently diverged as distinct species in different, isolated forest refugia, and subsequently, when the forests expanded again, developed overlapping ranges so that today one species (*H. cydno*) is essentially a forest species while the other occurs at forest edges and in secondary succession.[329] Periods when forests expanded provided the opportunity for considerable range expansion of these butterflies, as seen today in the widespread distribution of some species over Central and South America. Many other species, however, exhibit restricted distributions in South America, believed to be the result of forest refuge cycles. Some areas support a very high number of species, apparently as the result of past geographical expansion of the heliconiine faunas of four or five surrounding and now-obliterated forest refugia.[(622–624)] Concurrent with such effects is the likely differentiation of larval host plants (Passifloraceae in this case) within different forest refugia, so that when faunas and floras subsequently expanded, areas of overlap received a high diversity of host plant species as well.[(329)] Gilbert and Smiley[(329)] stress the need to separate out such historical effects from ecological factors in attempting to explain local patterns of heliconiine butterfly diversity.

Brown *et al.*[(622)] provide data on the distribution of races in the butterfly genus *Heliconius* in South America that support previous studies regarding the hypothesis of forest refugia in the recent geological past determining the patterns of race formation in other organisms.[(469,621,772)] The various races of *H. melpomene* and *H. erato* exhibit rigid parallel variation throughout South America as do 10 other similarly patterned species, as the result of strong stabilizing selection favoring Mullerian mimicry. The striking discovery here is that the butterflies are largely confined to forest refugia of the recent geological past and that the precise mimetic patterns evolved in different refugia are expected to be different because ecological conditions are not the same in all refugia. The existence of multiple species Mullerian mimicry complexes, and the phenotypic differences among species or races of butterflies in the mimicry complexes of different original refugia, reflects the isolation of the refugia in the past and does not necessarily reflect a pattern of geographical variation *sensu stricto*.[(622)] The studies of *Heliconius* and ithomiine butterflies in South America[(622–626)] demonstrate how varying ecological conditions among different isolated refugia in the geological past have shaped the evolution of individual species, and whole mimicry complexes.[(772)] Brown[(622–624)] recognizes five to nine endemic centers of mimetic color patterns for *Heliconius* and ithomiines, reflecting long evolutionary periods of specialized ecological adaptation to local environmental factors (soil types, host plants, etc.). Complex ''subcenters,'' as many as seven on the slopes of the Andes also occur, showing high numbers of endemic species, explained by ecological differences rather than geographical patterns *per se*.[(622–624)]

Brown[(623)] also discusses that some species are good at dispersal, thus breaking down any previously existing discontinuous patterns of distribution conforming to the expected patterns from the forest refugia model. These species pass through barriers effective for other species, becoming ''corridors for dispersal'' and resulting in such species having less distinct boundaries associated with environmental barriers. The problem is complex because different species of *Heliconius* will exhibit different responses to the same barriers, a function of behavioral patterns as well as population growth parameters also being different. Brown[(622,623)] noted an inverse relationship between diversity of species and degree of endemism, suggesting that contemporary ecological rather than historical factors often determine patterns of co-occurrence among species. Thus the so-called paleoecological refuge model cannot be used to explain all distributional patterns, even when the organisms are species that exhibit some patterns conforming to the inferred past distribution of forest refugia.

Brown[478] has analyzed the geographical patterns of subspecies and species of the ithomiine butterfly genera *Melinaea* and *Mechanitis,* showing that mimetic affinities and patterns of localization are clear for *Melinaea* and less so for *Mechanitis*. A detailed summary of the analysis of the evolutionary radiation of *Heliconius* butterflies in relation to the forest refugia model is given in Brown,[468] in which it is recognized that some species and races are explained by the inferred positions of forest refugia and core areas with geographical isolation producing these forms, while others have arisen as the direct result of secondary contact and creation of zones of hybridization following isolation in refugia.

The examination of the geographical distribution patterns of subspecies and species of *Heliconius,* together with information on life cycles, larval host plant needs, and behavior,[203] represents a synthetic approach to understanding the role of biogeographical and ecological factors in molding distributions of species and races spatially. In the case of many species of *Heliconius,* the maintenance of local and geographically varying polymorphisms within a species has been due to strong local stabilizing selection favoring Mullerian mimetic association with other species and ithomiines.[204] Thus *Heliconius numata* is unpalatable and highly polymorphic over much of its extensive geographical range in South America, with as many as seven distinct morphs co-occurring in the same population. Such a pattern of local variation is not expected for a Mullerian mimic since strong stabilizing selection would favor one phenotype or form to be associated with mimics of other species. But Brown and Benson[204] explain this unusual condition by the existence of many species of abundant ithomiines, also unpalatable, but with local habitats having sufficiently abundant species to interact mimetically with the various morphs of *H. numata*. Thus the complex polymorphism of *H. numata* is maintained by the very heterogeneous assemblages of ithomiine butterflies. The system provides an example of how a rare species of unpalatable Mullerian mimic is maintained through polymorphism favored by interaction with abundant and unpalatable ithomiines.

Although not all groups of insects attain their greatest taxonomic diversity in the tropical regions, as exemplified by Michener's recent biogeographical analysis of bees,[167] that most insects do attain great species and generic richness in the tropics is likely the result both of historical and of contemporary ecological conditions. The most convincing piece of evidence relating to why insects are so diverse in the tropics has to do with the great diversity of tropical floras and the timing of major periods of evolutionary radiation in relation to insect evolution. The tropics support the greatest diversity of pollinating organisms and flower-visiting organisms of any region, a factor that may contribute to the lower representation of bees in the tropics (many plants are pollinated in the tropics by butterflies, moths, flies, beetles, bats, hummingbirds, mice, etc.). Yet insects also represent the largest group of herbivores in the tropics. Small body size, high population growth rates, short generation times, genetic and phenotypic flexibility—these are probably some of the major properties of insects making them successful herbivores in the tropics. A plethora of studies of plant–insect interactions in the tropics in recent years has demonstrated the role of such interactions in shaping the structure of tropical plant communities. And although hypotheses such as the paleoecological refuge model of regional variations in the generic, species, and subspecific richness of tropical insect (and other organisms) faunas help to explain why such taxonomically rich areas of the tropics are often discontinuous geographically, equal emphasis, in attempting to explain the great taxonomic diversity of most insect groups in the tropics, should be given to the analysis of contemporary ecological interactions, particularly through the lenses of present-day and past coevolutionary properties of species interactions.

10.5. SOME MECHANISMS OF ECOLOGICAL AND EVOLUTIONARY DIFFERENTIATION

Mayr[773] discusses the role of genetic and phenotypic variation in determining the ability of a species to thrive in a range of environments. A consideration of why species of tropical insects are widespread across certain kinds of envrionmental gradients while others, even ecologically similar ones, are far more restricted in distribution, necessitates an understanding of the relationship between adaptive or phenotypic flexibility within breeding populations of the species, the range of environments available for colonization, and the evolutionary history of the species. The ability of a species to become a resident of a given community within a particular habitat will determine the overall geographical distribution of the species, or its distribution relative to core areas or refugia if that sort of analysis is appropriate. The reason for this statement is that the geographical range of a species is really a compilation of different habitats capable of being occupied by the species to varying degrees. The degree (i.e., a percentage rating of successful colonizing episodes might be one measure of "degree") to which a species becomes a member of a given community within a potential or realized geographical range depends upon (1) the constellation of physical envrionmental parameters affecting the ability of different genotypes to feed, mate, and survive, (2) the interactions of the species with others in the community, including competitors, predators, and mutualists, and (3) the amount of genetic and phenotypic flexibility represented by the range of individuals in the population as a means of "fitting into" the community. The degree to which the niche of the invading species is already fully occupied by another species, the degree to which the incumbent species is a specialist for that niche, and the degree to which the invading species is a specialist for the same niche will determine the ability for the species to enter into the community. The success of invading species may also be a function of the relationship between competitive ability and density or abundance of participating species: sometimes competitive ability is inversely related to density of competing species.[81,82]

Heterozygous loci have been considered the basic mechanism by which a species can adapt to changing conditions of the environment, both spatially and temporally.[773] Constant environments may result in a gradual loss of heterozygosity in *Drosophila,*[774] although it is not known if, in general, populations in uniform tropical environments are essentially homozygous. The problem of defining a uniform or constant environment is largely one of the units of measurements chosen: the environment might appear to be uniform overall, but finer resolution may reveal patterns of microheterogeneity.[775] Zones of hybridization between species may provide new genetic variation that permits one or the other of the species to enter into a new adaptive zone.[76] When formerly allopatric species come into secondary contact, distinct phenotypic patterns will be expressed within that part of the geographical range where the contact is occurring[776]; such patterns can usually only be interpreted through the inference of secondary contact, which may or may not eventually result in parapatric speciation. Given sufficient genetic variation, selection can modify ecological traits such as competitive ability or developmental time[232,777] as well as the ability to adapt to specific temperature conditions.[386] The data of Levins[386] illustrated how different species of *Drosophila* partition the environment in Puerto Rico in terms of differential genetic and behavioral preferences for ambient temperature. Levins showed that broad-niched species are successful in adapting to a range of temperature conditions by high individual flexibility rather than by the genetic or evolutionary divergence of populations across the range of environments studied. Widely separated breeding populations of the same insect species are not necessarily genetically different,[764] presumably since behavioral and physiological systems of flexibility buffer populations against environmental fluctuations.

The pattern of resources in the environment, both temporally and spatially, will determine the degree to which two or more ecologically similar species are juxtaposed into competition if the resource is limited in supply, the two or more species are specialists on it, and the geographical ranges of the species overlap. Where many ecologically similar species co-occur in the same area, as seen for many groups of insects in the tropics, it is necessary to examine patterns of niche partitioning arising from competition and related biotic interactions and those arising from the interaction of phenotypic constraints within local populations of each species with physical features of the environment.

Thus the evolution of thermal acclimation mechanisms in subtropical or tropical *Drosophila*[386] may be an adaptive response to the phsyical environment having little or nothing to do with interactions among species. If a species is capable of exploiting a range of resources locally or regionally, the observed restriction of the species to a portion of the resources may have resulted from "interference" competition, in which the presence of other species on some of the resource patches leads to an active displacement among species resulting in microallopatry,[778] instead of direct exploitation-type competition,[779] in which each species actively removes resources that would otherwise be used by the other species.

The structuring of insect communities in the tropics is highly dependent upon the organization of guilds of closely related genera or species exploiting the same resources or a narrow spectrum of resources within habitats. Some tropical communities are characterized by the co-occurrence of several closely related species on the same resources, and such assemblages[780] may be organized by the level of competitive interaction between species if resources are limiting and populations approach the carrying capacity. Mayr[773] discusses the various kinds of mechanisms that allow several closely related species to co-occur without the formation of inviable hybrids from cross-matings. In some insects, reproductive isolation is mediated by visual discrimination systems precluding copulations,[781] while in others ecological mechanisms and other forms of behavioral mechanisms reduce the likelihood of cross-matings. Much of the integrity of species richness in tropical insect communities depends upon the effective operation of reproductive isolation mechanisms in areas where species are sympatric or microsympatric.

It is the interplay of ecological interactions between species in contemporary time, incorporation of such ecological traits into whole communities over evolutionary time, and the impact of geological history that determines gradients in species richness in the tropics. When environments are stressful or rigorous in terms of physical fluctuations in certain parameters or rather uniform conditions of a harsh nature, biotic communities are thought to be physically controlled,[104] a result of which is the variable success for invading species to establish themselves in such communities.[632] Such situations may be generally characteristic of the very high latitudes and also high altitudes in the tropics. When environmental conditions are benign and predictable, there may be an evolutionary accumulation of highly coevolved species in local communities, as exemplified by the association of insects with the forest understory tree *T. cacao,* but with an overall increase in species richness being asymptotic,[634] One of the major "problems" with the control of insect pests on tropical crops such as cacao and coffee is the continual availability of a large pool of species associated with various kinds of natural vegetation, providing for replacement of specific pests as they are ''eliminated'' through ineffective (in the long term) pest control programs. Some regions of the tropics tend to accumulate a greater number of species of insects than expected in terms of predicted levels of species saturation in communities: such areas are ideal candidates for continual outbreaks of insect species on crops (Amazonia, Southeast Asia, tropical Australia). It is the exploitation by native, coevolved insect herbivores that limits the distribution and abundance of native plant species (populations).

Thus the geographical distribution of plant species can be greatly influenced by the intensity of herbivory from insect species that are specialized feeders on particular plants. In the mid-elevation and lowland tropics, such effects are expected to be intense for most dicotyledons. Both insect and vertebrate herbivore species may exhibit considerable variation in plant diet along altitudinal gradients in the tropics[(782)] not necessarily related to competitive exclusion or ecological release in certain zones along the gradient. Plant communities in montane localities in the tropics are sometimes greatly influenced by periodic natural catastrophes over long periods of time (i.e., landslides and volcanic eruptions) that cause large-scale changes in the floral composition of forests and postponing the natural sequence of succession.[(783)]

It is therefore the highly-coevolved interaction of insects and other herbivores with plants, highly coevolved with certain groups, geological and climatic history, and the periodic reoccurrence of regional catastrophes, that determines the composition of plant communities within the montane section of altitudinal gradients in the tropics. Other ecological processes, such as interspecies competition and character displacement, place upper limits on the overall species richness found in a certain kind of habitat,[(15)] especially when populations of component species within a given trophic level are approaching equilibrium conditions. Switches in host plant exploitation patterns within a widely distributed insect herbivore species along an altitudinal gradient or other form of environmental gradient may be promoted by shifts in community structure, emphasizing or de-emphasizing competition, shifts in distribution and abundance of the host plant, changes in the physical environment affecting the activity patterns of the insect, and large-scale alterations in the habitat.

Whether or not the insect can successfully switch to another host plant species is a function of the degree to which feeding behavior is regulated primarily by phenotypic flexibility (behavioral, physiological) or by genotype. The existence of a ''host selection gene'' may determine feeding flexibility in the sense that the homozygous condition may promote feeding on only one kind of host plant, the alternate homozygous condition (assuming a two-allele system) favoring feeding on the second kind of plant, and the heterozygous condition favoring the flexible feeding condition.[(391)] An insect population may contain different ''plant feeding'' genotypes in the sense that each genotype is capable of promoting feeding on a particular plant species. When an introduced plant species (an exotic) is juxtaposed in the same region or habitat with other native species, genotypes associated with the native species may shift to the introduced species if it is closely related to the native species. G. L. Bush suggests the occurrence of "host plant races" that may differ, in the specific case of apple fruit flies associated with fruit tree species, in terms of evolutionary divergence in the intestinal bacteria flora in the guts of fly larvae. In some groups of Old World tropical ants, polymorphic genera such as *Dorylus* in the subfamily Dorylinae exhibit considerably higher behavioral flexibility than closely related monomorphic genera such as *Aenictus* in terms of the degree of specialized foraging behavior or lack thereof.[430] Species of *Aenictus* are specialized predators on other ants in tropical habitats.[430] Thus for tropical insects in different trophic roles, polymorphism tends to result in more flexible behavior in terms of the breadth of resources utilized, but such patterns are expected to vary considerably in different communities, habitats, and along environmental gradients over large areas of terrain, because selection pressures will also vary.

10.6. SOME EXAMPLES OF DIFFERENTIAL ADAPTATION TO TROPICAL LOWLAND AND MONTANE AREAS BY INSECTS

The Neotropical carabid beetles (Carabidae) have provided the opportunity for the development of a rather comprehensive model explaining the complex relationships between evolutionary radiation within the family, dispersal, and climatic shifts.[417] Historically, carabids were of particular interest to biogeography because of the tendency toward flightlessness in some mid-elevation, montane, and high-latitude forms.[753] Flightlessness in carabids is believed to be related to the degree of stability of particular geographical regions, since otherwise this phenotypic trait would lead to extinction in regions of environmental instability.[417]

Flightlessness is found in insects occupying some isolated but stable habitats such as mountain tops. The likelihood for the evolution of flightlessness in insects under such conditions may involve a switch in the allocation of resources for other phenotypic traits, particularly those related to reproductive potential. An insect population may segregate out from each generation a certain portion of "dispersal prone" and "nondispersing" genotypes. As the number of dispersal-prone genotypes is reduced by emigration and if no new immigrants arrive in the population, there is a net decrease in the abundance of dispersing individuals and population size declines.[733] The result is an increase in the proportion of nondispersing genotypes, and if the environment is stable, directional selection favors the nondispersing genotypes and subsequent stabilizing selection promotes the evolution of phenotypic traits that optimize adaptation to the local and stable environment. There is a switch in the allocation of energy to the development of flight systems to other traits such as reproduction if selection favors the evolution of phenotypic traits that optimize local adaptation. Depending upon the degree of isolation, the size of the stable environment, the size and distribution of the insect population, and the distribution and abundance of resources, selection may favor the retention of some dispersal-prone genotypes, since under some conditions local dispersal would be selected for. The retention of such genotypes, even at low frequency, would provide the mechanism for the species to emigrate should the environment suddenly change. Slater[(424)] presents evidence from African populations of lygaeid bugs that flightless forms within flight-polymorphic species are more frequently encountered in stable habitats. More recently Noonan[(745)] found that individuals of *N. peruviana* have short wings where habitats are favorable and continuous and longer wings where otherwise. The data support the "checkerboard" hypothesis of Darlington.[(784)]

Thiele[(785)] has summarized in some detail the major adaptive features of carabid beetles, with an emphasis on the relation of body form to foraging habits, habitat associations, and trophic positions in communities. The great evolutionary and ecological diversification of the group, particularly in the tropical regions, attests to the versatility of carabids to exploit a range of environments. The carabids are hypothesized to have arisen in the Mesozoic with a "rapid pulse of radiation and diversification" with the early lineages being associated with river and stream edge habitats.[(417)]

The periodic and predictable cycles of inundation in Amazonian forests provided subsequent selection for invasion in evolutionary time of mesic forests, leading to another pulse of adaptive radiation, and other pulses in diversification paralleled stages in the evolution of tropical forests, notably in terms of vertical structuring of forests and the invasion of arboreal microhabitats (and canopies) by carabids.[(417)] Subsequent geological events altering the climate of Amazonia, such as cycles of Pleistocene climatic change, coupled with the dispersal abilities of lineages, resulted in complex geographical distribution patterns.[(417)] Thus the general pathway of carabid beetle evolution in South America consisted of an origin in tropical wet lowlands where the insects, as ecological generalists, were placed in competition with ecologically similar but unrelated forms such as ants, with subsequent radiation and diversification primarily into (1) mesic forests with exploitation of arboreal microhabitats and from here into mid-elevation (highland) and montane forests and mountain tops, and (2) as specialists into forest environments at other latitudes from the wet lowlands and also into other kinds of habitats. Erwin[(417)] points out that carabid beetles are most diverse generically in the tropical lowlands and mesic forests, with greatest species richness occurring in specialized environments such as mountain tops and tropical forest canopies and in north and south cold temperate habitats. Erwin's model for the evolution and diversification of carabid beetles stresses the concept of generic and species richness being driven by fluctuations in the physical environment, coupled with ecological processes in local communities, resulting in strong local selection pressures that leads to within- and between-habitat diversification in taxa. His model may be more generally applicable for other groups of insects with definitive Neotropical origins but with subsequent highly cosmopolitan distributions of genera and species.

Erwin's model accounts for the tropics being rich in genera and species of carabid beetles, even though populations may be very patchy in space, particularly in lowland and mid-elevation zones. Darlington[(753)] recognized the importance of local ecological interactions affecting carabids as a major source of selection pressure in terms of how different genera and species respond to temperature and humidity gradients in tropical wet forests. Darlington[(753)] originally developed an application of the taxon cycle to the evolution of carabid beetles in the tropics, stressing the origin of forms in the tropical wet lowlands with good powers of dispersal into other habitats (as discussed much later by Erwin[(417)] and the eventual evolution of other dominant forms that subsequently colonize other habitats or reinvade lowland areas where previous forms have become extinct. Both Darlington[(753)] and Erwin[(417)] have developed conceptual models to explain the geographical distribution, local and regional diversity, and patterns of morphological, behavioral, and ecological differentiation as related to patterns of climatic change and changes in forest communities.

Jeanne[(410)] has found that ant species in the tropics exhibit a great degree of microhabitat specificity and that many species in lowland rain forests are canopy specialists. Such results reveal that ants have been a major source of selection pressure for other groups of insects, e.g., carabid beetles, in similar microhabitats. Tropical soil arthropod faunas are very complex in terms of species richness,[(786)] more so than in the Temperate Zone. Erwin[(417)] mentions the impact of ants in tropical forests on the evolution of carabid beetles, and one might go so far as to say that in the lowland tropics, where there is a high level of generic and species richness for many groups of invertebrates, there are probably many examples of the ''joint evolution'' of whole groups such as ants and carabid beetles: the ants are a selection pressure in the sense of competing for the same microhabitats as carabids, and the carabid beetles are a likely source of selection pressure for ants, although probably less so than vice versa. The high species number of carabids at higher altitudes in the tropics[(417)] and the corresponding decline in ant species along such gradients may be indicative of a compensatory replacement of whole groups in the context that carabids have been more successful at colonizing high-altitude habitats than ants

As for most groups of insects in the tropics, we need to examine the adpative significance of subtle differences in morphology, behavior, and ecological traits between taxa of carabid beetles occupying the same communities and to extend such an approach to the analysis of how such communities change geographically.[(787)] The fact that carabids exhibit gradual reductions in mean body size from the lowlands to mountain tops in the tropics[739] has to be examined in terms of the relationship of body size, energy allocation strategies, and effects of the physical environment and the interactions with other organisms that occur along such gradients. Explaining gradients in body size strictly in terms of physiological constraints associated with temperature and moisture does not really elucidate the adaptive significance of such morphological patterns to interactions with other organisms in communities. If carabid beetle and other insect populations on tropical mountain tops are characterized by a relatively larger size (abundance) than populations of the same or closely related species at lower elevations, we need to critically examine the strategy of resource allocation in such differing populations. This strategy in each community relates to the distribution and abundance of resources; the quality of resources as affecting fitness; and the selection pressures arising from competition, predation, and other ecological interactions in each habitat. It is necessary to determine the role of heterogeneous gradients of abiotic factors along with biotic factors in determining the local assemblages of groups such as carabids and how these factors change along altitudinal gradients. Specialized feeders such as termites in the lowland tropics, insects for which a specialized feeding habit and moisture constraints may limit geographical range,[788] play a major role in nutrient cycling.

The degree to which an insect entering into a new adaptive zone such as a tropical mountain top can differentiate in response to local selection pressures is a function of the degree of genetic and phenotypic flexibility within the population[789] and the amount of time that has passed since the invasion occurred. Perhaps some of the most interesting data on such properties of adaptive response, and constraints upon it, come from the recent studies of Shapiro[472,759–761] on Andean pierine butterflies. As Shapiro[472] points out, little is known about the ecology and evolutionary history of the unique genera of pierines associated with the high Andes. Shapiro[759] has been most interested in examining the hypothesis that the Andean pierine fauna was derived from Nearctic stock during the Pleistocene glaciation period. From his studies of life histories in some selected genera in the Andes[472,761] he has discovered some evidence suggesting divergence from the Nearctic stock. Most interesting, however, is the relationship between geographical region and the capacity of pierines to exhibit wing color pattern polyphenisms in response to changes in photoperiod regimes.[760]

Most extratropical pierine butterflies are multivoltine and exhibit distinctive seasonal phenotypes and a facultative diapause, both controlled by photoperiod. Some univoltine and monophenic Holarctic species exhibit polyphenisms when experimentally induced under laboratory conditions, a phenomenon Shapiro calls "latent polyphenisms." Such phenotypic flexibility suggests that these species are derived from multivoltine and polyphenic ancestors.[760]

In the Andes of South America, some endemic pierine butterflies are both multivoltine and monophenic but cannot be induced to exhibit polyphenisms under experimental conditions as can some monophenic Holarctic forms.[759] Yet it is believed that these Andean forms are derived from Nearctic stock during the Pleistocene. The lack of phenotypic flexibility suggests that either they have secondarily lost the ability to exhibit plasticity in response patterns to photoperiod or they never had the condition which calls for a reexamination of the hypothesis that these pierines are derived from Nearctic forms.

Inherent in the first explanation is the idea that since Pleistocene glaciation is relatively recent, the time for complete differentiation from the Nearctic stock should have been sufficient and that some portion of the phenotype should exhibit relict traits such as latent polyphenism in response to changes in photoperiod. Since the apparent invasion of the Andes by pierines occurred at about the same time as the invasion of the Andes by their larval host plants, the Cruciferae,[790] one is tempted to accept the first explanation, although more data are needed on the ecology of Andean pierines and that of other groups of insects with similar evolutionary and biogeographical histories.

The examination of phenotypic traits of butterflies such as voltinism and life history and of responses of the phenotype to environmental factors such as photoperiod is a methodology that can be used to assess the degree of evolutionary and ecological differentiation in other groups of insects along latitudinal and altitudinal gradients, with an emphasis on comparing temperate and tropical faunas. Scott[791] has discussed the apparent, recent partial differentiation of the butterfly fauna of the Antilles, with derivations from Central and South America and similarities with that of Florida. Comparative studies, in the spirit of Shapiro,[760] on the divergence in ecological traits and phenotypic flexibility or lack thereof, between the island and mainland populations for some of these genera and species would be particularly instructive in explaining the success of invasions into the archipelago.

The determination of the evolutionary pathways of divergence in relation to geographical patterns,[792] across either latitudinal or altitudinal gradients (both situations exposing species or whole complexes of species to sets of very heterogeneous environmental conditions) and including biotic and abiotic components of ecosystems, constitutes the approach needed to explain not only the origins of tropical insect faunas but indeed the origins of such groups on a world wide basis. Both latitudinal and altitudinal gradients expose species or whole complexes of species to heterogeneous environmental conditions, resulting in some phenotypic traits being greatly modified while others remain stable (relict). Natural-history and ecological studies emphasizing the subtle differences in form, behavior, and physiological response systems among co-occurring groups such as a species complex, supplement biogeographical analyses in undertaking such a task. Yet the data base from such an approach for most groups of insects in the tropics is very meager. We need detailed data on trophic organization in ecosystems, including the determination of rather precise resources needed by species in certain habitats at a given time of the year, as exemplified by Greathead.[793] Little is known about the impact of vertebrates on insect populations, and the data of Hespenheide[794] indicate that such predation can be heavy in some habitats. Such studies are needed to explain short-term fluctuations in tropical insect communities and how, along with fluctuations in the physical environment, specific kinds of biotic interactions change or remain stable geographically.

Much of the foregoing discussion, and the preceding chapters, emphasizes the need for biologists to consider both contemporary ecological factors and the historical or biogeographical effects in understanding the distribution and abundance of insects in the tropics. A comprehensive understanding of both the broad geographical distributions of species or whole groups of species, and the distribution patterns of species associated with different habitats, in the tropics, requires the total view, namely, that of the ecologist considering the historical perspective and the biogeographer gathering as much ecological data as possible.

REFERENCES

1. R. C. Lewontin, *The Genetic Basis of Evolutionary Change,* Columbia University Press, New York (1974).
2. R. C. Lewontin and J. L. Hubby, A molecular approach to the study of genic heterozygosity in natural populations. II. Amount of variation and degree of heterozygosity in natural populations of *Drosophila pseudoobscura, Genetics* **54:**595–609 (1966).
3. R. Levins, *Evolution in Changing Environments,* Princeton University Press, Princeton, N.J. (1968).
4. R. Patrick, A discussion of the results of the Catherwood Expedition to the Peruvian headwaters of the Amazon, *Verh. Int. Ver. Theor. Angew. Limnol.* **15:**1084–1090 (1964).
5. R. Patrick, The Catherwood Foundation Peruvian Amazon expedition, *Monogr. Acad. Nat. Sci. Philadelphia* (1966).
6. P. R. Ehrlich and P. H. Raven, Butterflies and plants: A study in coevolution, *Evolution* **18:**586–608 (1964).
7. D. R. Paulson, Oviposition in the tropical dragonfly genus *Micrathyria* (Odonata, Libellulidae), *Tombo* **12:**12–16 (1969).
8. A. E. Emerson, Termite studies in the Belgian Congo, *Rapp. Ann. Inst. Rech. S. Afr. Commis.,* 149–159 (1949).
9. A. E. Emerson, Geographical origins and dispersions of termite genera, *Fieldiana Zool.* **37:**465–521 (1955).
10. W. L. Peters and G. F. Edmunds, Jr., Revision of the generic classification of the eastern hemisphere Leptophlebiidae (Ephemeroptera), *Pac. Insects* **12:**157–240 (1970).
11. E. G. Linsley, Ecology of Cerambycidae, *Ann. Rev. Entomol.* **4:**99–138 (1959).
12. J. A. L. Watson, Taxonomy, ecology, and zoogeography of dragonflies (Odonata) from the north-west of western Australia, *Aust. J. Zool.* **17:**65–112 (1969).
13. O. Park, Pselaphid beetles of Jamaica, *Trans. Am. Microsc. Soc.* **79:**5–25 (1960).
14. R. H. MacArthur, Patterns of species diversity, *Biol. Rev.* **40:**510–533 (1965).
15. R. H. MacArthur, Patterns of communities in the tropics, *Biol. J. Linn. Soc.* **1:**19–30 (1969).
16. R. H. MacArthur, *Geographical Ecology: Pattern and Distribution of Species,* Harper & Row, New York and London (1972).
17. Th. Dobzhansky, Evolution in the tropics, *Am. Sci.* **38:**209–221 (1950).
18. E. J. Corner, in: *Evolution as a Process* (J. Huxley, A. C. Hardy, and E. B. Ford, eds.), pp. 34–45, G. Allen, London, (1954).
19. W. H. Gotwald, Jr., Comparative morphological studies of the ants, with particular reference to the mouthparts (Hymenoptera: Formicidae), *Cornell Univ. Agric. Exp. Stn. Mem.* No. 408 (1969).
20. C. D. Michener, Observations on the nests and behavior of *Trigona* in Australia and New Guinea (Hymenoptera, Apidae), *Am. Mus. Novit.* No. 2026 (1961).
21. A. Wille, Behavioral adaptations of bees for pollen collecting from *Cassia* flowers, *Rev. Biol. Trop. San Jose Costa Rica* **11:**205–210 (1963).

22. E. J. Fittkau, Remarks on limnology of central-Amazon rain-forest streams, *Verh. Int. Theor. Angew. Ver. Limnol.* **15:**1092–1096 (1964).
23. E. J. Fittkau, Artenmannigfältigkeit amazonischer Lebensräume aus ökologischer Sicht, *Amazoniana* **4:**321–340 (1973).
24. E. J. Fittkau, in: *Biogeography and Ecology in South America* (E. J. Fittkau, J. Illies, H. Klinge, G. H. Schwabe, and H. Sioli, eds.), Dr. W. Junk, The Hague (1969).
25. J. Illies, The invertebrate fauna of the Huallaga, a Peruvian tributary of the Amazon River, from the sources down to Tingo Maria, *Verh. Int. Theor. Angew. Verein. Limnol.* **15:**1077–1083 (1964).
26. C. W. Rettenmeyer, Observations on the biology and taxonomy of flies found over swarm raids of army ants (Diptera: Tachinidae, Conopidae), *Univ. Kans. Sci. Bull.* **42:**993–1066 (1961).
27. C. W. Rettenmeyer, The diversity of arthropods found with Neotropical army ants and observations on the behavior of representative species, *Proc. North Cent. Branch, Entomol. Society Am.* **17:**14–15 (1962).
28. S. P. Hubbell, Tree dispersion, abundance, and diversity in a tropical dry forest, *Science* **203:**1299–1309 (1979).
29. D. H. Janzen, Herbivores and the number of tree species in tropical forests, *Am. Nat.* **104:**501–529 (1970).
30. P. D. Hurd, Jr., and E. G. Linsley, A classification of the Squash and Gourd Bees *Peponapis* and *Xenoglossa* (Hymenoptera: Apoidea), *Univ. Calif. Publ. Entomol.*, Vol. 62 (1970).
31. L. E. Gilbert, Pollen feeding and reproductive biology of *Heliconius* butterflies, *Proc. Natl. Acad. Sci. U.S.A.* **69:**1403–1407 (1972).
32. E. G. Linsley and R. L. Usinger, Insects of the Galapagos Islands, *Proc. Calif. Acad. Sci.* **33:**113–196 (1966).
33. D. H. Janzen, Coevolution of mutualism between ants and acacias in Central America, *Evolution* **20:**249–275 (1966).
34. D. R. Paulson, New records of Bahamian Odonata, *Q. J. Fla. Acad. Sci.* **29:**97–110 (1966).
35. A. J. Kohn and G. H. Orians, Ecological data in the classification of closely related species, *Syst. Zool.* **11:**119–127 (1962).
36. J. Reiskind, Ant-mimicry in Panamanian clubionid and salticid spiders (Araneae: Clubionidae, Salticidae), *Biotropica* **9:**1–8 (1977).
37. T. Erwin and L. V. J. M. Erwin, Relationships of predaceous beetles to tropical forest wood decay. Part II. The natural history of Neotropical *Eurycoleus macularis* Chevrolat (Carabidae: Lebiini) and its implications in the evolution of ectoparasitoidism, *Biotropica* **8:**215–224 (1976).
38. B. Holldobler, Canopy orientation: A new kind of orientation in ants, *Science* **210:**86–88 (1980).
39. H. Takaoka, Pathogens of blackfly larvae in Guatemala and their influence on natural populations of three species of Onchocerciasis vectors, *Am. J. Trop. Med. Hyg.* **29:**467–472 (1980).
40. D. H. Janzen, Euglossine bees as long-distance pollinators of tropical plants, *Science* **171:**203–205 (1971).
41. D. H. Janzen, in: *Taxonomy and Ecology* (V. H. Heywood, ed.), pp. 201–211, Academic Press, New York (1973).
42. D. H. Janzen, Promising directions of study in tropical animal-plant interactions, *Ann. Mo. Bot. Gard.* **64:**706–736 (1977).
43. G. H. Orians, The number of bird species in some tropical forests, *Ecology* **50:**783–801 (1969).
44. C. W. Myers, The ecological geography of cloud forest in Panama, *Am. Mus. Novit.* No. 2396 (1969).
45. W. Schwerdtfeger (ed.), *Climates of Central and South America,* Elsevier, New York (1976).
46. D. H. Janzen, Why mountain passes are higher in the tropics, *Am. Nat.* **101:**233–249 (1967).

47. D. H. Janzen, Why are there so many species of insects?, *Proc. Int. Congr. Entomol.* **15:**84–94 (1976).
48. P. M. Marsh, Descriptions of new Braconidae (Hymenoptera) parasitic on the potato tuberworm and on related Lepidoptera from Central and South America, *J. Wash. (D.C.) Acad. Sci.* **69:**12–17 (1979).
49. J. H. Vandermeer, Niche theory, *Ann. Rev. Ecol. Syst.* **3:**107–132 (1972).
50. N. L. Stanton, Patterns of species diversity in temperate and tropical litter mites, *Ecology* **60:**295–304 (1979).
51. R. L. Wenzel and V. J. Tipton (eds.), *Ectoparasites of Panama,* Field Museum of Natural History, Chicago (1966).
52. D. Lack, *The Natural Regulation of Animal Numbers,* Clarendon Press, Oxford (1954).
53. D. Lack, *Ecological Isolation in Birds,* Blackwell, Oxford. (1971).
54. R. C. Lewontin and M. J. D. White, Interaction between inversion polymorphisms of two chromosome pairs in the grasshopper, *Moraba scurra, Evolution* **14:**116–129 (1960).
55. S. Wright, On the roles of directed and random changes in gene frequency in the genetics of populations, *Evolution* **2:**279–294 (1948).
56. S. Wright, *Evolution and the Genetics of Populations,* Vol. 2, University of Chicago Press, Chicago (1969).
57. D. H. Janzen, Interfield and interplant spacing in tropical insect control, in: *Proc. Ann. Tall Timbers Conf. Ecol. Anim. Contr. Habitat Manage.* pp. 1–6, Tall Timbers Research Station, Tallahassee (1972).
58. Thomas Park, Beetles, competition, and populations, *Science* **138:**1369–1375 (1962).
59. D. R. A. Wharton, J. E. Lola, and M. L. Wharton, Population density, survival, growth, and development of the American Cockroach, *J. Insect Physiol.* **13:**699–716 (1967).
60. D. Pimentel, Animal population regulation by the genetic feed-back mechanism, *Am. Nat.* **95:**65–79 (1961).
61. J. Le Gay Brereton, in: *The Evolution of Living Organisms,* Symposium of Royal Society of Victoria, Melbourne, Australia, 1959 (1962).
62. J. Haffer, Speciation in Amazonian forest birds, *Science* **165:**131–137 (1969).
63. E. R. Pianka, Latitudinal gradients in species diversity: A review of concepts, *Am. Nat.* **100:**33–46 (1966).
64. R. T. Paine, Food web complexity and species diversity, *Am. Nat.* **100:**65–76 (1966).
65. J. H. Connell, Diversity in tropical rain forests and coral reefs, *Science* **199:**1302–1310 (1978).
66. J. H. Connell and E. Orias, The ecological regulation of species diversity, *Am. Nat.* **98:**399–414 (1964).
67. M. L. Cody, A general theory of clutch size, *Evolution* **20:**174–184 (1966).
68. R. E. Ricklefs and G. W. Cox, Taxon cycles in the West Indian avifauna, *Am. Nat.* **106:**195–219 (1972).
69. H. Groot, The reinvasion of Colombia by *Aedes aegypti:* Aspects to remember, *Am. J. Trop. Med. Hyg.* **29:**330–338 (1980).
70. A. M. Young, Cicada ecology in a Costa Rican tropical rain forest, *Biotropica* **4:**152–159 (1972).
71. A. M. Young, Environmental partitioning in lowland tropical rain forest cicadas, *J. N.Y. Entomol. Soc.* **88:**86–101 (1980).
72. D. H. Janzen, Sweep samples of tropical foliage insects: Effects of seasons, vegetation types, elevation, time of day, and insularity, *Ecology* **54:**687–708 (1973).
73. R. F. Denno, Ecotope differentiation in a guild of sap-feeding insects on the salt marsh grass, *Spartina patens, Ecology* **61:**702–714 (1980).
74. Th. Dobzhansky and C. Pavan, Local and seasonal variations in relative frequencies of species of *Drosophila* in Brazil, *J. Anim. Ecol.* **19:**1–14 (1950).

75. A. B. Da Cunha and Th. Dobzhansky, A further study of chromosomal polymorphism in *Drosophila willistoni* in its relation to the environment, *Evolution* **8:**119–134 (1954).
76. R. C. Lewontin and L. C. Birch, Hybridization as a source of variation to new environments, *Evolution* **20:**315–336 (1966).
77. A. M. Young, Notes on the behavioral ecology of *Perrhybris lypera* (Pieridae) in northeastern Costa Rica, *J. Lepid. Soc.* **34:**36–47 (1980).
78. R. C. Lewontin, The units of selection, *Ann. Rev. Ecol. Syst.* **1:**1–18 (1970).
79. M. J. Wade and D. E. McCauley, Group selection: The phenotypic and genotypic differentiation of small populations, *Evolution* **34:**799–812 (1980).
80. E. O. Wilson, *The Insect Societies,* Belknap Press, Harvard University Press, Cambridge, Mass. (1971).
81. F. J. Ayala, Dynamics of populations: Factors controlling population growth and population size in *Drosophila pseudoobscura* and *D. melanogaster, Ecology* **48:**67–75 (1967).
82. F. Ayala, Competition between species: Frequency dependence, *Science* **171:**820–824 (1971).
83. W. W. Anderson, Genetic equilibrium and population growth under density-regulated selection, *Am. Nat.* **105:**489–498 (1971).
84. R. C. Lewontin, The effects of population density and competition on viability in *Drosophila melanogaster, Evolution* **9:**27–41 (1955).
85. R. C. Lewontin, The adaptations of populations to varying environments, *Cold Spring Harbor Symp. Quant. Biol.* **22:**395–408 (1957).
86. I. M. Lerner, *Resistance of populations to change, Proceedings of the 1956 International Genetics Symposia Cytol. Suppl.* pp. 520–521 (1957).
87. R. C. Lewontin, in: *The Genetics of Colonizing Species* (H. G. Baker and G. L. Stebbins, eds.), pp. 77–94, Academic Press, New York (1965).
88. A. M. Young, Notes on the biology of *Hypothyris euclea* (Lepidoptera: Nymphalidae: Ithomiinae) in Costa Rica, *Pan-Pac. Entomol.* **53:**104–113 (1976).
89. A. M. Young, Weather and the regulation of *Hypothyris euclea* (Lepidoptera: Nymphalidae: Ithomiinae) populations in northeastern Costa Rica. *J. Lepid. Soc.* **33:**68–69 (1979).
90. F. J. Ayala, Evolution of fitness. I. Improvement in the productivity and size of irradiated populations of *Drosophila serrata* and *Drosophila birchii, Genetics* **53:**883–895 (1966).
91. E. O. Wilson, The nature of the taxon cycle in the Melanesian ant fauna, *Am. Nat.* **95:**169–193 (1961).
92. M. Gadgil and W. H. Bossert, Life historical consequences of natural selection, *Am. Nat.* **104:**1–24 (1970).
93. D. H. Janzen, Host plants as islands in evolutionary and contemporary time, *Am. Nat.* **102:**592–595 (1968).
94. E. Mayr, *Animal Species and Evolution,* Belknap Press, Harvard University Press, Cambridge, Mass. (1963).
95. W. G. Wellington, Individual differences as a factor in population dynamics: The development of problem, *Can. J. Zool.* **35:**293–323 (1957).
96. W. G. Wellington, The synoptic approach to studies of insects and climate, *Ann. Rev. Entomol.* **2:**143–162 (1957).
97. W. C. Allee, A. E. Emerson, O. Park, T. Park, and K. P. Schmidt, *Principles of Animal Ecology,* Saunders, Philadelphia (1949).
98. P. M. Tuskes and L. P. Brower, Overwintering ecology of the monarch butterfly, *Danaus plexippus* L., in California, *Ecol. Entomol.* **3:**141–153 (1978).
99. L. B. Slobodkin and A. Rapoport, An optimal strategy of evolution, *Q. Rev. Biol.* **49:**181–200 (1974).
100. J. Grossfield, Species differences in light-influenced mating behavior in *Drosophila, Am. Nat.* **104:**307–309 (1970).

101. N. Smythe, Relationships between fruiting seasons and seed dispersal methods in a neotropical forest, *Am. Nat.* **104:**25–35 (1970).
102. P. Feeny, Defensive ecology of the Cruciferae, *Ann. Mo. Bot. Gard.* **64:**221–234 (1977).
103. R. Levins, Theory of fitness in a heterogeneous envrionment. II. Developmental flexibility and niche selection, *Am. Nat.* **97:**75–90 (1963).
104. L. B. Slobodkin and H. L. Sanders, in: *Diversity and Stability in Ecological Systems* (G. M. Woodwell and H. H. Smith, eds.), pp. 82–95, *Brookhaven Symp. Biol.* No. 22, Brookhaven National Laboratory, Brookhaven, N.Y. (1969).
105. K. A. J. Wise and J. L. Gressitt, Far southern animals and plants, *Nature (London)* **207:**101–102 (1965).
106. F. R. Schram, Arthropods: A convergent phenomenon, *Fieldiana Geol.* **39:**61–108 (1978).
107. A. Wille, The phylogeny and relationships between the insect orders, *Rev. Biol. Trop.* **8:**93–123 (1960).
108. H. E. Hinton, On the origin and function of the pupal stage, *Trans. R. Entomol. Soc. London* **99:**395–409 (1948).
109. C. Istock, in: *Evolution of Insect Migration and Diapause* (H. Dingle, ed.), pp. 171–190, Springer-Verlag, Berlin, (1978).
110. F. M. Carpenter, Geological history and evolution of the insects, *Proc. Int. Congr. Entomol.* **15:**63–70 (1976).
111. J. Maynard Smith, Optimization theory in evolution, *Ann. Rev. Ecol. Syst.* **9:**31–56 (1978).
112. D. E. McCauley and M. J. Wade, Group selection: The genetic and demographic basis for the phenotypic differentiation of small populations of *Tribolium castaneum, Evolution* **34:**813–821 (1980).
113. S. L. Swihart, Neural adaptations in the visual pathway of certain heliconiine butterflies, and related forms, to variations in wing coloration, *Zoologia* **52:**1–14 (1967).
114. T. Eisner, in: *Chemical Ecology* (E. Sondheimer and J. B. Simeone, eds.), pp. 157–217, Academic Press, New York (1970).
115. D. M. Wilson and R. R. Hoy, Optomotor reaction, locomotory bias, and reactive inhibition in the milkweed bug *Oncopeltus* and the beetle *Zophobas, Z. Vgl. Physiol.* **58:**136–152 (1968).
116. D. H. Janzen, Synchronization of sexual reproduction of trees within the dry season in Central America, *Evolution* **21:**620–637 (1967).
117. G. W. Frankie, in: *Coevolution of Animals and Plants* (L. E. Gilbert and P. H. Raven, eds.), pp. 192–209, University of Texas Press, Austin (1975).
118. G. W. Frankie, H. G. Baker, and P. A. Opler, Comparative phenological studies of trees in tropical wet and dry forests in the lowlands of Costa Rica, *J. Ecol.* **62:**881–919 (1974).
119. E. B. Edney, Water balance in desert arthropods, *Science* **156:**1059–1066 (1967).
120. G. Fraenkel and M. Blewett, The utilization of metabolic water in insects, *Bull. Entomol. Res.* **35:**127–139 (1944).
121. M. J. Berridge and B. L. Gupta, Fine-structural changes in relation to ion and water transport in the rectal papillae of the blowfly, *Calliphora,* J. Cell Sci. **2:**89–112 (1967).
122. H. A. Schneiderman and L. I. Gilbert, Control of growth and development in insects, *Science* **143:**325–333 (1964).
122a. V. B. Wigglesworth, *The Life Of Insects,* World Press, Cleveland, Ohio (1964).
122b. C. M. Williams, Physiology of insect diapause. IV. The brain and prothoracic glands as an endocrine system in the Cecropia silkworm, *Biol. Bull.* **103:**120–138 (1952).
123. L. C. G. Simoes and G. Uemura, Hormonal control in insect development, *Cienc. Cult. (Maracaibo)* **32:**331–335 (1979).
124. A. Krishnakumaran, S. J. Berry, H. Oberlander, and H. A. Schneiderman, Nucleic acid synthesis during insect development. II. Control of DNA synthesis in the Cecropia silkworm and other saturniid moths, *J. Insect Physiol.* **13:**1–57 (1967).

125. A. Spielman and V. Skaff, Inhibition of metamorphosis and of ecdysis in mosquitoes, *J. Insect Physiol.* **13:**1087–1095 (1967).
126. P. F. Roseler, I. Roseler, and A. Strambi, The activity of corpora allata in dominant and subordinate females of the wasp *Polistes gallicus, Insectes Soc.* **27:**97–107 (1980).
127. S. H. Clark and A. P. Platt, Influence of photoperiod on development and larval diapause in the Viceroy Butterfly, *Limenitis archippus, J. Insect Physiol.* **15:**1951–1957 (1969).
128. S. D. Beck, *Insect Photoperiodism,* Academic Press, New York (1968).
129. K. D. Kappus and C. E. Venard, The effects of photoperiod and temperature on the induction of diapause in *Aedes triseriatus* (Say), *J. Insect Physiol.* **13:**1007–1019 (1967).
130. V. B. Wigglesworth, *The Principles of Insect Physiology,* 7th ed., Chapman and Hall, London (1972).
131. Y. C. Siew, The endocrine control of adult reproductive diapause in the chrysomelid beetle *Galeruca tanaceti* (L.). II, *J. Insect Physiol.* **11:**463–479 (1965).
132. S. L. Goldson and R. M. Emberson, Relict diapause in an introduced weevil in New Zealand, *Nature (London)* **286:**489–490 (1980).
133. S. R. Sims, Diapause dynamics and host plant suitability of *Papilio zelicaon* (Lepidoptera: Papilionidae), *Am. Midl. Nat.* **103:**375–384 (1980).
134. D. H. Janzen and T. W. Schoener, Differences in insect abundance and diversity between wetter and drier sites during a tropical dry season, *Ecology* **49:**96–110 (1968).
135. A. M. Young, The population biology of Neotropical cicadas. III. Notes on the behavioral natural history of *Pacarina* cicadas in some Costa Rican grasslands, *Entomol. News* **85:**239–256 (1974).
136. A. M. Young, The population biology of neotropical cicadas. I. Emergences of *Procollina* and *Carineta* in a mountain forest, *Biotropica* **7:**248–258 (1975).
137. A. M. Young, Notes on the faunistic complexity of cicadas (Homoptera: Cicadidae) in northern Costa Rica, *Rev. Biol. Trop.* **24:**267–279 (1976).
138. A. M. Young, Habitat and seasonal relationships of some cicadas (Homoptera: Cicadidae) in central Costa Rica, *Am. Midl. Nat.* **103:**155–166 (1980).
139. A. M. Young, Seasonal adult emergences of cicadas (Homoptera: Cicadidae) in northwestern Costa Rica, *Milwaukee Publ. Mus. Contrib. Biol. Geol.* No. 40 (1981).
140. H. Dingle, Life history and population consequences of density, photoperiod, and temperature in a migrant insect, the milkweed bug *Oncopeltus, Am. Nat.* **102:**149–163 (1968).
141. H. Dingle, Migration strategies in insects, *Science* **175:**1327–1335 (1972).
142. H. Dingle, in: *Evolution of Insect Migration and Diapause* (H. Dingle, ed.), pp. 254–276, Springer-Verlag, Berlin (1978).
143. C. G. Johnson, *The Migration and Dispersal of Insects by Flight,* Methuen, London (1969).
144. E. O. Wilson and W. H. Bossert, *A Primer in Population Biology,* Sinauer Associates, Stamford, Conn. (1971).
145. T. R. E. Southwood, Migration of terrestrial arthropods in relation to habitat, *Biol. Rev.* **37:**171–214 (1962).
146. R. B. Root and S. J. Chaplin, The life-styles of tropical milkweed bugs. *Oncopeltus* (Hemiptera: Lygaeidae) utilizing the same hosts, *Ecology* **57:**132–140 (1976).
147. A. M. Young, Notes on a migration of *Urania fulgens* (Lepidoptera: Uraniidae) in Costa Rica, *J. N.Y. Entomol. Soc.* **78:**60–70 (1970).
148. W. Stubblebine, J. H. Langenheim, and D. Lincoln, Vegetative response to photoperiod in the tropical leguminous tree *Hymenaea courbaril* L., *Biotropica* 10:18–29 (1978).
149. G. W. Ankersmit and P. L. Adkisson, Photoperiodic responses of certain geographical strains of *Pectinophora gossypiella* (Lepidoptera), *J. Insect Physiol.* **13:**553–564 (1967).
150. A. M. Young, Temporal selection for communicatory optimization: The dawn-dusk chorus as an adaptation in tropical cicadas, *Am. Nat.* **117:**826–829 (1980).

151. W. H. Whitford, E. Depree, and P. Johnson, Foraging ecology of two Chihuahuan desert ant species: *Novomessor cockerelli* and *Novomessor albisetosus, Insectes Soc.* **27:**148–156 (1980).

152. H. A. Hespenheide, Dispersion and the size composition of the aerial insect fauna, *Ecol. Entomol.* **2:**139–141 (1977).

153. W. F. Snow, Field estimates of the flight speed of some West African mosquitoes, *Ann. Trop. Med. Parasitol.* **74:**239–242 (1980).

154. J. A. Downes, Arctic insects and their environment, *Can. Entomol.* **96:**279–307 (1964).

155. D. R. Oliver, Adaptations of Arctic Chironomidae, *Ann. Zool. Fenn.* **5:**111–118 (1968).

156. A. M. Young, Some differences between temperate and tropical populations of Monarch (*Danaus plexippus*) and Queen (*Danaus gilippus*) butterflies (Lepidoptera: Danaidae), *Entomol. News* **85:**116–126 (1974).

157. J. A. Mutchmor, Temperature adaptation in insects, in: *Molecular Mechanisms of Temperature Adaptation,* pp. 165–176, American Association for the Advancement of Science, Washington, D.C. (1967).

158. G. Alexander, Cold hardiness in overwintering juvenile grasshoppers, *Entomol. News* **78:**147–154 (1967).

159. J. A. Mutchmor and A. G. Richards, Low temperature tolerance of insects in relation to the influence of temperature on muscle apyrase activity, *J. Insect Physiol.* **7:**141–158 (1961).

160. W. Nachtigall and D. M. Wilson, Neuro-muscular control of dipteran flight, *J. Exp. Biol.* **47:**77–97 (1967).

161. N. S. Church, Heat loss and the body temperatures of flying insects. I. Heat loss by evaporation of water from the body, *J. Exp. Biol.* **37:**171–185 (1960).

162. N. S. Church, Heat loss and the body temperatures of flying insects. II. Heat conduction within the body and its loss by radiation and convection, *J. Exp. Biol.* **37:**186–213 (1960).

163. A. E. Kammer, Motor patterns during flight and warm-up in Lepidoptera, *J. Exp. Biol.* **48:**89–109 (1968).

164. J. E. Rawlins, Themoregulation by the Black Swallowtail Butterfly, *Papilio polyxenes* (Lepidoptera: Papilionidae), *Ecology* **6:**345–357 (1980).

165. B. Heinrich, Temperature regulation in the bumblebee *Bombus vagans:* A field study, *Science* **175:**185–187 (1972).

166. B. Heinrich and P. H. Raven, Energetics and pollination ecology, *Science* **176:**597–602 (1972).

167. C. D. Michener, Biogeography of the bees, *Ann. Mo. Bot. Gard.* **66:**277–347 (1979).

168. J. S. Weir, The effect of temperature variations on queen oviposition and colony foundation in *Myrmica, J. Insect Physiol.* **1:**352–360 (1958).

169. P. L. Miller, Factors altering spiracle control in adult dragonflies: Water balance, *J. Exp. Biol.* **41:**331–343 (1964).

170. P. L. Miller, Factors altering spiracle control in adult dragonflies: Hypoxia and temperature, *J. Exp. Biol.* **41:**345–357 (1964).

171. K. P. Lamb, Some effects of fluctuating temperatures on metabolism development, and rate of population growth in the cabbage aphid, *Brevicoryne brassicae, Ecology* **42:**740–745 (1961).

172. K. P. Lamb, Water loss from aphids at moderate temperatures, *J. Insect Physiol.* **9:**703–711 (1963).

173. B. J. Wall, Effects of dehydration and rehydration on *Periplaneta americana, J. Insect Physiol.* **16:**1027–1042 (1970).

174. C. Papageorgis, Mimicry in neotropical butterflies, *Am. Sci.* **63:**522–532 (1975).

175. A. M. Young, The comparative ethology and ecology of several species of *Morpho* butterflies in Costa Rica, *Stud. Neotrop. Fauna* **8:**17–50 (1973).

176. A. M. Young and A. Muyshondt, Ecological and geographical expansion in tropical butterflies of the genus *Morpho* in evolutionary time, *Rev. Biol. Trop.* **20:**231–263 (1972).
177. T. Pliske, Courtship behavior and use of chemical communication by males of certain species of ithomiine butterflies (Nymphalidae: Lepidoptera), *Ann. Entomol. Soc. Am.* **68:**935–942 (1975).
178. T. Pliske, Pollination of pyrrolizidine alkaloid-containing plants by male Lepidoptera, *Environ. Entomol.* **4:**474–479 (1975).
179. T. Pliske, Attraction of Lepidoptera to plants containing pyrrolizidine alkaloids, *Environ. Entomol.* **4:**455–473 (1975).
180. L. P. Brower and M. A. Jones, Precourtship interaction of wing and abdominal sex glands in male *Danaus* butterflies, *Proc. R. Entomol. Soc. London Ser. A* **40:**147–151 (1965).
181. R. I. Vane-Wright, Pre-courtship activity and a new scent organ in butterflies, *Nature (London)* **239:**338–340 (1972).
182. J. R. Aldrich, M. S. Blum, and S. S. Duffey, Male specific natural products in the bug, *Leptoglossus phyllopus:* Chemistry and possible function, *J. Insect Physiol.* **22:**1201–1206 (1976).
183. J. R. Aldrich and M. S. Blum, Aposematic aggregation of a bug (Hemiptera: Coreidae): The defensive display and formation of aggregations, *Biotropica* **10:**58–61 (1978).
184. R. Thornhill, Sexual selection within mating swarms of the lovebug, *Plecia nearctica* (Diptera: Bibionidae), *Anim. Behav.* **28:**405–412 (1980).
185. R. Thornhill, Competition and coexistence among *Panorpa* scorpionflies (Mecoptera: Panorpidae), *Ecol. Monogr.* **50:**179–197 (1980).
186. A. M. Young, Community ecology of some tropical rain forest butterflies, *Am. Midl. Nat.* **87:**146–157 (1972).
187. A. M. Young, Feeding behavior of *Morpho* butterflies in a seasonal tropical environment, *Rev. Biol. Trop.* **23:**101–123 (1975).
188. A. M. Young, The evolution of eyespots in tropical butterflies in response to feeding on rotting fruit: An hypothesis, *J. N.Y. Entomol. Soc.* **87:**66–77 (1979).
189. A. M. Young, The interaction of predators and "eyespot butterflies" feeding on rotting fruits and soupy fungi in tropical forests: Variations on a theme developed by the Muyshondts and Arthur M. Shapiro, *Entomol. Rec. J. Var.* **90:**63–70 (1980).
190. A. M. Young and J. H. Thomason, Demography of a confined population of the butterfly *Morpho peleides* during a tropical dry season, *Stud. Neotrop. Fauna* **9:**1–34 (1974).
191. B. W. Cornaby, Carrion reduction by animals in contrasting tropical habitats, *Biotropica* **6:**51–63 (1974).
192. M. G. Naumann, Swarming behavior: Evidence for communication in social wasps, *Science* **189:**642–644 (1975).
193. A. M. Young, Attacks by the army ant *Eciton burchelli* on nests of the social paper wasp *Polistes erythrocephalus* in northeastern Costa Rica, *J. Kans. Entomol. Soc.* **52:**759–768 (1979).
194. M. J. West-Eberhard, The evolution of social behaviour by kin selection, *Q. Rev. Biol.* **50:**1–33 (1975).
195. R. M. Crewe and M. S. Blum, Alarm pheromones of the Attini: Their phylogenetic significance, *J. Insect Physiol.* **18:**31–42 (1972).
196. M. S. Blum, R. M. Crewe, W. E. Kerr, L. H. Keith, A. W. Garrison, and M. M. Walker, Citral in stingless bees: Isolation and functions in trail-laying and robbing, *J. Insect Physiol.* **16:**1637–1648 (1970).
197. C. R. Carroll and D. H. Janzen, Ecology of foraging by ants, *Ann. Rev. Ecol. Syst.* **4:**231–258 (1973).
198. L. M. Riddiford and C. M. Williams, The effects of juvenile hormone analogues on the embryonic development of silkworms, *Proc. Natl. Acad. Sci. U.S.A.* **57:**595–601 (1967).

199. F. Engelmann, Endocrine control of reproduction in insects, *Ann. Rev. Entomol.* **13:**1–26 (1968).
200. W. K. Reisen and M. Aslamkhan, A release-recapture experiment with the malaria vector, *Anopheles stephensi* Liston, with observations on dispersal, survivorship, population size, gonotrophic rhythm and mating behaviour, *Ann. Trop. Med. Parasitol.* **73:**251–269 (1979).
201. B. B. Simpson, Pleistocene changes in the flora of the high tropical Andes, *Paleobiology* **1:**273–291 (1975).
202. A. H. Ehrlich and P. R. Ehrlich, Reproductive strategies in the butterflies. I. Mating frequency, plugging, and egg number, *J. Kans. Entomol. Soc.* **51:**666–697 (1978).
203. K. S. Brown, Jr., and O. H. H. Mielke, The heliconians of Brazil (Lepidoptera: Nymphalidae). II. Introduction and general comments, with a supplementary revision of the tribe, *Zoologica* **1972:**1–40 (1972).
204. K. S. Brown, Jr., and W. W. Benson, Adaptive polymorphism associated with multiple Mullerian mimicry in *Heliconius numata* (Lepid. Nymph.), *Biotropica* **6:**205–228 (1974).
205. K. S. Brown, Jr., and J. V. Neto, Predation on aposematic ithomiine butterflies by tanagers (*Pipraeidea melanonota*), *Biotropica* **8:**136–141 (1976).
206. L. P. Brower, W. H. Calvert, L. E. Hedrick, and J. Christian, Biological observations on an overwintering colony of Monarch butterflies (*Danaus plexippus*, Danaidae) in Mexico, *J. Lepid. Soc.* **31:**232–242 (1977).
207. A. M. Young, Wing coloration and reflectance in *Morpho* butterflies as related to reproductive behavior and escape from avian predators, *Oecologia* **7:**209–222 (1971).
208. A. J. Alexander, A study of the biology and behavior of the caterpillars and emerging butterflies of the subfamily Heliconiinae in Trinidad, West Indies. I. Some aspects of larval behavior, *Zoologia* **46:**1–24 (1961).
209. J. M. Ringo, Sexual behavior in Hawaiian *Drosophila* (Diptera: Drosophilidae), *J. N.Y. Entomol. Soc.* (Abstract) **88:**68 (1980).
210. H. R. Hermann, Crespuscular and nocturnal activities of *Paraponera clavata* (Hymenoptera: Formicidae: Ponerinae), *Entomol. News* **86:**94–98 (1975).
211. A. M. Young, Notes on the foraging of the giant tropical ant *Paraponera clavata* (Formicidae: Ponerinae) on two plants in tropical wet forest, *J. Ga. Entomol. Soc.* **11:**41–51 (1976).
212. A. M. Young and H. R. Hermann, Notes on foraging of the giant tropical ant *Paraponera clavata* (Hymenoptera: Formicidae: Ponerinae), *J. Kans. Entomol. Soc.* **53:**35–55 (1980).
213. E. S. McCluskey, Circadian rhythms in male ants of five diverse species, *Science* **150:**1037–1039 (1965).
214. T. J. Walker, Cryptic species among sound-producing ensiferan Orthoptera (Gryllidae and Tettigoniidae), *Q. Rev. Biol.* **39:**345–355 (1964).
215. A. M. Young, Observations on the aggregation of adult cicadas (Homoptera: Cicadidae) in tropical forests, *Can. J. Zool.* **58:**711–722 (1980).
216. H. G. Baker and I. Baker, in: *Taxonomy and Ecology* (V. H. Heywood, ed.), pp. 243–264, Academic Press, New York (1973).
217. H. G. Baker and I. Baker, in: *Coevolution of Animals and Plants* (L. E. Gilbert and P. H. Raven, eds.), pp. 100–140, University of Texas Press, Austin (1975).
218. H. G. Baker, P. A. Opler, and I. Baker, A comparison of the amino acid complements of floral and extrafloral nectars, *Bot. Gaz. (Chicago)* **139:**322–332 (1978).
219. W. L. Brown, Jr., and E. O. Wilson, The evolution of the dacetine ants, *Q. Rev. Biol.* **34:**278–294 (1959).
220. A. M. Young, Possible evolution of mutualism between *Mechanitis* caterpillars and an ant in northeastern Costa Rica, *Biotropica* **10:**77–78 (1977).
221. A. M. Young and M. W. Moffett, Studies on the population biology of the tropical butterfly *Mechanitis isthmia* in Costa Rica, *Am. Midl. Natl.* **101:**309–319 (1979).
222. A. M. Young and M. W. Moffett, Behavioral regulatory mechanisms in populations of the

butterfly *Mechanitis isthmia* (Lepidoptera: Nymphalidae: Ithomiinae) in Costa Rica: Adaptations to host plants in secondary and agricultural habitats, *Dtsch. Entomol. Z.* **26:**21–38 (1979).
223. R. W. Matthews, Nesting biology of the social wasp *Microstigmus comes* (Hymenoptera: Sphecidae, Pemphredoninae), *Psyche* **75:**23–45 (1968).
224. R. W. Matthews, *Microstigmus comes:* Sociality in a sphecid wasp, *Science* **160:**787–788 (1968).
225. L. L. Rockwood, Distribution, density, and dispersion of two species of *Atta* in Guanacaste Province, Costa Rica, *J. Anim. Ecol.* **42:**803–817 (1973).
226. L. L. Rockwood, The effects of seasonality on foraging in two species of leaf-cutting ants (*Atta*) in Guanacaste Province, Costa Rica, *Biotropica* **7:**176–193 (1975).
227. J. O. Schmidt, in: *Pesticides and Venom Neurotoxicity* (D. L. Shankland, R. M. Hollingworth, and T. Smyth, eds.), pp. 247–263, Plenum, New York (1978).
228. H. R. Herman and M. S. Blum, The morphology and histology of the hymenopterous poison apparatus. I. *Paraponera clavata* (Formicidae), *Ann. Entomol. Soc. Am.* **59:**397–409 (1966).
229. A. M. Young, A human sweat-mediated defense against multiple attacks by the wasp *Polybia diguetana* in northeastern Costa Rica, *Biotropica* **10:**73–74 (1977).
230. A. Wille and C. D. Michener, The nest architecture of stingless bees with special reference to those of Costa Rica (Hymenoptera: Apidae), *Rev. Biol. Trop. Spec. Suppl.* **21**(1) (1973).
231. T. Park, E. V. Miller and C. Z. Lutherman, Studies in population physiology. IX. The effect of imago population density on the duration of the larval and pupal stages of *Tribolium confusum* Duval, *Ecology* **20:**365–373 (1939).
232. P. S. Dawson, Developmental rate and competitive ability in *Tribolium*. II. Changes in competitive ability following further selection for developmental time, *Evolution* **21:**292–298 (1967).
233. F. J. Ayala, M. L. Tracey, L. G. Barr, and J. G. Ehrenfeld, Genetic and reproductive differentiation of the subspecies, *Drosophila equinoxalis caribbensis, Evolution* **28:**24–41 (1974).
234. W. C. Allee, Measurement of environmental factors in the tropical rain forest of Panama, *Ecology* **7:**273–302 (1926).
235. C. H. Fraser-Rowell, Experiments on aggregation of *Phymateus purpurascens* (Orthoptera, Acrididae, Pyrgomorphinae), *J. Zool.* **152:**179–193 (1967).
236. J. Myers and J. N. M. Smith, Head flicking by tent caterpillars: A defensive response to parasite sounds, *Can. J. Zool.* **56:**1628–1631 (1978).
237. D. Otte and A. Joern, Insect territoriality and its evolution: Population studies of desert grasshoppers on creosote bushes, *J. Anim. Ecol.* **44:**29–54 (1975).
238. C. T. Brues, The specificity of food plants in the evolution of phytophagous insects, *Am. Nat.* **58:**127–144 (1924).
239. V. G. Dethier, Evolution of feeding preferences in phytophagous insects, *Evolution* **8:**33–54 (1954).
240. V. G. Dethier, in: *Chemical Ecology* (E. Sondheimer and J. B. Simeone, eds.), pp. 83–102, Academic Press, London (1970).
241. V. G. Dethier, Evolution of receptor sensitivity to secondary plant substances with special reference to deterrents, *Am. Nat.* **115:**45–66 (1980).
242. G. Fraenkel, Insect nutrition, *J. R. Coll. Sci.* **13:**59–69 (1943).
243. G. Fraenkel, The raison d'être of secondary plant substances, *Science* **129:**1466–1470 (1959).
244. R. T. Yamamoto and G. S. Fraenkel, The specificity of the tobacco hornworm, *Protoparce sexta,* to solanaceous plants, *Ann. Entomol. Soc. Am.* **53:**503–507 (1960).
245. J. A. Downes, The food habits and description of *Atrichopogon pollinivorus* sp.n. (Diptera: Ceratopogonidae), *Trans. R. Entomol. Soc. London* **106:**439–453 (1955).

246. L. M. Schoonhoven and V. G. Dethier, Sensory aspects of host-plant discrimination by lepidopterous larvae, *Arch. Neerl. Zool.* **16:**497–530 (1966).
247. L. M. Schoonhoven, in: *Insect/Plant Relationships* (H. F. van Emden, ed.), pp. 87–99, Blackwell, Oxford (1973).
248. P. Jaisson, Environmental preference induced experimentally in ants (Hymenoptera: Formicidae), *Nature (London)* **286:**388–384 (1980).
249. B. Heinrich, The effect of leaf geometry on the feeding behaviour of the caterpillar of *Manduca sexta* (Sphingidae), *Anim. Behav.* **19:**119–124 (1971).
250. P. W. Miles, Interaction of plant phenols and salivary phenolases in the relationship between plants and Hemiptera, *Entomol. Exp. Appl.* **12:**736–744 (1969).
251. M. H. Sweet, On the original feeding habits of the Hemiptera (Insecta), *Ann. Entomol. Soc. Am.* **72:**575–579 (1979).
252. C. W. Schaefer and R. O'Shea, Host plants of three coreine tribes (Hemiptera: Heteroptera: Coreidae), *Ann. Entomol. Soc. Am.* **72:**519–523 (1979).
253. J. W. Webb, Life history and population dynamics of tree-dwelling Homoptera, *J. N.Y. Entomol. Soc.* (Abstract) **88:**80 (1980).
254. R. Levins and R. MacArthur, An hypothesis to explain the incidence of monophagy, *Ecology* **50:**910–911 (1969).
255. W. W. Benson, Evidence for the evolution of unpalatability through kin selection in the Heliconiinae (Lepidoptera), *Am. Nat.* **105:**213–226 (1971).
256. D. A. Levin, The chemical defenses of plants to pathogens and herbivores, *Ann. Rev. Ecol. Syst.* **7:**121–159 (1976).
257. G. A. Rosenthal and D. H. Janzen (eds.), *Herbivores: Their Interaction with Secondary Plant Metabolites,* Academic Press, New York (1979).
258. P. Feeny, Seasonal changes in oak leaf tannins and nutrients as a cause of spring feeding by winter moth caterpillars, *Ecology* **51:**565–581 (1970).
259. P. P. Feeny, K. L. Paauwe, and N. J. Demong, Flea beetles and mustard oils: Host plant specificity of *Phyllotreta cruciferae* and *P. striolata* adults (Coleoptera: Chrysomelidae), *Ann. Entomol. Soc. Am.* **63:**832–841 (1970).
260. J. M. Erickson and P. Feeny, Sinigrin: A chemical barrier to larvae of the black swallowtail butterfly, *Papilio polyxenes, Ecology* **55:**103–111 (1974).
261. S. S. Rehr, D. H. Janzen, and P. P. Feeny, L-Dopa in legume seeds: A chemical barrier to insect attack, *Science* **181:**81–82 (1973).
262. S. S. Rehr, P. P. Feeny, and D. H. Janzen, Chemical defence in Central American non-ant acacias, *J. Anim. Ecology* **42:**405–416 (1973).
263. S. S. Rehr, E. A. Bell, D. H. Janzen, and P. P. Feeny, Insecticidal amino acids in legume seeds, *Biochem. Syst.* **1:**63–67 (1973).
264. R. I. Krieger, P. P. Feeny, and C. F. Wilkinson, Detoxication enzymes in the guts of caterpillars: An evolutionary answer to plant defenses?, *Science* **172:**579–581 (1971).
265. N. E. Stamp, Egg deposition patterns in butterflies: Why do some species cluster their eggs rather than deposit them singly?, *Am. Nat.* **115:**367–380 (1980).
266. A. W. Ghent, A study of the group-feeding behaviour of larvae of the Jack Pine Sawfly, *Neodirprion pratti banksianae* Roh, *Behaviour* **16:**110–148 (1960).
267. P. D. Coley, Effects of leaf age and plant life history patterns on herbivory, *Nature (London)* **284:**545–546 (1980).
268. M. J. Raupp and R. F. Denno, Interactions among willow herbivores as a determinant of feeding and ovipositional patterns of the imported willow leaf beetle, *Plagiodera versicolora* Laich. (Coleoptera: Chrysomelidae) *J. N.Y. Entomol. Soc.* (Abstract) **88:**67 (1980).
269. C. R. Carroll and C. A. Hoffman, Chemical feeding deterrent mobilized in response to insect herbivory and counter adaptation by *Epilachna tredecimnotata, Science* **209:**414–416 (1980).

270. A. M. Young, The biology of the butterfly *Aeria eurimedea agna* (Nymphalidae: Ithomiinae: Oleriini) in Costa Rica, *J. Kans. Entomol. Soc.* **51:**1–10 (1978).
271. L. P. Brower and J. V. Z. Brower, Birds, butterflies, and plant poisons: A study in ecological chemistry, *Zoologica* **49:**137–159 (1964).
272. L. E. Gilbert and P. R. Ehrlich, The affinities of the Ithomiinae and Satyridae (Nymphalidae), *J. Lepid. Soc.* **24:**297–300 (1970).
273. A. M. Young, On the evolutionary distance between asclepiadaceous-feeding danaids and apocyanaceous-feeding ithomiids, *J. Res. Lepid.* **18:**251–254 (1980).
274. D. J. Futuyuma, Food plant specialization and environmental predictability in Lepidoptera, *Am. Nat.* **110:**285–292 (1976).
275. A. M. Young, Evolutionary responses by butterflies to patchy spatial distributions of resources in tropical environments, *Acta Biotheor.* **29:**37–64 (1980).
276. D. F. Rhoades, in: *Herbivores—Their Interaction with Secondary Plant Metabolites* (G. A. Rosenthal and D. H. Janzen, eds.), pp. 3–47, Academic Press, New York (1979).
277. L. B. Brattsten, C. F. Wilkinson, and T. Eisner, Herbivore-plant interactions: Mixed-function oxidases and secondary plant substances, *Science* **196:**1349–1352 (1977).
278. V. G. Dethier, Egg-laying habits of Lepidoptera in relation to available food, *Can. Entomol.* **91:**554–561 (1959).
279. V. G. Dethier, Food-plant distribution and density and larval dispersal as factors affecting insect populations, *Can. Entomol.* **91:**581–596 (1959).
280. R. Straatman, Notes on certain Lepidoptera ovipositing on plants which are toxic to their larvae, *J. Lepid. Soc.* **16:**99–103 (1962).
281. M. S. McClure, Biology of *Erythroneura lawsoni* (Homoptera: Cicadellidae) and coexistence in the sycamore leaf-feeding guild, *Environ. Entomol.* **3:**59–68 (1974).
282. W. W. Benson, Resource partitioning in passion vine butterflies, *Evolution* **32:**493–518 (1978).
283. W. W. Benson, K. S. Brown, Jr., and L. E. Gilbert, Coevolution of plants and herbivores: Passion flower butterflies, *Evolution* **29:**659–680 (1975).
284. A. Muyshondt, Notes on the life cycle and natural history of butterflies of El Salvador. II. *Anaea (Zaretis) itys* (Nymphalidae), *J. Lepid. Soc.* **27:**294–302 (1973).
285. A. Muyshondt, Notes on the life cycle and natural history of butterflies of El Salvador. III. *Anaea (Consul) fabius* (Nymphalidae), *J. Lepid. Soc.* **28:**81–89 (1974).
286. A. Muyshondt, Notes on the life cycle and natural history of butterflies of El Salvador. V. *Anaea (Memphis) morvus* Boisduvali (Nymphalidae), *J. Lepid. Soc.* **29:**32–39 (1975).
287. A. Muyshondt, Notes on the life cycle and natural history of butterflies of El Salvador. VI. *Anaea (Memphis) pithyusa* (Nymphalidae), *J. Lepid. Soc.* **29:**168–175 (1975).
288. A. Muyshondt, A. M. Young, and A. Muyshondt, Jr., The biology of the butterfly *Dione juno huascama* (Nymphalidae: Heliconiinae) in El Salvador, *J. N.Y. Entomol. Soc.* **81:**137–151 (1973).
289. A. M. Young, Mimetic associations in natural populations of tropical butterflies. I. Life history and structure of a tropical dry forest breeding population of *Battus polydamus*, *Rev. Biol. Trop.* **19:**210–240 (1971).
290. A. M. Young, On the life cycle and natural history of *Hymenitis nero* (Lepidoptera: Ithomiidae) in Costa Rica, *Psyche* **79:**284–294 (1972).
291. A. M. Young, Breeding success and survivorship in some tropical butterflies, *Oikos* **23:**318–326 (1972).
292. A. M. Young, Notes on the life cycle and natural history of *Dismorphia virgo* (Pieridae: Dismorphiinae) in Costa Rica, *Psyche* **79:**165–178 (1972).
293. A. M. Young, A contribution to the biology of *Itaballia caesia* (Lepidoptera: Pieridae) in a Costa Rican mountain ravine, *Wasmann J. Biol.* **30:**43–70 (1972).

294. A. M. Young, The ecology and ethology of the tropical nymphaline butterfly, *Victorina epaphus*. I. Life cycle and natural history, *J. Lepid. Soc.* **26:**155–170 (1972).
295. A. M. Young, Notes on the biology of the butterfly, *Heliconius cydno* (Lepidoptera: Heliconiinae) in Costa Rica, *Wasmann J. Biol.* **31:**337-350 (1973).
296. A. M. Young, Notes on the life cycle and natural history of *Parides arcas mylotes* (Papilionidae) in a Costa Rican premontane wet forest, *Psyche* **80:**1–21 (1973).
297. A. M. Young, Notes on the biology of *Phycioies (Eresia) eutropia* (Lepidoptera: Nymphalidae) in a Costa Rican mountain forest, *J. N.Y. Entomol. Soc.* **81:**87–100 (1973).
298. A. M. Young, The life cycle of *Dircenna relata* (Ithomiidae) in Costa Rica, *J. Lepid. Soc.* **27:**258-267 (1973).
299. A. M. Young, On the biology of *Hamadryas februa* (Lepidoptera: Nymphalidae: Nymphalinae) in Guanacaste, Costa Rica, *Z. Angew. Entomol.* **75:**380-393 (1974).
300. A. M. Young, Further observations on the natural history of *Philaethria dido* (Lepidoptera: Nymphalidae: Heliconiinae), *J. N.Y. Entomol. Soc.* **82:**30–41 (1974).
301. A. M. Young, A natural historical account of *Oleria zelica pagasa* (Lepidoptera: Ithomiidae) in a Costa Rican mountain rain forest, *Stud. Neotrop. Fauna* **9:**123–140 (1974).
302. A. M. Young, Notes on the biology of *Pteronymia notilla* (Ithomiidae) in a Costa Rican mountain forest, *J. Lepid. Soc.* **28:**257–268 (1974).
303. A. M. Young, Notes on the natural history of rare *Adelpha* butterfly (Lepidoptera: Nymphalidae) in Costa Rican high country, *J. N.Y. Entomol. Soc.* **82:**235–244 (1974).
304. A. M. Young, On the biology of *Godyris zavaleta caesiopicta* (Lepidoptera: Nymphalidae: Ithomiinae), *Entomol. News* **85:**227-238 (1974).
305. A. M. Young, Observations on the life cycle of *Heliconius hecale zuleika* (Hewitson) in Costa Rica, *Pan-Pac. Entomol.* **51:**76–85 (1975).
306. A. M. Young, Notes on the life cycle of the butterfly *Hypanartia kerfersteini* (Nymphalidae: Nymphalinae: Nymphalini) in Costa Rica, *Brenesia* **9:**61–69 (1976).
307. A. M. Young, Studies on the biology of *Heliconius charitonius* L. (Nymphalidae: Heliconiinae) in Costa Rica, *Pan-Pac. Entomol.* **52:**291–303 (1976).
308. A. M. Young, Studies on the biology of *Parides iphidamas* (Papilionidae: Troidini) in Costa Rica, *J. Lepid. Soc.* **31:**100–108 (1977).
309. A. M. Young, Notes on the defoliation of coconut palm (*Cocos nucifera*) by the butterfly *Opsiphanes quiteria quirinus* (Lepidoptera: Brassolidae) in northeastern Costa Rica, *Dtsch. Entomol. Z.* **24:**353-365 (1977).
310. A. M. Young, Notes on the biology of the butterfly *Hypoleria cassotis* (Bates) (Nymphalidae: Ithomiinae) in northeastern Costa Rica, *Brenesia* **10:**97–108 (1977).
311. A. M. Young, Studies on the interactions of *Morpho peleides* (Morphidae) with Leguminosae, *J. Lepid. Soc.* **32:**65–74 (1978).
312. A. M. Young, Disappearances of eggs and larvae of *Heliconius* butterflies (Nymphalidae: Heliconiinae) in northeastern Costa Rica, *Entomol. News* **89:**81–89 (1978).
313. A. M. Young, Spatial properties of niche separation among *Eueides* and *Dryas* butterflies (Lepidoptera: Nymphalidae: Heliconiinae) in Costa Rica, *J. N.Y. Entomol. Soc.* **86:**2–19 (1978).
314. A. M. Young, Over-exploitation of larval host plants by the butterflies *Heliconius cydno* and *Heliconius sapho* (Lepidoptera: Nymphalidae: Heliconiinae: Heliconiini) in Costa Rica?, *J. N.Y. Entomol. Soc.* **88:**217–227 (1980).
315. A. M. Young, Notes on the seasonal distribution of *Anaea* butterflies (Nymphalidae) in tropical dry forests, *Acta Oecol.* **2:**17–30 (1981).
316. A. M. Young and A. Muyshondt, The biology of *Morpho peleides* in Central America, *Caribb. J. Sci.* **13:**1–49 (1973).
317. A. M. Young and A. Muyshondt, Ecological studies of the butterfly *Victorina stelenes*

(Lepidoptera: Nymphalidae) in Costa Rica and El Salvador, *Stud. Neotrop. Fauna* **9:**155–176 (1974).

318. A. M. Young and A. Muyshondt, Studies on the natural history of Central American butterflies in the family cluster Satyridae-Brassolidae-Morphidae (Lepidoptera: Nymphaloidea). III. *Opsiphanes tamarindi* and *Opsiphanes cassina* in Costa Rica and El Salvador, *Stud. Neotrop. Fauna* **10:**19–56 (1975).
319. A. M. Young and A. Muyshondt, Biology of *Morpho polyphemus* in El Salvador, *J. N.Y. Entomol. Soc.* **80:**18–42 (1972).
320. C. Wiklund, Oviposition preferences in *Papilio machaon* in relation to the host plants of the larvae, *Entomol. Exp. Appl.* **17:**189–198 (1974).
321. C. Wiklund, The evolutionary relationship between adult oviposition preferences and larval host plant range in *Papilio machaon* L., *Oecologia* **18:**185–197 (1975).
322. A. M. Young and D. Stein, Studies on the evolutionary biology of the neotropical nymphalid butterfly *Anartia fatima* in Costa Rica, *Milwaukee Publ. Mus. Contrib. Biol. Geol.* No. 8 (1976).
323. A. M. Young, Oviposition of the butterfly *Battus belus varus* Kollar (Papilionidae: Papilioninae: Troidini), *J. Lepid. Soc.* **33:**56–57 (1979).
324. A. M. Young, Oviposition by *Heliconius hecale* (Nymphalidae: Heliconiinae) on a grass inflorescence in Costa Rica, *J. Lepid. Soc.* **34:**66–68 (1980).
325. J. M. Pires, Th. Dobzhansky, and G. A. Black, An estimate of the number of species of trees in an Amazonian forest community, *Bot. Gaz.* (Chicago) **114:**467–477 (1953).
326. L. L. Rockwood, The effect of defoliation on seed production of six Costa Rican tree species, *Ecology* **54:**1363–1369 (1973).
327. D. E. Breedlove and P. R. Ehrlich, Coevolution: Patterns of legume predation by a lycaenid butterfly, *Oecologia* **10:**99–104 (1972).
328. K. Nakano, A model for the problem of the effects of defoliation on plant growth, *Jpn. J. Ecol.* **27:**227–247 (1977).
329. L. E. Gilbert and J. T. Smiley, in: *Diversity of Insect Faunas* (L. A. Mound and N. Waloff, eds.), pp. 89–104, Blackwell, London (1978).
330. B. J. Rathcke and R. W. Poole, Coevolutionary race continues: Butterfly adaptation to plant trichomes, *Science* **187:**175–176 (1975).
331. A. M. Young, The rearing of *Morpho peleides* (Morphinae) on peanuts, *J. Lepid. Soc.* **28:**90–99 (1974).

331a. A. M. Young, Notes on the natural history of *Morpho granadensis polybaptus* Butler (Lepidoptera: Nymphalidae: Morphinae), and its relation to that of *Morpho peleides limpida* Butler, *J. N.Y. Entomol. Soc.* **90:** in press (1982).

332. D. H. Janzen, Tropical blackwater rivers, animals, and mast fruiting by the Dipterocarpaceae, *Biotropica* **6:**69–103 (1973).
333. D. McKey, P. G. Waterman, C. N. Mbi, J. S. Gartlan, and T. T. Struhsaker, Phenolic content of vegetation in two African rain forests: Ecological implications, *Science* **202:**61–64 (1978).
334. L. Van Der Pijl, Evolutionary action of tropical animals on the reproduction of plants, *Biol. J. Linn. Soc.* **1:**85–96 (1969).
335. K. Faegri and L. Van Der Pijl, *The Principles of Pollination Ecology,* Pergamon Press, New York (1971).
336. D. J. Carter and J. C. Demming, *Azygophleps albovittata* Bethune-Baker (Lepidoptera: Cossidae) attacking groundnuts in northern Nigeria, with descriptions of the immature and imaginal stages, *Bull. Entomol. Res.* **70:**399–405 (1980).
337. G. S. Hartshorn, in: *Tropical Ecological Systems: Trends in Terrestrial and Aquatic Research* (F. B. Golley and E. Medina, eds.), pp. 41–51, Springer-Verlag, New York (1975).
338. L. E. Gilbert, in: *Coevolution of Animals and Plants* (L. E. Gilbert and P. H. Raven, eds.), pp. 210–240, University of Texas Press, Austin (1975).

339. W. W. Benson, Natural selection for mullerian mimicry in *Heliconius erato* in Costa Rica, *Science* **176:**936–939 (1972).

340. D. Simberloff, Species turnover and equilibrium island biogeography, *Science* **194:**572–578 (1976).

341. R. H. MacArthur and E. O. Wilson, *The Theory of Island Biogeography,* Princeton University Press, Princeton, N.J. (1967).

342. L. P. Brower, J. V. Z. Brower, and C. T. Collins, Experimental studies of mimicry. 7. Relative palatability and Mullerian mimicry among neotropical butterflies of the subfamily Heliconiinae, *Zoologica* **48:**65–84 (1963).

343. M. Rothschild, in: *Insect/Plant Relationships* (H. F. van Emden, ed.), pp. 59–83, Blackwell, Oxford (1972).

344. M. Rothschild and D. N. Kellett, Notes on the reactions of various predators to insects storing heart poisons (cardiac glycosides) in their tissues, *J. Entomol. Ser. A* **46**(2):103–110 (1972).

345. M. Rothschild, J. Von Euw, and T. Reichstein, Some problems connected with warningly coloured insects and toxic defense mechanisms, *Mitt. Basler Afrika Bibliogr.* **4–6:**135–158 (1972).

346. M. Rothschild, J. von Euw, and T. Reichstein, Cardiac glycosides (heart poisons) in the polka-dot moth *Syntomeida epilais* Walk. (Ctenuchidae: Lep.) with some observations on the toxic qualities of *Amata* (=*Syntomis*) *phegea* (L.), *Proc. R. Soc. London Ser. B* **183:**227–247 (1973).

347. F. H. Pough and L. P. Brower, Predation by birds on Great Southern White butterflies as a function of palatability, sex, and habitat, *Am. Midl. Nat.* **98:**50–58 (1977).

348. L. P. Brower and C. M. Moffitt, Palatability dynamics of cardenolides in the Monarch Butterfly, *Nature (London)* **249:**280–283 (1974).

349. L. P. Brower, J. V. Z. Brower, and J. M. Corvino, Plant poisons in a terrestrial food chain, *Proc. Natl. Acad. Sci. U.S.A.* **57:**893–898 (1967).

350. L. P. Brower, W. N. Ryerson, L. I. Coppinger, and S. C. Glazier, Ecological chemistry and the palatability spectrum, *Science* **161:**1349–1351 (1968).

351. L. P. Brower, M. Edmunds, and C. M. Moffitt, Cardenolide content and palatability of a population of *Danaus chrysippus* butterflies from West Africa, *J. Entomol.* **49:**183–196 (1975).

352. L. P. Brower, P. B. McEvoy, K. L. Williamson, and M. A. Flannery, Variation in cardiac glycoside content of monarch butterflies from natural populations in Eastern North America, *Science* **177:**426–429 (1972).

353. S. S. Duffey and G. G. E. Scudder, Cardiac glycosides in North American Asclepiadaceae, a basis for unpalatability in brightly coloured Hemiptera and Coleoptera, *J. Insect Physiol.* **18:**63–78 (1972).

354. F. A. Vaughan, Effect of gross cardiac glycoside content of seeds of common milkweed, *Asclepias syriaca,* on cardiac glycoside uptake by the milkweed bug *Oncopeltus fasciatus,* J. Chem. Ecol. **5:**89–100 (1979).

355. A. M. Young, Notes on the interaction of the Neotropical bug *Paryphes blandus* Horvath (Hemiptera: Coreidae) with the vine *Anguria warscewiczii* Hook. f. (Cucurbitaceae), *Brenesia* **17:**27–42 (1980).

356. D. F. Owen, *Tropical Butterflies,* Clarendon Press, Oxford (1971).

357. D. A. Clark, Age- and sex-dependent foraging strategies of a small mammalian omnivore, *J. Anim. Ecol.* **49:**549–563 (1980).

358. B. C. Campbell and S. S. Duffey, Tomatine and parasitic wasps: Potential incompatibility of plant antibiosis with biological control, *Science* **205:**700–702 (1979).

359. J. H. Vandermeer, Relative isolation and seed predation in *Calliandra grandiflora,* a mimosaceous legume from the highlands of Guatemala, *Biotropica* **6:**267–268 (1974).

360. J. M. Kenyi, Seasonal defoliation of thickets by larvae of *Belenois aurota, Afr. J. Ecol.* **18:**123–126 (1980).
361. S. D. Hendrix, An evolutionary and ecological perspective of the insect fauna of ferns, *Am. Nat.* **115:**171–196 (1980).
362. D. H. Janzen, Seed-eaters versus seed size, number, dispersal and toxicity, *Evolution* **23:**1–27 (1969).
363. D. H. Janzen, Escape of *Cassia grandis* L. beans from predators in time and space, *Ecology* **52:**964–979 (1971).
364. D. H. Janzen, Escape of juvenile *Dioclea megacarpa* (Leguminosae) vines from predators in a deciduous tropical forest, *Am. Nat.* **105:**97–112 (1971).
365. D. H. Janzen, Escape in space by *Sterculia apetala* seeds from the bug *Dysdercus fasciatus* in a Costa Rican deciduous forest, *Ecology* **53:**350–361 (1972).
366. J. White and C. E. Strehl, Xylem feeding by periodical cicada nymphs on tree roots, *Ecol. Entomol.* **3:**323–327 (1978).
367. R. F. Denno, M. J. Raupp, D. W. Tallamy, and C. F. Riechelderfer, Migration in heterogeneous environments: Differences in habitat selection between wing forms of the dimorphic planthopper, *Prokelisia marginata* (Homoptera: Delphacidae), *Ecology* **61:**859–867 (1980).
368. S. McNeill and T. R. E. Southwood, in: *Biochemical Aspects of Plant and Animal Coevolution* (J. B. Harborne, ed.), pp. 77–98, Academic Press, London (1978).
369. D. Horsfield, Relationships between feeding of *Philaenus spumarius* (L.) and the amino acid concentration in the xylem sap, *Ecol. Entomol.* **2:**259–266 (1977).
370. D. P. Janos, Mycorrhizae influence tropical succession, *Biotropica* **12** (Supplement):56–64 (1980).
371. V. C. Sgarbieri, Estudo do conteudó e de algumas caracteristicas das proteinas em sementes de plantas da familia Leguminosae, *Cienc. Cult. (Sao Paolo)* **32:**78–84 (1980).
372. D. H. Janzen, Swollen-thorn acacias of Central America, *Smithson. Contrib. Bot.* No. 13 (1974).
373. D. H. Janzen, Protection of *Barteria* (Passifloraceae) by *Pachysima* ants (Pseudomyrmecinae) in a Nigerian rain forest, *Ecology* **53:**885–892 (1972).
374. D. H. Janzen, Allelopathy by myrmecophytes: The ant *Azteca* as an allelopathic agent of *Cecropia, Ecology* **50:**147–153 (1969).
375. D. H. Janzen, Dissolution of mutualism between *Cecropia* and its *Azteca* ants, *Biotropica* **5:**15–28.
376. S. Risch, M. McClure, J. Vandermeer, and S. Waltz, Mutualism between three species of tropical piper (Piperaceae) and their ant inhabitants, *Am. Midl. Nat.* **98:**433–444 (1977).
377. B. L. Bentley, The protective function of ants visiting the extrafloral nectaries of *Bixa orellana* (Bixaceae), *J. Ecol.* **65:**27–38 (1977).
378. K. H. Keeler, The extrafloral nectaries of *Ipomoea leptophylla* (Convolvulaceae), *American Journal of Botany* **67:**216–222 (1980).
379. K. H. Keeler and R. B. Kaul, Morphology and distribution of petiolar nectaries in *Ipomoea* (Convolvulaceae), *Am. J. Bot.* **66:**946–952 (1979).
380. S. Koptur, Facultative mutualism between weedy vetches bearing extrafloral nectaries and weedy ants in California, *Am. J. Bot.* **66:**1016–1020 (1979).
381. R. Levins, in: *Genetics and Evolution Symposium* (E. T. Drake, ed.), pp. 325–340, Yale University Press, New Haven, Conn. (1968).
382. J. S. Jones and R. F. Probert, Habitat selection maintains a deleterious allele in a heterogeneous environment, *Nature (London)* **287:**632–633 (1980).
383. R. C. Richmond, in: *Ecological Genetics: The Interface* (P. F. Brussard, ed.), pp. 127–142, Springer-Verlag, New York (1978).
384. R. H. Richardson and J. S. Johnston, Ecological specialization in Hawaiian *Drosophila*. Habitat selection in Kipuka Puaulu, *Oecologia* **21:**193–204 (1975).

385. K. Y. Kaneshiro, Ethological isolation and phylogeny in the *plantibia* subgroup of Hawaiian *Drosophila,* Evolution **30:**740–745 (1976).

386. R. Levins, Thermal acclimation and heat resistance in *Drosophila* species, *Am. Nat.* **103:**483–499 (1969).

387. J. R. Powell, Genetic polymorphisms in varied environments, *Science* **174:**1035–1036 (1971).

388. J. F. McDonald and F. J. Ayala, Genetic response to environmental heterogeneity, *Nature (London)* **250:**572–574 (1974).

389. C. E. Taylor and J. R. Powell, Microgeographic differentiation of chromosomal and enzyme polymorphisms in *Drosophila persimilis, Genetics* **85:**681–695 (1977).

390. J. S. Jones, Can genes choose habitats?, *Nature (London)* **286:**757–758 (1980).

391. G. L. Bush, Sympatric host race formation and speciation in frugivorous flies of the genus *Rhagoletis* (Diptera, Tephritidae), *Evolution* **23:**237–251 (1969).

392. G. B. Johnson, Relationship of enzyme polymorphism to species diversity, *Nature (London)* **242:**193–194 (1973).

393. M. Edmunds, Polymorphism in the mimetic butterfly *Hypolimnas misippus* L. in Ghana, *Heredity* **24:**281–302 (1969).

394. P. W. Richards, *The Tropical Rain Forest,* Cambridge University Press, Cambridge (1952).

395. L. R. Holdridge, *Life Zone Ecology,* Tropical Science Center, San Jose, Costa Rica (1967).

396. L. R. Holdridge, W. C. Grenke, W. H. Hatheway, T. Liang, and J. A. Tosi, Jr., *Forest Environments in Tropical Life Zones: A Pilot Study,* Pergamon Press, Oxford (1971).

397. J. L. Wolfson, Oviposition response of *Pieris rapae* to environmentally-induced variation in *Brassica nigra,* Entomol. Exp. Appl. **27:**223–232 (1980).

398. H. B. N. Hynes, Notes on the aquatic Hemiptera-Heteroptera of Trinidad and Tobago, B.W.I., with a description of a new species of *Martarega* B. White (Notonectidae), *Trans. R. Entomol. Soc. London* **99:**341–360 (1948).

399. J. L. Gressitt, J. Sedlack, and J. J. H. Szent-Ivany, Flora and fauna on backs of large Papuan moss-forest weevils, *Science* **150:**1833–1835 (1965).

400. R. Silberglied and A. Aiello, Camouflage by integumentary wetting in bark bugs, *Science* **207:**773–775 (1980).

401. L. F. Jiron and R. Zeledon, El genero *Anastrepha* (Diptera; Tephritidae) en las principales frutas de Costa Rica y su relación con pseudomiasis humana, *Rev. Biol. Trop.* **27:**155–161 (1979).

402. J. H. Bruga and R. J. Olembo, Plant food preferences of some sympatric drosophilids of tropical Africa, *Biotropica* **3:**151–158 (1971).

403. P. Walter, Comportement de recherche et d'exploitation d'une masse steriorale chez quelques coprophages Afro-tropicaux (Col., Scarabaeidae), *Ann. Soc. Entomol. Fr.* **16:**307–323 (1980).

404. J. C. Schuster, Saltatorial Orthoptera as common visitors to tropical flowers, *Biotropica* **6:**133–140 (1974).

404a. H. Fraser Rowell, Food plant specificity in neotropical rain-forest acridids, *Entomol. Exp. Appl.* **24:**451–462 (1978).

405. A. M. Young, The ineffectiveness of the stingless bee *Trigona jaty* (Hymenoptera: Apidae: Meliponinae) as a pollinator of cocoa (*Theobroma cacao* L.), *J. Appl. Ecol.* **18:**149–155 (1981).

406. P. G. Bystrak and W. W. Wirth, The North American species of *Forcipomyia,* subgenus *Euprojoannisia* (Diptera: Ceratopogonidae), *U.S. Dept. Agr. Tech. Bull.* No. 1591 (1978).

407. A. M. Laessle, A micro-limnological study of Jamaican bromeliads, *Ecology* **42:**499–517 (1961).

408. M. H. Robinson and B. Robinson, Associations between flies and spiders: Bibiocommensalism and dipsoparasitism, *Psyche* **84:**150–157 (1977).

408a. A. M. Young, Feeding and oviposition in the giant tropical damselfly *Megaloprepus coerulatus* (Drury) in Costa Rica, *Biotropica* **12:**237–239 (1980).

409. D. M. Windsor, Nesting association between two neotropical polybiine wasps (Hymenoptera, Vespidae), *Biotropica* **4:**1–3 (1972).

410. R. L. Jeanne, A latitudinal gradient in rates of ant predation, *Ecology* **60:**1211–1224 (1979).

411. R. K. Colwell, Competition and coexistence in a simple tropical community, *Am. Nat.* **107:**737–760 (1973).

412. D. H. Janzen, M. Ataroff, M. Farinas, S. Reyes, N. Rincon, A. Soler, P. Soriano, and M. Vera, Changes in the arthropod community along an elevational transect in the Venezuelan Andes, *Biotropica* **8:**193–203 (1976).

413. D. J. Futuyma and F. Gould, Associations of plants and insects in a deciduous forest, *Ecol. Monogr.* **1979:**33–50 (1979).

414. W. W. Murdoch, F. C. Evans, and C. H. Peterson, Diversity and pattern in plants and insects, *Ecology* **53:**819–828 (1972).

415. P. Becker, Island colonization by carnivorous and herbivorous Coleoptera, *J. Anim. Ecol.* **45:**893–906 (1976).

416. T. W. Schoener and D. H. Janzen, Notes on environmental determinants of tropical versus temperate insect size patterns, *Am. Nat.* **102:**207–224 (1968).

417. T. L. Erwin, in: *Carabid Beetles: Their Evolution, Natural History, and Classification* (T. L. Erwin, G. E. Ball, D. R. Whitehead, and A. Halpern, eds.), pp. 539–592, Dr. W. Junk The Hague. (1979).

418. R. A. Beaver, Host specificity of temperate and tropical animals, *Nature (London)* **281:**139–141 (1979).

419. M. A. Sharp and D. R. Parks, Plant resources and butterfly habitat selection, *Ecology* **55:**870–875 (1974).

420. M. D. Rausher, Larval habitat suitability and oviposition preference in three related butterflies, *Ecology* **60:**503–511 (1979).

421. P. M. Dolinger, P. R. Ehrlich, W. L. Fitch, and D. E. Breedlove, Alkaloid and predation patterns in Colorado lupine populations, *Oecologia* **13:**191–204 (1973).

422. F. S. Chew, Coevolution of pierid butterflies and their cruciferous foodplants. II. The distribution of eggs on potential foodplants, *Evolution* **31:**568–579 (1977).

423. J. A. Slater, Lygaeid bugs (Hemiptera: Lygaeidae) as seed predators of figs, *Biotropica* **4:**145–151 (1972).

424. J. A. Slater, The incidence and evolutionary significance of wing polymorphism in lygaeid bugs with particular reference to those of South Africa, *Biotropica* **9:**217–229 (1977).

425. J. H. Lawton, in: *Diversity of Insect Faunas* (L. A. Mound and N. Waloff, eds.), Blackwell, London (1978).

426. U. Irmler and K. Furch, Production, energy and nutrient turnover of the cockroach *Epilampra irmleri* Rocha e Silva & Aguiar in a Central-Amazonian inundation forest, *Amazoniana* **6:**497–520 (1979).

427. J. C. Miller, Niche relationships among parasitic insects occurring in a temporary habitat, *Ecology* **61:**270–275 (1980).

428. E. O. Wilson, A social ethogram of the neotropical arboreal ant *Zacryptocerus* varians (Fr. Smith), *Anim. Behav.* **24:**354–363 (1976).

429. W. H. Gotwald, Jr., Predatory behavior and food preferences of driver ants in selected African habitats, *Ann. Entomol. Soc. Am.* **67:**877–886 (1974).

430. W. H. Gotwald, Jr., Behavioral observations on African army ants of the genus *Aenictus* (Hymenoptera: Formicidae), *Biotropica* **8:**59–65 (1976).

431. W. H. Gotwald, Jr., Trophic ecology and adaptation in tropical Old World ants of the subfamily Dorylinae (Hymenoptera: Formicidae), *Biotropica* **10:**161–169 (1978).

432. M. Yasuno, The study of the ant population in the grassland at Mt. Hakkoda. II. The distribution pattern of ant nests at the Kayano grassland, *Sci. Rep. Tohoku Univ. Ser. 4* **30:**43–55 (1964).

433. L. K. Johnson and S. P. Hubbell, Aggression and competition among stingless bees: Field studies, *Ecology* **55:**120–127 (1974).
434. L. C. Cole, The population consequences of life history phenomena, *Q. Rev. Biol.* **29:**103–137 (1954).
435. J. H. Connell, in: *The Dynamics of Populations* (P. J. Den Boer and G. R. Gradwell, eds.), pp. 298–312, Oosterbeek, Utrecht, The Netherlands (1971).
436. D. H. Janzen, When is it coevolution?, *Evolution* **34:**611–612 (1980).
437. R. B. Root, Organization of a plant-arthropod association in simple and diverse habitats: The fauna of collards (*Brassica oleracea*), *Ecol. Monogr.* **43:**95–124 (1973).
438. Y. Ito, A. Shibazaki, and O. Iwahashi, Biology of *Hyphantria cunea* Drury (Lepidoptera: Arctiidae) in Japan. IX. Population dynamics, *Res. Popul. Ecol.* **11:**211–228 (1969).
439. H. G. Andrewartha and L. C. Birch, *The Distribution and Abundance of Animals,* University of Chicago Press, Chicago, (1954).
440. L. B. Slobodkin, F. E. Smith, and N. G. Hairston, Regulation in terrestrial ecosystems, and the implied balance of nature, *Am. Nat.* **101:**109–124 (1967).
441. M. Lloyd, Self-regulation of adult numbers by cannibalism in two laboratory strains of flour beetles (*Tribolium castaneum*), *Ecology* **49:**245–259 (1968).
442. A. Milne, The natural control of insect populations, *Can. Entomol.* **84:**193–213 (1957).
443. Y. Ito, Factors that affect the fluctuations of animal numbers, with special reference to insect outbreaks, *Bull. Natl. Inst. Agric. Sci. (Japan) Ser. C* **13:**57–89 (1961).
444. L. M. Cook, K. Frank, and L. P. Brower, Experiments on the demography of tropical butterflies. I. Survival rate and density in two species of *Parides, Biotropica* **3:**17–20 (1971).
445. R. Silberglied and A. Aiello, Disruptive coloration in butterflies: Lack of support in *Anartia fatima, Science* **209:**617–619 (1980).
446. W. W. Murdoch, Population stability and life history phenomena, *Am. Nat.* **100:**5–11 (1966).
447. D. Pimentel, The influence of plant spatial patterns on insect populations, *Ann. Entomol. Soc. Am.* **54:**61–69 (1961).
448. S. Wright, General, group and special size factors, *Genetics* **17:**603–619 (1932).
449. S. Wright, Tempo and mode in evolution: A critical review, *Ecology* **26:**415–419 (1945).
450. R. A. Fisher and E. B. Ford, The spread of a gene in natural conditions in a colony of the moth *Panaxia dominula* L., *Heredity* **1:**143–174 (1947).
451. P. R. Ehrlich and L. E. Gilbert, Population structure and dynamics of the tropical butterfly *Heliconius ethilla, Biotropica* **5:**69–82 (1973).
452. L. M. Cook, E. W. Thomason, and A. M. Young, Population structure, dynamics and dispersal of the tropical butterfly *Heliconius charitonius, J. Anim. Ecol.* **45:**851–863 (1976).
453. A. M. Young, A communal roost of the butterfly *Heliconius charitonius* L. in Costa Rican premontane tropical wet forest, *Entomol. News* **89:**235–243 (1978).
454. A. M. Young and J. H. Thomason, Notes on communal roosting of *Heliconius charitonius* in Costa Rican high country, *J. Lepid. Soc.* **29:**243–255 (1975).
455. A. M. Young and M. E. Carolan, Daily instability of communal roosting in the neotropical butterfly *Heliconius charitonius* (Lepidoptera: Nymphalidae: Heliconiinae), *J. Kans. Entomol. Soc.* **49:**346–359 (1976).
456. D. F. Owen and D. O. Chanter, Population biology of tropical African butterflies. Sex ratio and genetic variation in *Acraea encedon, J. Zool.* **157:**345–374 (1969).
457. W. K. Reisen and P. F. L. Boreham, Host selection patterns of some Pakistan mosquitoes, *Am. J. Trop. Med. Hyg.* **28:**408–421 (1979).
458. J. T. Landahl and R. B. Root, Differences in the life tables of tropical and temperate milkweed bugs, genus *Oncopeltus* (Hemiptera: Lygaeidae), *Ecology* **50:**734–737 (1969).
459. R. A. Fisher, *The Genetic Theory of Natural Selection,* Clarendon Press, Oxford (1930).
460. U. Irmler, Population-dynamics and physiological adaptation of *Pentacomia egregia* Chaud. (Col. Cicindelidae) to the Amazonian inundation forest, *Amazoniana* **4:**219–227 (1973).

461. W. M. Schaffer, Optimal reproductive effort in fluctuating environments, *Am. Nat.* **108:**783–790 (1974).
462. j. T. Giesel, Reproductive strategies as adaptations to life in temporally heterogeneous environments, *Ann. Rev. Ecol. Syst.* **7:**57–79 (1976).
463. J. A. Wiens, Population responses to patchy environments, *Ann. Rev. Ecol. Syst.* **7:**81–120 (1976).
464. J. A. Wiens, On competition and variable environments, *Am. Sci.* **1977:**590–597 (1977).
465. R. H. MacArthur and R. Levins, The limiting similarity, convergence and divergence of coexisting species, *Am. Nat.* **101:**377–385 (1967).
466. E. L. Charnov and W. M. Schaffer, Life-history consequences of natural selection: Cole's result revisited, *Am. Nat.* **107:**791–793 (1973).
467. R. H. MacArthur and E. R. Pianka, On optimal use of a patchy environment, *Am. Nat.* **100:**603–609 (1966).
468. K. S. Brown, Jr., Geographical patterns of evolution in Neotropical Lepidoptera. Systematics and derivation of known and new Heliconiini (Nymphalidae: Nymphalinae), *J. Entomol. Ser. B* **44:**201–242 (1975).
469. J. Haffer, in: *The South American Herpetofauna: Its Origin, Evolution, and Dispersal* (W. E. Duellman, ed.), pp. 107–140, Museum of Natural History, University of Kansas Monograph No. 7, Lawrence (1979).
470. B. A. Fuseini and R. Kumar, Ecology of cotton stainers (Heteroptera: Pyrrhocoridae) in southern Ghana, *Biol. J. Linn. Soc.* **7:**113–146 (1975).
471. H. A. Hespenheide, Ecological inferences from morphological data, *Ann. Rev. Ecol. Syst.* **4:**213–229 (1973).
472. A. M. Shapiro, The life history of an Equatorial montane butterfly, *Tatochila xanthodice* (Lepidoptera: Pieridae), *J. N.Y. Entomol. Soc.* **86:**51–55 (1978).
473. R. C. Fisher, A study in insect multiparasitism. II. The mechanism and control of competition for possession of the host, *J. Exp. Biol.* **38:**605–628 (1961).
474. L. L. Rockwood, Seasonal changes in susceptibility of *Crescentia alata* leaves to the flea beetle, *Oedionychus* sp., *Ecology* **55:**142–148 (1974).
475. R. M. Andrews, Evolution of life histories: A comparison of *Anolis* lizards from matched island and mainland habitats, *Breviora,* No. 454, 1–51 (1979).
476. D. Leston, Entomology of the cocoa farm, *Ann. Rev. Entomol.* **15:**273–294 (1970).
477. J. M. Scriber and P. P. Feeny, Growth of herbivorous caterpillars in relation to feeding specialization and to the growth form of their food plants, *Ecology* **60:**829–850 (1979).
478. K. S. Brown, Jr., Geographical patterns of evolution in Neotropical Lepidoptera: Differentiation of the species of *Melinaea* and *Mechanitis* (Nymphalidae, Ithomiinae), *Syst. Entomol.* **2:**161–197 (1977).
479. J. R. G. Turner, Experiments on the demography of tropical butterflies. II. Longevity and home-range behavior in *Heliconius erato, Biotropica* **3:**21–31 (1971).
480. S. B. J. Menken, Allozymes and the population structure of *Yponomeuta cagnagellus* (Hubner) (Lepidoptera), *Neth. J. Zool.* **30:**228–242 (1980).
481. R. Levins and R. H. MacArthur, Maintenance of genetic polymorphism in a spatially heterogeneous environment: variation on a theme by Howard Levene, *Am. Nat.* **100:**585–590 (1966).
482. O. Halkka, in: *Diversity of Insect Faunas* (L. A. Mound and N. Waloff, eds.), pp. 41–55, Blackwell, London (1978).
483. U. N. Safriel and U. Ritte, Criteria for the identification of potential colonizers, *Biol. J. Linn. Soc.* **13:**287–297 (1980).
484. P. W. Price, Colonization of crops by arthropods: Nonequilibrium communities in soybean fields, *Environ. Entomol.* **5:**605–611 (1976).

485. J. N. Thompson and M. F. Willson, Disturbance and the dispersal of fleshy fruits, *Science* **200:**1161–1163 (1978).

486. A. W. Harvey, Hybridization studies in the *Schistocera americana* complex. I. The specific status of the Central American Locust, *Biol. J. Linn. Soc.* **12:**349–355 (1979).

487. P. DeBach, Weather and the success of parasites in population regulation, *Canadian Entomologist* **97:**848–863 (1965).

488. H. J. L. Heesen, Egg production of *Pterostichus oblongopunctatus* (Fabricius) (Col., Carabidae) and *Philonthus decorus* (Gravenhorst) (Col., Staphylinidae), *Neth. J. Zool.* **30:**306–314 (1980).

489. R. D. Alexander, Life cycle origins, speciation, and related phenomena in crickets, *Q. Rev. Biol.* **43:**1–41 (1968).

490. G. W. Miskimen, Environmental factors affecting soldier beetle distribution and coloration in Colombia, *Biotropica* **4:**85–92 (1972).

491. L. A. Lacey and J. D. Charlwood, On the biting activities of some anthropophilic Amazonian Simulidae (Diptera), *Bull. Entomol. Res.* **70:**495–509 (1980).

492. H. Ebert, On the frequency of butterflies in eastern Brazil with a list of the butterfly fauna of Pecos de Caldas, Minas Gerais, *J. Lepid. Soc. Suppl.* **3:**1–48 (1969).

493. H. K. Clench, A list of the butterflies of Andros, Bahamas, *Ann. Carnegie Mus.* **46:**173–193 (1977).

494. D. A. T. Baldry and D. H. Molyneux, Observations on the ecology and trypanosome infections of a relict population of *Glossina medicorum* Austen in the Komoe Valley of Upper Volta, *Ann. Trop. Med. Parasitol.* **74:**79–91 (1980).

495. W. W. K. Cheung and A. T. Marshall, Water and ion regulation in cicadas in relation to xylem feeding, *J. Insect Physiol.* **19:**1801–1816 (1973).

496. U. Irmler, Die Struktur der Carabiden- und Staphylinidengesellschaften in zentralamazonischen Überschwemmungswäldern, *Amazoniana* **6:**301–326 (1978).

497. U. Irmler, Abundance fluctuations and habitat changes of soil beetles in Central Amazonian inundation Forests (Coleoptera: Carabidae, Staphylinidae), *Stud. Neotrop. Fauna Environ.* **14:**1–16 (1979).

498. H. Puig, Production de litière en forêt guyanaise: Résultats préliminaires, *Bull. Soc. Hist. Nat. Toulouse* **115:**338–346 (1980).

499. G. W. Frankie, P. A. Opler, and K. S. Bawa, Foraging behaviour of solitary bees: Implications for outcrossing of a neotropical forest tree species, *J. Ecol.* **64:**1049–1057 (1976).

500. Th. H. Fleming, Do tropical frugivores compete for food?, *Am. Zool.* **19:**1157–1172 (1979).

501. R. I. M. Dunbar, Demographic and life history variables of a population of Gelada Baboons (*Theropithecus gelada*), *J. Anim. Ecol.* **49:**485–506 (1980).

502. T. Alerstam and P. H. Enckell, Unpredictable habitats and evolution of bird migration, *Oikos* **33:**228–232 (1979).

503. G. P. Waldbauer, in: *Evolution of Insect Migration and Diapause* (H. Dingle, ed.), pp. 127–144, Springer-Verlag, Berlin (1978).

504. K. Vepsalainen, in: *Evolution of Insect Migration and Diapause* (H. Dingle, ed.), pp. 218–253, Springer-Verlag, Berlin (1978).

505. C. A. Istock, The evolution of complex life cycle phenomena: An ecological perspective, *Evolution* **21:**592–605 (1967).

506. S. Masaki, in: *Evolution of Insect Migration and Diapause* (H. Dingle, ed.), pp. 72–100, Springer–Verlag, Berlin (1978).

507. H. A. Mooney, P. R. Ehrlich, D. E. Lincoln, and K. S. Williams, Environmental controls on the seasonality of a drought deciduous shrub, *Diplacus aurantiacus* and its predator, the Checkerspot Butterfly, *Euphydryas chalcedona, Oecologia* **45:**143–146 (1980).

508. P. J. Den Boer, Exclusion or coexistence and the taxonomic or ecological relationship between species, *Neth. J. Zool.* **30:**278–306 (1980).
509. J. J. M. Van Alphen and H. Boer, Avoidance of scramble competition between larvae of the Spotted Asparagus Beetle, *Crioceris duodecimpunctata* L. (Chrysomelidae) by discrimination between unoccupied and occupied Asparagus berries, *Neth. J. Zool.* **30:**136–143 (1980).
510. H. Wolda, Seasonal fluctuations in rainfall, food and abundance of tropical insects, *J. Anim. Ecol.* **47:**369–381 (1978).
511. H. Wolda, Fluctuations in abundance of tropical insects, *Am. Nat.* **112:**1017–1045 (1978).
512. H. Wolda, Seasonal fluctuations in rainfall, food and abundance of tropical insects, *J. Anim. Ecol.* **47:**369–381 (1978).
513. P. D. N. Hebert, Moth communities in montane Papua New Guinea, *J. Anim. Ecol.* **49:**593–602 (1980).
514. D. F. Owen, Species diversity and seasonal abundance in tropical Sphingidae, *Proc. R. Entomol. Soc. London Ser. A* **44:**162–168 (1969).
515. D. F. Owen and D. O. Chanter, Species diversity and seasonal abundance in tropical Ichneumonidae, *Oikos* **21:**142–144 (1970).
516. D. F. Owen, J. Owen, and D. O. Chanter, Seasonal changes in relative abundance and estimates of species diversity in a family of tropical butterflies, *Oikos* **23:**200–205 (1972).
517. G. G. Montgomery (ed.), *The Ecology of Arboreal Folivores,* Smithsonian Institution Press, Washington, D.C. (1978).
518. H. W. Bates, *The Naturalist on the River Amazons: A Record of Adventures, Habits of Animals, Sketches of Brazilian and Indian Life, and Aspects of Nature under the Equator, during Eleven Years of Travel,* J. Murray, London (1892).
519. C. S. Elton, Conservation and the low population density of invertebrates inside neotropical rainforest, *Biol. Conserv.* **7:**3–15 (1975).
520. H. M. Brasell, G. L. Unwin, and G. C. Stocker, The quantity, temporal distribution and mineral-element content of litterfall in two forest types at two sites in tropical Australia, *J. Ecol.* **68:**123–139 (1980).
521. A. K. Singh and R. S. Ambasht, Production and decomposition rate of litter in a teak (*Tectona grandis*) plantation at Varanasi (India), *Rev. Ecol. Biol. Sol* **17:**13–22 (1980).
522. J. A. Winder, Cacao pollination: Microdiptera of cacao plantations and some of their breeding places, *Bull. Entomol. Res.* **61:**651–655 (1972).
523. R. E. Buskirk and W. H. Buskirk, Changes in arthropod abundance in a highland Costa Rican forest, *Am. Midl. Nat.* **95:**288–298 (1976).
524. D. Denlinger, The developmental response of flesh flies (Diptera: Sarcophagidae) to tropical seasons: Variation in generation time and diapause in East Africa, *Oecologia* **35:**105–107 (1978).
525. D. L. Denlinger, Seasonal and annual variation of insect abundance in the Nairobi National Park, Kenya, *Biotropica* **12:**100–106 (1980).
526. M. Bates, Observations on climate and seasonal distribution of mosquitoes in eastern Colombia, *J. Anim. Ecol.* **14:**17–25 (1945).
527. A. M. Young, Feeding and oviposition in the giant tropical damselfly *Megaloprepus coerulatus* (Drury) in Costa Rica, *Biotropica* **12:**237–239 (1980).
528. C. S. Pittendrigh, The Bromeliad-*Anopheles*-Malaria complex in Trinidad. I. The Bromeliad flora, *Evolution* **2:**60–89 (1948).
529. H. T. Dalmat, Ecology of simuliid vectors of onchocerciasis in Guatemala, *Am. Midl. Nat.* **52:**175–196 (1954).
530. H. T. Dalmat, Biology and control of simuliid (Diptera) vectors of onchoceriasis in Central America, *Proc. Int. Congr. Entomol.* **3:**517–533 (1958).

531. H. Sioli, General features of the limnology of Amazonia. *Verh. Intern. Verein. Limnol.* **15:**1053–1058 (1965).
532. J. Bowden, Weather and the phenology of some African Tabanidae, *J. Entomol. Soc. South. Afr.* **39:**207–245 (1976).
533. W. A. Lamborn, Some features in the life history of Tabanidae in Nyasaland, *East Afr. Med. J.* **15:**46–50 (1938).
534. J. P. Dempster, The population dynamics of grasshoppers and locusts, *Biol. Rev.* **38:**490–529 (1963).
535. W. A. Sodeman, Jr. G. E. Rodrick, and A. L. Vincent, Lampyridae larva: A natural predator of schistosome vector snails in Liberia, *Am. J. Trop. Med. Hyg.* **29:**319 (1980).
536. J. J. Petersen, H. C. Chapman, O. R. Willis, and T. Fukuda, Release of *Romanomermis culicivorax* for the control of *Anopheles albimanus* in El Salvador. II. Application of the nematode, *Am. J. Trop. Med. Hyg.* **27:**1268–1273 (1978).
537. M. J. Klowden and A. O. Lea, Blood meal size as a factor affecting continued host-seeking by *Aedes aegypti* (L.), *Am. J. Trop. Med. Hyg.* **27:**827–831 (1978).
538. J. A. Seawright, D. G. Haile, M. G. Rabbani, and D. E. Weidhaas, Computer simulation of the effectiveness of male-linked translocations for the control of *Anopheles albimanus* Wiedemann, *Am. J. Trop. Med. Hyg.* **28:**155–160 (1979).
539. J. T. Whitlaw, Jr., and B. N. Chaniotis, Palm trees and Chagas' disease in Panama, *Am. J. Trop. Med. Hyg.* **27:**873–881 (1978).
540. R. C. Rainy, Weather and movements of locust swarms: A new hypothesis, *Nature (London)* **168:**1057–1060 (1951).
541. E. S. Brown and G. Swaine, New evidence on the migration of moths of the African Armyworm, *Spodoptera exampta* (Wlk.) (Lepidoptera, Noctuidae), *Bull. Entomol. Res.* **56:**671–684 (1966).
542. J. J. Matthee, The induction of diapause in eggs of *Locustana pardalina* Walker (Acrididae) by high temperature, *J. Entomol. Soc. South. Afr.* **41:**25–30 (1978).
543. W. L. Nutting, in: *Biology of Termites,* Vol. 1 (K. Krishna and F. M. Wessner, eds.), pp. 233–282, Academic Press, New York (1969).
544. W. Leuthold and O. Bruinsma, Pairing behavior in *Hodotermes mossambicus* (Isoptera), *Psyche* **84:**109–119 (1977).
545. S. M. Omer and J. L. Coudsley-Thompson, Dry season biology of *Anopheles gambiae* Giles in the Sudan, *Nature (London)* **217:**879–880 (1968).
546. M. J. Morris, The influence of constant and changing photoperiods on imaginal diapause in the red locust (*Nomadacris septemfasciata* Serv.), *J. Insect Physiol.* **11:**1105–1119 (1965).
547. E. J. Usua, Induction of diapause in the maize stem borer, *Busseola fusca, Entom. Exp. Appl.* **16:**322–328 (1973).
548. D. B. Carlisle, P. E. Ellis, and E. Betts, The influence of aromatic shrubs on sexual maturation in the desert locust *Schistocera gregaria, J. Insect Physiol.* **11:**1541–1558 (1965).
549. D. C. D. Happold, Seasonal distribution of adult dragonflies at Khartoum, Sudan, *Rev. Zool. Bot. Afr.* **77:**50–61 (1968).
550. D. C. D. Happold, Dragonflies of Jebel Marra, Sudan, *Proc. R. Entomol. Soc. London* **41:**87–91 (1966).
551. P. S. Corbet, The life-histories of two summer species of dragonfly (Odonata: Coenagriidae), *Proc. Zool. Soc. London* **128:**403–418 (1957).
552. P. S. Corbet, Temperature in relation to seasonal development of British dragonflies (Odonata), *Proc. Int. Congr. Entomol.* **2:**755–758 (1958).
553. P. S. Corbet, Notes on the insect food of the Nile crocodile in Uganda, *Proc. R. Entomol. Soc. London Ser. A* **34:**17–21 (1959).

554. D. R. Paulson and C. E. Jenner, Population structure in overwintering larval Odonata in North Carolina in relation to adult flight season, *Ecology* **52:**96–107 (1971).
555. T. T. Macan, The Odonata of a Moorland fishpond, *Int. Rev. Gesamten Hydrobiol.* **49:**325–360 (1964).
556. T. W. Donnelly, The Odonata of Dominica British West Indies, *Smithson. Contrib. Zool.* No. 37 (1970).
557. D. A. Maelzer, The effect of temperature and moisture on the immature stages of *Aphodius tasmaniae* Hope (Scarabaeidae) in the lower south-east of south Australia, *Aust. J. Zool.* **9:**173–202 (1961).
558. G. Halffter and E. G. Matthews, The natural history of dung beetles of the subfamily Scarabeinae (Coleoptera, Scarabaeidae), *Folia Entomol. Mex.* Nos. 12–14 (1966).
559. G. S. Rajulu and G. Krishnan, The epicuticle of millipedes belonging to the genera *Cingalobolus* and *Aulacobolus* with special reference to seasonal variations, *Z. Naturforsch.* **23b:**845–851 (1968).
560. R. J. Hoffman, Environmental control of seasonal variation in the butterfly *Colias eurytheme:* Effects of photoperiod and temperature on pteridine pigmentation, *J. Insect Physiol.* **20:**1913–1924 (1974).
561. R. J. Hoffman, Environmental uncertainty and evolution of physiological adaptation in *Colias* butterflies, *Am. Nat.* **112:**999–1015 (1978).
562. A. M. Young, Aggregative behavior of *Anartia fatima* (Nymphalidae: Nymphalinae) in Guanacaste Province, Costa Rica during the dry season, *J. Lepid Soc.* **33:**58–60 (1979).
563. T. C. Emmel and C. F. Leck, Seasonal changes in the organization of tropical rain forest butterfly populations in Panama, *J. Res. Lepid.* **8:**133–152 (1970).
564. A. M. Young, D. Tyrrell, and D. MacLoed, *Entomophthora echinospora,* a fungus pathogenic on the tropical cicada, *Procollina biolleyi, J. Invertebr. Pathol.* **21:**87–90 (1972).
565. P. de T. Alvim, Eco-physiology of the cacao tree, *Proc. Conf. Int. Rech. Agron. Cacao,* Abidjan, 23–35 (1967).
566. F. I. O. Nwoke, Effects of number of photoperiodic cycles on induction and development of flowers and fruits in *Corchorus olitorius* L. (var. *oniyaya* Epen.), *Ann. Bot.* **45:**569–576 (1980).
567. A. Raw, Seasonal changes in the numbers and foraging activities of two Jamaican *Exomalopsis* species (Hymenoptera, Anthophoridae), *Biotropica* **8:**270–277 (1976).
568. R. L. Jeanne, Aspects of the biology of *Stelopolybia areata* (Say) (Hymenoptera: Vespidae), *Biotropica* **5:**183–198 (1973).
569. G. W. Frankie and R. Coville, An experimental study on the foraging behavior of selected solitary bee species in the Costa Rican dry forest (Hymenoptera: Apoidea), *J. Kans. Entomol. Soc.* **52:**591–602 (1979).
570. C. D. Michener, M. L. Winston, And R. Jander, Pollen manipulation and related activities and structures in bees of the family Apidae, *Univ. Kans. Sci. Bull.* **51:**575–601 (1978).
571. R. A. Schlising, Sequence and timing of bees foraging in flowers of *Ipomoea* and *Aniseia* (Convolvulaceae), *Ecology* **51:**1061–1067 (1970).
572. H. G. Baker, Studies in the reproductive biology of West African Rubiaceae, *J. West Afr. Sci. Assoc.* **4:**9–24 (1958).
573. H. G. Baker, Evolutionary mechanisms in pollination biology, *Science* **139:**877–883 (1963).
574. P. Feinsinger, Ecological interactions between plants and hummingbirds in a successional tropical community, *Ecol. Monogr.* **48:**269–287 (1978).
575. A. M. Young, Notes on a community ecology of adult sphinx moths in Costa Rican lowland tropical rain forest, *Caribb. J. Sci.* **12:**151–163 (1972).
576. S. Laroca and O. H. H. Mielke, Ensaios sobre ecologia de comunidade em Sphingidae na Serra Do Mar, Parana, Brasil (Lepidoptera), *Rev. Brasil Biol.* **35:**1–19 (1975).

577. E. S. Morton, On the evolutionary advantages and disadvantages of fruit eating in tropical birds, *Am. Nat.* **107:**8–22 (1973).

578. P. Utrio and K. Eriksson, Volatile fermentation products as attractants for Macrolepidoptera, *Ann. Zool. Fenn.* **14:**98–104 (1977).

579. J. A. Derr, Coevolution of the life history of a tropical seed-feeding insect and its food plants, *Ecology* **61:**881–892 (1980).

580. P. M. Barrer and J. M. Cherrett, Some factors affecting the site and pattern of leaf-cutting activity in the ant *Atta cephalotes* L., *J. Entomol. Ser. A* **47:**15–27 (1972).

581. J. M. Cherrett, The foraging behaviour of *Atta cephalotes* L. (Hymenoptera, Formicidae). I. Foraging pattern and plant species attacked in tropical rain forest, *J. Anim. Ecol.* **37:**387–403 (1968).

582. B. Shorrocks, The distribution and abundance of woodland species of British *Drosophila* (Diptera: Drosophilidae), *J. Anim. Ecol.* **44:**851–864 (1975).

583. H. L. Carson and W. E. Johnson, Genetic variation in Hawaiian *Drosophila*. I. Chromosome and allozyme polymorphism in *D. setostimentum* and *D. ochrobasis* from the island of Hawaii, *Evolution* **29:**11–23 (1975).

584. S. B. Pipkin, R. L. Rodriguez, and J. Leon, Plant host specificity among flower-feeding Neotropical *Drosophila* (Diptera: Drosophilidae), *Am. Nat.* **100:**135–156 (1966).

585. E. Duffey, Ecological studies on the Large Copper Butterfly *Lycaena dispar* Haw. *batavus* Obth. at Woodwalton Fen National Nature Reserve, Huntingdonshire, *J. Appl. Ecol.* **5:**69–96 (1968).

586. A. M. Young, Mimetic association in natural populations of tropical papilionid butterflies, *J. N.Y. Entomol. Soc.* **79:**210–244 (1971).

587. L. G. Saunders, Methods for studying *Forcipomyia* midges with special reference to cacao-pollinating species (Diptera, Ceratopogonidae), *Can. J. Zool.* **37:**33–51 (1959).

588. J. A. Winder, Recent research on insect pollination of cocoa, *Cocoa Growers' Bull.* **26:**11–19 (1977).

589. J. A. Winder, Cocoa flower Diptera; Their identity, pollinating activity and breeding sites, *PANS* **24:**5–18 (1978).

590. A. M. Young, Études expérimentales sur le terrain quant a l'écologie du comportement et à la biologie du peuplement des chironomides pollinisants du cacaoyer (Diptera: Ceratopogonidae) au Costa Rica, Conference Internationale sur la Recherche Cacaoyere, Douala, Cameroun (1979).

590a. A. M. Young, Effects of shade cover and availability of midge breeding sites on pollinating midge populations and fruit set in two cocoa farms, *J. Appl. Ecol.* **19:**46–67 (1982).

591. J. G. Myers, Notes on wild cacao in Surinam and British Guiana, *Kew Bull.* **1930:**1–10 (1930).

592. F. Hardy, The chemical and ecological researches on cacao, *Trop. Agric. (Trinidad)* **12:**175–178 (1935).

593. F. Hardy, Some aspects of cacao soil fertility in Trinidad, *Trop. Agric. (Trinidad)* **13:**315–317 (1936).

594. A. Thomas, The dry season in the Gold Coast and its relation to the cultivation of cacao, *J. Ecol.* **20:**263–269 (1932).

595. J. Cuatrecasas, Cacao and its allies—A taxonomic revision of the genus *Theobroma, Contrib. U.S. Natl. Mus.* **35:**379–614 (1964).

596. B. Haines and R. B. Foster, Energy flow through litter in a Panamanian forest, *J. Ecol.* **65:**147–155 (1977).

597. M. J. N. Okwakol, Estimation of soil and organic matter consumption by termites of the genus *Cubitermes, Afr. J. Ecol.* **18:** 127–131 (1980).

598. An. A. Federov, The structure of the tropical rain forest and speciation in the humid tropics, *J. Ecol.* **54:**1–11 (1966).

599. T. Uchida and T. Nakashima, Some observations on the population density of the spruce barkbeetle, *Ips typographus* Linne, during the period of the outbreak in forests devastated by typhoon of September, 1954, in Hokkaido, *Res. Bull. Coll. Exp. For. Hokkaido Univ.* **21:**149–168 (1961).
600. R. W. Thorp, Structural, behavioral, and physiological adaptations of bees (Apoidea) for collecting pollen, *Ann. Mo. Bot. Gard.* **66:**788–812 (1979).
601. R. L. Dressler, Pollination in euglossine bees, *Evolution* **22:**202–210 (1968).
602. S. Salas and L. F. Jiron, Simbiosis entre cochinillas de cola (Homoptera: Margarodidae) y otros insectos. II. Estructura del sistema y significado ecologico, *Brenesia* **10/11:**57–64 (1977).
603. H. E. Evans and R. W. Matthews, Observations on the nesting behavior of South American sand wasps (Hymenoptera), *Biotropica* **6:**130–134 (1974).
604. R. E. Coville and P. L. Coville, Nesting biology and male behavior of *Trypoxylon (Trypargilum) tenoctitlan* in Costa Rica (Hymenoptera: Sphecidae), *Ann. Entomol. Soc. Am.* **73:**110–119 (1980).
605. W. J. Mattson and N. D. Addy, Phytophagous insects as regulators of forest primary production, *Science* **190:**515–522 (1975).
606. S. T. A. Pickett, Non-equilibrium coexistence of plants, *Bull. Torrey Bot. Club* **107:**238–248 (1980).
607. D. C. Culver, Analysis of simple cave communities: Niche separation and species packing, *Ecology* **51:**949–958 (1970).
608. D. C. Culver, Species packing in Caribbean and north temperate ant communities, *Ecology* **55:**974–988 (1974).
609. J. R. Karr, Structure of avian communities in selected Panama and Illinois Habitats, *Ecol. Monogr.* **41:**207–233 (1971).
610. D. H. Morse, The insectivorous bird as an adaptive strategy, *Ann. Rev. Ecol. Syst.* **2:**177–200 (1971).
611. P. W. Price, *Insect Ecology,* John Wiley, New York (1975).
612. G. E. Hutchinson, Homage to Santa Rosalia, or why are there so many kinds of animals?, *Am. Nat.* **93:**145–159 (1959).
613. A. M. Young, Notes on a population outbreak of the beetle *Uroplata* sp. (Coleoptera: Chrysomelidae) on the tree *Bunchosia pilosa* (Malpighiaceae) in Costa Rica, *J. N.Y. Entomol. Soc.* **87:**289–298 (1979).
614. D. R. Strong, Jr., Rolled-leaf hispine beetles (Chrysomelidae) and their Zingiberales host plants in Middle America, *Biotropica* **9:**156–169 (1977).
615. J. A. Slater, Monocots and chinch bugs: A study of host plant relationships in the lygaeid subfamily Blissinae (Hemiptera: Lygaeidae), *Biotropica* **8:**143–165 (1976).
616. D. E. Wilson and D. H. Janzen, Predation on *Scheelea* palm seeds by bruchid beetles: Seed density and distance from the parent palm, *Ecology* **53:**954–959 (1972).
617. B. A. Drummond, III, Comparative ecology and mimetic relationships of ithomiine butterflies in eastern Ecuador, Ph.D. dissertation, University of Florida, Gainesville (1976).
618. S. Iwao, Some effects of grouping in lepidopterous insects. L'effect de groupe chez les animaux, *Colloq. Int. C. Nl. R. S.* **173:**185–212 (1967).
619. K. Fujisaki, Breakup and re-formation of colony in the first-instar larvae of the Winter Cherry Bug, *Acanthocorus sordidus* Thunberg (Hemiptera: Coreidae), in relation to the defense against their enemies, *Res. Popul. Biol. (Kyoto)* **16:**252–264 (1975).
620. B. J. Rathcke, Competition and coexistence within a guild of herbivorous insects, *Ecology* **57:**76–87 (1976).
621. J. R. G. Turner, Adaptive radiation and convergence in subdivisions of the butterfly genus *Heliconius* (Lepidoptera: Nymphalidae), *Zool. J. Linn. Soc.* **58:**297–308 (1976).

622. K. S. Brown, Jr., P. M. Sheppard, and J. R. G. Turner, Quarternary refugia in tropical America: Evidence from race formation in *Heliconius* butterflies, *Proc. R. Soc. London Ser. B* **187:**369–378 (1974).

623. K. S. Brown, Jr., Geographical patterns of evolution in neotropical Lepidoptera. Systematics and derivation of known and new Heliconiini (Nymphalidae: Nymphalinae), *J. Entomol. Ser. B*. **44:**201–242 (1976).

624. K. S. Brown, Jr., in: *Biogéographie et Évolution en Amérique Tropicale* (H. Descimon, ed.), pp. 118–160, Publications du Laboratoire de Zoologie de l'École Normale Supérieure, Paris (1976).

625. K. S. Brown, Jr., Centros de evolucao, refugios quaternarios e conservacao de patrimonios geneticos na regiao neotropical: Padroes de diferenciacao em Ithomiinae (Lepidoptera: Nymphalidae), *Acta Amazonica* **7:**75–137 (1977).

626. K. S. Brown, Jr., *Ecologia Geografica E Evolucao Nas Florestas Neotropicais*—Part VI Na Serie "*Padroes Geograficos De Evolucao Em Lepidopteros Neotropicais*", Universidade Estadual de Campinas, Sao Paulo, Brazil (1979).

627. T. G. Whitham, Habitat selection by *Pemphigus* aphids in response to resource limitation and competition, *Ecology* **59:**1164–1176 (1978).

628. T. G. Whitham, The theory of habitat selection: Examined and extended using *Pemphigus* aphids, *Am. Nat.* **115:**449–466 (1980).

629. R. E. Silberglied and O. R. Taylor, Jr., Ultraviolet reflection and its behavioral role in the courtship of the sulfur butterflies *Colias eurytheme* and *C. philodice* (Lepidoptera, Pieridae), *Behav. Ecol. Sociobiol.* **3:**203–243 (1978).

630. J. R. G. Turner, in: *Evolutionary Biology*, Vol. 10 (M. K. Hecht, W. C. Steere, and B. Wallace, eds.), Plenum, New York (1977).

631. D. J. Stradling, The nature of the mimetic patterns of the brassolid genera, *Caligo* and *Eryphanis*, *Ecol. Entomol.* **1:**135–138 (1976).

632. D. Pimentel, Population ecology of insect invaders of the maritime provinces, *Can. Entomol.* **98:**887–894 (1966).

633. R. L. Metcalf, R. A. Metcalf, and A. M. Rhodes, Cucurbitacins as kairomones for diabroticite beetles, *Proc. Natl. Acad. Sci. U.S.A.* **77:**3769–3772 (1980).

634. D. R. Strong, Rapid asymptotic species accumulation in phytophagous insect communities: The pests of cacao, *Science* **185:**1064–1066 (1974).

635. R. T. Holmes, J. C. Schultz, and P. Nothnagle, Bird predation on forest insects: An exclosure experiment, *Science* **206:**462–463 (1979).

636. D. W. Inouye and O. R. Taylor, Jr., A temperate region plant-ant-seed predator system: Consequences of extra floral nectar secretion by *Helianthella quinquenervis*, *Ecology* **60:**1–7 (1979).

637. A. M. Young, Interactions of *Phileathria dido* (Heliconiinae) and *Victorina steneles* (Nymphalinae) at *Stachytarpheta* flowers in Costa Rica: Evidence against mimetic association, *Acta Biol. Venez.* **8:**1–17 (1972).

638. A. M. Young, Mimetic associations in populations of tropical butterflies. II. Mimetic interactions of *Battus polydamus* and *Battus bellus*, *Biotropica* **4:**17–27 (1972).

638a. K. S. Bawa, Breeding systems of tree species of a lowland tropical community, *Evolution* **28:**85–92 (1974).

639. K. S. Bawa and P. A. Opler, Dioecism in tropical trees, *Evolution* **29:**167–179 (1975).

640. D. A. Levin, in: *Topics in Plant Population Biology* (C. T. Solbrig, S. Jain, G. B. Johnson, and P. H. Raven, eds.), pp. 131–156, Columbia University Press, New York (1979).

641. D. A. Levin and W. W. Anderson, Competition for pollinators between simultaneously flowering species, *Am. Nat.* **104:**455–467 (1970).

642. D. W. Schemske, Pollinator specificity in *Lantana camara* and *L. trifolia* (Verbenaceae), *Biotropica* **8:**260-264 (1976).

643. E. M. Barrows, Floral maturation and insect visitors of *Pachyptera hymenaea* (Bignoniaceae), *Biotropica* **9:**133-134 (1977).

644. J. W. Purseglove, *Tropical Crops: Dicotyledons*. John Wiley, New York (1968).

645. J. P. Parvais, P. de Reffye, and P. Lucas, Observations sur la pollinisation libre chez *Theobroma cacao:* Analyse mathémathique des données et modelisation, *Cafe Cacao The* **21:**253-262 (1977).

646. F. Massaux, C. Tchiendji, C. Misse, and B. Decazy, Étude du transporte du pollen de cacaoyer par marquage au ^{32}P, *Cafe Cacao The* **20:**163-172 (1976).

647. T. Kaufmann, Preliminary observations on cecidomyiid midge and its role as a cocoa pollinator in Ghana, *Ghana J. Agric. Sci.* **6:**193-198 (1973).

648. T. Kaufmann, Ecology and behavior of cocoa pollinating Ceratopogonidae in Ghana, W. Africa, *Environ. Entomol.* **4:**347-351 (1975).

649. F. Privat, Les Bromeliaces, lieu de développement de quelques insectes pollinisateurs des fleurs de cacao, *Brenesia* **16:**197-212 (1979).

650. D. R. Glendinning, Natural pollination of cacao, *Nature (London)* **193:**1305 (1962).

651. P. Dessart, Contribution a l'étude des Ceratopogonidae (Diptera), *Bull. Agric. Congo Belge* **52:**525-540 (1961).

652. S. de J. Soria, and W. W. Wirth, Ciclos de vida dos polinizadores do cacaueiro *Forcipomyia* spp. (Diptera, Ceratopogonidae) e algumas anotacoes sobre o comportamento das larvas no laboratorie, *Rev. Theobroma* **5:**3-22 (1975).

653. S. de J. Soria, W. W. Wirth, and H. A. Besemer, Breeding places and sites of collection of adults of *Forcipomyia* spp. midges (Diptera, Ceratopogonidae) in cacao plantations in Bahia, Brazil: A progress report, *Rev. Theobroma* **8:**21-29 (1978).

654. D. Fish and S. de J. Soria, Water-holding plants (phytotelmata) as larval habitats for ceratopogonid pollinators of cacao in Bahia, Brazil, *Rev. Theobroma* **8:**133-146 (1978).

655. J. De la Cruz and S. de J. Soria, Estudio de fluctaciones de polinizacion del cacao por las mosquitas *Forcipomyia* spp. (Diptera, Ceratopogonidae), en Palmira, Valle, Colombia, *Acta Agron. (Palmira)* **23:**1-17 (1973).

656. F. W. Cope, The mechanism of pollen incompatability and compatability in *Theobroma cacao* L., *Heredity* **17:**157-182 (1962).

657. A. F. Posnette, Natural pollination of cocoa, *Theobroma leiocarpa* Bern. on the Gold Coast II, *Trop. Agric. (Trinidad)* **19:**188-191 (1942).

658. A. F. Posnette, Pollination of cacao in Trinidad, *Trop. Agric. (Trinidad)* **21:**115-118 (1944).

659. D. Leston, Entomology of the cocoa farm, *Ann. Rev. Entomol.* **15:**273-294 (1970).

660. D. G. Gibbs and D. Leston, Insect phenology in a forest cocoa-farm locality in West Africa, *J. Appl. Ecol.* **7:**519-548 (1970).

661. D. G. Gibbs, A. D. Pickett, and D. Leston, Seasonal population changes in cocoa capsids (Hemiptera, Miridae) in Ghana, *Bull. Entomol. Res.* **58:**279-293 (1968).

662. W. M. Wheeler, Studies of neotropical ant-plants and their ants, *Bull. Mus. Comp. Zool. Harv. Univ.* **110(1)** (1942).

663. B. Hocking, in: *Coevolution of Animals and Plants* (L. E. Gilbert and P. H. Raven, eds.), pp. 78-90, University of Texas Press, Austin (1975).

664. J. B. Free, *Insect Pollination of Crops,* Academic Press, London (1970).

665. E. L. Charnov, J. Maynard Smith, and J. J. Bull, Why be an hemaphrodite?, *Nature (London)* **263:**125-126 (1976).

666. T. J. Givnish, Ecological constraints on the evolution of breeding systems in seed plants: Dioecy and dispersal in gymnosperms, *Evolution* **34:**959-972 (1980).

667. M. R. Mesler, J. D. Ackerman, and K. L. Lu, The effectiveness of fungus gnats as pollinators, *Am. J. Bot.* **67:**564–567 (1980).
668. H. G. Baker, Evolution in the tropics, *Biotropica* **2:**101–111 (1970).
669. J. Roughgarden, Resource partitioning among competing species—A coevolutionary approach, *Theor. Popul. Biol.* **9:**388–424 (1976).
670. A. H. Gentry, Bignoniaceae of southern Central America: Distribution and ecological specificity, *Biotropica* **8:**117–131 (1976).
671. E. M. Barrows, Nectar robbing and pollination of *Lantana camara* (Verbenaceae), *Biotropica* **8:**132–135 (1976).
672. E. M. Barrows, Robbing of exotic plants by introduced carpenter and honey bees in Hawaii, with comparative notes, *Biotropica* **12:**23–29 (1980).
673. E. O. Wilson, Patchy distribution of ant species in New Guinea rain forests, *Psyche* **65:**26–38 (1958).
674. A. M. Young, Nectar robbing of "four-o-clocks," *Mirabilis jalapa* (Nyctaginaceae) by the social paper wasp *Epipona querini* (Vespidae: Polistinae: Polybiini) in Costa Rica, *Biotropica* **12:**222–223 (1980).
675. L. A. McDade and S. Kinsman, The impact of floral parasitism in two neotropical hummingbird-pollinated plant species, *Evolution* **34:**944–958 (1980).
676. C. E. King, Resource specialization and equilibrium population size in patchy environments, *Proc. Natl. Acad. Sci. U.S.A.* **68:**2634–2637 (1971).
677. W. W. Anderson, Genetic equilibrium and population growth under density-regulated selection, *Am. Nat.* **105:**489–498 (1971).
677a. A. M. Young, Nectar and pollen robbing of *Thunbergia grandiflora* by *Trigona* bees in Costa Rica, *Biotropica* **14:** (in press) (1982).
678. D. W. Roubik, Competitive interactions between neotropical pollinators and Africanized honey bees, *Science* **201:**1030–1032 (1978).
679. D. H. Janzen, Tropical agroecosystems, *Science* **182:**1212–1219 (1973).
680. A. J. Nicholson, The balance of animal populations. *J. Anim. Ecol.* **2:**138–178 (1933).
681. L. C. Birch, The intrinsic rate of natural increase of an insect population, *J. Anim. Ecol.* **17:**15–26 (1948).
682. J. S. Denslow, in: *Tropical Succession,* Vol. 12, (J. Ewel, ed.), pp. 47–55, *Biotropica Suppl.* (1980).
683. J. O. Tahvanainen and R. B. Root, The influence of vegetational diversity on the population ecology of a specialized herbivore, *Phyllotreta cruciferae* (Coleoptera: Chrysomelidae), *Oecologia* **10:**321–346 (1972).
684. R. G. Cates and G. H. Orians, Successional status and the palatability of plants to generalized herbivores, *Ecology* **51:**410–418 (1975).
685. R. S. Miller, Pattern and process in competition, *Adv. Ecol. Res.* **4:**1–74 (1967).
686. R. S. Miller, in: *Diversity and Stability in Ecological Systems,* pp. 63–70, Brookhaven Symposia in Biology (1969).
687. J. N. Raybould and H. K. Mhiddin, Studies on the immature states of the *Simulium neavei* Roubaud complex and their associated crabs in the Eastern Usambara Mountains in Tanzania. III. Investigations on development and survival and their relevance to control, *Ann. Trop. Med. Parasitol.* **72:**177–187 (1978).
688. L. E. Gilbert and M. C. Singer, Butterfly ecology, *Ann. Rev. Ecol. Syst.* **6:**365–398 (1975).
689. L. E. Gilbert, Butterfly-plant coevolution: Has *Passiflora adenopoda* won the selectional race with heliconiine butterflies?, *Science* **172:**585–586 (1971).
690. P. P. Calvert, Studies on Costa Rican Odonata. II. The habits of the plant-dwelling larva of *Mecistogaster modestus, Entomol. News* **22:**402–411 (1910).

691. J. H. Vandermeer, J. Addicott, A. Andersen, J. Kitasako, D. Pearson, C. Schnell, and H. Wilbur, Observations on *Paramecium* occupying arboreal standing water in Costa Rica, *Ecology* **53:**291–293 (1972).
692. D. H. Janzen, The unexploited tropics, *Bull. Ecol. Soc. Am.* **51:**4–7 (1970).
693. D. S. Hayman, Mycorrhiza and crop production, *Nature (London)* **287:**487–488 (1980).
694. C. R. Carroll, Biotic interactions in tropical agroecosystems, *Am. Zool.* **19:**1057–1064 (1979).
695. C. B. Heiser, Jr., Cultivated plants and cultural diffusion in nuclear America, *Am. Anthropol.* **67:**930–949 (1965).
696. R. Daubenmire, Some ecologic consequences of converting forest to savanna in northwestern Costa Rica, *Trop. Ecol.* **13:**31–51 (1972).
697. L. K. Vats and J. S. Singh, Population, biomass and secondary net production of above ground insects in a tropical grassland, *Trop. Ecol.* **19:**51–64 (1978).
698. D. Pimentel and F. A. Stone, Evolution and population ecology of parasite-host systems, *Can. Entomol.* **100:**655–662 (1968).
699. M. B. Usher, Studies on a wood-feeding termite community in Ghana, West Africa, *Biotropica* **7:**217–233 (1975).
700. H. W. Mielke, Termitaria and shifting cultivation: The dynamic role of the termite in soils of tropical wet-dry Africa, *Trop. Ecol.* **19:**117–122 (1978).
701. D. F. Owen, Adaptation to man-made environments and phenology in *Precis* butterflies, *Rev. Zool. Bot. Afr.* **83:**361–373 (1971).
702. J. B. Free and I. H. Williams, *The Pollination of Crops by Bees,* Apimondia and International Bee Research Association Special Publication, Bucharest (1977).
703. T. R. Soderstrom and C. E. Calderon, Insect pollination in tropical rain forest grasses, *Biotropica* **3:**1–16 (1971).
704. I. H. Williams, Behaviour of insects foraging on pigeon pea (*Cajanus cajan* (L.) Millsp.) in India, *Trop. Agric. (Trinidad)* **54:**353–363 (1977).
705. D. Leston, The natural history of some West African insects. 14. Entomological changes accompanying degradation of the forest, *Entomol. Mon. Mag.* **115:**105–112 (1979).
706. T. Lewis, J. M. Cherrett, I. Haines, B. Haines, and P. L. Mathias, The crazy ant (*Anoplolepis longipes* (Jerd.) (Hymenoptera, Formicidae) in Seychelles, and its chemical control, *Bull. Entomol. Res.* **66:**97–111 (1976).
707. E. H. Medina, The value of utilizing existing shade in the growing of vanilla, *J. Agric.* **40:**117–124 (1943).
708. C. Pennington, F. A. Jimenez, and T. Theis, A comparison of three methods of vanilla culture in Puerto Rico, *Turrialba* **4:**79–87 (1954).
709. R. W. Smith and J. B. Whittaker, Factors affecting *Gastrophysa viridula* populations (Coleoptera: Chrysomelidae) in different habitats, *J. Anim. Ecol.* **49:**537–548 (1980).
710. R. A. Bean and R. F. Denno, The effects of mowing on the guild of sap-feeding insects associated with Kentucky bluegrass, *J. N.Y. Entomol. Soc.* (Abstract) **8:**36–37 (1980).
710a. J. M. Scriber and P. P. Feeny, New foodplant and oviposition records for *Battus philenor* (Papilionidae), *J. Lepid. Soc.* **30:**70–71 (1976).
711. D. A. Maelzer, The effect of other organisms on the survival rate of *Aphodius tasmaniae* Hope (Scarabaeidae) in the lower south-east of South Australia, *Aust. J. Zool.* **10:**617–629 (1962).
712. D. A. Maelzer, The emergence as a pest of *Aphodius tasmaniae* Hope (Scarabaeidae) in pastures in the lower south-east of South Australia, *Aust. J. Zool.* **10:**95–112 (1962).
713. D. A. Maelzer, Rainfall and changes in abundance of *Aphodius tasmaniae* Hope (Coleoptera: Scarabaeidae) in the lower south-east of South Australia, *Aust. J. Zool.* **12:**263–278 (1964).
714. P. B. Carne, An ecological study of the pasture scarab *Aphodius howitti* Hope, *Aust. J. Zool.* **4:**259–314 (1956).

715. I. Hanski, Movement patterns in dung beetles and in the dung fly, *Anim. Behav.* **28:**953–964 (1980).
716. H. F. Howden and V. G. Nealis, Effects of clearing in a tropical rain forest on the composition of the coprophagous scarab beetle fauna (Coleoptera), *Biotropica* **7:**77–83 (1975).
717. U.S.D.A., Beetle biocontrol in livestock diseases, *U.S.D. Agric. Res.* **29:**15 (1980).
718. B. D. Gouck, Release and establishment of the baculovirus disease of *Oryctes rhinoceros* (L.) (Coleoptera: Scarabaeidae) in Papua New Guinea, *Bull. Entomol. Res.* **70:**445–453 (1980).
718a. A. M. Young, Notes on seasonality and habitat associations of tropical cicadas (Homoptera: Cicadidae) in premontane and montane tropical moist forests in Costa Rica, *J. N.Y. Entomol. Soc.* **89:**123–142 (1981).
719. M. Lloyd and J. White, On the oviposition habits of 13-year vs. 17-year periodical cicadas of the same species, *J. N.Y. Entomol. Soc.* **84:**148–155 (1976).
720. R. Silberglied, A. Aiello, and G. Lamas, Neotropical butterflies of the genus *Anartia:* Systematics, life histories, and general biology (Lepidoptera: Nymphalidae), *Psyche* **86:**219–260 (1979).
721. R. Levins and M. Wilson, Ecological theory and pest management, *Ann. Rev. Entomol.* **25:**287–308 (1980).
722. A. D. Hinckley, Ecology of the coconut rhinoceros beetle, *Oryctes rhinoceros* (L.) (Coleoptera: Dynastidae), *Biotropica* **5:**111–116 (1973).
723. R. Bardner, K. E. Fletcher, and J. H. Stevenson, Pre-flowering and post-flowering insecticide applications to control *Aphis fabae* on field beans: Their biological and economic effectiveness, *Ann. Appl. Biol.* **88:**265–271 (1978).
724. N. Waloff and P. Thompson, Census data of populations of some leafhoppers (Auchenorrhyncha, Homoptera) of acidic grassland, *J. Anim. Ecol.* **49:**395–416 (1980).
725. W. M. Rogoff, Behavior modification for insect management by competitive displacement, *Bull. Entomol. Soc. Am.* **26:**121–125 (1980).
726. T. Burnett, A model of host-parasite interaction, *Proc. 10th Internat. Congress of Entomology,* Vol. *2,* 679–686 (1958).
727. T. Burnett, Aspects of the interaction between a chalcid parasite and its aleurodid host, *Can. J. Zool.* **45:**539–578 (1967).
728. M. P. Hassell, Foraging strategies, population models and biological control: A case study, *J. Anim. Ecol.* **49:**603–628 (1980).
729. K. J. Beek and D. L. Bramo, in: *Biogeography and Ecology in South America* (E. J. Fittkau, J. Illies, H. Klinge, G. H. Schwabe, and H. Sioli, eds.), pp. 82–112, Dr. W. Junk, The Hague (1968).
730. G. F. Edmunds, Jr., Phylogenetic biogeography of mayflies, *Ann. Mo. Bot. Gard.* **62:**251–263 (1975).
731. R. Schuster, in: *Biogeography and Ecology in South America* (E. J. Fittkau, J. Illies, H. Klinge, G. H. Schwabe, and H. Sioli, eds.), pp. 741–763, Dr. W. Junk, The Hague (1968).
732. G. H. Schwabe, in: *Biogeography and Ecology in South America* (E. J. Fittkau, J. Illies, H. Klinge, G. H. Schwabe, and H. Sioli, eds.), pp. 113–136, Dr. W. Junk, The Hague (1968).
733. G. T. Prance, The origin and evolution of the Amazon flora, *Interciencia* **3:**207–222 (1978).
734. W. B. Heed, in: *Ecological Genetics: The Interface* (P. F. Brussard, ed.), pp. 109–126, Springer-Verlag, New York (1978).
735. P. R. Ehrlich, D. D. Murphy, M. C. Singer, C. B. Sherwood, R. R. White, and I. L. Brown, Extinction, reduction, stability, and increase: The responses of Checkerspot Butterfly (*Euphydryas*) populations to the California drought, *Oecologia* **46:**101–105 (1980).
736. R. I. Vane-Wright, in: *Diversity of Insect Faunas* (L. A. Mound and N. Waloff, eds.), pp. 56–70, Blackwell, London (1978).

737. J. G. Brittnacher, S. R. Sims, and F. J. Ayala, Genetic differentiation between species of the genus *Speyeria* (Lepidoptera: Nymphalidae), *Evolution* **32:**199–210 (1978).
738. W. Geiger, Observations eco-faunistiques sur les lépidoptères de la Tourbière du Cachot (Jura Neuchatelois). I. Méthodes, faunistiques et caractèristiques de peuplement, *Bull. Soc. Neuchatel. Sci. Nat.* **103:**11–27 (1980).
739. M. S. Mani, *Ecology and Biogeography of High Altitude Insects,* Dr. W. Junk, The Hague (1968).
740. M. S. Mani and S. Singh, Entomological survey of Himilaya. Part XXVI. A contribution to our knowledge of the geography of the high altitude insects of the nival zones from the north-west Himalaya. Part 3, *J. Bombay Nat. Hist. Soc.* **59:**77–99 (1962).
741. M. J. D. White, *Modes of Speciation,* W. H. Freeman, San Francisco (1978).
742. H. Heatwole and R. Levins, Biogeography of the Puerto Rican bank: Species-turnover on a small cay, Cayo Ahogado, *Ecology* **54:**1042–1055 (1973).
743. J. Terbough, Distribution on environmental gradients: Theory and a preliminary interpretation of distributional patterns in the avifauna of the Cordillera Vilcabamba, Peru, *Ecology* **52:**23–40 (1971).
744. A Trawoger, Der Alpenwollafter, *Eriogaster arbusculae* FRR. Ein Beitrag zur Kenntnis und Erforschung der alpinen Schmetterlingsfauna (Insect: Lepidoptera, Lasiocampidae), *Ber Nat. Med. Ver. Innsbruck* **64:**107–132 (1977).
745. Noonan, G. R., South American species of the subgenus *Anisotarsus* Chaudoir (Genus *Notiobia* Perty, Carabidae, Coleoptera, Insecta). Part I. Systematics and natural history, *Milwaukee Publ. Mus. Contrib. Biol. Geol.* No. 44 (1981).
746. P. E. Vanzolini and E. E. Williams, South American anoles: Geographic differentiation and evolution of the *Anolis chrysolepis* species group (Sauria, Iguanidae), *Arq. Zool. (Sao Paulo)* **19:**1–298 (1970).
747. P. J. Darlington, Jr., Carabidae of mountains and islands: Data on the evolution of isolated faunas, and on atrophy of wings, *Ecol. Monogr.* **13:**37–61 (1943).
748. P. J. Darlington, Jr., Australian carabid beetles, V. Transition of wet forest faunas from New Guinea to Tasmania, *Psyche* **68:**1–26 (1961).
749. P. A. Opler, Oaks as evolutionary islands for leaf-mining insects, *Am. Sci.* **62:**67–73 (1974).
750. V. F. Eastop, in: *Diversity of Insect Faunas* (L. A. Mound and N. Waloff, eds.), pp. 71–88, Blackwell, London (1978).
751. H. Schreiber, Dispersal Centres of *Sphingidae (Lepidoptera)* in the Neotropical Region, Dr. W. Junk, The Hague (1978).
752. K. S. Brown, Jr. and W. W. Benson, Evolution in modern Amazonian non-forest islands: *Heliconius hermathena, Biotropica* **9:**95–117 (1977).
753. P. J. Darlington, Jr., *Zoogeography: The Geographical Distribution of Animals,* John Wiley, New York (1957).
754. W. L. Brown, Jr., and E. O. Wilson, Character displacement, *Syst. Zool.* **5:**49–64 (1956).
755. E. J. Fittkau, J. Illies, H. Klinge, G. H. Schwabe, and H. Sioli (eds.), *Biogeography and Ecology in South America,* Vols. 1 and 2, *Monogr. Biol,* **18–19:**1–946 (1969).
756. J. Illies, in: *Biogeography and Ecology in South America* (E. J. Fittkau, J. Illies, H. Klinge, G. H. Scwabe, and H. Sioli, eds.), pp. 685–708, Dr. W. Junk, The Hague (1969).
757. G. Lamas, A preliminary check-list of the butterflies (Lepidoptera) of Peru west of the Andes, *Rev. Cienc. Universidad Nacional Mayor de San Marcos (Lima)* **70:**59–77 (1977).
758. G. Lamas, Mariposas y conservacion de la naturaleza en el Peru, *Bol. Colon. Suiza Peru 1978:* 61–66 (1978).
759. A. M. Shapiro, Evidence for obligate monophenism in *Reliquia santamarta,* a neotropical-alpine pierine butterfly (Lepidoptera: Pieridae), *Psyche* **84:**183–190 (1977).
760. A. M. Shapiro, Developmental and phenotypic responses to photoperiod and temperature in

an Equatorial montane butterfly, *Tatochila xanthodice* (Lepidoptera: Pieridae), *Biotropica* **10:**297–301 (1978).

761. A. M. Shapiro, The life history of *Reliquia santamarta,* a neotropical alpine pierine butterfly (Lepidoptera: Pieridae), *J. N.Y. Entomol. Soc.* **86:**45–50 (1978).

762. R. C. Buckley, R. T. Corlett, and P. J. Grubb, Are the xeromorphic trees of tropical upper montane rain forests drought-resistant?, *Biotropica* **12:**124–136 (1980).

763. A. H. Gentry, Floristic knowledge and needs in Pacific tropical America, *Brittonia* **30:**134–153 (1978).

764. R. D. Schrier, M. J. Cullenward, P. R. Ehrlich, and R. R. White, The structure and genetics of a montane population of the Checkerspot Butterfly, *Chlosyne palla, Oecologia* **25:**279–289 (1976).

765. N. A. Campbell and J. M. Dearn, Altitudinal variation in, and morphological divergence between, three related species of grasshoppers, *Praxibulus* sp., *Kosciuscola cognatus* and *K. usitatus* (Orthoptera: Acrididae), *Aust. J. Zool.* **28:**103–118 (1980).

766. P. A. Parsons and I. R. Bock, The population biology of Australian *Drosophila, Ann. Rev. Ecol. Syst.* **10:**229–245 (1979).

767. A. Wille, Observations on the behavior of a tropical rain forest dung beetle, *Megathoposoma candezei* (Harold) (Coleoptera: Scarabaeidae), *Rev. Biol. Trop.* **21:**41–57 (1973).

768. F. Vuilleumier, Insular biogeography in continental regions. The northern Andes of South America, *Evolution* **23:**688–702 (1970).

769. D. A. Livingstone, History of the tropical rain forest, *Paleobiology* **6:**243–244 (1980).

770. R. W. Fairbridge, Effects of Holocene climatic change on some tropical geomorphic processes, *Quat. Res. (N.Y.)* **6:**529–556 (1976).

771. R. G. Klein, Environmental and ecological implications of large mammals from Upper Pleistocene and Holocene sites in southern Africa, *Ann. S. Afr. Mus.* **81:**223–283 (1980).

772. J. R. G. Turner, in: *Biogéographie et Évolution en Amérique Tropicale* (H. Descimon, ed.), pp. 98–117, Publications du Laboratoire de Zoologie de l'École Normale Supérieure, Paris (1976).

773. E. Mayr, *Animal Species and Evolution,* Harvard University Press, Belknap Press, Cambridge, Mass. (1963).

774. R. C. Lewontin, Studies on heterozygosity and homeostasis. II. Loss of heterosis in a constant environment, *Evolution* **12:**494–503 (1958).

775. D. A. Roff, Population stability and the evolution of dispersal in a heterogeneous environment, *Oecologia* **19:**217–237 (1975).

776. A. M. Lucas, Clinal variation in pattern and colour in coastal populations of the butterfly *Tisiphone abeona* (Donovan) (Lepidoptera: Satyrinae), *Aust. J. Zool.* **17:**37–48 (1969).

777. T. Park and M. Lloyd, Natural selection and the outcome of competition, *Am. Nat.* **89:**235–240 (1955).

778. M. V. Brian, Exploitation and interference in interspecies competition, *J. Anim. Ecol.* **25:**339–347 (1956).

779. D. H. Morse, Resource partitioning in bumble bees: The role of behavioral factors, *Science* **197:**678–680 (1977).

780. R. E. Moreau, Ecological isolation in a rich tropical avifauna, *J. Anim. Ecol.* **17:**113–126 (1948).

781. J. K. Waage, Reproductive isolation and the potential for character displacement in the damselflies, *Calopteryx maculata* and *C. aequabilis* (Odonata: Calopterygidae), *Syst. Zool.* **24:**24–36 (1975).

782. J. J. Pizzimenti and R. DeSalle, Dietary and morphometric variation in some Peruvian rodent communities: The effect of feeding strategy on evolution, *Biol. J. Linn. Soc.* **13:**263–285 (1980).

783. T. T. Veblen, F. M. Schlegel, and B. Escobar, Structure and dynamics of old-growth *Nothofagus* forests in the Valdivian Andes, Chile, *J. Ecol.* **68:**1–31 (1980).

784. P. J. Darlington, Jr., The carabid beetles of New Guinea. Part IV. General considerations; analysis and history of fauna; taxonomic supplement, *Bull. Mus. Comp. Zool. Harv. Univ.* **142** (1971).

785. H. U. Thiele, *Carabid Beetles in Their Environments. A Study on Habitat Selection by Adaptations in Physiology and Behaviour,* Springer-Verlag, Berlin (1977).

786. J. A. Wallwork, *The Distribution and Diversity of Soil Fauna,* Academic Press, New York (1976).

787. A. S. Corbet, The distribution of butterflies in the Malay Peninsula (Lepid.), *Proc. R. Entomol. Soc. London Ser. A* **16:**101–116 (1941).

788. T. Abe, Studies on the distribution and ecological role of termites in a lowland rain forest of West Malaysia. 4. The role of termites in the process of wood decomposition in Pasoh Forest Reserve, *Rev. Ecol. Biol. Sol* **17:**23–40 (1980).

789. H. L. Carson, D. E. Hardy, H. T. Spieth, and W. S. Stone, in: *Essays in Evolution and Genetics in Honor of Theodosius Dobzhansky,* pp. 437–543, Appleton-Century-Crofts, New York (1970).

790. P. H. Raven and D. I. Axelrod, Angiosperm biogeography and past continental movements, *Ann. Mo. Bot. Gard.* **61:**539–673 (1974).

791. J. A. Scott, Biogeography of Antillean butterflies, *Biotropica* **4:**32–45 (1972).

792. M. G. Emsley, Speciation in *Heliconius* (Lep. Nymphalidae): Morphology and geographic distribution, *Zoologica* **50:**191–254 (1965).

793. D. J. Greathead, Observations on two species of *Systoechus* (Diptera Bombyliidae) preying on the Desert Locust, *Schistocerca gregarid* (Forskal) in eastern Africa, *Entomophaga* **3:**1–22 (1958).

794. H. A. Hespenheide, Selective predation by two swifts and a swallow in Central America, *Ibis* **117:**82–99 (1975).

INDEX

Abdominal hair pencils, 91
Acacia, 165–166, 356–357, 390
  karroo, 117
Acanthaceae, 215, 267, 272, 442
Acizzia russellae, 117
Acraea encedon, 242
Acrididae, 310
Activating hormone, 71, 72
Adaptations to mating, 52
Adaptive radiation, 163, 350
Aedes aegypti, 30
Aenictus, 455
Aeria eurimedea, 131
Africa, 3, 80, 82, 83, 179, 213, 257, 258, 261, 290, 310, 321, 323, 332, 335, 364, 374, 391, 410, 439
Aggregative behavior, 111, 154, 258, 320. 347
Agriculture, 178, 430
  livestock, 409
  monoculture, 210, 388, 389, 394, 420
  polyculture, 210, 389, 394, 420
Agroecosystems, 280, 390, 422
  corn–bean–squash, 390
Aiello, A., 182
Alcohols, 96
Aldrich, J. R., 91, 154
Alkaloids, 120, 122
Allopatry, 262
  micro-, 262, 453
Altitudes, high, 57, 231
Amazon forest, 207, 249, 292, 308, 309, 327, 333, 337, 431, 432, 436, 438, 447, 456–457
Anaea, 321
  (*Memphis*) *morvus,* 300, 322
  (*Zaretis*) *itys,* 300, 321
Anartia, 417, 442–443
  amathea, 442
  chrysopeles, 442
  fatima, 267, 320
  lytrea, 442
Anastrepha, 183
Anderson, W. W., 48
Andes, 12, 57, 58, 159, 233, 287, 432, 435, 437, 443–444, 447
Andira inermis, 360
Andrewartha, H. G., 229
Andromonoecism
  functional, 366–367, 395, 399
  morphological, 367
Angiosperm host plants, 160
Anguria, 154
  warscewiczii, 88–89
Anole lizards, 352
  Anolis, 395
Anopheline mosquito, 10
Anopheles, 34, 83, 311
  gambiae, 312
Anoplolepis longipes, 426
Antarctica, 57
Antilles, 459
Ant mosaic, 395
Ants, 138, 303, 346, 382–385, 457
  army, 93, 106, 210, 343
  community shift, 278
  crazy, 426
  dacetine, 105
  desert, 83
  distribution, 390
  diversification of, 6, 9
  driver, 213–214, 385
  foraging behavior, 96, 101, 210–213, 345, 371, 377
  ground-dwelling, 213
  host plants, 166–169, 356–357
  leaf-cutter, 110, 115, 210–213, 329
  mutualisms, 356–357
  nesting sites, 117, 189–191
  predation on eggs, 156
  venoms, 110
Aphelandra, 372
Aphids, 88, 352

Aphodius
  howitti, 407
  tasmaniae, 318, 407
Apocyanaceae, 131, 150, 345
Aposematism
  color, 150
  display, 154
Archipelagos, 436
  taxon cycle, 47, 165
Arctic region, 83, 84
Area-effect, 434
Aristolochia
  constricta, 348
  as host plant, 203, 406
Arthropods, 96, 320
  abundance, 304
  evolutionary success, 62–67
  foliage-inhabiting, 303
  as hosts, 377
  predatory, 319, 395
  seasonal habitat change, 69
Ascia monusta, 150
Asclepiadaceae, 131, 150
Asia, 107, 410
Asilidae, 260
Asynchrony, between plants and pollinators, 333
Atta, 329, 399
Attinae, 96
Aulacobolus, 319
Australia, 3, 301, 407, 445
Avoidance behavior, of predators, 93
Ayala, F. J., 40, 43, 111
Azteca, 206, 357
Azygophleps albovittata, 146

Bahamas, 8, 289
Baits, 426
  artificial, 215, 380–385
  fruit, 183–184, 383, 401
Baker, H. G., 12, 104, 145, 216, 357, 362
Baker, I., 104, 145, 216, 362
Barriers, evolutionary, 434
Barro Colorado Island, 126, 183, 327
Barrows, E. M., 361, 371
Barteria, 166
Bates, M., 306
Battus, 126
  philenor, 406
Bawa, K. S., 357, 364, 366, 368
Bees, 145, 215–216, 279, 300, 323–324, 326, 383, 450
  bumblebee, 86–87
  honeybee, 374
  orchid, 340
  pollination by, 340, 358–361, 367, 370, 393, 399
  robber, 96
  solitary foraging, 325, 360
  stingless, 96, 111, 263, 361, 372, 374, 392, 403
Beetles, 150, 196, 233, 262, 328, 451
  adaptations to desert environment, 70
  coconut rhinoceros, 421
  cophrophagous scarabs, 409–410
  density of, 346
  dung, 184, 318, 409–410, 446
  flea, 350
  habitat selection, 203, 216
  hispine, 344
  ladybird, 197
  migration, 80, 308–309
  outbreaks, 318
  pasture scarabs, 318, 407–409
  as pollinators, 324, 360–361
  rolled-leaf hispine, 343
  seed-eating, 195
  soldier, 445
  sprucebark, 339
Belostomidae, 11, 382
Belenois aurota, 159, 319
Belgian Congo, 3
Beltian bodies, 166
Benson, W. W., 132, 141, 148, 292, 349, 350, 438–439, 447
Bentley, B. L., 167
Bignoniaceae, 361
Biogeographic events, 350, 430, 460
Biogeography, 2, 56, 148, 258
Birch, L. C., 35, 229
Birds, 278
Bixa, 356
  orellana, 167
Black pepper, 417
Black water, 308
Blum, M. S., 91, 96, 154
Body size in insects, 198–199, 278, 285, 304, 329

Bombyliidae, 310
Bossert, W. H., 48, 229
Braconidae, 232, 235, 244, 278
Brassolidae, 184, 353
Brazil, 163, 292, 300
Breeding sites, 364
Breeding systems, 278, 357-358
  outcrossing, 87, 326, 340
Breedlove, D. E., 140, 203
Brereton, J. Le Gay, 25
British Isles, 8, 331
Bromeliads, 179, 186-189, 378
  artificial, 382
  as breeding sites, 189, 307
Brower, L. P., 54, 91, 98, 150
Brown, K. S., Jr., 98, 150, 292, 431, 436, 438-439, 447, 449
Brown, W. L., Jr., 105
Bruchidae, 164, 344
Brues, C. T., 114
Bugs
  African cotton stainers, 257, 328
  ambush, 261
  blissine, 344
  coccoid, 340
  lovebugs, 91
  milkweed, 246-248
  plant hoppers, 163, 429
  seed-eating, 195
  spittle, 268
  triatomine, 312
Bush, G. L., 176, 455
Buskirk, R. E., 303
Buskirk, W. H., 303
Butterfly
  abundance, 235, 293, 321, 338
  coloration, 150
  color sensitivity, 68
  communal roosting in, 259, 267
  courtship behavior, 88-93, 97-100
  daily activity pattern, 83, 101
  differentiation of species, 254
  dispersal, 225
  disruptive wing coloration, 235
  distribution, 3, 4
  diversity, 391
  feeding habits, 19, 116, 154-156, 216, 298, 328, 336, 353, 408
  generation time, 280
Butterfly (*cont'd*)
  habitat selection, 203-205, 321, 417
  host plant interactions, 135, 319, 324, 359
  mate recognition, 352-353
  migration, 78, 80
  monarch, 53-54, 294, 437
  mortality of, 235, 294
  oviposition, 126-132, 157, 181, 204, 231, 289, 299
  as pollinators, 279, 370, 451
  population structures, 256
  predation on, 244
  seasonal activity, 332
  sex ratios, 241-242
  speciation, 431
  thoracic temperatures, 84
  troidine swallowtail, 348
  vertical structuring of, 89
  viceroy, 77
  whole faunas, 443, 459
Bystrak, P. G., 364

Cacao, 261, 394
  flowers, 362-363, 367
  plantations, 186-188, 262-264, 332, 380-382, 394-404
  pollination, 363-367
  shade cover, 399
Caesaria, 321
California, 77
Caligo, 101
Calpodes ethlius, 43, 46
Calvert, P. P., 378
Cameroon, 142
Campbell, B. C., 158
Canna edulis, 43
Cannibalism, 231
Capparis tomentosa, 159, 319
Capsidae, 399
Carabidae, 58, 157, 201, 210, 237, 298, 308-309, 408, 442, 455-458
Cardiac glycosides, 120, 151
Caribbean islands, 167, 197
Caribbean watershed, 289
Carineta postica, 415
Carpenter, F. M., 63
Carrion colonization, 409
Carroll, C. R., 96, 168, 213, 262, 390, 422
Cassia, 215

Caste systems, 105
Catastrophes
  natural, 227, 269, 296, 297, 343, 420, 430, 454
    fires, 34
    floods, 178, 292, 303, 307-309, 337, 341, 388
    hurricanes, 388
    mud slides, 388
  typhoons, 339
Cecidomyiidae, 93, 382, 393
Cecropia, 206, 357
Central America, 6, 74, 79, 80, 132, 147, 161, 164, 165, 179, 206, 255, 265, 279, 290, 307, 322, 328, 336, 390, 408, 409, 417, 437
Cerambycidae, 3
Ceratopogonidae, 93, 116, 186, 261, 303, 332-333, 362, 382, 396-398
Chagas' disease, 312
Chalcidae, 43
Chanter, D. O., 312
Chaplin, S. J., 81
Characters, 239, 254
  displacement, 183, 259
  divergence, 259
  plesiomorphic, 117
  symplesiomorphic, 117
Chemical defenses, of plants, 117, 141, 182, 202
Chew, F. S., 205
Chironomidae, 84
Chironomous, chromosomes, 34
Chromosome inversion frequencies, 177
Chrysomelidae, 32-33, 77, 131, 194, 299, 346, 355, 405
Cicada, 50, 83, 93, 97, 101, 114
  aggregations, 207
  body size, 199-200
  chorus, 50, 102-104, 286, 290, 417
  emergence, 289-291, 293, 313-317, 327, 343, 411, 416
  food of, 117, 413
  generation time, 280
  habitat, 147, 184, 199-201, 208-209, 322-323
  host plants of, 161-163, 323, 413
  life cycle length, 220
  niche overlap, 30
  niche shift, 265
Cicada (*cont'd*)
  nymphs, 162, 291, 414-420
  oviposition, 413, 416
  seasonal emergence, 78-79
Cicindelidae, 308
Cingalobolus, 319
Circadian rhythms, 287
Climatic predictability, 351
Clinal shift, 432
Clinal variation, 447
Clutch size, 27, 97, 114
Coaction, 376
  definition, 223
Cockroaches, 24, 70, 88, 101, 111, 207, 309, 382
  molt induction, 73
Coconut palm, 410
Cody, M. L., 27, 97
Coevolution, 25, 86, 132-133, 138, 141, 181, 338, 341, 355, 358, 438
Coffee plantations, 411
Cole, L. C., 230
Coleoptera, 32, 151, 277, 278, 382, 383, 384, 398, 405
Coley, P. D., 126, 146-147
Colias eurytheme, 320
Collembolans, 303, 335
Colombia, 30, 81, 306, 409
Colonizing episodes, 41, 48, 176, 218-219, 224, 226, 229, 245, 270
Colonizing species, 248, 257
Colony
  behavior in, 105
  mortality in, 106
Colwell, R. K., 190-191, 215
Commensal associations, 189
Communal displays, 98
Communal feeding, 126
Communication
  acoustical, 287
Community, 2, 19, 43, 49, 69-70, 97, 114, 436
  biologically accommodated, 239, 255
  foliage-inhabiting, 290
  litter, 113, 335-336
  physically controlled, 239, 255
  pollinator, 157
Competition, 26, 32, 41, 43, 86, 96, 138, 228, 259-266, 355, 453
  diffuse, 437

Competition (*cont'd*)
exploitation-type, 453
"interference", 453
Competitive ability, 226
frequency-dependent, 343
Compositae, 404
Connell, J. H., 26, 138, 178, 338
Cook, L. M., 294, 332
Cordillera central, 418
Core area, 434–436, 438, 447
Coreidae, 88, 91, 154
Corner, E. J., 6
Corpora allata, 71–72
Corpora cardiaca, 71
Cossidae, 146
Costa Rica, 16, 30, 41, 52, 78, 88, 102, 110, 140, 142, 147, 154, 162, 167, 183, 184, 189, 196, 201, 205, 230, 255, 262, 267, 269, 275, 282, 289, 292, 293, 303, 307, 313, 321–323, 326–327, 329, 331, 344, 346, 348, 361, 371, 372, 378, 381–383, 389, 399, 411, 415, 417, 419, 442
Courtship patterns, 153
Courtship pheromones, 91
Crabs, 377
Cretaceous, 63
Crewe, R. M., 96
Crickets, 283, 297–298
Crops
cash, 389
food, 14, 388–391, 420
pests of, 424–427
Cruciferae, 146, 205, 459
Cryptic behavior, 120
Cryptic coloration, 91
Cuatrecasas, J., 364, 394
Cubitermes, 335
Cucurbitaceae, 8, 88, 258, 355
Curcurbitacins, 355
Curcurbita moschata, 131
Culex, 83
Culicidae, 93
Cuticle, 84, 319
Cyanogenic substances, 122
Cycles
climatic, 431
daily, 286–287, 290, 327
fragmentation, 438
glacial, 446
of inundation, 456
Cycles (*cont'd*)
morphisms, 249–250, 288, 296, 297
population, 157, 307, 318, 329, 342
seasonal, 289, 299–322, 423

DaCunha, A. B., 175
Danaidae, 91, 120, 131
Danaus
chrysippus, 150
plexippus, 54–55, 294
Darlington, P. J., 436, 456–457
Darwin, C., 115
Darwin's theory of evolution, 38
Dawson, P. S., 111
Defensive behavior, 63, 78, 242
Defogging, 196
Defoliation, 137, 140, 158–160, 298, 319, 341, 394
Demes, 255, 266
selection, 288
Denlinger, D. L., 312
Denno, R. F., 32
Density, 22, 40–43, 81–82, 111, 114
dependence, 229, 230, 232–233, 311, 407, 408, 426
independence, 226, 229, 232–233, 261, 280, 282, 299, 339, 348, 388–389, 407, 409, 427
Denslow, J. S., 376
Dethier, V. G., 115, 116, 377
Detoxification, 123
enzymes of, 123, 131, 139
Development
and environmental cues, 72
hemimetabolous, 62
holometabolous, 62
integration by hormones, 71–74
molt cycles, 71–72
rate of, 59, 61, 218–220
Diapause, 71, 76–79, 146, 279, 298, 306, 309, 312, 329
facultative, 330
pupal, 330
reproductive, 197, 268, 278, 288, 294, 300, 322, 330
Diel rhythms, 104
Dietary overlap, 330
Diets, and diapause, 82
Dimorphism, 296
sexual, color, 50

Dingle, H., 79–81, 246
Dioecism, 364, 368
Diplacus aurantiacus, 298
Diptera, 359, 378, 382, 444
Dismorphia
  egg placement, 157
  virgo, 38
Dispersal, 5, 25, 49, 79, 80, 111, 224–226, 240, 257, 269
  barriers to, 433
  genotypes, 456
  home range movements, 353, 370, 377
  seasonal, 292
Dispersal agents, of seed or fruit, 138, 161, 328, 341, 358
Dispersion, 61
  definition, 225
Displacement, ecological, 453
Distephana, 141
Distribution
  body size, 291
  patch, 88, 266, 268, 328
    generalist, 349
    specialist, 147, 148, 368
    structure, 135, 256, 266, 295, 341, 377, 388
  and spatial heterogeneity, 250, 254–255, 269
  and temporal heterogeneity, 250, 268
Diurnal cycle
  of activity, 101
  of temperature, 83, 281
Dobzhansky, T. H., 6, 26, 34, 175
Dominica, 262
Dominance
  ecological, 138, 343
DNA synthesis
  during molt cycle, 72
L-dopa, 160
Dorylinae, 455
Dorylus, 455
Downes, J. A., 83–84, 398
Dragonflies, 11, 88, 225, 316–318
Drift, genetic, 239, 267–268, 288, 447
Drosophila, 54, 98, 218, 445, 452–453
  flower-feeding, 330
  fungus-feeding, 331
  genetics, 34–35, 177
  grimshawi, 98
  habitat, 174–175, 330
  persimilis, 175
  population density, 40, 41, 183, 300
Drosophila (cont'd)
  pseudoobscura, 35
  resource partitioning, 194
  simulans, 35
  willistoni, 175
Drosophilidae, 189
Drummond, B. A., III, 345
Duffey, S. S., 150, 158
Dysdercus, 257, 328, 329
  bimaculatus, 329
  fasciatus, 257–258
  nigrofasciatus, 258
  superstitiosus, 257

Ecdysone, 72
Eciton, 93
Ecological barrier, 141, 265, 287, 346
Ecological colonization, 47, 49, 256
Ecological differentiation, local rain forest forms, 432, 445
Ecological diversity, 100, 148, 341, 434
Ecological release, 309
Ecological screen, 358
Ecological shifts, 56, 163, 349, 422
Ecosystems
  trophic organization of, 278
  tropical, 13, 335
Ectatomma, 169, 399
Ecuador, 93, 101, 233, 345
Egg
  maturation, 362
  predation on, 156, 231
  production, 59, 216
Ehrlich, A. H., 97
Ehrlich, P. R., 3, 97, 131, 140, 156, 182, 203, 241, 332
Eisner, T., 96
Elevational gradients, 180, 239, 255, 286, 356
El Salvador, 311, 320, 352
Emergence
  allochronic, 314, 324, 327, 343
  cicada, 184–185, 313–317, 343
  polymodal, 295
  seasonal, 323
Emerson, A. E., 3
Emigration, density-dependent, 311
Emmel, T. C., 321
Endemism, 48, 433, 447, 450
Endocrine control systems, and social evolution, 73–74
Endogenous activity patterns, 101

Endopterygota, 62
Energetics, foraging, 86
Energy budget, 96
Environments
  carrying capacity of, 175, 226
  coarse-grained, 174, 252, 256, 258-260, 295
  fine-grained, 174, 252, 255, 256, 258-260
  nonseasonal, 233
  physically controlled, 234
  predictable, 56, 58, 229
  seasonal, 233
  spatial patchiness of, 250
  tropical lowland, 234-235
  unpredictable, 56, 58
Enzyme systems, 73, 256
Erickson, J. M., 123
Erwin, T. L., 9, 201, 456-457
Erythrina, 187-188, 398
Esterifying acids, 146
Ethiopia, 294
Eueides
  aliphera, 205
  lybia, 205
Eugenol, 383
Euglossinae, 86, 342
Euphydryas chalcedona, 298
Euprojoannisia, 362
Euptychia, 348
Eurema daira, 78, 295, 296
Eusociality, 164
Evapotranspiration, 286
Evolution, arthropod, 62-67
Evolutionary history, 433
Ewel, J., 303
Exocrine control systems, 73
Exopterygota, 62
Extinction, 156, 254, 256
  rates, 338
  and species abundance, 6
Exudates
  of homopterous insects, 190
Eyespot, markings, 98, 353

Facultative associations, 168
Facultative shifts, 261
Farnesol, 72
Fecundity, 40, 61, 219, 245
  definition, 218
Fecundity schedule, age-specific, 217, 221
Feeding aggregates, 91
Feeding behavior, 71, 93-95, 202, 408
Feeding categories, 114-115
Feeding habits, 63, 101, 105, 164, 257, 320, 346
  facultative switches, 341
  obligatory switches, 341
Feeding stimulants, 355
Feeding strategies, 256, 408
Feeny, P. P., 56, 123, 146, 157, 264, 406
Ferns, as host plants, 160
Fidicina
  mannifera, 102, 322, 418
  pronoe, 102-103, 184, 417
  semilata, 411, 416
  sericans, 102-103, 162
Fig seeds, 206
Fisher's Fundamental Theorem of Natural Selection, 219-221
Fitness, 2, 40, 67, 120, 131, 154, 219-221, 245, 320, 322, 371
Fittkau, E. J., 6, 439, 440
Flacourticeae, 300
Flies, 93, 97, 176, 216, 249, 267, 328, 340, 360-361, 383, 410, 451, 455
  black, 9, 311
  fungus gnats, 269
  tsetse, 290
Flight, heat production, 84-87
Flightlessness, 455-457
Flight speeds, 83
Florida, 8
Flower
  cauliflorous, 366
  color change syndrome, 361
  drop, 299, 400
  morphological specializations, 87, 358
  pollination of, 69
  production, 368
  resources, 279
Flowering, 275, 426
  seasonal periodicity, 282, 323, 333
Flowering strategies
  diel, 86, 104
  mass, 324, 325-327, 360-361, 371
Foliage
  as food, 325
  quality, 405
Folivores, 294, 299
Food, spermatophores as, 97
Foraging behavior, 164, 210, 287, 329, 370
  density-dependent, 371

Foraging behavior (*cont'd*)
  diversification of, 96
  diurnal pattern of, 101
Forcipomyia 364–365
Forest
  Amazon inundation, 207, 292
  canopy, 180, 290, 308, 326
    flowers, 10
    specialists, 457
  islands, 431
  light gap, 233, 278, 331, 339, 376
  moist pockets, 277
  tropical dry, 7, 266, 279, 282, 301, 321–324, 329, 331, 390, 430
  tropical rain, 7, 8, 110, 167, 196, 266, 301, 313, 327, 332, 336, 430
Foundress, 106
Four-o'clocks, 372–373
Frankie, G. W., 12, 69, 145, 157, 324–325, 327, 357
Fraser-Rowell, C. H., 111
French Guyana, 13, 292
Fruit
  removal, 278
  rotten, 336
  set, 332, 360, 367, 369, 394, 399, 401
  shadow of, 328
Fruit-inhabiting flies, 176
Fruit investment strategy, 368
Fruiting
  mast, 375
  seasonal, 327–328
Fungus gardens, 329

Gadgil, M., 48, 229
Galapagos Islands, 8
Galeruca tanaceti, 77
Gene
  flow, 239
  frequency, 177
  host selection, 454
  pool, 22, 77
Generalists, 57, 140, 201, 228, 234–235, 456
  feeding habitats, 253
  habitat of, 210
Genetic variation, 1, 41, 49, 79, 219, 255, 268, 338, 340
  homozygosity, 34, 256
Genotype, 47, 48, 218
  enzyme, 175
  frequencies, 182
Gentry, A. H., 444
Geographical differentiation, 183, 443
Geographical distribution, 454
Geographical patchiness, 432
Geographical range, 77
Geographical variation, 150
Geological histories, 430, 453
Gerris, 296
Ghana, 329, 395, 398, 401
Ghent, A. W., 126
Gilbert, L. E., 8, 131, 132, 141, 148, 216, 241, 332, 345, 347–348, 350, 355
Givnish, T. J., 368
Glaciers, 440
Gliridicia nigra, 417
Glossina medicorum, 290, 377
Goethalsia meiantha, 162, 413
Gold Coast, 333
Gotwald, W. H., Jr., 6, 213, 385
Gradient
  altitudinal, 433, 435, 454, 459
  body size, 441
  clinal, 279
  ecotonal, 336
  geographic, 336, 432, 451
  sunlight, 195, 336
  thermal, 195
Granadilla, 141
Grasshopper, 112, 150, 221, 340, 444
Groundnut tree, 146
Guanacaste Province, 78, 102, 140, 154, 230, 275
Guatemala, 307, 311
Guild, insect, 123, 235, 266, 336

Habitat, 5, 49, 56, 67, 97, 104, 178–195
  associations, 431
  carrying capacity, 338
  complexity, 262, 401
  correlation, 251
  definition of, 61, 178
  divergence, 176
  diversity, 8, 65, 101, 142, 239, 387
  foraging, 105
  invasion, 59, 70, 391
  partitioning, 83
  selection, 174–178, 202–207, 216, 250, 345, 376
Habitat types
  agricultural, 47, 165, 301
  arboreal, 309, 456

Habitat types (*cont'd*)
cow pats, 409-410
desert, 57, 70, 74, 78
fields, 388-389
garden, 321, 391
grasslands, 390, 393, 407
lowland hot and wet, 81, 83, 432
marginal, 163
micro, 175-176, 343
pasture, 265, 410
riverine, 5
secondary, 41
successional, 391
water, 5
Haffer, J., 435
Haldane, J. B. S., 43
Halffter, G., 318
Happold, D. C. D., 316
Hardy-Weinberg equilibrium, 39
Hartshorn, G. S., 161, 376
Hawaii, 98, 175, 177
Heinrich, B., 85-86, 117, 145
Heliconiines, 120, 169
Heliconinae, 141, 205, 349-351
Heliconius, 8, 98, 100, 117, 132, 141, 154-156, 216, 242, 267, 299, 362, 431, 436, 438, 448-449, 451
charitonius, 241-243, 332
cydno, 448
erato, 449
ethilla, 241, 332
hermathena, 439
melpomene, 448-449
numata, 292
Hemiptera, 30, 32, 93, 117, 150, 151, 258, 269
coloration, 154
migration, 80
Herbivores, 8, 17, 32, 49, 97, 114, 115, 137, 138, 140, 146, 149, 231, 275, 295, 337, 338
deterrants, 356
guild, 227, 426
and host plant interactions, 158-160, 319, 350, 410, 423
population size, 24-25
seasonal changes in populations, 76
Heritability, measure of, 220
Hermann, H. R., 101, 189, 211
Hermaphroditism, 364, 367
simultaneous, 369
Hespenheide, H. A., 460
Heteroptera, 183
Hevea brasiliensis, 362, 394, 398, 399
Hibiscus, 329
Hoffman, R. J., 320
Holdridge, L. R., 217
Holldobler, B., 9
Homoptera, 30, 32, 115, 117, 120, 147, 163, 191, 196, 269, 299, 342, 343
Host plant
defensive systems, 49, 357
distribution, 410
diversity, 350
egg placement on, 231
insect colonization of, 49, 117, 132, 135, 142, 323
larval, 195, 244, 294, 295, 321, 448
range, 344
selection, 406
specialization, 345
specificity, 176, 429
variation, 150
Hubbell, S. P., 7, 11, 215
Humic acid, 308
Humidity, and body water content, 70-71
Hummingbirds, 191, 327
Hurd, P. D., Jr., 8
Hybridization, zones of, 450, 452
Hymenaea courbaril, 82
Hymenoptera, 105, 110, 215, 261, 382, 389
parasitic, 65, 305, 444
social evolution of, 73
Hypercontingent species, 190-191, 194
Hyperodes bonariensis, 77
Hypothuris 121, 127, 346-347
euclea, 346
colonizing episodes, 41-45

Ichneumonidae, 312
Illies, J., 6, 440
India, 319
Indo-Malaysia, 179
Inga 38, 157, 398, 418
Inouye, D. W., 356
Insecticide application, 426
Interstitial species, 191
Intertropical Convergence Zone, 316-317
Ipomoea, 326
Ips typographus, 233, 339
Isoptera, 105
Ithomiidae, 90-92, 98, 121, 127, 131, 146, 292, 345, 352, 436, 449

Jaisson, P., 117
Janzen, D. H., 8, 13, 14, 22, 26, 32, 49, 64-67, 69, 86, 88, 96, 104, 117, 123, 138, 140, 142, 146, 156, 161, 162, 163, 165, 166, 168, 177, 195-198, 202, 206, 213, 216, 217, 230, 262, 266, 275, 277, 279, 291, 300, 303, 323, 327, 331, 337, 338, 344, 346, 351, 355-357, 364, 369, 374-375, 387-391, 398, 420, 432, 433
Jeanne, R. L., 190, 457
Johnson, C. G., 80
Johnson, L. K., 215
Jones, M. A., 91
Jurassic, 62
Juvenile hormone, 72, 79

Kellett, D. N., 150
Kenyi, J. M., 159, 319
Ketones, 96
Khartoum, 316-317
Kinsman, S., 372
Kohn, A. J., 9
Krieger, R. I., 131

Lack, D., 19
Lamas, G., 443
Lampyridae, 311
Lantana camara, 371
Latitude, 79, 81, 82
Lauraceae, 300, 322
Lawton, J. H., 277
Leaf hopper, 182, 342, 405
Leck, C. F., 321
Leguminosae, 117, 142, 163, 413, 419
  seeds, 163-164
Lepidoptera, 84, 99, 117, 127, 161, 195, 216, 227, 244, 262, 268, 277, 288, 312, 319, 347, 358, 361, 377, 398, 435
Lerner, I. M., 41, 43
Leston, D., 398
Levins, R., 56, 121, 175, 177, 295, 421, 452-453
Lewontin, R. C., 1, 35, 40, 41, 47, 59, 111, 218, 221, 228
Liberia, 535
Life cycle
  patterns, 62-63
  stages, 122, 156, 229, 280, 294
Life history, 81, 217, 257, 289, 323, 353, 375
  synchronization, 283, 294
  traits, 339
Life spans, 242
Life style, definition, 7, 61
Life tables, 2, 59, 61, 217, 224, 243, 395
Limenitis archippus, 77
Limiting factors
  for growth, 333
  for host plants, 345
  for insect populations, 339, 408
  for pollination, 332, 368, 394
Lineages, phylogenetic, 442
Linsley, E. G., 3, 8
Lipids
  and diapause, 82
  in diet, 82, 169
Litter
  leaf, 292, 301, 303, 336
  organisms, 335
Lloyd, M., 413
Locusts, 79, 80, 82, 279, 310-312
Longevity, 59, 73
Lumbering, 420
  as a perturbation, 178, 430
Lycaena dispar, 331
Lycaenidae, 140
Lygaeidae, 205, 456

MacArthur, R. H., 6, 9, 27, 49, 63, 74, 120, 148, 219, 295, 338, 346, 435, 436
McCauley, D. E., 40
McDade, L. A., 372
Machaerium seemani, 269-270
McKey, D., 142
Maelzer, D. A., 318, 407
Manduca sexta, 85
Mating
  strategies, 93, 96, 287, 340
Matthews, E. G., 318
Matthews, R. W., 110, 189
Mayflies, 377
  taxonomic diversity, 3
Mayr, E., 49, 451, 453
Mechanitis, 121, 264-265, 450
  isthmia, 106-107, 141, 231, 346-347
Mecistogaster, 378
  coerulatus, 307
Megachile, 393
Megaloprepus, 378
  coerulatus, 378
Melinaea, 451
Meristems, 132, 169, 203, 260, 277, 279, 293, 405, 406

Mesozoic, 62
Metamorpha, 417
  epapha, 442
  stelenes, 320, 442
Mexico, 43, 93, 98, 203, 361, 437
Michener, C. D., 6, 111
Micrathyria, 3
Microsomal mixed-function oxidases, 123
Microstigmus comes, 110, 189
Midges
  pollinating, 261, 332–333, 367, 394–404
Migratory behavior, 54, 71, 80–82, 257, 295, 310–311, 329, 330
Milkweed plants, 79, 81, 151, 294
Millipedes, 319
Mimicry, 98, 120, 235, 336, 352–353, 431, 449
Mirabilis jalapa, 372–373
Mischocyttarus immarginatus, 189
Mites, nectarivorous, 191
Moffett, M. W., 126, 141, 150, 263
Monoecism, 368
Monophagy, 120, 132, 140, 176, 196, 237, 256, 330, 348
Monophenism, 458
Monoterpenes, 353
Monteverde cloud forest, 12, 327
Moore, T. E., 290
Moraba, 221
  scura, 22
Morphidae, 353
Morpho, 90–91, 98, 99, 116, 142, 184, 241, 406, 443
  achilles, 101
  amathonte, 98
  granadensis, 142, 348
  peleides
    abundance, 322
    aggregates, 93–95
    daily activity, 101
    food of, 128, 184
    habitats, 270
    host plants of, 142, 348
    phenotypic variation, 254–255
    population dynamics, 240–241
    sexual dimorphism, 50, 53
Mosquitoes, 306–307, 311, 323, 382
Moths, 324–325, 346, 360, 451
  armyworm, 160
  gypsy, 137
Moths (*cont'd*)
  sphinx, 85, 87
  spruce budworm, 137
  tent caterpillar, 137
  tobacco hornworm, 117
Mountain
  insect faunas, 288, 432
  tops, 70, 74, 84, 430, 433, 455–458
Mucuna, 160
  urens, 164, 270
Mustard oil glycosides, 122
Mutchmor, J. A., 84
Mutualism, 138, 165, 202, 210, 356
  dissolution of, 206
Muyshondt, A., 132, 156, 443
Mycorrhizal fungi, 163, 389

Nairobi National Park, 305
Natural history studies, 195, 229, 257, 260, 301, 346, 399, 424, 459
Naumann, M. G., 93
Nectandra, 322
Nectar
  nutrient concentrations, 145, 216, 327, 361, 362
  rewards, 340, 370–372
  thief, 194, 371–372
  time of peak production, 104
Nectaries
  extrafloral, 138, 167–170, 189, 356–357
  foliar, 166
Nephila clavipes, 378
Nesting habits, 164, 189–190, 213, 340
Neto, J. V., 150
Net reproductive rate, 219, 221, 245
New World tropics, 17
New Zealand, 77
Niche, 5, 237, 358
  breadth, 31, 48, 251, 252, 256
  overlap, 194
  and physiological adaptations, 67–71
  shifts, 263
Nigeria, 146, 166
Nitrogen
  and egg production, 59
  in plant tissues, 117, 163, 407, 408, 413
Nomadacris septempfasciata, 312
Noonan, G. R., 435, 443–444, 456
North America, 3, 137, 413
Notiobia peruviana, 435, 456

Novomessor
  albisetosus, 83
  cockerelli, 83
Nymphalidae, 300, 320

Oak trees, 157, 275
Ocotea, 36, 38
Odonata, 3, 5, 8, 17, 93, 189, 307, 318, 378
Oleria zelica pagasa, 142
Oligocene, 63
Oligolectic behavior, 86
Onchocerciasis, 311
Oncopeltus, 69, 79–82, 246–247
  fasciatus, 80
Onthophagus gazella, 410
Opler, P. A., 78, 295, 296
Optimization theory, 67
Optomotor response systems, 69
Orchidaceae, 398
Organization for Tropical Studies, 12
Orians, G. H., 9
Orthoptera, 101, 157, 184, 186, 196, 251, 269, 277, 288, 292, 299, 300, 346, 347
Oryctes rhinoceros, 410, 421
Osa Peninsula, 215, 293
O'Shea, R., 117
Overwintering behavior, 294
Oviposition, 43, 93, 98, 126, 133, 169, 299, 310
  cluster, 126, 231, 259
  single, 347
  sites, 413
Oviposition behavior, 3, 132, 135, 348, 407–408
Owen, D. F., 312, 321
Oxidases, mixed function, 123, 131, 154

Pachyptera hymenaea, 361
Pachysima, 166
Paine, R. T., 26
Palatability, 141, 150, 154, 259
Paltothyreus, 9
Panama, 12, 147, 299, 321, 327
Papilio
  machaon, 133
  polyxenes, 85
  zelicaon, 77
Papua New Guinea, 183, 410, 444
Paraponera, 9
  clavata, 101, 110, 189, 211
Parasites, 19, 48, 65, 78, 114, 207–210, 254, 278, 426
  population dynamics of, 76, 238, 243
Parasite types
  ecto-, 440
  endo-, 194, 261
  klepto-, 189
Parasitic ants, 231
Parasitic flies, 16, 228, 230, 231
Parasitic wasps, 16, 159, 228, 230, 231, 244, 261
Parasitism
  floral, 371–372
  multi-, 261
Parides, 50, 235, 294, 332, 406
Park, O., 5
Park, T., 24, 43, 111
Pars intercerebralis, 71
Paryphes blandus, 88–89, 154
Passiflora
  as food, 350
  vitifolia, 205
Passifloraceae, 141, 169, 351, 448
Patrick, R., 3
Paulson, D. R., 3, 8
Pectinophora gossypiella, 82
Pemphigus, 352
Pentaclethra, 30, 189, 418
  macroloba, 147, 161, 336, 418
Pentacomia egregia, 308
Periplaneta americana, 24
Permian, 62
Perrhybris, 38
  lypera, 36, 98, 231
Peru, 443
Pesticide application, 390, 422
Peters, W. L., 3
Petiolar cavities, 167
Phasmids, 101
Pheidole, 167
Phenolics, 122
Phenotype
  definition, 34
  flexibility of, 132, 255, 338, 341
  frequencies, 182
  seasonal, 458
Pheromones
  alarm, 96
  sex, 120
Philaenus, 268

Philaethria dido, 443
Photoperiod, 73, 81, 82, 279, 312, 320
  and activity, 285, 290
  and diapause, 79
  and migration, 69
Phymateus purpurascens, 112
Pianka, E. R., 25, 27
Pieridae, 12, 36, 78, 98, 233, 296, 458-459
Pieris, 205
  rapae, 181
Piegon pea, 393
Pimentel, D., 24, 353
Piperaceae, 167
Piper nigrum, 417
Plantations, cacao, 186-188
Plant defense mechanisms, 13, 49, 71, 117, 132, 137-140, 145, 156, 182, 243, 273-276, 337, 406
Plate tectonics, 438
Platypodidae, 202
Plecoptera, 440
Pleistocene, 25, 149, 254, 431, 438, 456, 458-459
Pliske, T., 90-91, 120, 146
Polistes, 73, 93, 106
Pollen
  balls, 362
  collecting, 326
  donor, 370, 395
  feeding, 116, 215, 216, 241, 300, 398
  interference, 364
  nitrogen, 408
  production cost, 368
  rewards, 371
  robber, 371, 372, 403
  starch content, 362
Pollinating guild, 325, 370-371
Pollination
  cross-, 371
  systems, 138, 149, 323-327, 329, 340, 358-375, 391-394
    self-compatible, 360
    self-incompatible, 360, 364, 394
Pollinator
  abundance, 332-333, 368
  asynchrony with host plant, 401
  availability, 138, 364
  limitation of fruit set, 367
  long-distance, 266, 342
Polybia, 93, 110
  occidentalis, 189
Polybiine wasp, 189
Polymorphism, 34-35, 175, 177, 249, 255, 267, 292, 296, 450, 455
Polyphagy, 121, 132, 140, 154, 176, 196, 231, 237, 256, 348
Polyphenisms
  latent, 458-459
Pompilidae, 107
Ponerinae, 110, 211
Poole, R. W., 141
Population
  census history, 15, 22-24, 233, 239, 276
  control mechanisms
    abiotic, 14, 228
    biotic, 14, 217, 224, 228, 232, 234-235, 237, 244, 248, 257, 260, 281, 288, 297, 319, 341, 347, 355, 356, 389, 390, 410, 424
    physiological, 313
  differentiation, 40, 220, 254
  dynamics, 22, 54, 221, 226, 232, 240, 327
  outbreaks, 137, 238, 281, 318-319, 335, 339, 348, 355, 399, 410, 421, 453
  regulation, 25, 229-239, 242
  size, 240, 242-246, 271, 332, 377
  stability, 241, 254
  structure, 96, 241, 256
Populus, 352
Pough, F. H., 150
Prance, G. T., 438, 440
Predation, 19, 32, 48, 65, 78, 86, 96-97, 114, 254
Predator
  avoidance, 287
  population dynamics of, 238, 243
  seasonal changes in populations, 76
  seeds, 160-164
  vertebrate, 235, 258
Predator-parasitoid complex, 388
Prey preferences, 340, 385
Price, P. W., 273
Procollina biolleyi, 343, 414-415
Productivity
  harvestable, 210, 214, 249, 250, 261, 350, 388, 399
  hypothesis, 26
Protective coloration, 235
Protective guard insects, 361

Prothoracic gland, 71-72
Psephalidae, 5
Pseudomyrmex, 165, 166, 356-357, 390
Psyllidae, 117
Pteridines, 320
Puerto Rico, 175, 206, 398, 452
Pulsations, glacial, 441
Puntarenas Province, 102
Purseglove, J. W., 366, 394
Pursuit invitation behavior, 352
Pyrrolizidine alkaloids, 120, 146

Quesada gigas, 102, 184-185, 313-314

Rainfall
  and butterfly mortality, 235-236, 294
  and diapause, 79
  and insect density, 348
  seasonal patterns, 317, 318, 322
Rain shadows, 362, 441
Range
  geographical, 434, 452-454
  topographical, 77
Range expansions, 180, 431
Rapoport, A., 54
Rathcke, B. J., 141
Rausher, M. D., 203
Raven, P. H., 3, 156, 182
Refugia
  evolutionary, 434
  glacial forest, 350
  model, 439, 448-451
  moist, 248, 427
  moist forest, 78, 196, 275, 278, 288, 407
  mountain forest, 148
  natural habitat, 422
  Quaternary, 351, 431, 441, 446, 448
  riparian forest, 312
Region, definition, 254-255
Rehr, S. S., 123, 160, 164, 356
Reproductive effort, 229
Reproductive isolation, 182
Reproductive output, 332
Reproductive patterns of plants, 169
Reproductive rate, 247
Reproductive value, 248
Resource
  distribution, 65
  limitation, 135, 260, 340
  partitioning, 194, 345, 349-351
Resource (*cont'd*)
  patchy, 260
  utilization, 28-29, 43
  variety, 351
Resource abundance, 207
Resource allocation, 229
Rettenmeyer, C. W., 6
Rhoades, D. F., 133, 139
Rhodnius, 72, 77
Rhythms
  activity, 100-101, 104
  daily, 285-287
Richards, P. W., 179, 366
Ringo, J. M., 98
Risch, S., 167
RNA synthesis, 72
Robinson, B., 189
Robinson, M. H., 189
Rockwood, L. L., 110, 140, 262, 329
Roosting, communal, 259, 267
Root
  clover, 407, 408
  crowns, 30, 117, 291, 413, 415
  systems, 117, 189, 407
Root, R. B., 81
Rosenthal, G. A., 123, 140
Rothschild, M., 150
Rowell, H. F., 186
Ruwenzori National Park, 319

Salticidae, 395
Sanders, H. L., 70
Sap-feeding insects, 163
Satiation, of predators and parasites, 230
Saturniidae, 295
Satyridae, 91, 184, 320, 353
Satyrinae, 348
Savanna zone, 329
Scavenging insects, 278
Scent organs, 153
Schaefer, C. W., 117
Scheelea, 312
Schistocarpha oppositifolia, 186, 404
Schistocera gregaria, 312
Schistosoma, 311
Schmidt, J. O., 110
Schoener, T. W., 88, 195, 196, 198, 217, 230, 275, 291, 331
Schram, F. R., 62
Schwerdtfeger, W., 74

Scolytidae, 202
Scorpion flies, 377
Scott, J. A., 459
Scriber, J. M., 264, 406
Scudder, G. G. E., 150–151
Season
  dry, 3, 52, 69, 73, 74–75, 78, 79, 81, 82, 86, 146, 244, 248, 250, 261, 267, 278–280, 283, 292, 297, 306–312, 318–333, 382, 389, 401, 409, 427
    fauna density in, 180, 196–199, 275
  growing, 73, 426
  wet, 75, 78, 197, 248, 261, 269, 277, 286–289, 305
    and insect abundance, 275, 293, 297
Seasonal changes, in insect populations, 74, 75, 299–322
Seasonality, 12–13, 26–27, 57, 64, 96, 138, 235, 248, 273–333, 341, 388, 437
Seed
  predation, 160–164, 195, 206, 231, 266, 327, 329, 339, 341, 357, 369
  production costs, 368
Seed crop, 328
  production, 161
Seed dispersal agents, 10, 358–359
Seed mortality agents, 10
Selection, 41, 54, 59, 217–221, 239
  density-dependent, 295
  directional, 36–38, 49, 238, 296, 353, 420, 431, 455–457
  disruptive, 38–40, 267, 296
  frequency-dependent, 288
  individual, 221, 245
  K, 226–228, 231, 234–237, 266
  models, 36–40
  r, 226–228, 231, 235, 266
  seasonal, 289
  stabilizing, 36–37, 56, 89, 169, 338, 353, 420, 449–450, 455–456
  units of, 220
Sequestering mechanisms, 120–121, 151
Sex ratios, 240–241
Sexual selection, 96
Seychelles, 426
Sgarbieri, V. C., 163
Shapiro, A. M., 12, 233, 443, 458–459
Sheltering behavior, 352
Sida, 329
Silberglied, R., 183, 442
Simberloff, D., 148
Simpson, B. B., 97, 447
Simulidae, Amazonian, 287
Simulium, 307
  neavi, 377
Slash-and-burn clearing, 79
Slater, J. A., 206, 344, 456
Slobodkin, L. B., 54, 70
Smiley, J. T., 141, 345, 347, 348
Smith, J. M., 177
Smythe, N., 54, 327
Social behavior, 63, 93, 96, 104–111
Solanaceae, 345
Solanum rugosum, 41, 346
Songs
  species-specific, 101
  timing of chorus, 102–104, 290
South Africa, 117, 206
South America, 25, 74, 79, 101, 148, 179, 255, 265, 340, 348, 351, 394, 429, 431, 435, 437
Southwood, T. R. E., 80
Specialists
  ecological, 57, 65, 140, 201, 239, 331, 342, 456
  grain, 191
  habitat of, 210
  sequential, 191
Speciation, 27, 49, 430, 431
  and species abundance, 6
  isolating mechanisms, 279
Species
  diversity, 278
  richness, 338, 384
Speyeria, 432
Sphecidae, 107, 340
Spiders, orb weavers, 189, 378
Sphingidae, 86, 313, 325, 327, 359
Stanton, N. L., 16
Staphylinidae, 308–309, 383
Stem-boring guilds, 348
Sterculia, 329
  apetala, 328
Stradling, D. J., 353
Strehl, C. E., 162
Strong, D. R., Jr., 344, 355
Stubblebine, W., 82
Succession, 86, 339
  seasonal, 327–329
  stages, 388

Surinam, 284
Survivorship, 40, 240, 246
  age-specific, 221
Swarm raids, 342
Sweat, human, 108–110
Swihart, S. L., 68
Swollen-thorn acacia, 165
Symphyla, 62
Synchronization
  of flowering, 86, 279, 292, 323, 368, 371
  of life cycles, 230, 283, 289, 308, 312, 323
  of populations, 290, 329

Tabanidae, 309–311
Tabanus
  guineensis, 310
  taeniola, 310
Tachinidae, 207, 231, 232, 235, 278
Tannins, 157, 273
Tanzania, 377
Taxon cycle, 47–49, 165, 296, 457
Taylor, O. R., Jr., 356
Temperate Zone, 3, 6, 17, 32, 79, 80, 83–84, 100
  changes in climate, 64
  insect populations, 1, 14, 15, 75–80
Temperature
  body, 81–88
  daily variations in, 284–285
Tephritidae, 183
Termites, 105, 335, 391
Tertiary, mid-, 438
Texas, 82
Thasus acutangulus, 154
Theobroma, wild species, 366
Theobroma cacao, 394, 399
  insects on, 453
  pollination of, 261, 325, 332–333, 362–365
Thermal acclimation, 453
Thermoregulation, 85–86, 320, 352
Thermoregulatory stress, 320
Thomason, J. H., 93, 240
Thornhill, R., 93
Thrips, 393
Thunbergia grandiflora, 215, 372
Thyrnnidae, 408
Tiliaceae, 162, 413
Time stability hypothesis, 26
Tobago, 183
Tomatine, 158
Toxic secondary substances, 146, 256, 258
Transient species, 336
Trapline home range movements, 370–371
Tree falls, 178, 337, 388
Tribolium, 24, 43, 231, 255
Trichome hairs
  evolution of, 141
Trigona, 86, 186, 326
  foraging behavior, 215
  fulviventris, 372
  nesting habits, 6
  jaty, 186, 403–404
Trinidad, 183, 186, 332, 348
Trophic level, 251, 278, 335
Trophic relationships, 6, 198
Trypoxylon, 340
Turner, J. R. G., 350, 352, 448
Tuskes, P. M., 54

Uganda, 112, 142, 159, 183, 243, 312, 319, 377
Ultraviolet reflection cues, 352
Upper Carboniferous, 62
Urania fulgens, 81
Usinger, R. L., 8

Vandermeer, J. H., 159
Van der Pijl, L., 145, 358
Vane-Wright, R. I., 91, 431
Vanilla fragrans, 398
Vanzolini, P. E., 435
Vaughn, F. A., 151
Vectors, disease, 335
Venezuela, 82, 284
Venoms, ant, 110
Vepsalainen, K., 296
Vespidae, 107, 325
Vicariance model, 434
Voltinism, 77, 86, 137, 276–277, 318, 408, 459

Wade, M. J., 40
Waldbauer, G. P., 295
Walker, T. J., 101
Wasps, 93, 111, 159, 189, 249, 300, 323, 324, 345–347, 360–361, 383
  paper, 73, 107, 343, 372
  sand, 340
  sawflies, 126

Watson, J. A. L., 3
Webb, J. W., 117
Weevils
  Argentine stem, 77
  flightless, 183
Wellington, W. G., 63
Wenzel, R. L., 17
West Africa, 150, 377
West-Eberhard, M. J., 93
Western Samoa, 410
West Indies, 49, 206
White, J., 162, 413
White, M. J. D., 221, 434
White water, 308
Wiens, J. A., 251, 253, 254
Wigglesworth, V. B., 71, 77
Wiklund, C., 135
Wille, A., 6, 62, 111, 215
Williams, C. M., 72
Williams, E. E., 435
Wilson, D. E., 344
Wilson, D. M., 69
Wilson, E. O., 40, 47, 93, 104, 105, 107, 148, 164, 213, 219
Wilson, M., 420
Winder, J. A., 395, 401
Windsor, D. M., 189
Wing
  damage patterns, 98
  evolution of, 63
  muscle degeneration, 257
  pigments, 320
  pockets, 91
Wirth, W. W., 364
Wisconsin, 84, 176
Wolda, H., 299
Wright, S., 239, 266

Young, A. M., 30, 36, 41, 50, 78, 83, 84, 93, 98, 101–104, 110, 114, 116, 117, 126, 131, 132, 135, 141, 142, 156, 157, 161, 162, 186, 189, 201, 205, 207, 211, 215, 240, 255, 263, 289, 290, 291, 300, 307, 313, 320, 323, 328, 332, 340, 343, 345, 346, 348, 372, 395, 401, 417, 442–443

Zacryptocerus varians, 213
Zaire, 184
Zammara
  smaragdina, 50–52, 161–162, 343, 413
  smaragdula, 184, 313, 315, 317
Zingiberales, 343–344
Zophobas, 69
Zygia longifolia, 416, 418